T0137310

Application of Microalgae in Wastewater Treatment

Sanjay Kumar Gupta • Faizal Bux
Editors

Application of Microalgae in Wastewater Treatment

Volume 1: Domestic and Industrial Wastewater Treatment

Editors
Sanjay Kumar Gupta
Environmental Engineering, Department of Civil Engineering
Indian Institute of Technology – Delhi
New Delhi, Delhi, India

Faizal Bux
Institute for Water and Wastewater Technologies
Durban University of Technology
Durban, South Africa

ISBN 978-3-030-13915-5 ISBN 978-3-030-13913-1 (eBook)
https://doi.org/10.1007/978-3-030-13913-1

This Springer imprint is published by the registered company Springer Nature Switzerland AG.
The registered company address is: Gewerbestrasse 11, 6330 Cham, Switzerland

Preface

In the past few decades, algal technologies have been one of the extensively studied fields of biological sciences for numerous environmental, biological, biomedical, and industrial applications. Microalgae are one of the simplest photosynthetic life-forms which have an amazing potential of growing in very harsh environmental conditions. Microalgae hold amazing potential for the sequestration of various nutrients from water to carbon dioxide from the air. These organisms hold great potential, and are desperately required for sustainable and renewable management of food, fodder, and fuels. Algal biomass can be used for food, bioremediation, biofuels, and a number of chemicals. Microalgae have a capacity to produce polymers, toxins, fatty acids, and enzymes, which can be useful for pharmaceutical, nutraceutical, and cosmeceutical developments. The present book, Application of Algal Technologies for Wastewater Treatment Volume I, deals with the application of the characteristic features of various types of microalgae, diatoms, and blue-green algae for the treatment of domestic and industrial wastewater. Chapter 1 of this book provides a brief introduction to the global perspective of phycoremediation technologies. The authors have provided an overview and current status of algae-based bioremediation, and the challenges towards achieving global sustainability.

Diatoms are heterokonts which are highly diverse and have significant evolutionary differences compared to green algae, but serve as a sink for greenhouse gas. About 20% of the total photosynthetic carbon dioxide fixation and 40% of annual marine primary productivity depend on diatoms. Diatoms have great potential as bio-indicators as their population diversity reflects the environmental conditions of their oceanic or riverine ecosystems. The ease of their detection and versatility across different eco-systems complements their sensitivity to many physicochemical and biological changes. Chapter 2 and Chapter 15 provide detailed information about the current research on the potential advantages and lacuna pertinent to the utilization of diatoms for domestic and industrial wastewater remediation.

Chapter 3 and Chapter 4 describes the suitable approach to sustainable wastewater treatment using different strains of microalgae as well as by developing designer microalgae consortia. This chapter elaborates how developed microalgae consortia can play a futuristic role for carbon capture and could be used as a cost-effective tool for the production of various chemicals, as well as for wastewater treatment. Both of the chapter covers the various aspects related to the utilization of algal-bacterial interaction in wastewater remediation from laboratory scale to pilot scale studies.

Despite promising research findings on microalgae for WWT at the laboratory scale, the large scale of microalgae-based WWT processes is reliable only in outdoor systems that still need further investigations. Chapter 5 provides an overview of the most up-to-date information on outdoor cultivation of microalgae for wastewater treatment and discusses the progress and the important operational factors facing the outdoor culture.

Endocrine disrupting chemicals interference in the endocrine system of an exposed organism are considered one of the most emerging pollutants. The removal of increasing level of EDCs from wastewater has become a major concern nowadays. Chapter 6 provides state-of-the-art information on algae mediated remediation of endocrine disrupting chemicals (EDCs) from wastewater. In this chapter, the authors have reviewed recent literature pertaining to the application of microalgae for remediation of EDCs and various practical avenues of this technology in the area of wastewater treatment. Several eco-friendly natural methods have emerged for bioremediation of wastewater. However, algae-based bioremediation offers dual advantages of wastewater treatment as well as biomass production from wastewater which has tremendous secondary and tertiary uses. Chapter 7 and Chapter 13 deal with bioremediation of municipal sewage water using microalgae and algal-bacterial consortia. The biomasss production using wastewater for various applications is another aspect thoroughly covered in these chapters.

Chapter 8 provides an overview of applications, challenges, and future prospects of phycoremediation of petroleum hydrocarbons polluted sites. In this chapter, the authors have mainly focused on present practical and technological constraints to employing sustainable methods for the removal of petroleum hydrocarbons from wastewater. This chapter discusses the biogeochemical pathway leads degradation of petrochemical polluted soils and groundwater using phycoremediation techniques.

Chapter 9 discusses the genetic transformation and transgenesis technologies that could be applied for algae, and highlights the potential use of transgenic algae in wastewater treatment. The authors have reviewed various genetic modifications and transgenesis technologies which can improve the physiological characteristics of algae, further enhancing the potential utilization of algae in wastewater treatment and other bioremediation applications.

Due to longer persistence and higher toxicity, mutagenicity, and carcinogenicity, persistent organic chemicals (POPs) and pharmaceutical compounds are the most widespread pollutants which affect both terrestrial and aquatic organisms. Conventional water treatment plants are not efficient enough to remove the POPs, including pharmaceutical compounds. Therefore, more effective and cost-effective waste treatment procedures are required for the removal of these chemicals. Chapter 10 discusses various micro algal-based systems for the removal of pharmaceutical compounds, application challenges and future prospects. Chapter 11 provides a detailed review of the molecular mechanisms involved in bioremediation and biotransformation of POPs. The limitations and various approaches to enhance phycoremediation is discussed in detail.

Chapter 12 presents a comprehensive overview of the feasibility of application of microalgae in pathogen removal from wastewater. The mechanisms involved in pathogens removal, factors affecting pathogens elimination and feasibility of algal technologies for pathogen removal are discussed in detail.

Textile effluents contain high levels of synthetic dyes, detergents, stain repellents, waxes, and biocides. The dyes are often non-biodegradable and carcinogenic. Therefore, treatment of such effluents before release into the environment is imperative. Several physicochemical treatment methods have been developed, but most of them are energy intensive. Application of algae for the bioremediation of textile effluents has emerged as an environmentally friendly and economic technique. Chapter 14 discusses the possibilities and constraints of phycoremediation of textile effluents.

Pesticides and pesticide residues have led to significant contamination of entire terrestrial and aquatic ecosystems, and are major causes of biodiversity loss. Most organic and inorganic pesticides pose a tremendous threat to humans an ecosystems. Microbial bioremediation has proven to be efficient, inexpensive, and eco-friendly. Chapter 16 presents a comprehensive overview about the feasibility of using bacterial-microalgal consortium for the bioremediation of common industrial, agricultural and domestic pesticides leading to soil and water contamination, while outlining a variety of remediation approaches to treat wastewater. Furthermore, this chapter includes a discussion on the factors affecting both bioaccumulation and biodegradation efficiencies, including limitationsassociated with approach, environment and microbial consortium.

The use of blue-green algae (BGA) started in the early twentieth century and has gained immense attention for its various applications in agricultural biotechnology, natural products, cosmetics, and the production of numerous secondary metabolites including vitamins, enzymes, and pharmaceuticals. Recent understandings of cellular and metabolic diversities of BGA have given a new hope for its application in wastewater treatment, which has started to gain popularity in the last few decades. Chapter 17 summarizes recent reports of BGA usage in wastewater treatment and its future applications in Phytoremediation.

Conventional open algal ponds are used for wastewater treatment, and are considered as a low-cost option for algal biomass production. However, such systems are dependent on the prevalent environmental factors and do not provide a sufficient level of control over the process, thus achieving a sub-optimal performance. Two chapters deal with the design and optimization aspects of algal systems for wastewater treatment. Chapter 18 provides an overview of photobioreactor technology, the inherent complexity of their application, and current technical advances leading to their large-scale application. Chapter 19 provides an overview of the design and basic limiting factors of algal cultivation systems. The design considerations include light irradiance/distribution, culture mixing/agitation, air-CO_2 mixture supply, heat and gas-liquid mass transfers, and energy inputs. A detailed description of several algal growing systems, viz., facultative waste stabilization ponds, shallow ponds, raceway, tubular photobioreactors, flat panel photobioreactors, and airlift photobioreactors, are provided in this chapter.

This book is intended to be a practical guide for scholars and experts working on the application of algal technologies for bioremediation. This book is divided into two volumes. The first volume contributes significant knowledge about various algal technologies using microalgae, diatoms and blue-green algae applied for the treatment of domestic and various types of industrial wastewater as well as phycoremediation of emerging pollutants, whereas the second volume comprises of various aspects of water and wastewater based algal biorefineries.

New Delhi, Delhi, India — Sanjay Kumar Gupta
Durban, South Africa — Faizal Bux

Acknowledgments

First and foremost, I thank the Almighty God for sustaining the enthusiasm with which I plunged into this endeavor.

I avail this opportunity to express my profound sense of sincere and deep gratitude to the many people who are responsible for the knowledge and experience I have gained during this book project.

First of all, I would like to extend my sincere gratitude to each of the authors for devoting their time and effort towards this book and for the contribution of their excellent pieces of academic work. Without the overwhelming support of several authors, we may not have been able to accomplish this mammoth task. The contributions of all authors are sincerely appreciated and gratefully acknowledged.

I would like to thank the Publisher, Springer, for providing continuous support and a platform to publish this book. I would also like to thank publishing editor Dr. Sherestha Saini, Senior Editor, Springer New York for her suggestions, and continuous support. This wonderful compilation get published thanks to her tremendous efforts. I would also like to thank the production team, especially, Ms. Chandhini Kuppusamy, M. Gabriele, Aaron Schiller and Susan Westendorf for their technical support. They worked with me throughout the course of this book project.

I express my sincere gratitude to all the faculties of Environmental Engineering, Department of Civil Engineering, Indian Institute of Technology Delhi, for their wholehearted support and encouragement for this book and all the academic works.

From the bottom of my heart, I would like to thank my colleagues, who continually and convincingly conveyed a spirit of adventure, whenever I felt low, which kept me agile during this endeavor.

I do not have words to express my thankfulness to my mother Smt. Manju Gupta who has always been a source of motivation in every walk of my life. I would like to express my deepest appreciation to Ajay and Maushami for their endless support.

Last but not least, it would be unfair on my part if I failed to record the support, encouragement, and silent sacrifices of my wife Preeti and my lovely kids Shubhangi & Adweta. Without their persistent support, this book would not be possible. I would like to dedicate this book to Master Arjan, the cutest junior member of my family.

New Delhi, Delhi, India Sanjay Kumar Gupta

Contents

About the Editors

Sanjay Kumar Gupta is Technical Superintendent, Environmental Engineering, Department of Civil Engineering at the Indian Institute of Technology, Delhi, India. Dr. Gupta started his research carrier in 1999 at CSIR-Indian Institute of Toxicology Research, Lucknow India. His PhD was awarded in 2010 from Dr. Ram Manohar Lohia Avadh University, Faizabad. Later, he did his post-doctoral research from Durban University of Technology, South Africa. He has been recognized thrice as "One of the Top Publisher Post-Doc Fellow" in 2014, 2016 and 2017 for his active research contributions. Dr. Gupta has authored 65 articles in peer-reviewed journals and books, and has presented 20 papers in national and international conferences. He was an Editor for Journal of Ecophysiology and Occupational Health, and is a life member of many professional societies including the International Society of Environmental Botanists, Society of Toxicology, Academy of Environmental Biology, and the Indian Network for Soil Contamination Research. His research interests include ecotoxicological risk assessment, bioremediation of wastewater and industrial effluents, and algal biotechnology.

Faizal Bux is Director of the Institute for Water and Wastewater Technology at Durban University of Technology in South Africa. He received his B.Sc. from University of Durban-Westville in 1986, his M. Tech. in Biotechnology from Technikon Natal in 1997, and his Ph.D. in Biotechnology from Durban institute of Technology in 2003. He has edited 5 books, written 121 articles in peer-reviewed journals, authored 17 book chapters, and is an Editor for the journals Environmental Science and Health, Biofuels Research Journal, and Water Science and Technology. His research interests include biological nutrient removal in wastewater treatment, algal biotechnology, and bioremediation of industrial effluents. He has more than 20 years of experience at higher Education Institutes and has received numerous institutional awards including the Vice Chancellor's and University Top Senior researcher awards. Prof Bux is ranked as the most published researcher at Durban University of Technology. He has supervised over 50 Masters and Doctoral students, and 10 Post doctoral fellows served their tenure under his guidance. He was an editor for CLEAN–Soil, Air, Water (John Wiley & Sons, Germany), Environmental Science and Health Part A (Taylor Francis, USA), and served as a reviewer for 28 International Journals. His citations are in excess of 5159 with an H Index of 33. Prof Bux is an invited Member of the Management Committee of the International Water Association IWA) specialist group (Microbial Ecology and Water Engineering, MEWE) and is actively involved in coordinating activities of MEWE globally.

Phycoremediation Technology: A Global prospective

Sumedha Nanda Sahu, Narendra Kumar Sahoo, and Satya Narayana Naik

1 Introduction

Water crisis is realized as one of the major issues and global threat, even though sufficient water and land resources are available (CA 2007). According to United Nations World Water Development Report (2014), more than two million tons of sewage, agricultural, and industrial wastes is dumped untreated into lakes, rivers, and other waterbodies in developing countries that is eventually polluting the usable water supply. Almost all waterbodies globally are highly polluted because of release of various industrial as well as domestic wastewaters. This untreated wastewater provides various organic and inorganic nutrients such as nitrogen (N) and phosphorus (P), for the autotrophs which in turn leads the process of eutrophication in waterbodies (Schindler et al. 2008).

The art of utilization of algae (macro- or microalgae) in removal, biotransformation, or mineralization of various nutrients, heavy metals, and xenobiotics from wastewater and carbon dioxide from waste air (Olguin and Sanchez-Galvan 2012) is known as phycoremediation. During this treatment, carbon, nitrogen, phosphorus, and other salts are used by algae as nutrients, from the wastewater or air as the case may be. Other pollutants and xenobiotics are even taken care of by the organisms by various cellular mechanisms. This is an eco-friendly process as there is no secondary pollution if the biomass produced is harvested for utilization (Mulbry et al. 2008). Literature reveals that algal bioremediation (phycoremediation) technology is highly relevant and has immense potential for future applications in various waste removal strategies. In the past few decades, extensive research has been made in algal biotechnological advancement and has successfully established the

S. N. Sahu · N. K. Sahoo · S. N. Naik (✉)
Centre for Rural Development and Technology, Indian Institute of Technology Delhi, New Delhi, India
e-mail: snn@rdat.iitd.ac.in

S. K. Gupta, F. Bux (eds.), *Application of Microalgae in Wastewater Treatment*,
https://doi.org/10.1007/978-3-030-13913-1_1

system of wastewater remediation using algae, microalgae in particular, in reduction of an array of organic, inorganic nutrients, and some highly toxic chemicals (Beneman et al. 1980; Thomas et al. 2016).

The agents of phycoremediation, algae, are photosynthetic organisms, capable of growing in extremely harsh and difficult environments. In addition, there are various research reports on microalgal sequestration of various heavy metals in their cell walls through process of adsorption or ion exchange, as a means of bioremediation of heavy metals (Priyadarshani et al. 2011). While microalgae are microscopic, macroalgae are visible to naked eye. Phycoremediation can serve many purposes such as (i) utilization of nutrients from wastewater; (ii) transformation, degradation, or removal of xenobiotics; (iii) remediation of acidic and metal rich wastewaters; (iv) CO_2 sequestration; and (v) biosensor-based detection of toxic compounds (Gani et al. 2015). By taking in carbon dioxide from the atmosphere and giving out oxygen through photosynthesis, not only they purify the air, but their interplay with the pollutants reduces the load from entering the waterbodies. However, it is still a challenge to develop and optimize processes to treat industrial effluent as well as to restore polluted rivers and lakes through the process of phycoremediation. In addition, there are various research reports on microalgal sequestration of various heavy metals in their cell walls through process of adsorption or ion exchange, as a means of bioremediation of heavy metals (Priyadarshani et al. 2011).

As algae are emerging as a potential biofuel candidate due to its productivity and other beneficial characteristics, successful pilot-/field-scale trials are now coming into existence. These production systems for biofuels can be exploited for phycoremediation to make it more profitable and eco-friendly. Such approach will help in making the biofuel technology economically feasible. In the recent day, technological advancement has explored the scope of microalgae to mitigate various hazardous pollutants in the environments. Moreover, phycoremediation strategy coupled with energy production is well established; however algal biofuel technology is not feasible commercially because of higher energy inputs. Additionally, modification in the cultivation system, harvesting systems, extraction technology, and biomass utilization approaches (biochemical and thermochemical) could be adopted to cope with sustainability issue via an integrated/biorefinery approach as demonstrated in Fig. 1. Let us discuss the various important issues of phycoremediation in details starting from cultivation itself.

2 Different Algal Systems Used for Bioremediation

The cultivation systems for algal biomass production coupling with remediation of wastewater are basically open systems and closed systems (photobioreactor). Other than the suspension culture, attached cultivation is also frequently implemented both in open systems and closed systems. Among these, open pond (raceway ponds) algae culturing and turf scrubbers are the most popular systems for algae cultivation. On the other hand, the closed systems for algae cultivation (photobioreactors) have more

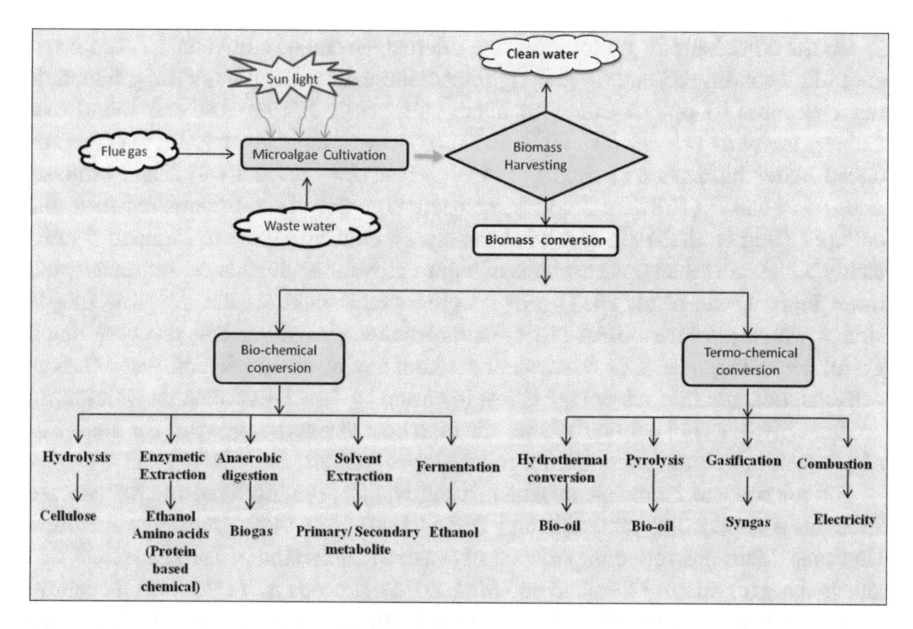

Fig. 1 Integrated biomass conversion flow chart

diversity based on its shapes and configurations of the bioreactor. The most frequently used closed systems for algae cultivation are tubular, bubble column, airlift, and flat panel (Richardson et al. 2012). The algae employed in these systems may act in monoculture, consortia, or natural assemblage of algal community. However, other two modes are considered more sustainable than monoculture mode.

2.1 *Open and Closed Culture Systems*

Open ponds for algae cultivation are the oldest and simplest form of cultivation systems for microalgae biomass production (Handler et al. 2012). In the past decade, stabilization ponds have been found to be used for the treatment of urban wastewater (Caldwell 1946); however to be more efficient, it requires a lot of land. In general high rate algal ponds (HRAPs) are shallow-type open raceway system with a single or multiple loops, and to obtain a water velocity of 0.15–0.3 m/s, it uses paddle wheel (Park et al. 2011). The depth of the systems is in between 0.2 and 0.4 m (sometimes up to 1 m) where CO_2 can be added in a sump of about 1.5 m depth. It is found that high rate algal ponds reduce the surface needed in comparison to stabilization ponds by a factor of 5 (Picot et al. 1992) and achieving a three-fold improvement in biomass productivity with a yield from 10 ton/year/ha (Craggs et al. 2011). As compared to activated sludge systems, the capital costs and operational costs are significantly reduced in case of HRAPs (Craggs et al. 2011).

On the other hand in photobioreactor, microalgae can be cultivated in axenic and controlled conditions, and there is significant increase in the volumetric productivities compared to open systems of algae cultivation. Earlier research found that *Chlamydomonas reinhardtii* cultivation using wastewater in photobioreactor produced better biomass and lipid (+144% and +271%, respectively), and removal rates of N and P (+38% and +15%, respectively) were found compared to flasks culture (Kong et al. 2010). However, the closed culturing systems demand significantly higher cost than open systems of algae cultivation which is approximately ten times high (Davis et al. 2011). For keeping axenic cultures and to grow fragile strains which produce potent bioactive molecules, closed systems are very much useful. However, in case of wastewater medium having a huge diversity of microorganisms, this precious advantage is lost in closed system. Moreover, the volumetric productivity does not counterbalance the high cost of photobioreactor for treatment of urban or agricultural wastewater for algae production.

For microalgae culturing in attached cultivation, immobilized microalgae are fixed on a supporting material, and that is immersed in the nutrients medium. However, there are few comparisons of wastewater treatment for suspended and attached algal systems (Kesaano and Sims 2014). It needs more research on certain factors which affect growth, nutrient mass transport, selection of species, algal-bacterial mutualistic interactions, and upscaling of laboratory research. The attached cultivation systems have provided promising results with certain wastewaters. It has been reported that use of dairy manure for cultivation of benthic algae in an attached cultivation system would require 26% less land area for equivalent nitrogen uptake compared to the conventional corn/rye rotation process (23% for phosphorus) (Wilkie and Mulbry 2002). In addition, biofilm rotating disk reactor is one of the efficient attached cultivation system for microalgae cultivation using wastewater with better biomass productivity. Earlier studies reveal that biomass productivities between 20 and 31 $g/m^2/day$ with nutrient reduction rates as high as 14.1 $g/m^2/day$ for nitrogen and 2.1 $g/m^2/day$ for phosphorus (Christenson and Sims 2012). Using rotating biological contactor-based photobioreactor, an average biomass productivity of 20.1 ± 0.7 $g/m^2/day$ was obtained over a period of 21 weeks without reinoculation (Blanken et al. 2014). These reactors provide better surface area to volume ratio in comparison to HRAPs.

2.2 Use of Monoculture and Consortia

Microorganisms usually exist in nature as part of organized communities and consortia, which gain benefits from cohabitation to keep invaders away, tackle risk of contamination, and simultaneously improve productivity and product diversity. In contrary, most of the cultivation trials are attempting a monoculture of selected species with advantageous traits. Promising genera/species/strains with specialized characters are generally employed under algal technology either as monocultures or consortia or as natural community depending upon the purpose. When the intention is biological manufacturing, i.e., to harvest high-value items from mass cultivation of biological organisms, axenic monocultures are mainly used (Mcneil et al. 2013). But, such monocultures are highly prone to contamination and losses. They require controlled conditions, high degree of sophistication hence, and higher throughput for the cultivation.

Controlled, symbiotic co-cultures possess features to overcome these bottlenecks, and co-cultures have shown improvements in yields of biomass, lipids (Yen et al. 2014), and high-value products (Dong and Zhao 2004). Maintaining axenic cultures has also proved as expensive and labor intensive, given the recurrent problem of contamination by bacteria, viruses, protozoa, fungi yeast, fungi, and microplasma (Langer 2008). Moreover, parasites or grazers can outcompete the working cell culture and influence production outputs and cell health.

With a wide range of thallus organization, algae could be found in a diverse habitat ranging from fresh to marine environments. Some commonly studied species for phycoremediation include *Botryococcus*, *Chlamydomonas*, *Chlorella*, *Phormidium*, *Haematococcus*, *Spirulina*, *Oscillatoria*, *Dunaliella*, *Desmodesmus*, *Arthrospira*, *Nodularia*, *Nostoc*, *Cyanothece*, *Scenedesmus*, etc. (Dubey et al. 2015; Rawat et al. 2011). Different species of algae possess different phycoremediation attributes such as growth rates, photosynthesis, total biomass production, biotransformation of certain molecule, faster uptake of certain heavy metal or nutrient, etc. Therefore, a mixture (consortia) of selected algal species will show a better performance than any individual species. Different species with distinct attributes and without any antagonistic effects could be taken together to form a functionally distinct community. However, for the formation of consortia, the screening of various traits of each constituent monoculture is necessary, as each species bear distinct inherent traits to make it superior than others. Association of eukaryotic algae with other organisms like bacteria, yeast, or cyanobacteria may be beneficial in production outputs. So, symbiotic/synergistic/mutualistic association of organisms in artificial co-cultures may produce marketable products and allow a biorefinery mode of production (Markou and Nerantzis 2013; Gebreslassie et al. 2013). A mixture of microorganisms possessing different metabolic activities and adapted to various environmental conditions develops a healthy biological system that can operate under different nutrient loads and environmental conditions (Johnson and Admassu 2013; Boonma et al. 2014). Moreover, cooperative interactions can be established between the microorganisms integrating the consortia, which can increases nutrient uptake rates (Renuka et al. 2013).

The important factors for stability of consortia are the initial inoculum size of constituent species, duration of log phase of each species, carrying capacities, non-allelopathy (toxin/antibiotic nonproducing) features, and maintenance of their original features in the consortia (Patel et al. 2017). The distinct features of the constituent species could be nitrogen fixation, luxury uptake of phosphates, heavy metal detoxification, high CO_2 sequestration, easy harvesting feature, etc. In such consortia, while the nitrogen fixers will alleviate nitrogen limitation, phosphates from anthropogenic sources will help to enhance their growth, and heavy metal detoxifying strains such as *Chlorella* can provide a better growth condition. This means, both cyanobacteria and green algae as constituent in consortia may provide a successful functional community. More diverse community means a higher stability and more biomass production (Cardinale 2011). Even if open ponds are inoculated with different algal species, high degree of contamination by pathogens is a possibility which leads to change in the original community structure. But at the same time, the inherent diversity and interaction within the consortia protect the individuals from pathogen or high light intensity. Therefore, the structural diversity and functional stability of consortia help in decreasing overall throughput and increasing the sustainability of the phycoremediation.

3 The Agents with Special Attributes

Owing to rapid industrialization and rapid growth of human population, resulting environmental degradation is very alarming. To deal with such situation, we need to follow unique approaches. Among various strategies for waste mitigation, today by means of algal strains with special characteristics, the nutrient removal has been shown to be more efficient. They possess various desired attributes like extreme temperature tolerance, producing high-value molecules, quick sedimentation behavior, mixotrophic growth potential, etc. A *Phormidium* sp. that was isolated from polar environment is capable of removing nutrients more efficiently than a community of green algae at temperatures below 10 °C. This strain was appropriate for wastewater treatment in cold climates during spring and autumn (Tang et al. 1997). On the other hand, *Phormidium bohneri* is a high-temperature alga for treating wastewater in addition to its quick sedimentation behavior (Talbot and De la Noüe 1993).

Some marine seaweed, green macroalgae, and their alginate derivatives show high affinity for various metal ions (Mani and Kumar 2014). Alginate plays a vital role in metal biosorption process by brown algae. So, there should be an attempt by scientist and industries to make the microalgal technology more eco-friendly and cost-effective by focusing specific uses. It can meet most of the problems and lead toward global sustainability.

3.1 Cyanophycean as Bioremediators

The cyanobacteria are able to fix atmospheric N_2, catalyze the cycling of various nutrient elements, purify soil and water by discouraging growth of pathogenic microbes, and decompose organic substances. They could remediate heavy metals and detoxify pesticides and other xenobiotics to promote soil and water reclamation. They contribute to agriculture by improving soil quality and promoting plant growth by production of enzymes, vitamins, hormones, and other bioactive compounds (Higa and Wididana 1991). The restored soil fertility, land reclamation, nutrient cycling, and reduced agrochemical uses not only contribute to agricultural sustainability but also provide environmental protection and pollution prevention (Shukia et al. 2008). Some researcher found that the evolution of greenhouse gases such as CH_4 from the soils of various ecosystems is minimized to a great extent by the association of methanotrophs and cyanobacteria (Tiwari et al. 2015). It may be noted that CH_4 is 28 times more potential GHG than CO_2 (IPCC Fifth Assessment Report 2014 (AR5)).

The oxygen released by cyanobacteria during photosynthesis creates an aerobic environment in the rhizosphere, simultaneously reduces the methane genesis, and enhances the aerobic methane oxidation (Prasanna et al. 2002). As cyanobacteria minimizes methane flux without compromising the productivity of the flooded rice field, the cyanobacteria could be employed as a practical option for minimizing global warming potential and enhancing nitrogen fixation potential of paddy fields (Prasanna et al. 2002). Increased diversity of cyanobacteria, methanotrophs, and other organisms in the wet crop fields promotes higher production and reduced CH_4 emissions (Singh 2015). Furthermore, the cyanobacterial N fixation resulting in reduced fertilizer use makes the land restoration cost-effective, sustainable, and safer (Pandey et al. 2014) which also conserves the diversity of methanotrophs and CH_4 consumers. However, the cyanobacterial genetic and metabolic engineering in the future are expected to make the phycoremediation more effective in mitigating environmental pollution and empowering global sustainability.

3.2 *Role of Algae for Detoxification of Organic Pollutants and Heavy Metal*

The extensive occurrence of various toxic pollutants like heavy metals and other hazardous contaminants in the environment is a serious concern today. Several removal methods have been proposed and implemented in an eco-friendly manner to address various environmental pollution issues. Recent study by Oregon State University reveals that the marine plants and seaweeds in shallow coastal ecosystems can give a major role in increasing the effects of ocean acidification. Moreover, researchers found that seaweed, green macroalgae, and their alginate derivatives show high affinity for various metal ions (Mani and Kumar 2014). It is know that alginate plays a significant role in metal biosorption by brown algae. The potential of microalgae to perform well at very low levels of contaminants without producing toxic sludge is easy to culture and maintain. Furthermore, microalgae have a very good binding affinity (because of its relatively high specific surface area and net negative charge) and also appropriate remediation strategies (Suresh Kumar et al. 2015). Some earlier studies on potential of microalgae for bioremediation of heavy metals are noted in Table 1. There is still need of research that leads to the development of new bioremediation technologies which use algae in engineered systems to mitigate toxic organic pollutants for a green environment.

Table 1 Some studies on algal application for bioremediation of heavy metals (Priyadarshani et al. 2011)

Microalgae	References	Metal studied
Tetraselmis chuii	Ayse et al. (2005)	Cu
Spirulina (Arthrospira) Platensis	Arunakumara et al. (2008)	Pb
Spirogyra sp., *Nostoc commune*	Mane et al. (2011)	Se
Anabaena variabilis, *Aulosira* sp., *Nostoc muscorum*, *Oscillatoria* sp., and *Westiellopsis* sp.	Parameswari et al. (2010)	Cr(VI), Ni (II)
Spirogyra hyalina	Nirmal Kumar and Cini (2012)	Cd, Hg, Pb, As, and Co
Scenedesmus bijuga, *Oscillatoria quadripunctulata*	Ajayan et al. (2011)	*Cu*, *Co*, *Pb*, *Zn*
Scenedesmus acutus, *Chlorella vulgaris*	Travieso et al. (1999)	*Cd*, *Zn*, and *Cr*
Spirulina platensis	Garnikar (2002)	*Cu*, *Hg*, and *Pb*
Chlorella minutissima	Singh et al. (2011)	*Cr(VI)*

3.3 Exploiting Harmful Algal Blooms (HABs)

Algal blooms are generally a resultant of nutrient enrichment of aquatic bodies. Bloom is a state of higher productivity and is desired most of the times in man-engineered systems. However, when such blooms in nature comprise harmful cyanobacteria, it becomes an environmental concern because of their toxins which compromise the safety of water usage (Smith and Daniels 2018). Owing to incidence of various illnesses in livestock and human form algal toxins, there is a worldwide attention on harmful algal blooms which are characterized by very fast growth and biomass accumulation of one or several species of algae (Chen et al. 2016). The management challenges these blooms pose are a thorough understanding of the aquatic food web dynamics, community ecology, and the links with other ecosystems, along with the socioeconomic welfare and the administrative issue (Qin et al. 2015; Sun et al. 2015; Brooks et al. 2016). Thus, short-term management strategies of bloom control and eradication and reducing their harmful effects can lead to unseen damages to aquatic ecosystems and thereby significant socioeconomic losses (Ahlvik and Hyytiäinen 2015). On the other hand, a thorough understanding of the toxicological potential of HABs (Brooks et al. 2016) particularly of costal ecosystems is required for our safeguard and a successful management of the algal blooms. Therefore, there is an urgent need to improve our understanding of the toxicological potential of HABs (Brooks et al. 2016), especially for coastal ecosystems (Halpern et al. 2008).

Research on algal bloom for exploring its potential has given rise to conversion of the biomass into commercial products and as natural renewable bioresources (Kim et al. 2015). Moreover, the natural ones, in comparison to culture biomass which are photoautotrophic, need a lower cost of raw material input. Further, the occasion of bloom formation can be exploited for the purpose of phycoremediation. For this purpose, we simply have to feed the existing bloom with wastewater. The bloom remediates the wastewater at a much faster rate due to high population density of the algae. In this context, Sahoo (2010) found that nutrient uptake rate is much faster when fed to a denser algal population than low density population. In case of planned cultivation program, also a bloom is desired but at the cost of artificial nutrient inputs. So, the natural bloom can explored for phycoremediation purpose. However, a proper management plan needs to be in place which takes care of wastewater input and harvest of the bloom biomass so that the ecological balance is maintained. Pandhal et al. (2018) worked on harvesting of natural bloom and reported an improved ecosystem functioning in response to maximum rate of harvesting. The biomass of natural algae blooms could offer an abundant feedstock for conversion into various biofuel products like bio-oil, biochar, biodiesel, and so on. As the eutrophication and bloom ordinarily is damaging to the local ecosystems as well as economy, the utilization of this abundant waste biomass would provide a feedstock for green bioenergy while still mitigating the environmental burdens (Zeng et al. 2013).

3.4 *Bioremediation of Soil Using Algae-Bacteria Consortia*

There is a global concern for environmental impacts of soil contamination. Bioremediation of such areas is a high-priority research topic for researchers worldwide. A consortium of microorganisms could be employed for such purpose. Consortia are generally symbiotic community of microalgae, cyanobacteria, and other associated aerobic or anaerobic microorganisms. They synergistically neutralize various toxic, organic, and inorganic pollutants. Microalgae and bacteria complement each other, and the synergy results in better remediation efficiency. Oxygen, and electron acceptor from algal photosynthesis, helps the degradation of organic matter by heterotrophic bacteria, and in turn, algae get their CO_2, nutrients, and other stimulatory substances from bacteria (Subashchandrabose et al. 2011).

Comparing to chemical technologies, the bioremediation processes involving algae-bacteria consortia techno-economically feasible self-sustaining approach (Bose and Das 2013). The effect that is obtained from the algae-bacteria synergy is hardly possible from employing any single species of (Escobar et al. 2008). In comparison to monoculture, consortia are very robust, are resistant against invasion, can bear environmental stress, and can maintain a more stable community. This ensures more stability in their growth and production. Chemical substances on their cell surface acting sometimes as allelochemicals help in enhanced bacterial degradation of wastes (Luo et al. 2014).

Bioremediation by such approach not only minimizes the material inputs in terms of energy, nutrients, and CO_2, but it produces biomass as by-product which could be utilized for various purposes (biofuel or biomaterial). Some valuable elements can even be recovered from the biomass. Such bioremediation process is sometimes more efficient in treatment of hazardous chemicals. Algae, because of their sensitivity, can serve as bioindicators to identify contamination, genotoxicity, and ecotoxicity in sediment as well as in soil (Bose and Das 2013). Taking the help of modern molecular techniques, selected consortia can provide better results for waste remediation and side by side produce some desired metabolites. Further improvements could be achieved by application of computation biology in addition to experimental biology by getting more insight into the algal-bacterial interaction at both molecular and metabolic levels.

4 Genetic Engineering and Phycoremediation

Research advancement on algal biotechnology is exploring the method like recombinant DNA technique to create constructs for both prokaryotes and eukaryotes that may replicate and possess novel research utility. It can also be applied to modification of algal metabolic pathways to targeted cellular activities of the photosynthetic cells by manipulating enzymatic, transport, and various important regulatory functions.

The genetically modified algae produce higher yields of the primary metabolites as well as by-products (Snow and Smith 2012). They do so depending on the desired characters for which the genes are introduced. This is a product of synthetic biology which provides a superior feedstock (Tabatabaei et al. 2011). The introduction of DNA into algal cells is done through various routes such as artificial transposons, particle bombardment, agitation of a cell suspension in the presence of DNA and glass beads, electroporation, agrobacterium infection, viruses, and silicon carbide whiskers. Out of these all methods electroporation and particle bombardment are the best ones (Rismani-Yazdi et al. 2011).

By modification of genomic DNA, desirable traits can be incorporated in algae to make them survive and show improved performance in harsh conditions. Techniques like DNA sequencing, metagenomics, hybridization, and enhanced evolution are being employed as tools for this purpose (Dana et al. 2012). Non-transgenic methods could even be employed to develop improved algal strains (Tabatabaei et al. 2011; Flynn et al. 2010). When an improved trait is incorporated, normally a trade-off occurs to make some other traits unfavorable (Hall and Benemann 2011). The major challenges of genetic engineering which influences the global commercialization of algae are the lower growth rate and gene quality (Tabatabaei et al. 2011).

However, there is a need of suicide genes to control an accidental escape and occurrence of any dangerous algal strain which possess high risk to environment (Quinn and Davis 2015). Although the bioremediation concept using algae to degrade pollutants in situ has lately attracted a lot of public attention, introducing the "genetically engineered" algae into the environment to enhance the process is yet to be demonstrated with success.

5 Algal Omics in Phycoremediation

An in-depth understanding of the role of different factors related to metabolism, growth, function, and dynamics of the microbial communities of the contaminated site is required for a successful bioremediation application. Proteomics, transcriptomics, genomics, and metabolomics tools together are providing a crucial insight into interactions in microbial communities and the bioremediation mechanisms and understanding of toxicity. It also helps in predicting the risks associated with environmental toxicity and bioprospecting of value-added products. This "omics" technology has become highly helpful in producing a complete description of nearly all components within a biological entity. Further, the technique and related data processing activity have a great value in ecotoxicological research (Spurgeon et al. 2010). The different omic techniques are providing information about the microbes involved in soil bioremediation and their metabolic responses. In addition, algal omics technology has also been extensively applied to the examination of algal bioremediation (Merchant et al. 2007). This advance technology is helping to unlock the full potential of microalgae feedstocks for multiple uses, through utility in an array of industrial biotechnology, biofuel, and biomedical applications (Guarnieri and Pienkos 2015). Thus, algae are emerging as highly attractive microbial cell factories in producing wide array of algal bio-products. The omic concept can help in driving bio-product discovery and optimization in microalgal systems. Moreover, multi-omic analyses of algal biology are evolving as a potential tool for development of biocatalyst and offering a powerful path toward hypothesis-driven strain-engineering strategies for enhanced TAG biosynthesis (Arora et al. 2018).

6 Global Challenges

Global issues like rapid climate change because of global warming are the major threat to ecosystem health and sustainable human welfare. Moreover, several anthropogenic stresses are hampering our day-to-day activities and ecosystems equilibrium. Therefore, there is need of mitigation strategies to solve this serious issue related to environmental pollution. Among various mitigating strategies, phycoremediation is a powerful tool for addressing global changes. Microalgae possess effective CO_2sequestration capacity compared to other photosynthetic organisms. Furthermore, microalgae can use CO_2 from flue gases and produce several high-value products. Production of biofuel from microalgae is a very promising technology.

6.1 Changing Land Uses and Consumption Pattern

Microalgae cultivation is emerging as an important research and investment area these days, because of its wide potential for fuels, foods, animal feed, pharmaceuticals, industrial applications, and environmental benefits. In addition, microalgae promises many environmental benefits compared to existing waste treatment and fuel technology. However, there are certain issues to overcome for a feasible and sustainable wastewater management, emissions reduction, and land use changing pattern (Usher et al. 2014). Moreover, microalgae cultivation seems more advantageous owing to their high growth rates and option to use marginal land for cultivation, thereby minimizing competition with food production as compared to other bioenergy crops. However, large-scale algae cultivation could have significant impacts on global energy scenario and agricultural and land markets, leading to considerable changes in global resource demands and greenhouse gas emissions (Efroymson et al. 2016).

Topographic and soil constraints limit the land availability for algal cultivation system in raceway pond as the large shallow ponds require relatively flat terrain. In addition, the soil porosity/permeability will also affect the need for pond lining and sealing (Lundquist et al. 2010). Solar radiation is one of the important factors influencing growth of algae to achieve higher production all over the year with little seasonal variation. For this reason, the most suitable locations for algal cultivation are warm countries close to the equator where insolation is not less than 3000 h $yr.^{-1}$ and with an average of 250 h $month^{-1}$ (Necton 1990; Verween et al. 2011). So far, the most commercial microalgae production has occurred in low-latitude regions such as Israel, Hawaii, and Southern California.

6.2 Climate Change Phenomena

The global warming and climatic disturbance as a consequence of environmental pollution have emerged as an important issue around the globe. These issues have drawn the attention of environmental biologists, pathologists, eco-chemists, eco-toxicologists, and researchers from diverse fields. Moderation of climate change phenomena is one of the important incentives for the algal energy field in addition to control of global environmental pollution issues. The greenhouse gas (carbon dioxide, methane, and nitrous oxide) mediated global warming, and climate change appears to be seriously disturbing the natural world. The main source of energy in India is coal which is currently contributing around 54% of electricity need and may reach to around 70% in the future (Arora 2013). The other important energy source is crude oil, about 70% of which is consumed by automobiles. It is obvious that fuel requirements and associated pollution will increase with increasing living standard and expanding population.

The compatibility of cyanobacteria with predicted global climate change is expected to be positive as concluded from research on their in situ dynamics, evolutionary history, and ecophysiology (Paul 2008; Paerl and Huisman 2009). Many systems for algae cultivation and production continue to be developed for moderate and hot climates (e.g., USA, Europe, and Australia) (Pankratz et al. 2017). Recently, algae cultivation has explored for the use in fixation of CO_2, which is of higher interest in greenhouse gas mitigation and in biofuels production. Furthermore, algae may provide key to scientists to achieve a negative emissions technology and to produce electricity, biofuels, value chemicals, and protein while simultaneously removing substantial amounts of carbon dioxide from atmosphere and reducing deforestation. Oswald and Golueke (1960) initially conceptualized the idea of phycoremediation involving a large-scale system of dozens of large (40 ha) high-rate algal ponds. The harvesting of biomass was done by a simple flocculation-settling procedure. Anaerobic digestion of the concentrated algal sludge produced biogas (methane and CO_2). Microalgae are able to sequester CO_2 from the ambient media and also from soluble carbonates (Chanakya et al. 2012). Many stationary industries such as cement plants, thermal power plants, refineries and petrochemical plants, and fertilizer plants are the main source of CO_2 with high concentrations localized in the ambient environment (Mildbrandt and Jarvis 2010). Being environmental friendly and without any secondary pollution as in case of chemical methods of wastewater treatment, application of phycoremediation technology to the wastewater and gas is being thought of these days for carbon capture.

7 Conclusion

Phycoremediation is an eco-friendly solution with no secondary pollution conditioned prior to harvest and utilization of the algae biomass. It has immense potential for future applications in various waste removal strategies. However, phycoremediation is still in growing stage; there is a need to develop and optimize the processes to treat industrial effluent. To address the sustainability issues of wastewater treatment as well as resource recovery, modification in the cultivation system, harvesting systems, extraction technology, and biomass utilization approaches (biochemical and thermochemical) via an integrated/biorefinery approach is necessary. Genetic engineering, synthetic ecology (synthetic consortia), and omics approach of algae are futuristic approaches to optimize the phycoremediation technology. This simple technology has the ability to address the global issues of land use changes and global climate change.

References

Ahlvik L, Hyytiäinen K (2015) Value of adaptation in water protection – economic impacts of uncertain climate change in the Baltic Sea. Ecol Econ 116:231–240

Ajayan KV, Selvaraju M, Thirugnanamoorthy K (2011) Growth and heavy metals accumulation potential of microalgae grown in sewage wastewater and Petrochemical Effluents. Pak J Biol Sci 14(16):805–811

Arora V (2013) Energy statistics 2013. Central Statistics Office, Ministry of Statistics and Programme Implementation, Government of India. [cited 4.12.2013]

Arora N, Pienkosb PT, Pruthia V, Poluria KM, Guarnieri MT (2018) Leveraging algal omics to reveal potential targets for augmenting TAG accumulation. Biotechnol Adv 36:1274. https://doi.org/10.1016/j.biotechadv.2018.04.005
Arunakumara KKIU, Zhang X, Song X (2008) Bioaccumulation of PB2+ and its effects on growth, morphology and pigment contents of *Spirulina (Arthrospira) platensis*. J Ocean Univ China 7(4):397–403
Ayse BY, Oya I, Selin S (2005) Bioaccumulation and toxicity of different copper concentrations in *Tetraselmis chuii*. EU J Fish Aquat Sci 22(3–4):297–304
Beneman JR, Foopman BL, Weissman JC, Eisenher DM, Oswald WJ (1980) Cultivation on sewage of microalgae harvestable by microstrainer. Progress Report. Sanitary Engineering Research Laboratory, University of California, Berkeley
Blanken W, Janssen M, Cuaresma M, Libor Z, Bhaiji T, Wijffels RH (2014) Biofilm growth of *Chlorella sorokiniana* in a rotating biological contactor based photobioreactor. Biotechnol Bioeng 111:2436–2445
Boonma S, Chaiklangmuang S, Chaiwongsar S, Pekkoh J, Pumas C, Ungsethaphand T, Tongsiri S, Peerapornpisal Y (2014) Enhanced carbon dioxide fixation and bio-oil production of a microalgal consortium. Clean Soil Air Water 43:761–766
Bose S, Das C (2013) Preparation and characterization of low cost tubular ceramic support membranes using sawdust as a pore-former. Mater Lett 110:152–155
Brooks BW, Lazorchak JM, Howard MDA, Johnson M-VV, Morton SL, Perkins DAK, Reavie ED, Scott GI, Smith SA, Steevens JA (2016) Are harmful algal blooms becoming the greatest inland water quality threat to public health and aquatic ecosystems? Environ Toxicol Chem 35:6–13
CA (Comprehensive Assessment of Water Management in Agriculture) (2007) Water for food, water for life: a comprehensive assessment of water management in agriculture. Earthscan/International Water Management Institute, Colombo/London
Caldwell DH (1946) Sewage oxidation pond performance, operation and design. Sew Works J 3:433–458. Available online: http://www.jstor.org/stable/25030250. Accessed on 12 Nov 2015
Cardinale BJ (2011) Biodiversity improves water quality through niche partitioning. Nature 472:86–89
Chanakya H, Mahapatra D, Ravi S, Chauhan V, Abitha R (2012) Sustainability of large-scale algal biofuel production in India. J Indian Inst Sci 92(1). [cited 16.1.204]. Available at: http://journal.iisc.ernet.in/index.php/iisc/article/view/23/23
Chen L, Chen J, Zhang X, Xie P (2016) A review of reproductive toxicity of microcystins. J Hazard Mater 301:381–399
Christenson LB, Sims RC (2012) Rotating algal biofilm reactor and spool harvester for wastewater treatment with biofuels by-products. Biotechnol Bioeng 109:1674–1684
Craggs RJ, Heubeck S, Lundquist TJ, Benemann JR (2011) Algal biofuels from wastewater treatment high rate algal ponds. Water Sci Technol 63:660–665
Dana GV, Kuiken T, Rejeski D, Snow AA (2012) Synthetic biology: four steps to avoid a synthetic-biology disaster. Nature 483:29
Davis R, Aden A, Pienkos PT (2011) Techno-economic analysis of autotrophic microalgae for fuel production. Appl Energy 88:3524–3531
Dong Q, Zhao X (2004) In situ carbon dioxide fixation in the process of natural astaxanthin production by a mixed culture of Haematococcus pluvialis and Phaffia rhodozyma. Catal Today 98:537–544
Dubey KK, Kumar S, Dixit D, Kumar P, Kumar D, Jawed A, Haque S (2015) Implication of industrial waste for biomass and lipid production in Chlorella minutissima under autotrophic, heterotrophic, and mixotrophic grown conditions. Applied biochemistry and biotechnology 176(6):1581–1595
Efroymson RA, Dale VH, Langholtz MH (2016) Socioeconomic indicators for sustainable design and commercial development of algal biofuel systems. GCB Bioenergy 1673:1–19

Escobar J, Brenner M, Whitmore TJ, Kenney WF, Curtis JH (2008) Ecology of testate amoebae (thecamoebians) in subtropical Florida lakes. J Paleolimnol 40(2):715–731
Flynn KJ, Greenwell HC, Lovitt RW, Shields RJ (2010) Selection for fitness at the individual or population levels: modelling effects of genetic modifications in microalgae on productivity and environmental safety. J Theor Biol 263:269–280
Gani P, Sunar NM, Matias-Peralta H, Latiff AAA, Parjo UK, Razak ARA (2015) Phycoremediation of wastewaters and potential hydrocarbon from microalgae: a review. Adv Environ Biol 9(20). Special):1–8
Garnikar D (2002) Accumulation of copper, mercury and lead in Spirulina platensis studied in Zarrouk's medium. J KMITNB 12(4):333–335
Gebreslassie BH, Waymire R, You FQ (2013) Sustainable design and synthesis of algae-based biorefinery for simultaneous hydrocarbon biofuel production and carbon sequestration. AICHE J 59:1599–1621
Guarnieri MT, Pienkos PT (2015) Algal omics: unlocking bioproduct diversity in algae cell factories. Photosynth Res 123:255–263
Hall CAS, Benemann JR (2011) Oil from algae? Bioscience 61:741–742
Halpern BS, Walbridge S, Selkoe KA, Kappel CV, Micheli F, D'Agrosa C, Bruno JF, Casey KS, Ebert C, Fox HE, Fujita R, Heinemann D, Lenihan HS, Madin EMP, Perry MT, Selig ER, Spalding M, Steneck R, Watson R (2008) A global map of human impact on marine ecosystems. Science 319:948–952
Handler RM, Canter CE, Kalnes TN, Lupton FS, Kholiqov O, Shonnard DR, Blowers P (2012) Evaluation of environmental impacts from microalgae cultivation in open-air raceway ponds: analysis of the prior literature and investigation of wide variance in predicted impacts. Algal Res 1:83–92
Higa T, Wididana GN (1991) Changes in the soil microflora induced by Effective Microorganisms. In: Parr JF, Hornick SB, Whitman CE (eds) Proceedings of the First International Conference on Kyusei Nature Farming. U.S. Department of Agriculture, Washington, DC, pp 153–162
Johnson KR, Admassu W (2013) Mixed algae cultures for low cost environmental compensation in cultures grown for lipid production and wastewater remediation. J Chem Technol Biotechnol 88:992–998
Kesaano M, Sims RC (2014) Algal biofilm based technology for wastewater treatment. Algal Res 5:231–240
Kim JK, Kottuparambil S, Moh SH, Lee TK, Kim Y-J, Rhee J-s, Choi E-M, Kim BH, Hu YJ, Yarish C, Han T (2015) Potential application of nuisance microalgae blooms. J Appl Phycol 27:1223–1234
Kong Q-X, Li L, Martinez B, Chen P, Ruan R (2010) Culture of microalgae *Chlamydomonas reinhardtii* in wastewater for biomass feedstock production. Appl Biochem Biotechnol 160:9–18
Langer ES (2008) Batch Failure Rates in Biomanufacturing. Genetic Engineering and Biotechnology News 28(14)
Lundquist TJ, Woertz IC, Quinn NWT, Benemann JR (2010) A realistic technology and engineering assessment of algae biofuel production. Energy Biosciences Institute, University of California. Available from: http://works.bepress.com/tlundqui/5
Luo S, Chen B, Lin L, Wang X, Tam NFY, Luan T (2014) Pyrene degradation accelerated by constructed consortium of bacterium and microalga: effects of degradation products on the microalgal growth. Environ Sci Technol 48:13917–13924
Mane PC, Bhosle AB, Jangam CM et al (2011) Bioadsorption of selenium by Pretreated Algal Biomass Advances in applied. Sci Res 2(2):202–207
Mani D, Kumar C (2014) Biotechnological advances in bioremediation of heavy metals contaminated ecosystems: an overview with special reference to phytoremediation. Int J Environ Sci Technol 11(3):843–872
Markou G, Nerantzis E (2013) Microalgae for high-value compounds and biofuels production: a review with focus on cultivation under stress conditions. Biotechnol Adv 31:1532–1542

Mcneil B, Giavasis I, Archer D, et al. (2013) Microbial production of food ingredients, enzymes and nutraceuticals. 1st ed. Mcneil B, Giavasis I, Archer D, et al., editors. Woodhead Publishing, Cambridge

Merchant SS et al (2007) The Chlamydomonas genome reveals the evolution of key animal and plant functions. Science 318:245

Mildbrandt A, Jarvis E (2010) Resource Evaluation and site selection for microalgae production in India. National Renewable Energy Laboratory [cited 9.12.2013]. Available at: http://www.nrel.gov/docs/fy10osti/48380.pdf

Mulbry W, Kondrad S, Pizarro C, Kebede-Westhead E (2008) Treatment of dairy manure effluent using freshwater algae: algal productivity and recovery of manure nutrients using pilot-scale algal turf scrubbers. Bioresour Technol 99:8137–8142

Myhre G, Shindell D., Bréon FM, Collins W, Fuglestvedt J, Huang J, Koch D, Lamarque JF, Lee D, Mendoza B, Nakajima T, Robock A, Stephens G, Takemura T, Zhang H (2013) Anthropogenic and natural radiative forcing. IPCC Fifth Assessment Report, 2014 (AR5)

Necton SA (1990) Estudo de localizac¸a˜o de unidade industrial para produc¸a˜o de microalgas em portugal. Olha˜o, Portugal: Necton, Companhia Portuguesa de Culturas Marinhas, S.A. p. 71

Nirmal Kumar JI, Cini O (2012) Removal of heavy metals by biosorption using freshwater alga *Spirogyra hyalina*. J Environ Biol 33:27–31

Olguin EJ, Sanchez-Galvan G (2012) Heavy metal removal in phytofiltration and phycoremediation: the need to differentiate between bioadsorption and bioaccumulation. New Biotechnol 30:3–8

Oswald WJ, Golueke CG (1960) Biological transformation of solar energy. In Advances in applied microbiology (Vol. 2, pp. 223–262). Academic Press

Paerl HW, Huisman J (2009) Climate change: a Catalyst for global expansion of harmful cyanobacterial blooms. Environ Microbiol Rep 1(1):27–37

Pandey VC, Singh JS, Singh DP, Singh RP (2014) Methanotrophs: promising bacteria for environmental remediation. Int J Environ Sci Technol 11:241–250. https://doi.org/10.2134/jeq2011.0179

Pandhal J, Choon WL, Kapoore RV, Russo DA, Hanotu J, Grant Wilson IA, Desai P, Bailey M, Zimmerman WJ, Ferguson AS (2018) Harvesting environmental microalgal blooms for remediation and resource recovery: a laboratory scale investigation with economic and microbial community impact assessment. Biology 7:4. https://doi.org/10.3390/biology7010004

Pankratz S, Oyedun AO, Zhang X, Kumar A (2017) Algae production platforms for Canada's northern climate. Renew Sust Energ Rev 80:109–120

Parameswari E, Lakshmanan A, Thilagavathi T (2010) Phycoremediation of heavy metals in polluted waterbodies. Elec J Env Agricult Food Chem Title 9(4):808–814

Park JBK, Craggs RJ, Shilton AN (2011) Wastewater treatment high rate algal ponds for biofuel production. Bioresour Technol 102:35–42

Patel VK, Sahoo NK, Patel AK, Rout PK, Naik SN, Kalra A (2017) Exploring microalgae consortia for biomass production: a synthetic ecological engineering approach towards sustainable production of biofuel feedstock. In Algal biofuels (pp. 109–126). Springer, Cham

Paul VJ (2008) Global warming and cyanobacterial harmful algal blooms. In: Hudnell HK (ed) Cyanobacterial harmful algal blooms: state of the science and research needs, Advances in experimental medicine and biology, vol 619. Springer, New York, pp 239–257

Picot B, Bahlaoui A, Moersidik S, Baleux B, Bontoux J (1992) Comparison of the purifying efficiency of high rate algal pond with stabilization. Pond Water Sci Technol 25:197–206

Prasanna R, Kumar V, Kumar S, Yadav AK, Tripathi U, Singh AK et al (2002) Methane production in rice soils is inhibited by cyanobacteria. Microbiol Res 157:1–6. https://doi.org/10.1078/0944-5013-00124

Priyadarshani I, Sahu D, Rath B (2011) Microalgal bioremediation: current practices and perspectives. J Biochem Technol 3(3):299–304

Qin B, Li W, Zhu G, Zhang Y, Wu T, Gao G (2015) Cyanobacterial bloom management through integrated monitoring and forecasting in large shallow eutrophic Lake Taihu (China). J Hazard Mater 287:356–363

Quinn JC, Davis R (2015) The potentials and challenges of algae based biofuels: a review of the techno-economic, life cycle, and resource assessment modeling. Bioresour Technol 184:444–452

Rawat I, Kumar RR, Mutanda T, Bux F (2011) Dual role of microalgae: phycoremediation of domestic wastewater and biomass production for sustainable biofuels production. Appl Energy 88:3411–3424

Renuka N, Sood A, Ratha SK, Prasanna R, Ahluwalia AS (2013) Evaluation of microalgal consortia for treatment of primary treated sewage effluent and biomass production. J Appl Phycol 25:1529–1537

Richardson JW, Johnson MD, Outlaw JL (2012) Economic comparison of open pond raceways to photo bio-reactors for profitable production of algae for transportation fuels in the Southwest. Algal Res 1:93–100

Rismani-Yazdi H, Haznedaroglu BZ, Bibby K, Peccia J (2011) Transcriptome sequencing and annotation of the microalgae *Dunaliella tertiolecta*: pathway description and gene discovery for production of next-generation biofuels. BMC Genomics 12:148

Sahoo NK (2010) Nutrient removal, growth response and lipid enrichment by a phytoplankton community. J Algal Biomass Util 1(3):1–28

Schindler DW, Hecky RE, Findlay DL, Stainton MP, Parker BR, Paterson MJ, Beaty KG, Lyng M, Kasian SEM (2008) Eutrophication of lakes cannot be controlled by reducing nitrogen input: results of a 37–year whole–ecosystem experiment. Proc Natl Acad Sci U S A 105(32):11254–11258

Shukia SP, Singh JS, Kashyap S, Giri DD, Kashyap AK (2008) Antarctic cyanobacteria as a source of phycocyanin: an assessment. Ind J Marine Sci 37:446–449

Singh JS (2015) Microbes: the chief ecological engineers in reinstating equilibrium in degraded ecosystems. Agric Ecosyst Environ 203:80–82. https://doi.org/10.1016/j.agee.2015.01.026

Singh SK, Bansal A, Jha MK et al (2011) An integrated approach to remove Cr(VI) using immobilized *Chlorella minutissima* grown in nutrient rich sewage waste water. Bioresour Technol:1–9

Smith GJ, Daniels V (2018) Algal blooms of the 18th and 19th centuries. Toxicon 142:42–44

Snow AA, Smith VH (2012) Genetically engineered algae for biofuels: a key role for ecologists. Bioscience 62:765–768

Spurgeon DJ, Jones OAH, Dorne J-LCM, Svendsen C, Swain S, Stürzenbaum SR (2010) Systems toxicology approaches for understanding the joint effects of environmental chemical mixtures. Sci Total Environ 408:3725–3734

Subashchandrabose SR, Balasubramanian R, Megharaj M, Venkateswarlu K, Naidu R (2011) Consortia of cyanobacteria/microalgae and bacteria: biotechnological potential. Biotechnol Adv 29(6):896–907

Sun PF, Lin H, Wang G, Lu LL, Zhao YH (2015) Preparation of a new-style composite containing a key bioflocculant produced by *Pseudomonas aeruginosa* ZJU1 and its flocculating effect on harmful algal blooms. J Hazard Mater 284:215–221. https://doi.org/10.1016/j.jhazmat.2014.11.025

Suresh Kumar K, Dahms H-U, Won E-J, Lee J-S, Shin K-H (2015) Microalgae – A promising tool for heavy metal remediation. Ecotoxicol Environ Saf 113:329–352

Tabatabaei M, Tohidfar M, Jouzani GS, Safarnejad M, Pazouki M (2011) Biodiesel production from genetically engineered microalgae: future of bioenergy in Iran. Renew Sust Energ Rev 15:1918–1927

Talbot P, De la Noüe J (1993) Tertiary treatment of wastewater with Phormidium bonheri (Schmidle) under various light and temperature conditions. Water Res 27:153–159

Tang JX, Hoagland KD, Siegfried BD (1997) Differential toxicity of atrazine to selected freshwater algae. Bull Environ Contam Toxicol 59:631–637

Thomas DG, Minj N, Mohan N, Rao PH (2016) Cultivation of microalgae in domestic wastewater for biofuel applications – An upstream approach. J Algal Biomass Util 7(1):62–70

Tiwari S, Singh JS, Singh DP (2015) Methanotrophs and CH4 sink: effect of human activity and ecological perturbations. Climate Change Environ Sustain 3:35–50. https://doi.org/10.5958/2320-642X.2015.00004.6

Travieso L, Canizares RO, Borja R et al (1999) Heavy metal removal by microalgae. Bull Environ Contam Toxicol 62:144–151

United Nation World Water Development Report. United Nations Educational, Scientific and Cultural Organization 2014. ISBN 978-92-3-104259-1

Usher PK, Ross AB, Camargo-Valero MA, Tomlin AS, Gale WF (2014) An overview of the potential environmental impacts of large-scale microalgae cultivation. Biofuels 5(3):331–349

Verween A, Queguineur B, Casteleyn G, Fernandez GA, Leu S, Biondi N, et al. (2011) Algae and aquatic biomass for a sustainable production of 2nd generation biofuels: deliverable 1.6-mapping of natural resources in artificial, marine and freshwater bodies [Internet]. Available from: http://www.aquafuels.eu/deliverables.html

Wilkie AC, Mulbry WW (2002) Recovery of dairy manure nutrients by benthic freshwater algae. Bioresour Technol 84:81–91

Yen H-W, Chen P-W, Chen L-J (2014) The synergistic effects for the co-cultivation of oleaginous yeast-Rhodotorula glutinis and microalgae-*Scenedesmus obliquus* on the biomass and total lipids accumulation. Bioresour Technol 184:148–152

Zeng Y, Zhao B, Zhu L, Tong D, Hu C (2013) Catalytic pyrolysis of natural algae from water blooms over nickel phosphide for high quality bio-oil production. RSC Advances 3(27): 10806–10816

The Diatoms: From Eutrophic Indicators to Mitigators

Aviraj Datta, Thomas Kiran Marella, Archana Tiwari, and Suhas P. Wani

1 Introduction

Human activities can bring negative effects in the environment. Any substance that can cause a negative effect in the environment is considered a pollutant which has to be controlled in order to reduce adverse impacts. Pollutants can come from different sources, although human activities like agriculture, change of land use, and others are one source of production of pollutants which can affect the environment (Gottschalk et al. 2011). Elevated nutrients contribute to poor lake ecosystem, which highlights the need for efficient nutrient removal strategies that enable us to protect or restore the water bodies from eutrophication. Biological elements, such as macroinvertebrate species, macrophytes, diatoms, and zooplankton, have been used to monitor nutrient changes (Lougheed et al. 2007). Diatoms are one of the most explored species for water quality assessment around the world, due to their sensitive time-dependent response (Stevenson 2014).

The main purpose of developing biological monitoring strategies is to enable researchers to assess water quality of lotic and lentic systems. This approach makes use of aquatic biota to evaluate complex and dynamic changes in water quality. Biotic communities are generally sensitive to inflow of chemicals and change in physical factors that bring about a change in their morphology and diversity which reflects the physiochemical conditions of the ecosystems. This approach uses biota to represent the general environmental conditions and assess environmental quality of the monitored ecosystem.

The biotic organisms of water include macroinvertebrates, phytoplankton, zooplankton, phyto-benthic macroinvertebrates, and the fish communities (De Pauw

A. Datta (✉) · T. K. Marella · S. P. Wani
International Crops Research Institute for the Semi-Arid Tropics, Patancheru, India

A. Tiwari
AMITY University, Noida, India

S. K. Gupta, F. Bux (eds.), *Application of Microalgae in Wastewater Treatment*,
https://doi.org/10.1007/978-3-030-13913-1_2

et al. 2000). The ecological indicators are used based on species diversity of these organisms to monitor water quality, hydrology, and the overall health of a water body. Indicators species are used to monitor the levels of toxins, physicochemical parameters, and the overall nature of the water resource (Nixon 2009).

The role played by algae is crucial in all water ecosystems. They are identified as strands or filaments in rivers and along the lake shorelines and act as a link between the biotic and abiotic environments. The algal community assemblage and abundance change in response to water quality fluctuations, and this can be attributed to their direct reliance on making them sensitive to water quality changes. The sensitivity of algae to water quality changes makes them useful as bioindicators of the physical and chemical properties in water environments. Diatoms are single-celled organisms and basically the lone member group of algal organisms applied in aquatic studies until recent years (Ruhland et al. 2008). They are represented by over 100,000 species all over the world and are identified in rivers and from the lake shorelines as brown, slimy covering on submerged substrates such as mud, sand, macrophytes, or rocks. The benefits of diatoms used as bioindicators include the following: they are easily identifiable under a microscope, and they have cell walls with each species having specific shape and morphological structure. Diatom classification is well detailed and defined, as well as various species tolerance to environmental changes. The species cell walls composed of silica from silicon resist decomposition, and so can be preserved, thus providing a permanent record whereby short- or long-term changes can be assessed (Cox 1991). Moreover, historical conditions of water can be projected by use of the species cell walls preserved in sediments at the bottom of the lakes (Lavoie and Campeau 2010).

Diatoms play a major role in biomass production and sinking of atmospheric greenhouse gas in oceans. Diatoms are responsible for about 20% of the total photosynthetic CO_2 fixation, which is equivalent to the photosynthetic activity of all rainforests combined and approximately 40% of annual marine biomass production (Falkowski et al. 1998). Diatoms are exceedingly robust and can inhabit virtually all photic zones from the equator to arctic where they are extensively studied for their usefulness as indicators of changes in physiochemical conditions due to their rapid response to any slight changes. Thus, diatoms show high degree of flexibility in varied culture conditions that could be useful for their use in biotechnological applications despite challenging conditions. Diatoms are sensitive to changes in their aquatic environments and are reliable indicators of the water quality. The reason for this is their reproduction rate, which allows for significant increase in population of a given species under favorable conditions while other species concurrently decrease or disappear.

Diatoms are members of the heterokont class of algae, which are highly different, and have a more complex evolutionary history than green algae and vascular plants. The evolutionary age of diatoms has been estimated from molecular genetic data as 165–240 M·ya (Kooistra and Medlin 1996), which is in reasonable agreement with the fossil record. Diatoms are secondary endosymbionts and part of the heterokont group, which includes other silica-forming algae. Diatom genomes are a complex mixture derived from combination of higher organisms of both plant and animal origin. This unique combination gave diatoms a peculiar metabolic profile and process which is different from other algae (Armbrust et al. 2004). The evolutionary success of diatoms is also connected to their cell wall which is made up of silica which needs lower energy requirement to build when compared with (Raven 1983).

Diatoms are divided into four major groups based on their cell wall structure, radial centrics, bipolar and multipolar centrics, araphid pennates, and raphid pennates (Fig. 1). All these groups have evolved under decreasing CO_2 levels during the Mesozoic era (Armbrust 2009). This has led to an advanced carbon-concentrating mechanism making them highly adaptable to changing CO_2 levels.

Silica cell wall gives diatom algae an advantage of enhanced sinking rate which results in increased carbon burial in shallow seas and continental margins (Smetacek 1999; Falkowski et al. 2005) and are known to be primary contributors to present nascent petroleum reserves.

Diatoms are useful indicators of water quality because of their diversity in varied environments, species richness, and dynamic response to changes in physicochemical conditions of surrounding ecosystem (Dixit et al. 1992; McCormick and Cairns 1994). Diatoms play a significant role in controlling and biomonitoring of organic pollutants, heavy metals, hydrocarbons, PCBs, pesticides, etc. in aquatic ecosystems. Although diatoms are extensively studied for their role as indicators of different kinds of water pollution, their application in phycoremediation of polluted water bodies has just started. In this chapter we explore the potential of diatoms as indicator species for pollution and their implications on wastewater treatment.

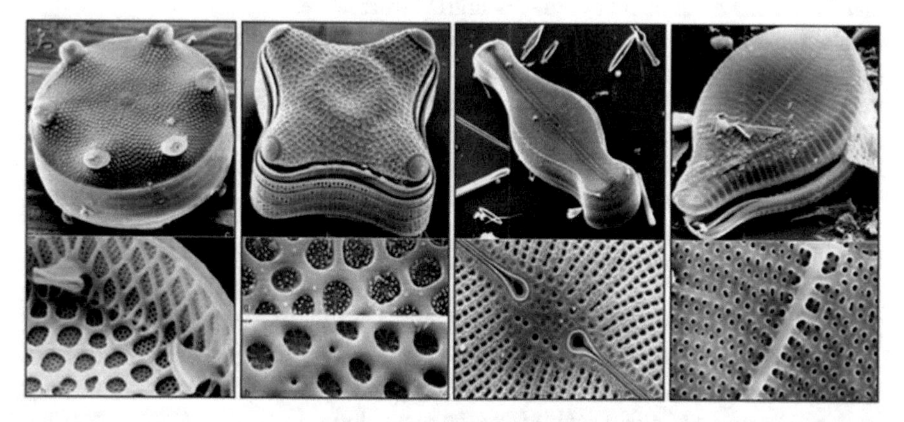

Fig. 1 Different silica frustule shapes and intrinsic frustule designs of diatoms *Aulacodiscus* sp., radial centric; *Amphitetras* sp., polar centric; *Didymosphenia* sp., raphid pennate; and *Podocystis* sp., araphid pinnate (Kröger and Poulsen 2008)

1.1 Why Diatoms as Bioindicators, Sensitivity of Diatoms to Physiochemical Changes

It is paramount to understand the biological, chemical, and physical processes of any water body in order to determine the mass balance of pollutants into and out of the system. The pollution of fresh water bodies from excess nutrients and hazardous chemicals is one of the greatest environmental issues of the developing world. For successful mitigation to these issues, along with treatment efficient monitoring approaches are needed. Ecosystem monitoring employs physical-, chemical-, and biological-based methods for routine monitoring. Although chemical and physical methods provide instant results, they do not provide us with information on previous dynamic changes of the ecosystem, but with biological monitoring we can get information on long-term effects on the ecosystem by different physicochemical fluxes. Therefore, complementing biological monitoring with physicochemical monitoring is the right way to monitor water quality.

In order to be considered as bioindicators, the species which are being monitored should show a strong correlation with a physiochemical parameter, should have a narrow tolerance range to that parameter, and should be commonly found in the sample. Diatoms meet all these criteria which make them ideal for biomonitoring water quality. Diatoms are present in all aquatic ecosystems due to which same species can be compared for assessment of different habitats like lakes, wetlands, oceans, streams, etc. Diatoms grow as attached biofilm on solid substrates so they can be monitored by sampling these substrates even when the water body is dry. Due to their faster growth rate compared to other species, they can give us an early warning to impeding pollution and water quality restoration. Diatom-based monitoring is cost-effective when compared with other methods, and they give an added advantage of retaining the samples for longer times for long-term studies. These attributes make diatom-based biomonitoring of water habitats an important parameter for habitat assessment in many countries worldwide.

Diatom-based water quality indices have been developed for monitoring water quality in many geographic areas. Nutrient influxes along with some physicochemical parameters are key factors which influence diatom growth and survival. Diatoms respond to nutrient influx by changing their community structure in terms of species response, where specific diatom species dominate nutrient-rich waters, whereas others prefer nutrient-depleted conditions. This dynamic response makes diatoms ideal indicators of nutrient enrichment. Physical and chemical monitoring methods where water samples are picked at one defined time cannot provide this dynamic nutrient influx data. Monitoring nonpoint source pollution of inorganic nutrients like phosphorus is quite difficult even with multiple sampling efforts due to its sudden fluctuations. With diatom-based monitoring, when diatom communities are exposed to cumulative nutrient, diatoms respond by changing their community structure leading to better monitoring efficiency (Table 1).

Nutrient monitoring based on diatoms is used widely since they are the major primary producers with an ability to strongly reflect their ecosystem nutrient

Table 1 Diatom-based monitoring of different parameters from varied ecosystems reported in literature

Ecosystem	Monitoring parameter/impact	References
Rivers, streams	Eutrophication	Lobo et al. (2004)
	Heavy metal contamination	Leguay et al. (2016)
Lakes	Eutrophication	Poulíčková et al. (2004)
	Heavy metal contamination	Cantonati et al. (2014)
Marine benthos	Various environmental parameters	Weckström and Juggins (2006)
Marine biofilm	Eutrophication	Cibic and Blasutto (2011)

concentrations by their community structure. Diatoms are useful for monitoring nutrient influx into lakes due to their relative abundance and richness which can provide a sensitive index for physicochemical changes (Black et al. 2011). Some diatom-based models measure interaction between diatom community dynamics, and nutrients can provide nutrient concentration information which will be useful to develop efficient management practices. Macroinvertebrates and fishes have also been used as biomonitors (Hering et al. 2006), but diatoms have an advantage due to their increased sensitivity (Leira and Sabater 2005). Benthic diatoms are known to be influenced more by local factors like major nutrients, pH, etc. than large-scale factors like climate and geology (Stevenson and Pan 1999; Leland 1995). Benthic diatoms also respond well to hydro-morphological modification and nutrient enrichment (Hering et al. 2006; Rott et al. 2003).

1.2 *Taxonomy of Indicator Species from Different Environments*

Diatoms are ecologically diverse and extensively distributed in both fresh and saline habitats. There are diatom species that are very tolerant with a wide ecological valence, yet other species have tolerance levels that are distinct and narrow optima for many environmental variables; these attributes enable them to be remarkably applied in quantifying environmental features with great precision (Dixit et al. 1992). Excess nutrient loading and organic contamination have been regularly monitored using diatoms and indices of various types developed to quantify the quality of water (Rott et al. 2003). Some of the extreme pollution-tolerant species are *Navicula atomus*, *Nitzschia palea*, *Gomphonema parvulum*, *Navicula cryptocephala*, and *Navicula minima*, and species sensitive to extreme pollution are *Achnanthes biasolettina*, *Cocconeis placentula*, and *Gomphonema minutum*. Heavy metal pollution can result in cell wall deformities and loss of diversity caused which are useful indicators to monitor heavy metal pollution (Walsh and Wepener 2009).

Intrinsic silica patterns on diatom frustule make diatoms unique in terms of taxonomic identification up to strain level compared to other algal species. Species diversity and biomass in terms of bio-volume are the two main criteria which are based solely on diatom-based monitoring. Sampling habitat plays a significant role in effectiveness of biomonitoring. Sampling of rocks and hard surfaces is recommended in the European Union (Kelly et al. 1998), whereas in US programs random sampling of any available substrate is recommended (Weilhoefer and Pan 2007). Species composition and biomass in terms of cell bio-volume are two of the key parameters on which diatoms can be differentiated from other algae and microbes (Table 2).

1.3 Diatom-Based Water Quality Indices

Water chemistry significantly influences diatom assemblage communities. Diatom development and structure respond extensively to eutrophication, organic pollution, fluctuations in conductivity and pH, and elevated levels of sediments suspended in water. Most researches have documented relationships between concentration of nutrients and assemblage of diatom communities and likened a high amount of the community difference proportionally to the recorded nutrients in the water bodies

Table 2 Indicator species of diatoms for different physiochemical parameters of wastewater

Species	Indicator for
Nitzschia palea	Organic pollution
Nitzschia fonticola	Organic pollution
Fragilaria capucina	Heavy metal
Achnanthidium minutissimum	Heavy metal
Fragilaria ulna var. *acus (Kütz.) Lange-Bert.*	High conductivity
Eunotia sp.	Eutrophication
Diatoma vulgare	Eutrophication
Eunotia exigua, Gomphonema angustum	pH
Amphora veneta, Gomphonema rautenbachiae	pH
Melosira nummuloides	Salinity
Melosira sp.	Flow rate
Cocconeis sp.	Flow rate
Gomphoneis herculeana, Achnanthidium sp., *Achnanthes subhudsonis* var. *kraeuselii*	Eutrophication
Gomphoneis herculeana, Achnanthidium sp., *Achnanthes subhudsonis* var. *kraeuselii*	Low total phosphate (TP)
Luticola goeppertiana, Navicula recens, Nitzschia inconspicua, Nitzschia palea, Rhopalodia sp., *Eunotia* sp.	High TP
A. minutissimum, Cocconeis placentula, Surella construens, Sorella pinnata	Nitrogen (autotrophic)
Gomphonema parvulum, Eolimna minima, and *Nitzschia palea*	Nitrogen (heterotrophic)
Navicula sp., *Nitzschia* sp., *Surirella* sp.	Silt tolerant
Eunotia sp., *Pinnularia* sp.	Acidic pH

(Torrisi et al. 2010). Others, on the other hand, have noted significant correlations among type of substrate, dissolved oxygen, and alkalinity (Blinn and Herbst 2003).

Lake classification based on algae is well documented in the literature (Stoermer 1978), which are listed in Table 3. Many classification systems employ diatoms to assess the water quality (Hecky and Kilham 1973; Carpelan 1978).

1.4 Studies on Water Quality Monitoring Using Diatoms

Anthropogenic pollution of surface waters in many countries has led to increased stress on water ecosystems. To understand and monitor its effect, we need to place more emphasis on developing trophic variables. In some European countries, several diatom-based indices have been employed and are being used routinely (Prygiel et al. 1999). The European Water Framework Directive (WFD) (Bennion and Battarbee 2007) has encouraged the application of ecological studies to understand the impact of anthropogenic pollution on fresh water ecosystems (Muxika et al. 2007). The WFD mandates the use of ecological monitoring of rivers and lakes based on biological indicators like microalgae, fish, invertebrates, macrophytes, etc. of which diatoms are most commonly used species (King et al. 2006). In Latvia, Furse et al. (2006) reported that diatom-based diversity indices correlated strongly with environmental variables when compared with macrophytes and fish. With conversion of community response to a particular gradient into a continuously monitored variable by using diatoms, we can simplify ecological monitoring of water bodies. Studies related to effects of eutrophication have shown that diatom metrics detect eutrophication more efficiently than other metrics studies (Hering et al. 2006). All these studies providing strong evidence of usefulness of diatoms as

Table 3 Diatom-based water quality indices

Abbreviation	Nomenclature	Reference
CEE	Commission for economical community metric	Descy and Coste (1991)
DESCY	Descy's pollution metric	Descy (1979)
EPID	Pollution metric based on diatoms	Dell'Uomo (1996)
IBD	Biological diatom index	Prygiel et al. (1999)
IDG	Generic diatom index	Prygiel et al. (1996)
IDAP	Indice Diatomique Artois-Picardie	Prygiel et al. (1996)
L&M	Leclercq and Maquet's pollution index	Leclercq and Maquet (1987)
ROOT	Trophic metric	Rott et al. (1999)
SLAD	Sla'decek's pollution index	Sládeček (1986)
TDI	Trophic diatom index	Kelly and Whitton (1995)
WAT	Watanabe et al. pollution metric	Watanabe (1988)
ABSS	Abundance of reference taxa	Delgado et al. (2010)

bioindicators resulted in their increased use as tools for efficient monitoring of water quality (Gómez and Licursi 2001; Wu and Kow 2002).

Although several authors (Stoermer and Yang 1970; Tilman et al. 1982) showed that diatoms are useful indicators of water quality, still the development of new indices is necessary for many geographic locations before their widespread application in monitoring studies. The development and wide use of software packages, such as Omnidia, which facilitates calculation of indices, is quite helpful (Eloranta and Soininen 2002; García et al. 2008). In North America, the use of diatom metrics based on sensitive and tolerant species is more widespread (Fore and Grafe 2002; Passy et al. 2004).

2 Role of Diatoms as Bioindicator in the Performance Evaluation of Constructed Wetland

Constructed wetlands are ecological systems which are influenced by a combination of physicochemical and biological processes. In order to maintain them, a balance between these processes is paramount. In a constructed wetland, ecological food web consists of planktonic and benthic algae, bacteria, and other higher trophic organisms, but the majority of the primary productivity is fueled by sunlight and nutrients available in influent wastewater. Algae are a part of any wet habitat, and they are an integral part of any wetland ecosystem. Many different species of algae inhabit CWL depending on the type of vegetation. In CWL many types of vegetation are promoted depending on the design like free floating, rooted floating, submerged aquatic, emergent aquatic, and shrubs. All these different vegetation techniques are used in the presence of wastewater at different depths; this leads to a congenial environment for benthic diatom algae which grow as colonies on submerged substrates and include epiphytic, epipsammic, epipelic, and epilithic forms. Structure and productivity of benthic diatom community is influenced by nutrient loading into CWL (Gaiser et al. 2014). So by monitoring the dynamics of diatoms on submerged and emerging plants and other substrates, we can access water quality and treatment efficiency of a CWL, and it can be a suitable alternative for physiochemical analysis to evaluate wetland performance and evaluation.

2.1 Assessment of Wastewater Characteristics Through Diatom Species Diversity

Using diatoms as indicators of wastewater quality can be attributed to their presence in diverse ecosystems, sensitivity to changes in nutrient and environmental conditions, and easiness to access their diversity. Their significance in water ecosystems is linked to their primary role in aquatic food webs and biogeochemical cycle (Lamberti 1996; Mulholland et al. 2008).

Diatoms are diverse species which are present in all wetland ecosystems throughout the world. They show dynamic sensitivity response to different range of water pollution. Their fast growth rate enables them to inhabit new habitats in rapid time which makes their species monitoring ideal for studying their response to environmental change. Their fast response in terms of species diversity and abundance gives them a competitive advantage over physicochemical sampling, where sudden spike in a parameter can lead to ecological significant fluctuations which cannot be monitored over time. Benthic diatoms are attached to substrates so they are confined to particular habitats with specific physiochemical characters which make them ideal for biomonitoring of those environments (De la Rey et al. 2004). The species-specific response of diatom to varied conditions can be studied by their increase in biomass and species diversity (Patrick 1961). Benthic diatoms have been increasingly used to monitor physiochemical changes such as pH, conductivity and organic nutrients, eutrophication, and global warming problems.

2.2 Importance of Diatoms in Water Quality Management and Natural Food Production in Aquaculture

Microalgae are major contributors to nutrition in natural marine and fresh water ecosystems and in aquaculture. Being major primary producers in oceans, diatoms are major food for many marine invertebrates. Diatoms also contribute as natural food in intensive aquaculture systems by forming the base of aquatic food web. Dissolved oxygen (DO) is one of the major factors influencing aquatic animal's metabolism and growth. Decreased dissolved oxygen in intensive aquaculture systems is a serious concern. Artificial aeration using electrical aerators adds additional costs and risks. By growing diatoms we can increase the DO levels in ponds very rapidly, and as diatoms move in the water column depending on light requirement, DO increase will be achieved even in the middle and bottom of the ponds which is not the case when we use mechanical aeration.

In aquaculture ponds, some phytoplankton species are considered undesirable, especially blue-green algae (BGA), sometimes called cyanobacteria, and are particularly troublesome. Due to their higher light requirement, they always grow as mats on top of the water column resulting in decreased mixing of atmospheric gases into the water leading to less DO. BGA also produce smelly compounds which give off odor to cultured organisms which can result in poor meat taste. Some BGA species like *microcystis* produce toxins that can kill fish and shrimp. By growing diatoms we can efficiently reduce the problems associated with BGA growth. BGA mainly dominate the ponds when there is high nutrient content and high pH; by growing diatoms which can utilize nutrients much faster than BGA and also help in lowering pH by maintaining water quality, we can eradicate BGA growth in aquaculture ponds.

Aquaculture ponds contain bacteria and viruses that can infect cultured organisms and thus potentially devastate aquaculture farms. These same Bacteria and viruses can also infect diatoms, so diatoms have developed self-defense mechanisms to protect themselves like secretion of compounds that inhibit bacterial growth or viral attachment so by transferring this compounds to feeding animals and also some species of microalgae especially diatoms grow on surface of the fish and shellfish there by it induces immunity to many harmful water born bacterial and viral pathogens in shrimp, fish and shell fish.

Diatoms possess many advantages as natural food in aquaculture. Their size and shape are ideal for ingestion and easy digestion; their biochemical composition is ideal for culture species and zooplankton with the right amount of carbohydrates, proteins, and fats. Diatoms also provide many phytonutrients like PUFAs – e.g., EPA, arachidonic acid (AA), and DHA.

3 Potential to Treat Effluents from Constructed Wetlands

The role of algae as primary producers and in nutrient cycling of wetlands is well established (Wu and Mitsch 1998). Diatom assemblages are increasingly used in lake bio-assessment and paleolimnological studies (Dixit et al. 1992). Weilhoefer and Pan (2007) found that diatoms growing on submerged macrophytes, sediment surface, and in the water column in wetlands are ideal for diatom-based wetland bio-assessment. Although much research is focused on using diatom-based biomonitoring of wetlands, their importance in mitigation of eutrophication through excess nutrient removal and natural oxygenation through photosynthesis is not well explored.

Phycoremediation using microalgae was considered as one of the effective ways to deal with water pollution because it causes no secondary pollution and has high efficiency and low cost (Olguın 2003). Furthermore, microalgae have the ability to use inorganic nutrients (N and P), metabolize organic compound, and remove heavy metals and toxic organic compounds, which were then converted into biomass (Renuka et al. 2015). The biomass may be harvested and used in various applications, which then assist in purification of the wastewater, besides reducing the biochemical oxygen demand (BOD), resulting from biodegradation of the dead cells in the treated water.

Wetlands are ideal environments to mitigate nutrient-enriched surface waters. Denitrification process in wetlands was known to happen at sediment interphase (Payne 1991). But recent research has shown that maximum denitrification rate was achieved in upper 3–5 cm of wetland sediment which is dominated by periphyton attached to natural substrate especially dominated by diatoms (Eriksson and Weisner 1997). Ishida et al. (2008) studied the potential relation between the algal community structure and bacterial cell densities and denitrification rates and found that elevated denitrification rates were found in periphyton with high relative diatom concentration but not with green or blue-green algae. Diatoms also contributed to increase in bacterial density; this might be due to specific relationship between diatom and bacterial community structure.

3.1 *Effect on Dissolved Oxygen Concentration*

Dissolved oxygen is one of the key factors which influence the survival rate of not only cultured organism but also aerobic bacteria in aquatic ecosystems. Due to high amount of oxygen in the atmosphere (21%–300 mg L^{-1} of air), terrestrial organisms rarely experience its depletion. But in aquatic environments, solubility of oxygen is less than 1% of its solubility in the air. This amount of oxygen solubility in water depends on factors like pH, temperature, and surface area. At an atmospheric pressure of 1, saturated DO concentrations can reach a maximum of 9 mg L^{-1} at 20 °C which is much less when compared with its concentration in the air (Wetzel 1981).

The significant O_2 generation from algal photosynthesis can offset the cost incurred by wastewater treatment plants and aquaculturists for mechanical aeration (Mallick 2002). Oxygenation due to algal photosynthesis in oxidation ponds facilitates enhanced breakdown of organic and inorganic compounds by aerobic bacteria (Munoz and Guieysse 2006). Algal photosynthesis provides dissolved oxygen for aerobic bacteria, while the bacteria provide carbon, nitrogen, and phosphorus needed by algae for growth. The technique has been widely utilized in treating agricultural, municipal, and industrial wastewater. Algae growth especially blue-green algae which grow on water surface can hinder light penetration and gaseous exchange in ponds with submerged vegetation leading to hindered growth and lower DO levels.

3.2 *Residual Nutrient Removal Efficiency*

Diatom algae can dominate under nutrient-limiting and excess conditions; it is shown that diatom species outcompete non-nitrogen-fixing cyanobacteria under low nitrogen concentration in a eutrophic lake (Amano et al. 2012). Enhanced carbon fixation ability and concomitant nutrient removal capability increase the applicability of diatoms for CO_2 mitigation and wastewater treatment. Diatom algae produce oxygen during photosynthesis which acts as stimulant for heterotrophic bacterial growth which in turn can enhance bacterial degradation of organic pollutants (de Godos et al. 2010). Growth of benthic diatom *Nitzschia* sp. has resulted in enhanced aerobic bacterial activity in sediment layer which can lead to accelerated decomposition of organic matter (Yamamoto et al. 2008). Phthalate acid esters (PAEs) are commonly occurring priority pollutants and endocrine disruptors. Marine benthic diatom *Cylindrotheca closterium* has shown increased PAE removal rate in surface sediments. In bottom sediment it helped in increase of aerobic bacterial growth by photosynthetic oxygen, thereby resulting in a combination of bacteria-diatom-dependent PAE removal (Li et al. 2015). Diatom *Stephanodiscus minutulus* under optimum nutrient availability has shown increased uptake of PCB integer 2,2′,6,6′-tetrachlorobiphenyl (Lynn et al. 2007). Polyaromatic hydrocarbon (PAH) phytoremediation has limited success rate due to their high toxicity, but diatoms *Skeletonema costatum* and *Nitzschia* sp. have shown accumulation and degradation of phenanthrone (PHE) and fluoranthene (PLA), two typical PAHs (Hong et al. 2008). Diatom algae-produced O_2 can help in bacterial degradation of PAHs, phenolics, and organic solvents in benthic environments. Diatom *Amphora coffeaeformis* is known to accumulate herbicide mesotrione (Valiente Moro et al. 2012). The potential of diatom algae in biodegradation and accumulation of pollutants is enormous, but till date little research is done in this field.

3.3 Role in Pathogenic Bacteria Removal

Phytoplankton and bacteria have coexisted in the environment for millions of years. There exists a positive and negative allelopathic interaction between the both. Microalgal photosynthesis can enhance pathogen removal by changing the water physical parameters like increased pH, dissolved oxygen, and temperature (Ansa et al. 2011). Diatoms develop natural defense mechanism to protect against bacteria which can harm them and are often harmful to humans and animals also. Effective control of some harmful bacteria can be achieved, if we can grow natural diatom populations in wastewaters which have innate defense mechanism to control their growth. Diatoms secrete volatile and nonvolatile substances like fatty acids, esters, and polysaccharides as antibacterial compounds to control their growth (Lebeau and Robert 2003). Many of these hydrophobic molecules act as deterrents to bacteria by disrupting their cell signaling mechanisms during their adhesion to diatom cells. In a study on diatom *Navicula delognei* by Findlay and Patil (1984), fatty acids and sterols have shown strong antibacterial effect against pathogens like *Staphylococcus epidermidis*, *Salmonella enterica*, etc. Diatom *Phaeodactylum tricornutum*-produced eicosapentaenoic acid (EPA) has shown to inhibit gram-positive bacteria (Desbois et al. 2009). The same diatom has also shown inhibitory effect on multiresistant staph aureus (MRSA) (Desbois et al. 2009). *Chetoceros* sp. a marine planktonic diatom when maintained at higher concentration in the aquaculture ponds has lead lowered pathogenic bacteria like *Vibrio vulnificus* and simultaneously reduced propagation of viruses in shrimp production system. Diatom *Skeletonema costatum* was shown to inhibit *Vibrio*, a pathogen of fish and shellfish (Naviner et al. 1999). Many pathogenic bacteria are anaerobes which cause many respiratory, digestive, and urinary tract infections which are waterborne. Walden and Hentges (1975) have shown that anaerobic pathogenic intestinal bacteria growth was inhibited in the presence of oxygen; so to counter this, many anaerobes grow at oxygen-deficient zones especially in the sediment layer of wastewater ponds. Mechanical aeration cannot provide enough oxygen to these zones leading to proliferation of harmful bacteria. This can be reversed if we can promote diatom growth in these ponds with high sediment accumulation as benthic diatoms are known to produce high amount of oxygen even inside the sediment leading to aerobic zones.

4 Diatom-Based Excess Nutrient Removal from Eutrophic Water Bodies

Diatoms can be grown using agricultural and municipal wastewater. Wastewater contains macronutrients like nitrate, phosphate, silica, and other trace metals which are essential for algal growth. Hence growing algae in wastewater can be economically and environmentally beneficial as it can lead to decreased water treatment cost with an option of generating value added (Oswald 1988). The combination of three roles of microalgae in CO_2 mitigation, wastewater treatment, and biofuel production has the potential to decrease the use of fresh water for biofuel production and on climate change through CO_2 removal; however many crucial challenges like isolation of algal strains with high growth and nutrient uptake, integration of algal growth system with wastewater treatment systems, improved algal harvesting, and life cycle analysis are to be further explored to maximize the enormous potential of algal biofuels. Benthic diatoms are the dominant algal community in wastewater bodies, and they contribute significantly to nutrient removal and primary productivity in water.

Any wastewater treatment plant had to remove high concentrations of N and P present; if not treated this will cause eutrophication to downstream waterbodies. P is very difficult to remove in a conventional STP as there are very few phosphorus-removing bacteria present than nitrate-removing bacteria, so it is primarily removed by chemical precipitation which cannot be recycled. Algae-based treatments are more efficient in removing excess P from wastewater than chemical treatments. Microalgae especially diatoms are efficient in utilizing N and P along with other metals present in wastewater for their growth through photosynthesis and play a significant role in excess nutrient mitigation. Furthermore, an algae-based bioremediation is more environmentally amenable and sustainable as it does not generate additional pollutants such as sludge; resultant algae biomass rich in nutrients can be used as low-cost fertilizer or as animal feed (Munoz and Guieysse 2006).

4.1 Diatom Physiological and Morphological Advantages for Efficient Nutrient Removal

Silica cell wall plays a significant role in carbon-concentrating mechanism (CCM) with diatom bio-silica acting as an effective pH buffer enabling increased carbonic anhydrase activity near cell surface which enables conversion of bicarbonate to CO_2

(Milligan and Morel 2002). Silica cell wall gives diatom algae an advantage of enhanced sinking rate which results in increased carbon burial in shallow seas and continental margins (Falkowski et al. 2005) and are major contributors to nascent petroleum reserves.

Diatoms possess larger storage vacuole compared to other algae which is one of the main factors for their dominance in oceans (Raven 1987). Nutrient utilization which is an important factor influencing growth is dependent on surface to volume ratio were smaller cells have an advantage but diatoms with their large storage vacuole can store nutrients inside the cell thus nullifies this factor even with large surface area. Thus in nutrient replete conditions, diatoms store nutrients, this enables them to perform several cell divisions even in deplete conditions, and this will further influence their dominance by preventing other algae to grow. Diatom algae consistently achieved growth rates in the range of two to four divisions per day which is much higher than other algae tested with the same size (Furnas 1990). Diatom algae can dominate other eukaryotic algae even under high turbulence, and mixing this makes them ideal for mass culturing under varied mixing regimes.

Diatom carbon fixing ability is greater than other algal groups in terms of productivity per unit of carbon. In comparison with *Chlorella vulgaris*, diatom *Phaeodactylum tricornutum* has shown two times more efficiency in converting light energy into biomass. This shows that diatoms have higher photosynthetic efficiency in low light conditions when compared with green algae (Smetacek 1999).

Diatoms lack α-carotene biosynthetic pathway which enables them to produce photo-protective and light harvesting pigments from the same precursors (Wagner et al. 2006). Diatom can perform both C3 and C4 biochemical fixation with a complete urea cycle (Armbrust et al. 2004). Diatoms store carbohydrate in the form of chrysolaminarin which is a soluble form of carbohydrate, whereas other classes of algae store in the form of starch in chloroplast. Although diatoms are not efficient in storing carbohydrates, relative energy required to utilizing soluble carbohydrate stored in CV to unsoluble carbohydrate stored in chloroplast is less (Hildebrand et al. 2012).

Diatoms synthesize their frustules with silica. The source of silica for diatoms is dissolved silicic acid which is absorbed in low quantities by silicic acid transporter proteins. The energy required to build silica cell wall is much less when compared with lignin or polysaccharide cell wall; this will also help in carbon saving as carbon in cell wall is replaced by silica and the carbon replaced is used for other cellular functions (Raven 1983). All these significant differences in cell structure and function might have contributed to the dominance of diatoms.

4.2 *Studies on the Use of Diatoms for Different Wastewater Treatment*

Integrating municipal wastewater treatment with microalgal cultivation can be a sustainable option for the existing STPs as it can reduce the high-maintenance costs and input cost for civil construction. Municipal wastewater contains ammonia, phosphate, and other essential nutrients which are required for microalgal growth. Over the past decade, many studies have been done on growing microalgae on different types of wastewaters like domestic wastewater, agricultural runoff, dairy wastewater, and industrial and municipal waste streams, and the success of these studies was dependent on biotic and abiotic factors. Majority of these studies concentrated on the use of green and blue-green algae, but in recent times, diatoms are increasingly recognized for their phycoremediation potential (Table 4).

In the 1950s, Oswald designed large-scale algae-based open pond systems called high-rate algal pond (HRAP). Algae photosynthesis was used to fulfill the oxygen demand to treat domestic wastewater which was a very efficient system for wastewater treatment (Olguın 2003). HRAP are shallow open ponds; under optimum

Table 4 Nitrogen and phosphorus removal efficiency of microalgae grown using different wastewaters

Algal strain	Wastewater source	Treatment time	Removal efficiency (%)		Reference
			Total nitrate (TN)	Total phosphate (TP)	
Chlorella vulgaris	MWW[a]	09	78	87	Ruiz-Marin et al. (2010)
Scenedesmus dimorphus	IWW[b]	08	70	55	González et al. (1997)
Scenedesmus obliquus	MWW	08	79	47	Ruiz-Marin et al. (2010)
Arthrospira platensis	IWW	15	96	87	Phang et al. (2000)
Oscillatoria sp.	MWW	14	100	100	Craggs et al. (1997)
Phaeodactylum tricornutum	MWW	14	100	100	Craggs et al. (1997)
Mixed culture	DWW[c]	15	96[d]	99	Woertz et al. (2009)
Mixed (diatom dominance)	MWW	23	82	88	Marella et al. (2015)
Diatom consortium	MWW	07	91	88	Marella et al. (2018)

[a]MWW – municipal wastewater
[b]IWW – industrial wastewater
[c]Dairy wastewater 25% dilution
[d]Total ammoniacal nitrogen (TAN)

conditions BOD removal rates were as high as 3500 mg m^2 d^{-1} with hydraulic retention time of 4–10 days. A modified version of this was advanced integrated wastewater pond systems (AIWPS) which are a series of facultative, settling, and maturation ponds. Diatoms can be harnessed for tertiary treatment for enhanced nitrogen and phosphorus removal. Diatoms utilize N and P thorough biotic and abiotic process. Diatoms incorporate N and P into their biomass in the form of protein, nucleic acids, and phospholipids, whereas the increased pH due to their photosynthesis will enhance ammonia and phosphate volatilization and precipitation.

In aquaculture, artificial feed and fish waste enrich the water with excess nutrients leading to unwanted BGA blooms which are detrimental to culture organism growth. Diatom *P. tricornutum* has shown 30–100% removal of ammonium and orthophosphate in batch and continuous modes using diluted effluent (Craggs et al. 1995). Diatom-dominated biofilms grown on artificial substrates in shrimp ponds led to 33% phosphate removal. The diatom-dominated biomass from these treated biofilms can be used as natural feed for filter feeding fish and bivalves. This fish- and bivalve-based aquaculture system could be effective to reduce cost of water treatment with simultaneous production of natural feed.

5 Other Applications

In spite of their dominance in world's oceans combined with their tremendous diversity and tropic flexibility compared with other algae, they are the least explored species for biotechnological applications. Most studies have focused on polyunsaturated fatty acids like eicosapentaenoic acid (EPA) and decosahexanoic acid (DHA) which is used for pharmaceutical applications. Applications for other molecules like amino acids for cosmetics, antioxidants, antibiotics, and antiproliferative agents are at the early stage of development (Lebeau and Robert 2003).

Diatom algae contain very interesting bioactive compounds which are highly sought after in pharmaceutical and nutraceutical industries. Diatoms are rich source of pigments, lipids, sterols, hydrocarbons, phenolic compounds, polysaccharides, alkaloids, and toxins with high bioactivity. Although diatoms contain a variety of active compounds, previous literature is predominantly dedicated to PUFA especially EPA.

Fucoxanthin is a major light harvesting pigment and carotenoid present in seaweeds and diatoms. Fucoxanthin is known to show strong antioxidant, anti-inflammatory, anti-obesity, antidiabetic, anticancer, and antihypertensive activities (Abidov et al. 2010). At present the main commercial source for fucoxanthin is seaweeds, but they have major drawbacks like slow growth, less fucoxanthin content, and contamination by heavy metals. In comparison diatoms contain fucoxanthin in the range of 0.2–2% of dry weight which is 100 times more than that of brown seaweed which is a primary industrial source (Kim et al. 2012).

EPA which is an omega-3 polyunsaturated fatty is de novo synthesized in diatoms. These are the richest primary sources of EPA. The major dietary sources of EPA and DHA for humans are fatty fishes, but advantage of diatom-derived EPA is that it will be a vegetarian source of nutritional fatty acid. Pennate diatom *Phaeodactylum tricornutum* which can accumulate high levels of EPA is presently explored as a potential source for its industrial production.

Microalgal fatty acids are an integral part of animal nutrition; as higher organism cannot synthesize polyunsaturated fatty acids; they can only acquire them through food (Yongmanitchai and Ward 1989). EPA (20:5 (n-3)) and DHA (22:6 (n-3)) are the two main PUFAs required by marine animals to maintain good growth and survival (Renaud et al. 1991).

Microalga as a source of fuel is gaining popularity. Every single microalgal cell can act as a lipid factory which is not the case with terrestrial plants which produce specialized oil-bearing organelles like seeds. Due to this unique ability of microalgae, they are targeted organisms for large-scale funding and scientific studies for biomass and bioenergy production.

6 Conclusions

Algae culture can be integrated within the present wastewater treatment facilities with no or little change to existing infrastructure. This approach will enable reduced capital, maintenance cost, and scalability issues with enhanced treatment efficiency. Although there is much research done on this aspect, research lacuna still exists in areas like photobioreactor design, harvesting technology, drying methods, and other downstream processes which if worked on can lead to effective commercial exploitation of this environmental energy-efficient technology.

Microalgal biotechnology especially for wastewater treatment has received more attention in recent years as a viable alternative to conventional wastewater treatment systems. Algal biomass produced during this process is a sustainable bioresource for biofuel, nutraceutical, biofertilizer, animal feed, poultry feed, and aqua feed industries. In spite of its attractiveness, there are still some obstacles to be solved for its mass-scale exploitation.

References

Abidov M, Ramazanov Z, Seifulla R, Grachev S (2010) The effects of Xanthigen™ in the weight management of obese premenopausal women with non-alcoholic fatty liver disease and normal liver fat. Diabetes Obes Metab 12(1):72–81

Amano Y, Takahashi K, Machida M (2012) Competition between the cyanobacterium Microcystis aeruginosa and the diatom Cyclotella sp. under nitrogen-limited condition caused by dilution in eutrophic lake. J Appl Phycol 24(4):965–971

Ansa E, Lubberding H, Ampofo J, Gijzen H (2011) The role of algae in the removal of Escherichia coli in a tropical eutrophic lake. Ecol Eng 37(2):317–324

Armbrust EV (2009) The life of diatoms in the world's oceans. Nature 459(7244):185–192

Armbrust EV, Berges JA, Bowler C, Green BR, Martinez D, Putnam NH, Zhou S, Allen AE, Apt KE, Bechner M (2004) The genome of the diatom Thalassiosira pseudonana: ecology, evolution, and metabolism. Science 306(5693):79–86

Bennion H, Battarbee R (2007) The European Union water framework directive: opportunities for palaeolimnology. J Paleolimnol 38(2):285–295

Black RW, Moran PW, Frankforter JD (2011) Response of algal metrics to nutrients and physical factors and identification of nutrient thresholds in agricultural streams. Environ Monit Assess 175(1–4):397–417

Blinn D, Herbst D (2003) Use of diatoms and soft algae as indicators of environmental determinants in the Lahontan Basin, USA. Annual report for California state water resources board Contract agreement 704558

Cantonati M, Angeli N, Virtanen L, Wojtal AZ, Gabrieli J, Falasco E, Lavoie I, Morin S, Marchetto A, Fortin C (2014) Achnanthidium minutissimum (Bacillariophyta) valve deformities as indicators of metal enrichment in diverse widely-distributed freshwater habitats. Sci Total Environ 475:201–215

Carpelan LH (1978) Revision of Kolbe's System der Halobien based on diatoms of California lagoons. *Oikos* 31:112–122

Cibic T, Blasutto O (2011) Living marine benthic diatoms as indicators of nutrient enrichment: a case study in the Gulf of Trieste. In: Diatoms: classification, ecology and life cycle. Nova Science Publishers, Inc, New York, pp 169–184

Cox EJ (1991) What is the basis for using diatoms as monitors of river quality? In: Whitton BA, Rott E, Friedrich G (eds) Use of Algae for Monitoring Rivers. Universität Innsbruck, Austria. pp 33–40

Craggs RJ, Smith VJ, McAuley PJ (1995) Wastewater nutrient removal by marine microalgae cultured under ambient conditions in mini-ponds. Water Sci Technol 31(12):151–160

Craggs RJ, McAuley PJ, Smith VJ (1997) Wastewater nutrient removal by marine microalgae grown on a corrugated raceway. Water Res 31(7):1701–1707

de Godos I, Vargas VA, Blanco S, González MCG, Soto R, García-Encina PA, Becares E, Muñoz R (2010) A comparative evaluation of microalgae for the degradation of piggery wastewater under photosynthetic oxygenation. Bioresour Technol 101 (14):5150-5158

De la Rey P, Taylor J, Laas A, Van Rensburg L, Vosloo A (2004) Determining the possible application value of diatoms as indicators of general water quality: a comparison with SASS 5. Water SA 30(3):325–332

De Pauw N, Beyst B, Heylen S (2000) Development of a biological assessment method for river sediments in Flanders, Belgium. Verh Int Ver Theor Angew Limnol 27(5):2703–2708

Delgado C, Pardo I, García L (2010) A multimetric diatom index to assess the ecological status of coastal Galician rivers (NW Spain). Hydrobiologia 644(1):371–384

Dell'Uomo A (1996) Assessment of water quality of an Apennine river as a pilot study for diatom-based monitoring of Italian watercourses. In: Use of algae for monitoring rivers. Eugen Rott, Innsbruck, pp 65–72

Desbois AP, Mearns-Spragg A, Smith VJ (2009) A fatty acid from the diatom Phaeodactylum tricornutum is antibacterial against diverse bacteria including multi-resistant Staphylococcus aureus (MRSA). Mar Biotechnol 11(1):45–52

Descy J (1979) A new approach to water quality estimation using diatoms. Nova Hedwingia, Beiheft 64:305–323

Descy J-P, Coste M (1991) A test of methods for assessing water quality based on diatoms. Verh Int Ver Theor Angew Limnol 24(4):2112–2116

Dixit SS, Smol JP, Kingston JC, Charles DF (1992) Diatoms: powerful indicators of environmental change. Environ Sci Technol 26(1):22–33

Eloranta P, Soininen J (2002) Ecological status of some Finnish rivers evaluated using benthic diatom communities. J Appl Phycol 14(1):1–7
Eriksson PG, Weisner SE (1997) Nitrogen removal in a wastewater reservoir: the importance of denitrification by epiphytic biofilms on submersed vegetation. J Environ Qual 26(3):905–910
Falkowski PG, Barber RT, Smetacek V (1998) Biogeochemical controls and feedbacks on ocean primary production. Science 281(5374):200–206
Falkowski PG, Katz ME, Milligan AJ, Fennel K, Cramer BS, Aubry MP, Berner RA, Novacek MJ, Zapol WM (2005) The rise of oxygen over the past 205 million years and the evolution of large placental mammals. Science 309(5744):2202–2204
Findlay JA, Patil AD (1984) Antibacterial constituents of the diatom Navicula delognei. J Nat Prod 47(5):815–818
Fore LS, Grafe C (2002) Using diatoms to assess the biological condition of large rivers in Idaho (USA). Freshw Biol 47(10):2015–2037
Furnas MJ (1990) In situ growth rates of marine phytoplankton: approaches to measurement, community and species growth rates. J Plankton Res 12(6):1117–1151
Furse M, Hering D, Moog O, Verdonschot P, Johnson RK, Brabec K, Gritzalis K, Buffagni A, Pinto P, Friberg N (2006) The STAR project: context, objectives and approaches. Hydrobiologia 566(1):3–29
Gaiser EE, Sullivan P, Tobias FA, Bramburger AJ, Trexler JC (2014) Boundary effects on benthic microbial phosphorus concentrations and diatom beta diversity in a hydrologically-modified, nutrient-limited wetland. Wetlands 34(1):55–64
García L, Delgado C, Pardo I (2008) Seasonal changes of benthic communities in a temporary stream of Ibiza (Balearic Islands). Limnetica 27(2):259–272
Gómez N, Licursi M (2001) The Pampean Diatom Index (IDP) for assessment of rivers and streams in Argentina. Aquat Ecol 35(2):173–181
González LE, Cañizares RO, Baena S (1997) Efficiency of ammonia and phosphorus removal from a Colombian agroindustrial wastewater by the microalgae Chlorella vulgaris and Scenedesmus dimorphus. Bioresour Technol 60(3):259–262
Gottschalk F, Ort C, Scholz R, Nowack B (2011) Engineered nanomaterials in rivers–exposure scenarios for Switzerland at high spatial and temporal resolution. Environ Pollut 159(12):3439–3445
Hecky RE, Kilham P (1973) Diatoms in alkaline, saline lakes: ecology and geochemical implications. Limnol Oceanogr 18(1):53–71
Hering D, Johnson RK, Kramm S, Schmutz S, Szoszkiewicz K, Verdonschot PF (2006) Assessment of European streams with diatoms, macrophytes, macroinvertebrates and fish: a comparative metric-based analysis of organism response to stress. Freshw Biol 51(9):1757–1785
Hildebrand M, Davis AK, Smith SR, Traller JC, Abbriano R (2012) The place of diatoms in the biofuels industry. Biofuels 3(2):221–240
Hong Y-W, Yuan D-X, Lin Q-M, Yang T-L (2008) Accumulation and biodegradation of phenanthrene and fluoranthene by the algae enriched from a mangrove aquatic ecosystem. Mar Pollut Bull 56(8):1400–1405
Ishida CK, Arnon S, Peterson CG, Kelly JJ, Gray KA (2008) Influence of algal community structure on denitrification rates in periphyton cultivated on artificial substrata. Microb Ecol 56(1):140–152
Kelly M, Whitton B (1995) The trophic diatom index: a new index for monitoring eutrophication in rivers. J Appl Phycol 7(4):433–444
Kelly M, Cazaubon A, Coring E, Dell'Uomo A, Ector L, Goldsmith B, Guasch H, Hürlimann J, Jarlman A, Kawecka B (1998) Recommendations for the routine sampling of diatoms for water quality assessments in Europe. J Appl Phycol 10(2):215–224
Kim SM, Jung Y-J, Kwon O-N, Cha KH, Um B-H, Chung D, Pan C-H (2012) A potential commercial source of fucoxanthin extracted from the microalga Phaeodactylum tricornutum. Appl Biochem Biotechnol 166(7):1843–1855

King L, Clarke G, Bennion H, Kelly M, Yallop M (2006) Recommendations for sampling littoral diatoms in lakes for ecological status assessments. J Appl Phycol 18(1):15–25

Kooistra WH, Medlin L (1996) Evolution of the diatoms (Bacillariophyta) IV A reconstruction of their age from small subunit rRNA coding regions and fossil record. Mol Phylogenet Evol 6(3):391–407

Kröger N, Poulsen N (2008) Diatoms-from cell wall biogenesis to nanotechnology. Annu Rev Genet 42:83–107

Lamberti GA (1996) The role of periphyton in benthic food webs. In: Stevenson RJ, Bothwell ML, Lowe LR (eds) Algal ecology: freshwater benthic ecosystems. Academic Press, California. pp 533–572

Lavoie I, Campeau S (2010) Fishing for diatoms: fish gut analysis reveals water quality changes over a 75-year period. J Paleolimnol 43(1):121–130

Lebeau T, Robert J (2003) Diatom cultivation and biotechnologically relevant products. Part II: Current and putative products. Appl Microbiol Biotechnol 60(6):624–632

Leclercq L, Maquet B (1987) Deux nouveaux indices chimique et diatomique de qualité d'eau courante: application au Samson et à ses affluents (Bassin de la Meuse Belge), comparaison avec d'autres indices chimiques, biocénotiques et diatomiques. Institut Royal des Sciences Naturelles de Belgique

Leguay S, Lavoie I, Levy JL, Fortin C (2016) Using biofilms for monitoring metal contamination in lotic ecosystems: the protective effects of hardness and pH on metal bioaccumulation. Environ Toxicol Chem 35(6):1489–1501

Leira M, Sabater S (2005) Diatom assemblages distribution in catalan rivers, NE Spain, in relation to chemical and physiographical factors. Water Res 39(1):73–82

Leland HV (1995) Distribution of phytobenthos in the Yakima River basin, Washington, in relation to geology, land use and other environmental factors. Can J Fish Aquat Sci 52(5):1108–1129

Li Y, Gao J, Meng F, Chi J (2015) Enhanced biodegradation of phthalate acid esters in marine sediments by benthic diatom Cylindrotheca closterium. Sci Total Environ 508:251–257

Lobo E, Bes D, Tudesque L, Ector L (2004) Water quality assessment of the Pardinho River, RS, Brazil, using epilithic diatom assemblages and faecal coliforms as biological indicators. Vie Milieu 54(2–3):115–126

Lougheed VL, Parker CA, Stevenson RJ (2007) Using non-linear responses of multiple taxonomic groups to establish criteria indicative of wetland biological condition. Wetlands 27(1):96–109

Lynn SG, Price DJ, Birge WJ, Kilham SS (2007) Effect of nutrient availability on the uptake of PCB congener 2, 2′, 6, 6′-tetrachlorobiphenyl by a diatom (Stephanodiscus minutulus) and transfer to a zooplankton (Daphnia pulicaria). Aquat Toxicol 83(1):24–32

Mallick N (2002) Biotechnological potential of immobilized algae for wastewater N, P and metal removal: a review. *Biometals* 15(4):377–390

Marella TK, Tiwari A, Bhaskar MV (2015) A new novel solution to grow diatom algae in large natural water bodies and its impact on CO_2 capture and nutrient removal. J Algal Biomass Util 6(2):22–27

Marella TK, Parine NR, Tiwari A (2018) Potential of diatom consortium developed by nutrient enrichment for biodiesel production and simultaneous nutrient removal from waste water. Saudi J Biol Sci 25 (4):704–709

McCormick PV, Cairns J (1994) Algae as indicators of environmental change. J Appl Phycol 6(5–6):509–526

Milligan AJ, Morel FM (2002) A proton buffering role for silica in diatoms. Science 297(5588):1848–1850

Muxika I, Borja A, Bald J (2007) Using historical data, expert judgement and multivariate analysis in assessing reference conditions and benthic ecological status, according to the European Water Framework Directive. Mar Pollut Bull 55(1–6):16–29

Mulholland PJ, Helton AM, Poole GC, Hall RO, Hamilton SK, Peterson BJ, Tank JL, Ashkenas LR, Cooper LW, Dahm CN (2008) Stream denitrification across biomes and its response to anthropogenic nitrate loading. Nature 452(7184):202

Munoz R, Guieysse B (2006) Algal–bacterial processes for the treatment of hazardous contaminants: a review. Water Res 40(15):2799–2815
Naviner M, Bergé J-P, Durand P, Le Bris H (1999) Antibacterial activity of the marine diatom Skeletonema costatum against aquacultural pathogens. Aquaculture 174(1):15–24
Nixon SW (2009) Eutrophication and the macroscope. Hydrobiologia 629(1):5–19
Olguın EJ (2003) Phycoremediation: key issues for cost-effective nutrient removal processes. Biotechnol Adv 22(1–2):81–91
Oswald WJ (1988) Large-scale algal culture systems (engineering aspects). In: Borowitzka MA, Borowitzka LJ (eds) Micro-algal biotechnology. Cambridge University Press, Cambridge, pp 357–394
Passy SI, Bode RW, Carlson DM, Novak MA (2004) Comparative environmental assessment in the studies of benthic diatom, macroinvertebrate, and fish communities. Int Rev Hydrobiol 89(2):121–138
Phang SM, Miah MS, Yeoh BG, Hashim MA (2000) Spirulina cultivation in digested sago starch factory wastewater. J Appl Phycol 12(3–5):395–400
Poulíčková A, Duchoslav M, Dokulil M (2004) Littoral diatom assemblages as bioindicators of lake trophic status: a case study from perialpine lakes in Austria. Eur J Phycol 39(2):143–152
Prygiel J, Lévêque L, Iserentant R (1996) A new Practical Diatom Index for the assessment of water quality in monitoring networks. J Water Sci 9(1):97–113
Prygiel J, Coste M, Bukowska J (1999) Review of the major diatom-based techniques for the quality assessment of rivers-state of the art in Europe. In: Prygiel J, Whitton BA, Bukowska J (eds) Use of algae for monitoring rivers, vol 3. Agences de l'Eau Artois-Picardie, Douai, pp 224–238
Patrick R (1961) A study of the number and kinds of species found in rivers of the Eastern Unisted States. Proc Acad Natl Sci Phila 113 : 215–258.
Payne WJ (1991) A review of methods for field measurements of denitrification. Forest Ecol Manag 44 (1):5-14
Raven JA (1983) The transport and function of silicon in plants. Biol Rev 58(2):179–207
Raven JA (1987) The role of vacuoles. New Phytol 106(3):357–422
Renaud S, Parry D, Thinh L, Kuo C, Padovan A, Sammy N (1991) Effect of light intensity on the proximate biochemical and fatty acid composition of Isochrysis sp. and Nannochloropsis oculata for use in tropical aquaculture. J Appl Phycol 3(1):43–53
Renuka N, Sood A, Prasanna R, Ahluwalia A (2015) Phycoremediation of wastewaters: a synergistic approach using microalgae for bioremediation and biomass generation. Int J Environ Sci Technol 12(4):1443–1460
Rott E, Pipp E, Pfister P (2003) Diatom methods developed for river quality assessment in Austria and a cross-check against numerical trophic indication methods used in Europe. Algol Stud 110(1):91–115
Rott E, Pipp E, Pfister E, van Dam H, Orther K, Binder N, Pall K (1999) Indikationslisten für Aufwuchsalgen in Österreichischen Fliessgewassern. Teil 2, Trophieindikation. Bundesministerium für Land, und Forstwirtschaft, Wien 248.
Ruhland K, Paterson AM, Smol JP (2008) Hemispheric-scale patterns of climate-related shifts in planktonic diatoms from North American and European lakes. Glob Chang Biol 14(11):2740–2754
Ruiz-Marin A, Mendoza-Espinosa LG, Stephenson T (2010) Growth and nutrient removal in free and immobilized green algae in batch and semi-continuous cultures treating real wastewater. Bioresour Technol 101(1):58–64
Sládeček V (1986) Diatoms as indicators of organic pollution. CLEAN–Soil, Air, Water 14(5):555–566
Smetacek V (1999) Diatoms and the ocean carbon cycle. Protist 150(1):25–32
Stevenson J (2014) Ecological assessments with algae: a review and synthesis. J Phycol 50(3):437–461

Stevenson RJ, Pan Y (1999) Assessing environmental conditions in rivers and streams with diatoms. In: Stoermer EF, Smol JP (eds) The diatoms: applications for the environmental and earth sciences, vol 1(4). Cambridge University Press, Cambridge
Stoermer E (1978) Phytoplankton assemblages as indicators of water quality in the Laurentian Great Lakes. Trans Am Microsc Soc 97:2–16
Stoermer EF, Yang JJ (1970) Distribution and relative abundance of dominant plankton diatoms in Lake Michigan. University of Michigan, Ann Arbor
Tilman D, Kilham SS, Kilham P (1982) Phytoplankton community ecology: the role of limiting nutrients. Annu Rev Ecol Syst 13(1):349–372
Torrisi M, Scuri S, Dell'Uomo A, Cocchioni M (2010) Comparative monitoring by means of diatoms, macroinvertebrates and chemical parameters of an Apennine watercourse of central Italy: the river Tenna. Ecol Indic 10(4):910–913
Valiente Moro C, Bricheux G, Portelli C, Bohatier J (2012) Comparative effects of the herbicides chlortoluron and mesotrione on freshwater microalgae. Environ Toxicol Chem 31(4):778–786
Wagner H, Jakob T, Wilhelm C (2006) Balancing the energy flow from captured light to biomass under fluctuating light conditions. New Phytol 169(1):95–108
Walden WC, Hentges DJ (1975) Differential effects of oxygen and oxidation reduction potential on the multiplication of three species of anaerobic intestinal bacteria. Appl Microbiol 30(5):781–785
Walsh G, Wepener V (2009) The influence of land use on water quality and diatom community structures in urban and agriculturally stressed rivers. Water SA 35(5):579–594
Watanabe T (1988) Numerical water quality monitoring of organic pollution using diatom assemblages. In: Proceedings of the 9th international diatom symposium, 1988. Biopress Limited, Koeltz Scientific Books, Bristol
Weckström K, Juggins S (2006) Coastal diatom–environment relationships from the Gulf of Finland, Baltic Sea. J Phycol 42(1):21–35
Weilhoefer C, Pan Y (2007) A comparison of diatom assemblages generated by two sampling protocols. J N Am Benthol Soc 26(2):308–318
Wetzel RG (1981) Longterm dissolved and particulate alkaline phosphatase activity in a hardwater lake in relation to lake stability and phosphorus enrichments. Verh Int Ver Theor Angew Limnol 21(1):369–381
Woertz I, Feffer A, Lundquist T, Nelson Y (2009) Algae grown on dairy and municipal wastewater for simultaneous nutrient removal and lipid production for biofuel feedstock. J Environ Eng 135(11):1115–1122
Wu J-T, Kow L-T (2002) Applicability of a generic index for diatom assemblages to monitor pollution in the tropical River Tsanwun, Taiwan. J Appl Phycol 14(1):63–69
Wu X, Mitsch WJ (1998) Spatial and temporal patterns of algae in newly constructed freshwater wetlands. Wetlands 18(1):9–20
Yamamoto T, Goto I, Kawaguchi O, Minagawa K, Ariyoshi E, Matsuda O (2008) Phytoremediation of shallow organically enriched marine sediments using benthic microalgae. Mar Pollut Bull 57(1):108–115
Yongmanitchai W, Ward OP (1989) Omega-3 fatty acids: alternative sources of production. Process Biochem 24(4):117–125

A Review of Micropollutant Removal by Microalgae

Sikandar I. Mulla, Ram Naresh Bharagava, Dalel Belhaj, Fuad Ameen, Ganesh Dattatraya Saratale, Sanjay Kumar Gupta, Swati Tyagi, Kishor Sureshbhai Patil, and Anyi Hu

1 Introduction

For the past two decades, medicines including antibiotics as well as antimicrobial agents have been used continuously to control or treat infections as well as diseases (Hom-Diaz et al. 2017; Mulla et al. 2018; Mulla et al. 2016a; Mulla et al. 2016b; Mulla et al. 2016c; Villar-Navarro et al. 2018). Because of their partial digestion by either humans or animals, these substances have been continuously released through excretion and persist at a level of nanograms to micrograms in the environment

S. I. Mulla (✉) · A. Hu
Key Laboratory of Urban Pollutant Conversion, Institute of Urban Environment, Chinese Academy of Sciences, Xiamen, China

R. N. Bharagava
Laboratory of Bioremediation and Metagenomics Research (LBMR), Department of Microbiology (DM), Babasaheb Bhimrao Ambedkar University (A Central University), Lucknow, India

D. Belhaj
University of Sfax-Tunisia, FSS, Department of Life Sciences, Laboratory of Biodiversity and Aquatic Ecosystems Ecology and Planktonology, Sfax, Tunisia

F. Ameen
Department of Botany and Microbiology, Faculty of Science, King Saud University, Riyadh, Saudi Arabia

G. D. Saratale
Department of Food Science and Biotechnology, Dongguk University-Seoul, Ilsandong-gu, Goyang-si, Republic of Korea

S. K. Gupta
Environmental Engineering, Department of Civil Engineering, Indian Institute of Technology – Delhi, New Delhi, Delhi, India

S. Tyagi · K. S. Patil
Division of Biotechnology, Chonbuk National University, Iksan, Republic of Korea

S. K. Gupta, F. Bux (eds.), *Application of Microalgae in Wastewater Treatment*,
https://doi.org/10.1007/978-3-030-13913-1_3

(Fig. 1) (Mulla et al. 2018; Mulla et al. 2016b; Sui et al. 2015). Lower concentrations of common and synthetic steroid hormones such as estrogens, progesterone, and norgestrel were also observed in aquatic systems and surrounding areas (Fig. 1) (Peng et al. 2014a; Ruksrithong and Phattarapattamawong 2017; Yu et al. 2013). Therefore, there is growing concern about the health of human beings, animals, and aquatic organisms because of the continuous and widespread contamination of ecological systems by these agents (Arnold et al. 2013; Ebele et al. 2017; Lapworth et al. 2012; Marcoux et al. 2013; Mulla et al. 2018; Norvill et al. 2016). Most of these contaminants have shown toxicity such as endocrine disruption, chronic effects, and induction of mutations at the genetic level (Mulla et al. 2018; Norvill et al. 2016; Yu et al. 2013). Increased resistance of pathogens towards such contaminants is also possible in neighboring organisms (Mulla et al. 2018; Weatherly and Gosse 2017).

Use of many of these lifesaving drugs is not ceasing; however, the controlling board can restrict the routine use of certain important drugs and thereby protect them from pathogens that can acquire resistance or transfer the genes responsible for resistance to neighboring organisms. Thus, in an emergency situation, the drugs can be used to save severely diseased persons from life-threatening conditions. From this aspect, it is clearly understandable that these contaminants will continue to be present in the environment, especially through wastewater treatment plants (WWTPs). However, studies have shown that several contaminants are not being completely removed in WWTPs, so that this is one of the important routes for the release of such contaminants into the environment (Fig. 1) (Christensen et al. 2006; Goeppert et al. 2015; Mulla et al. 2018). Hence, recently most of studies have

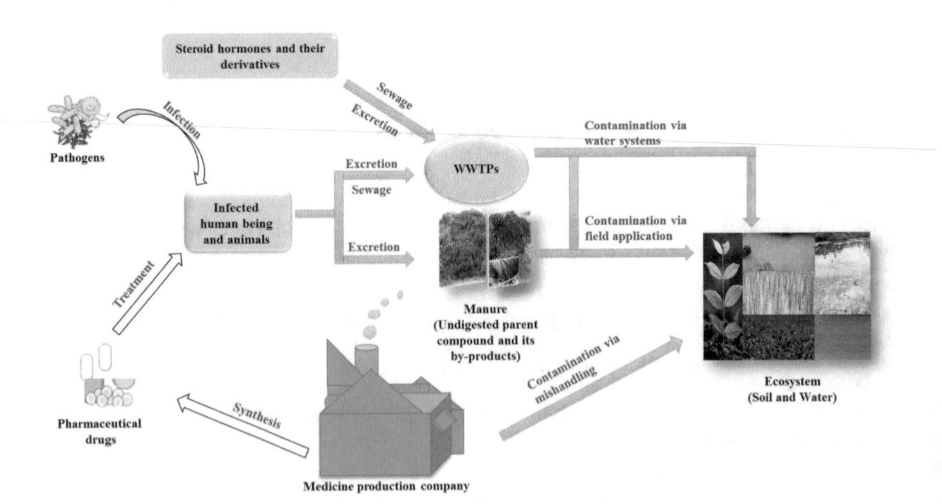

Fig. 1 A major route of medicine(s) and steroidal hormone(s) in the environment

focused on the quantification of contaminants such as pharmaceutical drugs, antimicrobial agents, and also steroidal hormones in the WWTPs (influent and effluent wastewater) (Ashfaq et al. 2017a; Ashfaq et al. 2018; Ashfaq et al. 2017b; Ashfaq et al. 2017c; Li et al. 2016; Wang et al. 2018). Various engineering techniques have been applied, including an advanced oxidation process to remove such pollutants from WWTPs, but these methods were found not to be useful because of their high cost (Mulla et al. 2018). Hence, biological removal of contaminants using algae either alone or in combination with other organisms was found to be economical, less expensive, convenient, and environmentally friendly (Fig. 2).

In addition, one positive outlook is that the use of algae can also help remove pathogens, nutrients, and biodegradable oxygen demand (BOD), especially in the sewage system and WWTPs (Heaven et al. 2003; USEPA 2011). However, there is not much information available on the removal of microcontaminants by algae (Norvill et al. 2016). Thus, in this chapter we mainly focus on the fate of steroidal hormones, nonsteroidal antiinflammatory drugs (Table 1), and quinolone antibiotics (Table 1), and also their removal by algae (Table 2).

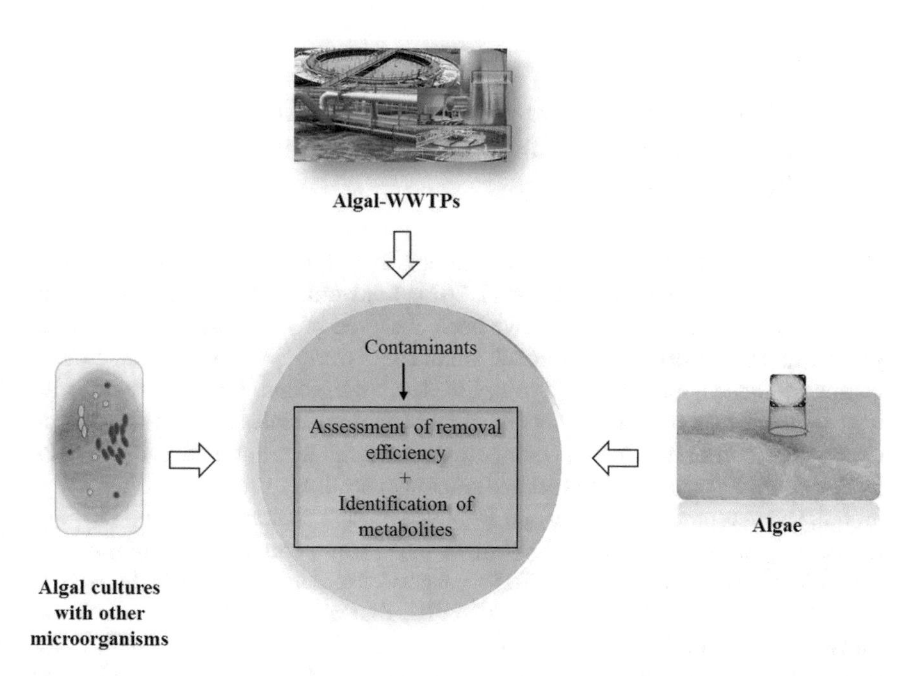

Fig. 2 Algae-mediated removal and degradation of contaminants

Table 1 Molecular structure and chemical formula of different medicines

Class	Antibiotic name (formula)	Molecular structure
Nonsteroidal antiinflammatory drugs	Ibuprofen ($C_{13}H_{18}O_2$)	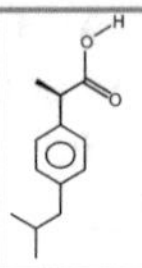
	Diclofenac ($C_{14}H_{11}Cl_2NO_2$)	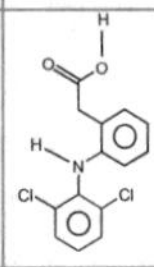
	Naproxen ($C_{14}H_{14}O_3$)	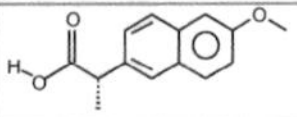
Quinolone (fluoroquinolone) antibiotics	Ciprofloxacin ($C_{17}H_{18}FN_3O_3$)	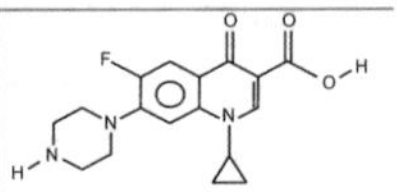
	Levofloxacin ($C_{18}H_{20}FN_3O_4$)	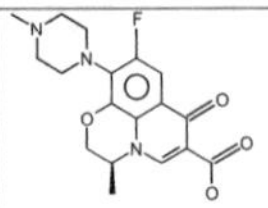
	Enrofloxacin ($C_{19}H_{22}FN_3O_3$)	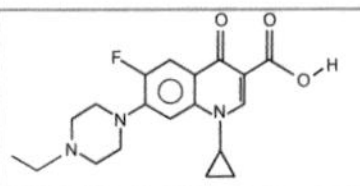

2 The Fate of Occurrence of Nonsteroidal Antiinflammatory Drugs, Quinolone Antibiotics, and Hormones in the Environment

There are reports of the existence of contaminants in water bodies, notably in wastewater and groundwater (Deblonde et al. 2011; Lapworth et al. 2012). Concentrated contaminants in aquatic systems have been observed in wastewater, surface water, groundwater, and drinking water all around the world (Austria, China, France, Germany, Greece, Italy, Korea, Spain, Sweden, Switzerland, USA). In this section we concentrate on only a few classes of persistent molecules such as steroid hormones, nonsteroidal antiinflammatory drugs, and quinolone antibiotics. In WWTPs, the concentrations of nonsteroidal antiinflammatory drugs such as diclofenac, ibuprofen, and ketoprofen were as much as 0.0942 mg/l, 0.603 mg/l, and 0.00856 mg/l, respectively in influent and 0.00069 mg/l, 0.055 mg/l, and 0.00392 mg/l, respectively, in effluent (Behera et al. 2011; Kasprzyk-Hordern et al. 2009; Loos et al. 2013; Santos et al. 2009; Singer et al. 2010; Stamatis et al. 2010; Stamatis and Konstantinou 2013; Terzic et al. 2008; Yu and Chu 2009; Zhou et al. 2010; Zorita et al. 2009). The occurrence of steroid hormones such as estrone, estradiol, and 17α-ethynylestradiol was observed up to 0.00017 mg/l, 0.00005 mg/l, and 0.000003 mg/l, respectively, in the influent (Behera et al. 2011; Janex-Habibi et al. 2009; Nie et al. 2012; Zorita et al. 2009). The occurrence of quinolone antibiotics

Table 2 Various algal cultures used for the removal of different contaminants in aqueous systems

Class	Substrate(s)	Organism(s)	References
Nonsteroidal antiinflammatory medicines	Ibuprofen, ketoprofen	*Stigeoclonium* sp., *Chlorella* sp., *Stigeoclonium* sp., *Monoraphidium* sp.	Matamoros et al. (2015)
	Ibuprofen	*Chlorella* sp., *Scenedesmus* sp.	Matamoros et al. (2016)
	Diclofenac	*Chlorella sorokiniana*, *Chlorella vulgaris*, *Scenedesmus obliquus*	Escapa et al. (2016)
	Diclofenac, ibuprofen	*Chlorella sorokiniana*	de Wilt et al. (2016)
	Naproxen	*Cymbella* sp., *Scenedesmus quadricauda*	Ding et al. (2017)
	Ibuprofen, diclofenac	*Coelastrum* sp.	Villar-Navarro et al. (2018)
	Ibuprofen	*Phaeodactylum tricornutum*	Santaeufemia et al. (2018)
Fluoroquinolone drugs	Levofloxacin	*Chlamydomonas mexicana*, *Chlamydomonas pitschmannii*, *Chlorella vulgaris*, *Ourococcus multisporus*, *Micractinium resseri*, *Tribonema aequale*	Xiong et al. (2017a)
	Levofloxacin	*Scenedesmus obliquus*	Xiong et al. (2017b)
	Enrofloxacin	*Scenedesmus obliquus*, *Chlamydomonas mexicana*, *Chlorella vulgaris*, *Ourococcus multisporus*, *Micractinium resseri*	Xiong et al. (2017c)
	Ciprofloxacin	*Chlamydomonas mexicana*	Xiong et al. (2017d)
Steroidal hormones	Estradiol, estrone, estriol, hydroxyestrone, estradiol, valerate, and ethinylestradiol (estrogens)	*Chlorella vulgaris*	Lai et al. (2002)
	17α-Estradiol and estrone	*Scenedesmus dimorphus*	Zhang et al. (2014)
	β-Estradiol and 17α-ethinylestradiol	*Selenastrum capricornutum*, *Chlamydomonas reinhardtii*	Hom-Diaz et al. (2015)
	Ethinylestradiol	*Chlorella* PY-ZU1 (mutant)	Cheng et al. (2018)
	17β-Estradiol	*Chlorella*	Parlade et al. (2018)
	17α-Ethinylestradiol	*Chlorella* sp., *Nitzschia acicularis*	Sole and Matamoros (2016)
	Progesterone and norgestrel	*Scenedesmus obliquus*, *Chlorella pyrenoidosa*	Peng et al. (2014a)

such as levofloxacin, ciprofloxacin, ofloxacin, and norfloxacin was observed at concentrations as great as 0.0068 mg/l, 0.099 μg/l, 1.287 μg/l, and 0.775 μg/l, respectively, in raw sewage (Ghosh et al. 2016; Jia et al. 2012). Further, this release of undigested WWTP effluent into the environment is a major cause of surface water pollution because of these contaminants (Kasprzyk-Hordern et al. 2009). On the other hand, with dilution by river water, such contaminant levels might be lower than in the effluent (Gros et al. 2007). In some situations, concentrations can be altered by rainfall, although not always. Moreover, in hot climates persisting contaminant concentrations were observed at a higher level compared to the rainy period (Luo et al. 2014). Also, a lower level of contamination was observed with respect to groundwater compared to surface water (Loos et al. 2010; Vulliet and Cren-Olivé 2011). Moreover, in a few cases a lower concentration was observed, for example, in leakage of septic tanks as well as garbage (landfill), contaminants up to 1–10 μg/l, respectively (Lapworth et al. 2012). Runoff from agricultural activity was also found to be one of the main routes for groundwater contamination (González-Rodríguez et al. 2011).

Hence, several researchers studied degradation of microcontaminants using biological agents such as algal cultures (Fig. 2). Generally, degradation of different contaminants by different genera or species of microalgae (*Selenastrum capricornutum*, *Selenastrum obliquus*, *Selenastrum quadricauda*, *Chlamydomonas mexicana*, and *Chlorella fusca*) involved different reaction mechanisms including glycosylation, oxidation, carboxylation, demethylation, hydroxylation, dehydrogenation, and hydrogenation, and also breaking the functional group bond linked to the side chains of substrates (Fisher et al. 1996; Matamoros et al. 2016; Peng et al. 2014a; Thies et al. 1996; Xiong et al. 2017d). In most cases, phase I (cytochrome P450 enzymes, including monooxygenase, dioxygenase, hydroxylase, carboxylase, and decarboxylase) and phase II (glutathione-*S*-transferases) families of enzymes are the main enzymes involved in the degradation and transformation of different classes of pollutants by microalgae (Fisher et al. 1996; Thies et al. 1996). The removal of antiinflammatory drugs, quinolone drugs, and steroidal hormones by algae is discussed in the following sections.

3 Removal of Nonsteroidal Antiinflammatory Drugs by Microalgae

A number of algae are able to remove antiinflammatory drugs. For example, Matamoros et al. (2015) evaluated the hydraulic retention time (HRT) and seasonality impact for destruction efficiency of several microcontaminants from urban wastewater in two pilot high-rate algal ponds (HRAPs). The HRAPs contained microalgae such as *Chlorella* sp., *Stigeoclonium* sp. (diatoms), and *Monoraphidium* sp., varying with respect to climate. The results showed that removal efficiency in the HRAPs varied from high to low; for example, more than 90% removal was observed for ibuprofen including other contaminants such as caffeine, acetaminophen, methyl dihydrojasmonate, and hydrocinnamic acid whereas low removal (i.e., 40–60%) was observed for diclofenac including other contaminants such as benzotriazole, *O,H*-benzothiazole, triphenyl phosphate, cashmeran, diazinon, benzothiazole, celestolide, 2,4-D, and atrazine. On the other hand, moderate to high removal (i.e., 60–90%) was observed for ketoprofen including other contaminants such as naproxen, triclosan, oxybenzone, 5-methyl/benzotriazole, galaxolide, tonalide, tributyl phosphate, bisphenol A, and octylphenol. It was observed that the biodegradation, photolysis, and adsorption to algal biomass were the main route in HRAPs, and such mechanisms were also found in engineered systems (WWTPs) such as constructed wetlands and activated sludge (Bell et al. 2013; Matamoros et al. 2015). In another study, the same research group demonstrated the removal of ibuprofen with other micropollutants in microalgae (especially *Chlorella* sp. and *Scenedesmus* sp.) wastewater (either urban or synthetic) under aerobic conditions, and nonaerated as well as darkness reactors were used as controls. The results showed that approximately 99% of microcontaminants were removed within 10 days whereas biodegradation was found to be an effective method for ibuprofen removal. In contrast, 40% of increased removal efficiency was observed for ibuprofen when algae were used (Matamoros et al. 2016). Additionally, during the process, ibuprofen was transformed into two intermediates identified as carboxy-ibuprofen and hydroxy-ibuprofen (Matamoros et al. 2016).

Escapa et al. (2016) used three different microalgae strains, namely, *Chlorella sorokiniana*, *Chlorella vulgaris*, and *Scenedesmus obliquus* for the removal of diclofenac from water. The removal rate of diclofenac (25 mg/l) was higher with *Scenedesmus obliquus*, that is, 3.4 fold higher than in *Chlorella sorokiniana* and 5.4 fold more than in *Chlorella vulgaris*. Overall, it was concluded that the removal rate is specifically related to cell size; for example, in the case of *Chlorella sorokiniana*, it was observed that the higher cells achieved a higher removal rate, but, because of the larger size of *Scenedesmus obliquus*, the organism was able to remove more diclofenac than the other two strains (Escapa et al. 2016).

Similarly, de Wilt et al. (2016) evaluated removal of six spiked pharmaceuticals (diclofenac, ibuprofen, paracetamol, metoprolol, carbamazepine, trimethoprim; final concentration, 100–350 μg/l) by an algal treatment system (*Chlorella sorokiniana*). The 60% to 100% removal of diclofenac, ibuprofen, paracetamol, and metoprolol was led by biodegradation and photodegradation.

Ding et al. (2017) studied the degradation mechanism and the toxicity of naproxen (1–100 mg/l) on algae (*Cymbella* sp. and *Scenedesmus quadricauda*). Naproxen at 100 mg/l was able to completely inhibit the growth of both algae within 24 h. Moreover, both strains showed varied trends with respect to removal efficiency of naproxen at 1 mg/l. *Cymbella* sp. was removed up to 97% within 30 days. During the degradation of naproxen in both strains, various by-products were generated, identified as 6-(2-amino-1-hydroxy-3-(4-hydroxyphenyl) propoxy)naphthalene-2,3-diol, 2-(1-hydroxy)-6-*O*-desmethylnaproxen, 6-*O*-desmethylnaproxen, 6-(2-(6,7-dimethoxynaphthalen-2-yl)-1-hydroxypropoxy)-3,4,5,6-tetrahydroxyhexanoic acid, 3,4,5-trihydroxy-6-((2-(7-hydroxy-6-ethoxynaphthalen-

2-yl)propanoyl)oxy) tetrahydro-2 *H*-pyran-2-carboxylic acid, 6-((2-(6,7-dimethoxynaphthalen-2-yl)propanoyl) oxy)-3,4,5-trioxotetrahydro-2 *H*-pyran-2-carboxylic acid, 2-(7-hydroxy-6-methoxynaphthalen-2-yl) propane-1,1,3-triol, 6,7-dihydroxynaphthalen-2-yl 2-amino-3-(4-hydroxyphenyl)propanoate, 2-(6-((carboxycarbonyl)oxy)-7-hydroxynaphthalen-2-yl)-3-hydroxypropanoic acid, 6-hydroxy-7-methoxynaphthalen-2-yl 2-amino-3-(4-hydroxyphenyl)propanoate, 1-(hydroxy(6-hydroxynaphthalen-2-yl) methyl)naphthalene-2,6-diol, and 6-hydroxy-7-methoxynaphthalen-2-yl 2-amino-3-(4-hydroxyphenyl)-3-(4-hydroxyphenyl) propanoate.

Villar-Navarro et al. (2018) applied microalgal treatment (dominant genus, *Coelastrum*) for the elimination of nutrients and contaminants from wastewater under different conditions for comparison. Ibuprofen removal was lower (less than 1 mg/l) compared to a conventional line; the removal of diclofenac was around 15–50% in HRAPs. Generally, it was found that the microalgal system efficiency for the removal of various contaminants including drugs was higher than that of conventional WWTPs. Hence, it can be considered as an alternative to an activated sludge system. For example, batch reactors consisting of freshwater green microalgae have proved the competency of these organisms to remove various substances including nutrients, inorganic components, active pharmaceutical drugs, and endocrine-disrupting compounds from wastewater (Norvill et al. 2016; Zhou et al. 2014). Furthermore, HRAP and dissolved air flotation treatments results showed removal efficiency at 55% and 71%, respectively, for diclofenac (Villar-Navarro et al. 2018). In the conventional line, the overall removal capacity of diclofenac was found to be approximately 19%, and it did not surpass 30% even after secondary treatment by active sludge (Collado et al. 2014; Kermia et al. 2016). Santaeufemia and coworkers recently studied ibuprofen removal by living or dead cells of *Phaeodactylum tricornutum*, with more than 99% of ibuprofen removal achieved in both dead and living cells of *P. tricornutum* (Santaeufemia et al. 2018).

Similarly, Ismail et al. (2016), studied ketoprofen degradation using a combination of *Chlorella* sp. with the K2 consortium. It was observed that the *Chlorella* sp. was resistant to ketoprofen. However, no degradation of ketoprofen was observed in MSM + ketoprofen under diurnal cycle conditions, and in MSM + ketoprofen or MSM + ketoprofen + *Chlorella* sp. under diurnal cycle conditions.

4 Removal of Quinolone Antibiotics by Microalgae

Xiong et al. (2017a) demonstrated the removal of levofloxacin in six different microalgal cultures: *Chlamydomonas mexicana*, *Chlamydomonas pitschmannii*, *Chlorella vulgaris*, *Ourococcus multisporus*, *Micractinium resseri*, and *Tribonema aequale*. Compared among all the strains, *Chlorella vulgaris* was found to be an efficient organism that removed up to 12% of levofloxacin at a given concentration of 1 mg/l; the removal rate was further significantly enhanced, to more than 80%, in the presence of sodium chloride (1%). In another study, the same research group demonstrated the degradation of levofloxacin degradation with *Scenedesmus obliquus* (a microalga) and found that in the presence of sodium chloride at 170 mM, the removal of levofloxacin (1 mg/l) was increased up to 93% whereas without sodium chloride (0 nM), only up to 4.53% removal was observed. Furthermore, six by-products were generated during the process (Xiong et al. 2017b).

Five pure microalgal cultures (*Scenedesmus obliquus, Chlamydomonas mexicana, Chlorella vulgaris, Ourococcus multisporus, Micractinium resseri*) and their consortium were used to study enrofloxacin removal in natural systems. The removal of enrofloxacin (1 mg/l) in *Scenedesmus obliquus, Chlamydomonas mexicana, Chlorella vulgaris, Ourococcus multisporus, Micractinium resseri*, and their consortium was as much as 23%, 25%, 26%, 18%, 20%, and 26%, respectively (Xiong et al. 2017c). Furthermore, abiotic (especially photolysis) removal of enrofloxacin was not observed during the process, which means removal proceeded through degradation and sorption to algal biomass (Xiong et al. 2017c).

Xiong et al. (2017d) also demonstrated ciprofloxacin (2 mg/l) removal in *Chlamydomonas mexicana* but found only 13% of removal after 11 days of incubation. Additionally, with various co-substrates used to check the removal efficacy of ciprofloxacin in algae, sodium acetate significantly improved the rate of removal of ciprofloxacin up to 32% compared to other co-substrates. Similar results were observed with co-substrates such as glucose, sodium acetate, sodium succinate, and sodium formate during the removal of different contaminants (Dawas-Massalha et al. 2014; Luo et al. 2008; Peng et al. 2014b; Reis et al. 2014). In another study, HRAP was used for the removal of ciprofloxacin (2 mg/l); photodegradation is a main mechanism for the removal of ciprofloxacin in daytime whereas at night the substrate was gathered on the surface of the biomass (Hom-Diaz et al. 2017).

5 Removal of Hormones by Microalgae

Lai et al. (2002) studied the removal of estrogens (estradiol, estrone, estriol, hydroxyestrone, estradiol valerate, and ethinylestradiol) at 500 ng each by *Chlorella vulgaris* under dark and light conditions and found that 50% of the estradiol was degraded into a unknown intermediate; a small amount was transformed to estrone under light conditions whereas in dark conditions estradiol degradation to estrone was the major route. Likewise, Zhang et al. (2014), demonstrated an alga-mediated (*Scenedesmus dimorphus*) removal of 17α-estradiol and estrone in bench-scale experiments and found removal rates up to 85% and 95%, respectively, for 17α-estradiol as well as estrone and 17β-estradiol as well as estriol at 8 days. Furthermore, this study also revealed that the sorption, photodegradation, and biotransformation are the main routes for the removal of such molecules. Overall, the reaction mechanisms involved hydroxylation, reduction, and oxidation during the removal of these molecules. The degradative pathway was initiated by transforming 17α-estradiol and 17β-estradiol to estrone and subsequently converted either to estriol or interconverted to 17α-estradiol and/or 17β-estradiol. Similar transformations were also observed in other studies (Lai et al. 2002; Pflugmacher et al. 1999).

Also, Hom-Diaz et al. (2015) demonstrated the degradation of β-estradiol and 17α-ethinylestradiol under anaerobic digestion using microalgal cultures, *Selenastrum capricornutum* and *Chlamydomonas reinhardtii*, with more than 88% and 60% removal for β-estradiol and 17α-ethinylestradiol, respectively, after 168 h. Recently, a mutant *Chlorella PY-ZU1* (with carbon dioxide fixation) was applied for the removal of ethinylestradiol (5 mg/l) with 94% removal of ethinylestradiol (Cheng et al. 2018). 17β-Estradiol removal was also studied using pilot-plant photobioreactors consisting of an algal–bacterial consortium under indoor and outdoor conditions and found that under harsh conditions (low temperatures and solar irradiation), 50% of 17β-estradiol removal by the established consortium was observed within 24 h. Furthermore, the removal rate was significantly enhanced, more than 94%, when favorable conditions were provided. During the entire process the alga *Chlorella* was found stable (Parladé et al. 2018). A by-product, estrone, was generated during 17β-estradiol treatment. In another study, Sole and Matamoros (2016) applied free (*Chlorella* sp. and *Nitzschia acicularis*) and immobilized microalgae (alginate beads) for the removal of six contaminants (including 17α-ethinylestradiol) and found that almost all contaminants were removed up to 80% by both free and immobilized microalgae. Additionally, under both free and immobilized microalgal treatments, total removal of NH_4-N was found up to 64% and 89%, respectively, whereas total phosphorus removal was found up to 90% and 96%, respectively, at 10 days of incubation. Various research groups also reported that the removal rate of different contaminants was significantly improved by immobilized (including alginate beads) enzymes as well as microorganisms (Edalli et al. 2016; Hoskeri et al. 2014; Mulla et al. 2012; Mulla et al. 2016; Mulla et al. 2013; Tallur et al. 2015).

Similarly, progesterone and norgestrel degradation was studied using *Scenedesmus obliquus* and *Chlorella pyrenoidosa*, with more than 95% removal of progesterone observed in both algal strains whereas 40% of norgestrel removal was achieved in *Chlorella pyrenoidosa*. On the other hand, almost all norgestrel was removed by *Scenedesmus obliquus* (Peng et al. 2014a). During alga-mediated biotransformation of progesterone, various intermediates such as 3β-hydroxy-5α-pregnan-20-one, 3,20-allopregnanedione, and 1,4-pregnadiene-3,20-dione, as well as six minor androgen compounds, were generated (Peng et al. 2014a).

6 Conclusion

During the past decade, research work around the world has established significant knowledge of detoxification/degradation, transformation, and/removal of various contaminants, especially micropollutants including pharmaceutical kinetic drugs and hormones, by using microorganisms that are alga-mediated. Pollutants such as drugs and hormones are released into the environment through various means such as agricultural activities and effluent from WWTPs. These contaminants (pollutants) were detected at levels ranging from 10^{-9} to 10^{-6}, and at these concentrations, they were found to be toxic to certain organisms. Additionally, most of them are endocrine-disrupting chemicals. Hence, several research groups used algal cultures as biological agents for the removal of various microcontaminants (including steroidal hormones, nonsteroidal antiinflammatory drugs, and quinolone antibiotics). A benefit of this treatment technique is that it can remove unwanted nutrients such as nitrogen and phosphorus from wastewater. Moreover, an engineering approach can further help to establish the removal of these contaminants from WWTPs.

References

Arnold KE, Boxall AB, Brown AR, Cuthbert RJ, Gaw S, Hutchinson TH, Jobling S, Madden JC, Metcalfe CD, Naidoo V, Shore RF, Smits JE, Taggart MA, Thompson HM (2013) Assessing the exposure risk and impacts of pharmaceuticals in the environment on individuals and ecosystems. Biol Lett 9:20130492. https://doi.org/10.1098/rsbl.2013.0492

Ashfaq M, Li Y, Wang Y, Chen W, Wang H, Chen X, Wu W, Huang Z, Yu CP, Sun Q (2017a) Occurrence, fate, and mass balance of different classes of pharmaceuticals and personal care products in an anaerobic-anoxic-oxic wastewater treatment plant in Xiamen, China. Water Res 123:655–667. https://doi.org/10.1016/j.watres.2017.07.014

Ashfaq M, Nawaz Khan K, Saif Ur Rehman M, Mustafa G, Faizan Nazar M, Sun Q, Iqbal J, Mulla SI, Yu CP (2017b) Ecological risk assessment of pharmaceuticals in the receiving environment of pharmaceutical wastewater in Pakistan. Ecotoxicol Environ Saf 136:31–39. https://doi.org/10.1016/j.ecoenv.2016.10.029

Ashfaq M, Noor N, Saif-Ur-Rehman M, Sun Q, Mustafa G, Nazar M, Yu C-P (2017c) Determination of commonly used pharmaceuticals in hospital waste of Pakistan and evaluation of their ecological risk assessment. Clean Water Air Soil 45:1500392

Ashfaq M, Li Y, Wang Y, Qin D, Rehman M, Rashid A, Yu CP, Sun Q (2018) Monitoring and mass balance analysis of endocrine disrupting compounds and their transformation products in an anaerobic-anoxic-oxic wastewater treatment system in Xiamen, China. Chemosphere 204:170–177

Behera S, Kim H, Oh J, Park H (2011) Occurrence and removal of antibiotics, hormones and several other pharmaceuticals in wastewater treatment plants of the largest industrial city of Korea. Sci Total Environ 409:4351–4360

Bell KY, Bandy J, Finnegan BJ, Keen O, Mauter MS, Parker AM, Sima LC, Stretz HA (2013) Emerging pollutants. Part II: treatment. Water Environ Res 85:2022–2071

Cheng J, Ye Q, Li K, Liu J, Zhou J (2018) Removing ethinylestradiol from wastewater by microalgae mutant *Chlorella* PY-ZU1 with CO_2 fixation. Bioresour Technol 249:284–289

Christensen AM, Ingerslev F, Baun A (2006) Ecotoxicity of mixtures of antibiotics used in aquacultures. Environ Toxicol Chem 25:2208–2215

Collado N, Rodriguez-Mozaz S, Gros M, Rubirola A, Barcelo D, Comas J, Rodriguez-Roda I, Buttiglieri G (2014) Pharmaceuticals occurrence in a WWTP with significant industrial contribution and its input into the river system. Environ Pollut 185:202–212

Dawas-Massalha A, Gur-Reznik S, Lerman S, Sabbah I, Dosoretz CG (2014) Co-metabolic oxidation of pharmaceutical compounds by a nitrifying bacterial enrichment. Bioresour Technol 167:336–342. https://doi.org/10.1016/j.biortech.2014.06.003

de Wilt A, Butkovskyi A, Tuantet K, Leal LH, Fernandes TV, Langenhoff A, Zeeman G (2016) Micropollutant removal in an algal treatment system fed with source separated wastewater streams. J Hazard Mater 304:84–92. https://doi.org/10.1016/j.jhazmat.2015.10.033

Deblonde T, Cossu-Leguille C, Hartemann P (2011) Emerging pollutants in wastewater: a review of the literature. Int J Hyg Environ Health 214:442–448. https://doi.org/10.1016/j.ijheh.2011.08.002

Ding T, Lin K, Yang B, Yang M, Li J, Li W, Gan J (2017) Biodegradation of naproxen by freshwater algae *Cymbella* sp. and *Scenedesmus quadricauda* and the comparative toxicity. Bioresour Technol 238:164–173. https://doi.org/10.1016/j.biortech.2017.04.018

Ebele A, Abdallah MAE, Harrad S (2017) Pharmaceuticals and personal care products (PPCPs) in the freshwater aquatic environment. Emerg Contam 3:1–16

Edalli VA, Mulla SI, Eqani SAMAS, Mahadevan GD, Sharma R, Shouche Y, Kamanavalli CM (2016) Evaluation of p-cresol degradation with polyphenol oxidase (PPO) immobilized in various matrices. 3 Biotech 6: 229

Escapa C, Coimbra R, Paniagua S, García A, Otero M (2016) Comparative assessment of diclofenac removal from water by different microalgae strains. Algal Res 18:127–134

Fisher M, Gokhman I, Pick U, Zamir A (1996) A salt-resistant plasma membrane carbonic anhydrase is induced by salt in *Dunaliella salina*. J Biol Chem 271:17718–17723

Ghosh G, Hanamoto S, Yamashita N, Huang X, Tanaka H (2016) Antibiotics removal in biological sewage treatment plants. Pollution 2:131–139

Goeppert N, Dror I, Berkowitz B (2015) Fate and transport of free and conjugated estrogens during soil passage. Environ Pollut 206:80–87. https://doi.org/10.1016/j.envpol.2015.06.024

González-Rodríguez R, Rial-Otero R, Cancho-Grande B, Gonzalez-Barreiro C, Simal-Gándara J (2011) A review on the fate of pesticides during the processes within the food-production chain. Crit Rev Food Sci Nutr 51:99–114

Gros M, Petrović M, Barcelo D (2007) Wastewater treatment plants as a pathway for aquatic contamination by pharmaceuticals in the Ebro River basin (northeast Spain). Environ Toxicol Chem 26:1553–1562

Heaven S, Lock AC, Pak LN, Rspaev MK (2003) Waste stabilisation ponds in extreme continental climates: a comparison of design methods from the USA, Canada, northern Europe and the former Soviet Union. Water Sci Technol 48:25–33

Hom-Diaz A, Llorca M, Rodriguez-Mozaz S, Vicent T, Barcelo D, Blanquez P (2015) Microalgae cultivation on wastewater digestate: β-estradiol and 17α-ethynylestradiol degradation and transformation products identification. J Environ Manag 155:106–113. https://doi.org/10.1016/j.jenvman.2015.03.003

Hom-Diaz A, Norvill ZN, Blanquez P, Vicent T, Guieysse B (2017) Ciprofloxacin removal during secondary domestic wastewater treatment in high rate algal ponds. Chemosphere 180:33–41. https://doi.org/10.1016/j.chemosphere.2017.03.125

Hoskeri R, Mulla S, Ninnekar H (2014) Biodegradation of chloroaromatic pollutants by bacterial consortium immobilized in polyurethene foam and other matrices. Biocatal Agric Biotechnol 3:390–396

Ismail M, Essam T, Ragab Y, Mourad F (2016) Biodegradation of ketoprofen using a microalgal-bacterial consortium. Biotechnol Lett 38:1493–1502

Janex-Habibi ML, Huyard A, Esperanza M, Bruchet A (2009) Reduction of endocrine disruptor emissions in the environment: the benefit of wastewater treatment. Water Res 43:1565–1576. https://doi.org/10.1016/j.watres.2008.12.051

Jia A, Wan Y, Xiao Y, Hu J (2012) Occurrence and fate of quinolone and fluoroquinolone antibiotics in a municipal sewage treatment plant. Water Res 46:387–394. https://doi.org/10.1016/j.watres.2011.10.055

Kasprzyk-Hordern B, Dinsdale RM, Guwy AJ (2009) The removal of pharmaceuticals, personal care products, endocrine disruptors and illicit drugs during wastewater treatment and its impact on the quality of receiving waters. Water Res 43:363–380. https://doi.org/10.1016/j.watres.2008.10.047

Kermia AEB, Fouial-Djebbar D, Trari M (2016) Occurrence, fate and removal efficiencies of pharmaceuticals in wastewater treatment plants (WWTPs) discharging in the coastal environment of Algiers. C R Chim 19:963–970

Lai KM, Scrimshaw MD, Lester JN (2002) Biotransformation and bioconcentration of steroid estrogens by *Chlorella vulgaris*. Appl Environ Microbiol 68:859–864

Lapworth DJ, Baran N, Stuart ME, Ward RS (2012) Emerging organic contaminants in groundwater: a review of sources, fate and occurrence. Environ Pollut 163:287–303. https://doi.org/10.1016/j.envpol.2011.12.034

Li M, Sun Q, Li Y, Lv M, Lin L, Wu Y, Ashfaq M, Yu CP (2016) Simultaneous analysis of 45 pharmaceuticals and personal care products in sludge by matrix solid-phase dispersion and liquid chromatography tandem mass spectrometry. Anal Bioanal Chem 408:4953–4964. https://doi.org/10.1007/s00216-016-9590-0

Loos R, Locoro G, Comero S, Contini S, Schwesig D, Werres F, Balsaa P, Gans O, Weiss S, Blaha L, Bolchi M, Gawlik BM (2010) Pan-European survey on the occurrence of selected polar organic persistent pollutants in ground water. Water Res 44:4115–4126. https://doi.org/10.1016/j.watres.2010.05.032

Loos R, Carvalho R, Antonio DC, Comero S, Locoro G, Tavazzi S, Paracchini B, Ghiani M, Lettieri T, Blaha L, Jarosova B, Voorspoels S, Servaes K, Haglund P, Fick J, Lindberg RH, Schwesig D, Gawlik BM (2013) EU-wide monitoring survey on emerging polar organic contaminants in wastewater treatment plant effluents. Water Res 47:6475–6487. https://doi.org/10.1016/j.watres.2013.08.024

Luo W, Zhao Y, Ding H, Lin X, Zheng H (2008) Co-metabolic degradation of bensulfuron-methyl in laboratory conditions. J Hazard Mater 158:208–214. https://doi.org/10.1016/j.jhazmat.2008.02.115

Luo Y, Guo W, Ngo H, Nghiem L, Hai F, Zhang J, Liang S, Wang X (2014) A review on the occurrence of micropollutants in the aquatic environment and their fate and removal during wastewater treatment. Sci Total Environ 473–474:619–641

Marcoux MA, Matias M, Olivier F, Keck G (2013) Review and prospect of emerging contaminants in waste. Key issues and challenges linked to their presence in waste treatment schemes: general aspects and focus on nanoparticles. Waste Manag 33(11):2147–2156. https://doi.org/10.1016/j.wasman.2013.06.022

Matamoros V, Gutierrez R, Ferrer I, Garcia J, Bayona JM (2015) Capability of microalgae-based wastewater treatment systems to remove emerging organic contaminants: a pilot-scale study. J Hazard Mater 288:34–42. https://doi.org/10.1016/j.jhazmat.2015.02.002

Matamoros V, Uggetti E, Garcia J, Bayona JM (2016) Assessment of the mechanisms involved in the removal of emerging contaminants by microalgae from wastewater: a laboratory scale study. J Hazard Mater 301:197–205. https://doi.org/10.1016/j.jhazmat.2015.08.050

Mulla S, Talwar M, Hoskeri RS, Ninnekar H (2012) Enhanced degradation of 3-nitrobenzoate by immobilized cells of *Bacillus flexus* strain XJU-4. Biotechnol Bioprocess Eng 17:1294–1299

Mulla SI, Talwar MP, Bagewadi ZK, Hoskeri RS, Ninnekar HZ (2013) Enhanced degradation of 2-nitrotoluene by immobilized cells of *Micrococcus* sp. strain SMN-1. Chemosphere 90(6):1920–1924. https://doi.org/10.1016/j.chemosphere.2012.10.030

Mulla SI, Bangeppagari M, Mahadevan GD, Eqani SAMAS, Sajjan DB, Tallur PN, Veena B, Megadi V, Ninnekar H (2016) Biodegradation of 3-chlorobenzoate and 3-hydroxybenzoate by polyurethane foam immobilized cells of *Bacillus* sp. OS13. J Environ Chem Eng 4:1423–1431

Mulla SI, Hu A, Wang Y, Sun Q, Huang SL, Wang H, Yu CP (2016a) Degradation of triclocarban by a triclosan-degrading *Sphingomonas* sp. strain YL-JM2C. Chemosphere 144:292–296. https://doi.org/10.1016/j.chemosphere.2015.08.034

Mulla SI, Sun Q, Hu A, Wang Y, Ashfaq M, Eqani SA, Yu CP (2016b) Evaluation of sulfadiazine degradation in three newly isolated pure bacterial cultures. PLoS One 11:e0165013. https://doi.org/10.1371/journal.pone.0165013

Mulla SI, Wang H, Sun Q, Hu A, Yu CP (2016c) Characterization of triclosan metabolism in *Sphingomonas* sp. strain YL-JM2C. Sci Rep 6:21965. https://doi.org/10.1038/srep21965

Mulla SI, Hu A, Sun Q, Li J, Suanon F, Ashfaq M, Yu CP (2018) Biodegradation of sulfamethoxazole in bacteria from three different origins. J Environ Manag 206:93–102. https://doi.org/10.1016/j.jenvman.2017.10.029

Nie Y, Qiang Z, Zhang H, Ben W (2012) Fate and seasonal variation of endocrine-disrupting chemicals in a sewage treatment plant with A/A/O process. Sep Purif Technol 84:9–15

Norvill ZN, Shilton A, Guieysse B (2016) Emerging contaminant degradation and removal in algal wastewater treatment ponds: identifying the research gaps. J Hazard Mater 313:291–309. https://doi.org/10.1016/j.jhazmat.2016.03.085

Parladé E, Hom-Diaz A, Blánquez P, Martínez-Alonso M, Vicent T, Gaju N (2018) Effect of cultivation conditions on β-estradiol removal in laboratory and pilot-plant photobioreactors by an algal-bacterial consortium treating urban wastewater. Water Res 137:86–96

Peng FQ, Ying GG, Yang B, Liu S, Lai HJ, Liu YS, Chen ZF, Zhou GJ (2014a) Biotransformation of progesterone and norgestrel by two freshwater microalgae (*Scenedesmus obliquus* and *Chlorella pyrenoidosa*): transformation kinetics and products identification. Chemosphere 95:581–58588. doi:https://doi.org/10.1016/j.chemosphere.2013.10.013.

Peng X, Qu X, Luo W, Jia X (2014b) Co-metabolic degradation of tetrabromobisphenol A by novel strains of *Pseudomonas* sp. and *Streptococcus* sp. Bioresour Technol 169:271–276. https://doi.org/10.1016/j.biortech.2014.07.002

Pflugmacher S, Wiencke C, Sandermann H (1999) Activity of phase I and phase II detoxification enzymes in Antarctic and Arctic macroalgae. Mar Environ Res 48:23–36

Reis PJ, Reis AC, Ricken B, Kolvenbach BA, Manaia CM, Corvini PF, Nunes OC (2014) Biodegradation of sulfamethoxazole and other sulfonamides by *Achromobacter denitrificans* PR1. J Hazard Mater 280:741–749. https://doi.org/10.1016/j.jhazmat.2014.08.039

Ruksrithong C, Phattarapattamawong S (2017) Removals of estrone and 17β-estradiol by microalgae cultivation: kinetics and removal mechanisms. Environ Technol. https://doi.org/10.1080/09593330.2017.1384068

Santaeufemia S, Torres E, Abalde J (2018) Biosorption of ibuprofen from aqueous solution using living and dead biomass of the microalga *Phaeodactylum tricornutum*. J Appl Phycol 30:471–482

Santos JL, Aparicio I, Callejon M, Alonso E (2009) Occurrence of pharmaceutically active compounds during 1-year period in wastewaters from four wastewater treatment plants in Seville (Spain). J Hazard Mater 164(2–3):1509–1516. https://doi.org/10.1016/j.jhazmat.2008.09.073

Singer H, Jaus S, Hanke I, Lück A, Hollender J, Alder A (2010) Determination of biocides and pesticides by on-line solid phase extraction coupled with mass spectrometry and their behaviour in wastewater and surface water. Environ Pollut 158:3054–3064

Sole A, Matamoros V (2016) Removal of endocrine disrupting compounds from wastewater by microalgae co-immobilized in alginate beads. Chemosphere 164:516–523. https://doi.org/10.1016/j.chemosphere.2016.08.047

Stamatis NK, Konstantinou IK (2013) Occurrence and removal of emerging pharmaceutical, personal care compounds and caffeine tracer in municipal sewage treatment plant in Western Greece. J Environ Sci Health B 48(9):800–813. https://doi.org/10.1080/03601234.2013.781359

Stamatis N, Hela D, Konstantinou I (2010) Occurrence and removal of fungicides in municipal sewage treatment plant. J Hazard Mater 175(1-3):829–835. https://doi.org/10.1016/j.jhazmat.2009.10.084

Sui Q, Cao X, Lu S, Zhao W, Qiu Z, Yu G (2015) Occurrence, sources and fate of pharmaceuticals and personal care products in the groundwater: a review. Emerg Contam 1:14–24

Tallur PN, Mulla SI, Megadi VB, Talwar MP, Ninnekar HZ (2015) Biodegradation of cypermethrin by immobilized cells of *Micrococcus* sp. strain CPN 1. Braz J Microbiol 46 (3):667-672

Terzic S, Senta I, Ahel M, Gros M, Petrovic M, Barcelo D, Muller J, Knepper T, Marti I, Ventura F, Jovancic P, Jabucar D (2008) Occurrence and fate of emerging wastewater contaminants in Western Balkan Region. Sci Total Environ 399(1-3):66–77. https://doi.org/10.1016/j.scitotenv.2008.03.003

Thies F, Backhaus T, Bossmann B, Grimme LH (1996) Xenobiotic biotransformation in unicellular green algae. Involvement of cytochrome P450 in the activation and selectivity of the pyridazinone pro-herbicide metflurazon. Plant Physiol 112(1):361–370

USEPA (2011) Principles of design and operations of wastewater treatment pond systems for plant operators, engineers, and managers. USA Patent
Villar-Navarro E, Baena-Nogueras R, Paniw M, Perales J, Lara-Martín P (2018) Removal of pharmaceuticals in urban wastewater: high rate algae pond (HRAP) based technologies as an alternative to activated sludge based processes. Water Res 139:19–29
Vulliet E, Cren-Olivé C (2011) Screening of pharmaceuticals and hormones at the regional scale, in surface and groundwaters intended to human consumption. Environ Pollut 159:2929–2934
Wang Y, Li Y, Hu A, Rashid A, Ashfaq M, Wang H, Luo H, Yu CP, Sun Q (2018) Monitoring, mass balance and fate of pharmaceuticals and personal care products in seven wastewater treatment plants in Xiamen City, China. J Hazard Mater 354:81–90. https://doi.org/10.1016/j.jhazmat.2018.04.064
Weatherly LM, Gosse JA (2017) Triclosan exposure, transformation, and human health effects. J Toxicol Environ Health B Crit Rev 20(8):447–469. https://doi.org/10.1080/10937404.2017.1399306
Xiong J-Q, Kurade M, Jeon BH (2017a) Biodegradation of levofloxacin by an acclimated freshwater microalga, *Chlorella vulgaris*. Chem Eng J 313:1251–1257
Xiong J-Q, Kurade M, Patil D, Jang M, Paeng K-J, Jeon BH (2017b) Biodegradation and metabolic fate of levofloxacin via a freshwater green alga, *Scenedesmus obliquus* in synthetic saline wastewater. Algal Res 25:54–61
Xiong J, Kurade M, Jeon BH (2017c) Ecotoxicological effects of enrofloxacin and its removal by monoculture of microalgal species and their consortium. Environ Pollut 226:486–493
Xiong JQ, Kurade MB, Kim JR, Roh HS, Jeon BH (2017d) Ciprofloxacin toxicity and its co-metabolic removal by a freshwater microalga *Chlamydomonas mexicana*. J Hazard Mater 323:212–219. https://doi.org/10.1016/j.jhazmat.2016.04.073.
Yu CP, Chu KH (2009) Occurrence of pharmaceuticals and personal care products along the West Prong Little Pigeon River in East Tennessee, USA. Chemosphere 75(10):1281–1286. https://doi.org/10.1016/j.chemosphere.2009.03.043
Yu CP, Deeb RA, Chu KH (2013) Microbial degradation of steroidal estrogens. Chemosphere 91(9):1225–1235. https://doi.org/10.1016/j.chemosphere.2013.01.112
Zhang Y, Habteselassie M, Resurreccion E, Mantripragada V, Peng S, Bauer S, Colosi L (2014) Evaluating removal of steroid estrogens by a model alga as a possible sustainability benefit of hypothetical integrated algae cultivation and wastewater treatment systems. ACS Sustain Chem Eng 2:2544–2553
Zhou H, Wu C, Huang X, Gao M, Wen X, Tsuno H, Tanaka H (2010) Occurrence of selected pharmaceuticals and caffeine in sewage treatment plants and receiving rivers in Beijing, China. Water Environ Res 82(11):2239–2248
Zhou GJ, Ying GG, Liu S, Zhou LJ, Chen ZF, Peng FQ (2014) Simultaneous removal of inorganic and organic compounds in wastewater by freshwater green microalgae. Environ Sci Process Impacts 16(8):2018–2027. https://doi.org/10.1039/c4em00094c
Zorita S, Martensson L, Mathiasson L (2009) Occurrence and removal of pharmaceuticals in a municipal sewage treatment system in the south of Sweden. Sci Total Environ 407(8):2760–2770. https://doi.org/10.1016/j.scitotenv.2008.12.030

Developing Designer Microalgae Consortia: A Suitable Approach to Sustainable Wastewater Treatment

Adi Nath, Kritika Dixit, and Shanthy Sundaram

1 Introduction

1.1 Natural Source of Microalgae Consortia

Recycling of microalgae polycultures for aqueous-phase co-products (ACP) is not only feasible but in fact increases biomass production through the recycling of nutrients. Mixed cultures outperformed even the most effective single species in terms of co-product tolerance and productivity at the time of utilization of ACP (Godwin et al. 2017). CO_2 effectively increased light energy conversion and storage into lipids through the interaction of heterotrophy in microalgae consortia (Sun et al. 2016). Consortia species have been evaluated for exact analysis of protein, amino acids, fatty acids, and pigments. The approximate values, especially protein, are as much as 45% of the entire mixture (Yahya et al. 2016). The growth of consortium of native protoctist (protist) strains isolated from dairy-farm effluents is appropriate for the production of biodiesel in addition to wastewater quality improvement (Hena et al. 2015). The drying procedure for a microalgae consortium is characterized by drying rate and moisture diffusion (Viswanathan et al. 2011). In the carpet business, combining stream effluent with a 10% to 15% waste matter mixture was found to produce a good growth medium for the cultivation of microalgae. Bioenergy

A. Nath
Department of Botany, Nehru Gram Bharati (Deemed University), Allahabad, Uttar Pradesh, India

Centre of Biotechnology, University of Allahabad, Prayagraj, Uttar Pradesh, India

K. Dixit
Defence Institute of Physiology and Allied Sciences, DRDO, New Delhi, Delhi, India

S. Sundaram (✉)
Centre of Biotechnology, University of Allahabad, Prayagraj, Uttar Pradesh, India

S. K. Gupta, F. Bux (eds.), *Application of Microalgae in Wastewater Treatment*,
https://doi.org/10.1007/978-3-030-13913-1_4

production using carpet factory effluent exploiting an algal consortium might make the management of sewage effluent purification a cost-effective business for future industry operations (Chinnasamy et al. 2010).

The microalgae consortium has the capability to intake *p*-chlorophenol under different sources of light energy. *p*-Chlorophenol adsorbed by zeolite has been frequently used in biodegradation (Lima et al. 2003). The mixed algal consortia enhance removal of nutrients within the scope of biofuel production (Mahapatra et al. 2014). The alga–bacterial community interaction could be widely used for the detoxification of commercial wastes to develop new treatment strategies, such as membrane photobioreactors to boost biomass management and separate edible algae from inedible algae by inhibition (Munoz and Guieysse 2006). The design pool for quick growth rate with lipid accumulation and separability from liquid parts is another facet of consortium development (Aneesh et al. 2015). The association of filamentous strains from native surroundings followed as a dominating priority in nutrient removal efficiency to increase biomass (Renuka et al. 2013). Agroforestry might enhance primary production in microbiome community ecosystems, although altered diversity effects are less well studied (Bruno et al. 2005). The utilization of resources and space occupied by a microalgae consortium indicates that species richness improves the consistency of the idea that the community is related to biomass stability (May 1974), although the stability of the aggregate community tends to increase, such as in biomass and effective density (Tilman 1996; Tilman and Lehman 2001). Moreover, in the case of enormous resource delimitation, replacement of dominant genera after competition or on a hotspot-specific vast scale, the availability of resources or their utilization remains completely unaffected through ecosystem processes. The mechanisms of physiological vital activities as well as molecular coherent identity have been identified for enormous production by interactive trait-dependent or trait-independent factors (Crooks 2002). To date, the taxonomic functional groups have been mostly sessile; the availability of resources and structural shape, size, and texture may be less well studied than in similar properties in photosynthetic microorganism groups, for example, mobile species. The research of interest suggested by Levine (2000) is that data-based synthesis, including experimental and simulation approaches, may additionally be profitable when examining the microalgae consortia in primary and secondary biome functions. According to Emmerson and Huxham (2002), observation suggests gaining insight into microalgae consortia function and improvement and regulative services relationships at a very large scale and in a diversity-rich community. This topic is similar to what is ascertained in allopatric and sympatric variations that at the time of invasions eventually cease to exist; however, most regions gain the total variety of species, except some resident species (Marshall et al. 1982; Vermeij 1991). Invasive species can affect ecosystem shape, size and texture, structure, density, and provisory and regulatory function, less well studied in this regard. Ecosystems lacking dominant filter-feeding photosynthetic microorganisms experience altered behavior at the time species are introduced (Alpine and Cloern 1992). Evaluation of the impact of invasion-mediated enhanced productivity with respect to diverse habitat and ecosystem processes would allow a more interesting experiment of the diversity–function associations (Stachowicz and Byrnes 2006). The ecological niche or weather-mediated partitioning and agonistic growth-dependent biotic and abiotic factors of one species over another, based on niche and predation, are relatively dominant in monocultures and with decreased productivity (Rakowski and Cardinale 2016). The effect of temperature change on biomass productivity with respect to phycocyanin analysis greatly affected the resulting change in diversity (Rakowski and Cardinale 2016; Burgner and Hillebrand 2011). The use of global carbon-mediated factors such as glucose, ethanol, and acetate will overcome less rich consortium growth rates on morning heating of the media caused by biomass loss during the nighttime (Londt and Zeelie 2013). The microalgae consortia address queries about structuring processes, diversity, dominance, relative abundance, stability, and bionomics in the range of various useful groups (Steneck and Dethier 1994). Microalgal biovolume is often calculated to assess the relative abundance of co-occurring algal variables in form or size for formulae applying the surface-to-volume quantitative relationship (Hillebrand et al. 1999). Cyanobacterium mats are an ecological indicator, in specific participating in diversity and connectivity between niche portioning across a range of vegetation (Micheli et al. 2014). The microalgae consortia with respect to bacteria cell factories are significant in complex coherent manipulation engineering of fundamental CO_2 pathways for nitrogen fixation for better bioenergy and biofuels for the future demands of humanity (Ortiz-Marquez et al. 2013).

1.2 Community Behavior

Species richness as added in a consortium has a directly proportional utilization of biological resources to meet human demands and provide new biomass. This combined role of microalgae consortia and assemblage interactions alters primary productivity, sustainability, and services. In this regard also showed that resource capture and biomass production by the consortia begin to become saturated at a low level of richness. Many microalgae consortia use an astonishing amount of resources and manufacture more biomass than their best species singly. Biodiversity–ecosystem function (BEF) experiments for the fundamental study of photosynthetic physiological pathways that have been ongoing for several generations show that the impacts of diversity on microalgae consortia biomass tend to grow stronger with time (Cardinale et al. 2007). The multifarious impact should grow stronger in additional realistic environments with greater spatial and temporal nonuniformity. The relationship of biomass richness using a large distribution of agroforestry maintains algal diversity and consortia biomass. Statistical analysis of the functional relationship of algal richness to develop consortia biomass shows positive values that remain. The interspecific interactions tend to steady consortia biomass in numerous communities (Gross et al. 2014). However, the impact of species richness on growth-level stability is ambiguous. The latest strategy magnifies the community relationship among a number of species across different levels of species richness (Gross et al. 2014). In general, with decrease in aggregate productivity of the selected sampling cluster, resource depletion also decreases. At the same time, the biomass stability and resources captured through the foremost genera in a rich polyculture do not show an exact discrepancy from that of the most efficient species alone employed in the known mechanism; results are most in keeping with what is referred to as the 'sampling effect.' Sampling impact happens when numerous communities of consortia are elucidated through the interaction of the foremost dominant genera. A polyculture system is more productive than a single best species (Weis et al. 2008).

2 Evolutionary Principle Behind Microalgae Consortia Development

2.1 *Selection Effect*

The total aggregate or primary production of different nutrient factors of the microalgae consortia is most probably decreased by similar resources being taken up by other organisms. In a community of microalgae consortia, the net primary productivity is similar to the individually most productive species in the community by utilization of resources at a given time. The mechanism behind gross primary productivity and net primary productivity could be determined by designing the community towards the number of species added in microalgae consortia; if consistent regarding change in diversity, this is called the sampling effect (Cho et al. 2017). The microalgae consortia of different niche habitats are dominant or more productive over the most productive single species in the community by the sampling effect. In another way, we can say that designing of microalgae consortia of highly diverse habitats produced less biomass than the highest producing individual species, called the selection effect. The selection effect is also known as sampling effect or selection probability (Fox 2004).

2.2 *Species Richness Effect*

The stepwise substitution of several species at a given time is most probably responsible for varying the future of microalgae consortia among different habitats. There are great opportunities in the fate of next-generation productivity because interactions among diverse microalgae consortia vary from the average individual producing species in ecosystem functioning (Zimmerman and Cardinale 2014) (Fig. 1).

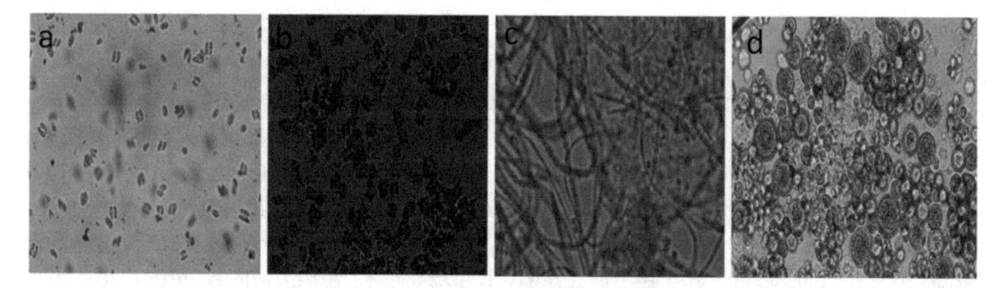

Fig. 1 Microscopic examination of monocultures and their consortia showing the species richness effect: (**a**) *Chlorella* sp., (**b**) *Scenedesmus abundans*, (**c**) *Anabaena variabilis*, (**d**) consortia

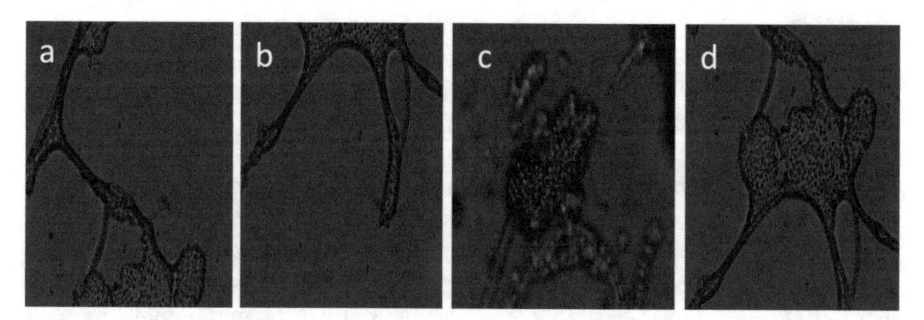

Fig. 2 Biofilm image of microalgae consortia shows identity effect: (**a**) consortia of *Lyngbya* sp., *Scytonema* sp., *Chlorella* sp.; (**b**) *Scytonema* sp., *Phormidium* sp.; (**c**) *Lyngbya* sp., *Oscillatoria* sp.; (**d**) *Oscillatoria* sp., *Scytonema* sp., *Phormidium* sp

2.3 Random Effect or Identity Effect

Discrimination in individual indigenous species through the microalgae consortia combination can be directly evaluated by primary productivity, sustainability, functioning, and services resulting from the randomness of identical species, shape size, characteristic features, and interactive properties (either premitotic phase or postmitotic phase interaction behavior) (Cardinale et al. 2012) (Fig. 2).

2.4 Complementarity Effect

In microalgae consortia, it is commonly assumed that the diverse habitats directly affect biomass productivity through species compatibility. This fact is the subject of controversy in applied experimental designing. The complementation in or within species at the time of niche partitioning by microalgae consortia of similar or dissimilar properties such as an organism discriminate by resource capturing capability and by compatibility interaction modes regarding time or space.

Trait-Independent Complementarity

The conversion of species in microalgae consortia produces higher productivity than individual species alone, called trait-independent complementarity (TIC). The TIC of an individual species growth rate and productivity is not determined by interactive species behavior. The enhanced value of TIC shows that dissimilar properties such as organism productivity are less because of similar properties such as heterogeneity and homogeneity behavior. Microbiome productivity in a niche habitat could elucidate the productivity of individual genera and microalgae consortia by portioning total abundance accessed through trait-dependent complementarity (TDC), trait-independent complementary (TIC), and dominant effect (DE).

Trait-Dependent Complementarity

If the recurring individual species gains rich production alone rather in a mixed culture, this is called trait-dependent complementarity (TDC). TDC occurs not only in interactions among community species. If the value of TDC increases, it demonstrates that the biomass of higher providing species is affected in coexistence rather than in irreversible interactions with a low biomass-producing individual species.

Dominant Complementarity

The exact productivity of consortia is directly affected at the premitotic or post-mitotic interaction stage. Species richness with respect to the exact productivity of other species is called dominant complementarity. The denominator indigenous individual species exhibit yields like other component species (Loreau 2000). The macroalgae consortia of diverse habitats did not overyield with respect to the highly efficient individual genera or the optimum yielding individual culture for primary production. The results of TIC are positive; however, effective intraspecific competition among genera recessively and in becoming workable with a community is enhanced. The enhanced value of trait-dependent complementarity (TDC) among the interactions of species and with their natural habitat for improved services were directly lower with dominance and alternative negative selection effects. The impact of interactive species for element effects will enhance gross primary production and net primary production proportionally. This change might suggest less identity impact in alternative environments and recommend improved evolutionary patterns to reverse efficiency against global warming impact within microalgae consortia interactions on species loss (Bruno et al. 2006).

2.5 Competition

The higher diverse microalgae consortia utilize available resources within their habitat. Some researchers believe that the consortia of higher dense microclimates are less productive than individuals or a community of a few species because of the invasion of new species in premitotic or postmitotic stage interaction. Interaction among microalgae consortia in the terrestrial environment is identified by experimental data and debatable by researchers: typically decreased competition and coaction occur with increasing species richness within an environment, and experimental evaluation shows alternative services (Davinson 1991). The great impact of sampling frequently elaborates the arrangement of environmental variation nonuniformity, which affects each microalgae consortia by increasing the natural habitat area (Zimmerman and Cardinale 2014).

2.6 Facilitation

Microalgae consortia diversity directly alter the structure participants of epiphytic algae and cyanobacteria, which inhabit by enhanced phototrophic productivity through greater primary production; in another way, participation of large and more structural and functional services affects the heterogeneity and homogeneity of behavior. Experimental evaluation of similar properties as in consortia is mixed. A few reports in this regard suggest the seagrass community has less interaction with associated lower nonchordates; however, it was strongly associated with species aggregations (Stachowicz 2006). In this regard, in diverse treatments dense shoots did have a significant role; one experiment reported an impact of variance on behalf of dominance for gross texture, structure, and compactness (Power and Cardinale 2009).

2.7 Coexistence

In coexistence, the stability of microalgae consortia is positively determined on the basis of sympatric and allopatric variation for the impact of producer richness. A study reported the impact of microalgae consortia on biomass stability within the interaction of macroalgal mats through modification of bioprocess engineering for stoichiometry. Participants of consortia showed durability of primary productivity along with weather with respect to applicable resource capturing among compatibility in participating microalgae consortia. In comparative analysis of microalgae consortia, experimental information resulted in an equivalent bridge connection with relevancy of duration and area (Bruno et al. 2005, 2006). Microalgae consortia have a tendency to achieve heterogeneity. Interaction-induced variation in total productivity was approximately increased for individual Firmicutes strains, with respect to average producing strains among highest species-rich microalgae consortia. In this approach, researchers apply microalgae grass growth kinetics with eventual low production within species-rich microalgae consortia rather than less species richness productivity with respect to sediment and temperature variation. Conclusively, we can say that marine microalgae, seaweeds, and ocean grasses showed decreased total productivity in microalgae consortia after genetic manipulation.

3 Designer Microalgae Consortia

3.1 Edible Algae Versus Nonedible Algae

In a previous study for the production of desired compounds and physiology of the organisms, every factor of specific individual algae thought to be important underwent separate experimentation (He et al. 2012; Weyhenmeyer et al. 2013; Weis et al. 2008). So, to preserve available resources, we considered independent variables at the same time by using statistical methods such as factorial, mixture optimal, central composite design (CCD), and response surface methodology (RSM). A number of other optimization methods are available but these have several limitations. Forecasting desired response with a point change in each variable, the concentrations of different variables were altered one at a time and the altered level of response or product ions was measured at each point through a computer device (Patel et al. 2014; Loreau et al. 2001; Tillman et al. 1982, 1996). To study the effect of different components of media BG11+ (important upstream parameters) on growth, enhancement of biomass, carbohydrate, and carbon fixation contents in eight different consortia in factorial design, a minimum and maximum level of nutrient concentration level L-1 for each medium component was chosen. The maximum concentrations were taken as described by Rippka et al. (1979); minimum concentrations were chosen randomly to a certain extent. For industrial productivity, we would refer to as "edible" algae, dominant over the "inedible" algae. Other Chlorophyta cannot grow on a bacteria-only diet, although alternative strains can do so (personal observation). The Cyanobacteria (members of Cyanophyta) and microalgae (members of Chlorophyta) commonly co-occur in nature but were not indicated to mimic any specific natural system.

3.2 Microalgae Consortia Functioning

The functioning of microalgae consortia was determined from a wide range of morphological and physiological categories to confirm some chance of community delineation and to assist association. Feeding trials indicated that microalgae consortia readily consume and grow best. Of seven better primary production yields, designer microalgae communities were a suitable size for intake by percent CO_2 and perhaps are expanded at higher levels. Occasionally microalgae consortia support growth on their own survivalists from their environment. The results also show highest heterogeneity could show better production. Inspection of population dynamic statistics indicated that rates of modification in algal biovolume usually slowed over the course of the experimentation, typically approaching a quasi-steady state when mean total consortium effectiveness varied considerably among microalgae compositions (block effects were not significant).

Primary Functioning

The aggregate productivity in each microalgae consortium of the most productive individual cultures significantly exceeded the premitotic phase (lag phase) in time. In individual cultures regarding postmitotic phase (stationary phase) interaction in time, assessments of the two most productive individual species and their consortia on the day of harvesting discriminated between them. The transgressed yield of microalgae consortia in individual cultures was directly related to overall productivity with respect to the consortia on the day of harvesting (Fig. 3).

Photosynthetic Quantum Yield

Developing microalgae consortia with different environmental factors suggests enormous resources through biodiversity by the evaluation of developing microalgae consortia participants for their better productivity and sustainability in their environment. The synthetic microalgae consortia could be designated in a targeted approach for heavy metal detoxification, water quality improvement, and

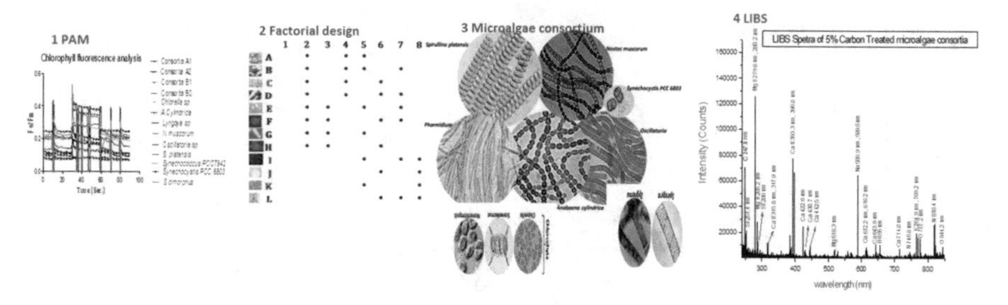

Fig. 3 Graphical diagram shows (1) pulse amplitude modulation chlorophyll fluorescence analysis; (2) factorial design through response surface methodology software via central composite design (CCD); (3) development of enhanced productivity consortia by optimal resource capture; (4) elemental characterization for responsible factor through laser-induced breakdown spectroscopy (LIBS)

fulfillment of human services. The microalgae consortia most often gain higher capacity than individual species. The enhanced photosynthetic functioning of the heterogeneity route of selected gene expression in developing microalgae consortia and the next generation of developed consortia is less well studied. To start, the biotechnology of microalgae consortia can be determined by one or two physiological trait factors for primary productivity that can be optimized for targeted achievements. The metabolic pathway among microalgae consortia cells with respect to different traits can be analyzed on the basis of the participants in the consortia. Fv/Fm (Fmax−Fmin/Fmax) was observed in all the cultures. Fv/Fm is a substantive means for assessing the potential of photosynthesis in microalgae consortia, chiefly for highlighting photoinhibition in excess illumination. Pulse-amplitude-morphometry (PAM) fluorometry results indicated that at optimum magnitudes, Fv/Fm ratios were completely different in various monocultures and consortia were found to be statistically different from each other ($p < 0.001$). In all cases, with increasing chlorophyll content, a lower Fv/Fm was determined with all the monocultures and consortia. This result indicated that the microalgal cells were growing and performing photosynthesis efficiently and were active within the photoinhibition process. The quenching analysis by PAM fluorometer showed the measure of photosynthetic efficiency in terms of quantum efficiency or quantum yield. Generally, the monocultures were poor in achieving maximum quantum yield, in comparison to all the consortia (Fig. 4).

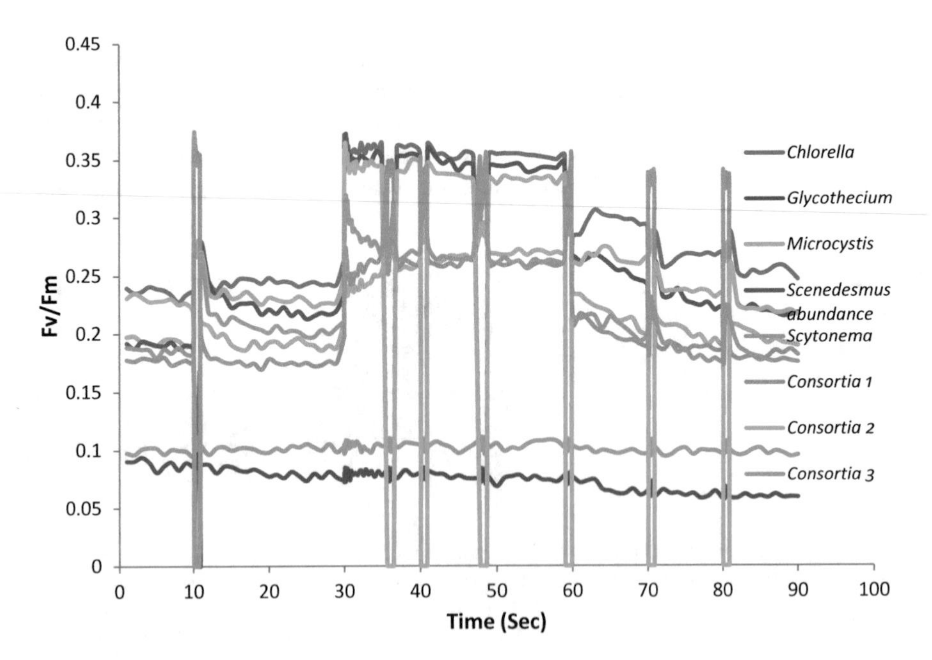

Fig. 4 Photosynthetic quantum yield in various monocultures and microalgae consortia

Photochemical and Nonphotochemical Quenching

According to the PAM graph obtained, we noticed that of all consortia *Scenedesmus dimorphus* was the foremost effective species (or the foremost effective strain) showing the greatest photosynthetic effectiveness. Equally, *Synechocystis* PCC6803, *Synchococcus* PCC7942, and *Spirulina platensis* had good photosynthetic ability compared to the alternative strains studied. As we know, nonphotosynthetic quenching provides a standard for complete quantum efficiency and is reciprocally related to it; here, *Anabaena cylindrica* is the least efficient strain regarding photosynthetic efficiency and shows the highest quenching. Intended for establishment of the purposeful arrangement for photosynthetic efficiency of photosynthetic blue-green algae, variable fluorescence magnitude relationships were monitored. The value of the initial fluorescence by the first electron acceptor from PS II antennae varied among blue-green algae samples. The maximum quantum yield of filamentous strains was at a low level. Altogether, consortia, *Spirulina platensis,* and *Chlorella* sp. displayed the maximum variable fluorescence value; next were *Anabaena cylindrica*, *Oscillatoria*, *Lyngbya*, *Nostoc muscorum*, *Synechococcus* PCC7942, and *Scenedesmus dimorphus*.

The higher Fv/Fm ratio for *Spirulina platensis* indicates higher photosynthesis assimilation for alternative blue-green algae. Fv/Fm (Fmax−Fmin/Fmax) was assessed in all individual strains and consortia cells, fully grown at completely different levels of light intensity. Fv/Fm is also a utile parametric quantity to gauge photosynthetic ability in algae and in the first place specializing in photoinhibition owing to excess light. PAM fluorometry outcomes indicate that at optimal light and contrasting optimum growing parametric quantity, the quantum efficacy is greatest. In all cases together with the accelerative chlorophyll content, a minimal value of variable fluorescence was scrutinized, pointing out that the microalgal cells were thriving and achieving photosynthesis and moreover were striving for total adaptation for survival in high-intensity light. The quenching investigation by PAM fluorometer gave the measure of photosynthetic potency in terms of quantum efficacy or yield. The Fv/Fm ordination depicts the evolutionary divulgence of these photosynthetic autotrophic organisms and indicates many attainable sources for improving potency. Fv/Fm (Fmax-Fmin/Fmax) was monitored in all the microalgal and consortia cultures. PAM fluorometry results indicated that at optimum intensity level, Fv/Fm ratios of totally different monocultures and consortia were observed to be significantly different from one another ($p < 0.001$).With increasing chlorophyll content, a lower Fv/Fm was ascertained in every case together with all the monocultures and consortia. The results made it clear that the cells were thriving and carrying out effective photosynthesis and were active in the photoinhibitory process. The quenching investigation by PAM fluorometry recorded the measure of photosynthetic potency in terms of quantum efficacy or yield. Broadly speaking, monocultures were poor in maximum quantum efficiency, in contrast to the consortia (Nath et al. 2017a, b) (Fig. 5).

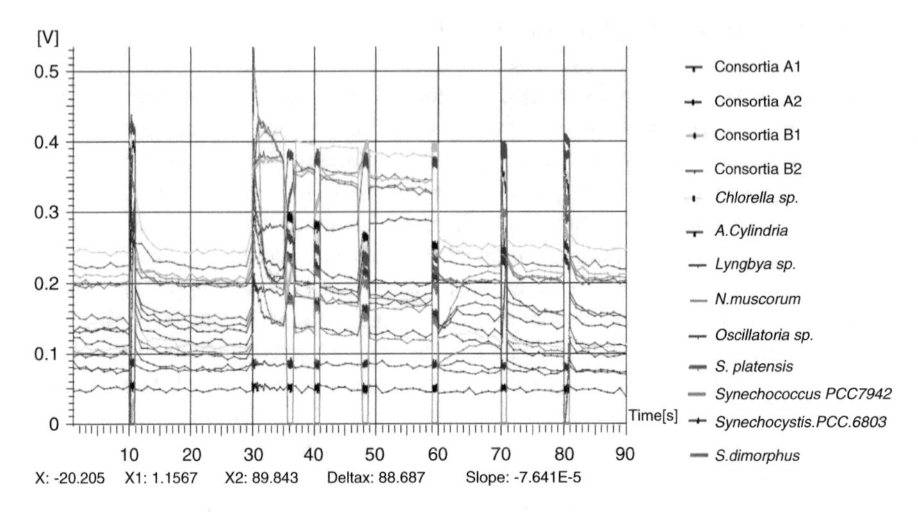

Fig. 5 Direct heat dissipation analysis in different individual cultures and their consortia

Fresh and Dried Biomass Productivity

Biomass productivity was successfully achieved by raising the temperature tolerance of microalgal consortia as well as an acceptable screening method. The delineated strategies should be applicable to induce a range of interesting characteristics in numerous microalgal consortia, which is often particularly engaging for systems wherever it is not obvious that genes need modification. Coupling contemporary production with the next-generation screening strategies would be enabling in choosing the strains with characteristics that do not yield a better fitness and cannot be chosen for in vitro experimentation. At the very least, random impact strategies will be used as a primary step during a combined approach with genetic alteration (which is analyzed through next-level strategies and techniques), to see appropriate target genes for many centered and comprehensive strategies for microalgal consortia improvement and alteration (Godwin et al. 2018; Nath et al. 2017b).

3.3 Microalgae Consortia Services

Carbon Regulatory Services

CO_2 effectively improved actinic light capacity transfer and storage into bioorganic compounds through higher concentration, that is mostly in a surprising way the properties of heterotrophy in microalgae consortia. Insight has not clearly identified through standardization a number of traits illustrating better proficiencies within the coupled dark biochemistry of bioenergy production and bioorganic molecule

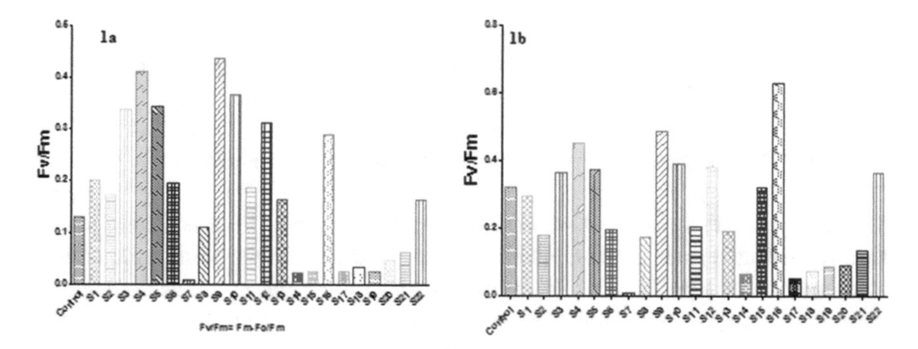

Fig. 6 Plant efficiency analyzed by a central composite designed from 22 trials from pulse amplitude modulation fluorescence

formation. The results progressively created associated economical instrumental quantification of photosynthesis for biodiesel formation at a high level, and carbon capture and storage from flue gases for carbon credit (Sun et al. 2016).

Nutrient Uptake Provisioning Services

To establish the functional position of photosynthesis rates, Fv/Fm ratio was measured in two different microalgae consortia with 22 runs of each in a test tube. This experiment imitated the highest photosynthetic quantum yield of PS II reaction centers of diverse dark-adapted microalgae culture conditions. In general, run filaments of each of the two microalgae consortia were identified for maximum rich quantum yield in the 16th run. The corresponding values of Fv/Fm of MC1(Scytonema sp., Microcystis aeruginosa, Gloeocapsa sp., Chlamydomonas reinhardtii, Scenedesmus abundance and Scenedesmus dimorphus) and MC2 (Calothrix sp., Westiellopsis prolifica, Aphanothece nageli, Chlamydomonas reinhardtii, Scenedesmus quadricauda, Chlorella sp) were 0.52 and 0.61, respectively. The lower Fv/Fm ratios in the microalgae consortia indicate their lower photorespiration rates resulting from lack of oxygen in competitive resources stress compared to other individual microalgae. The microalgae cell groups having a greater surface-to-volume proportion were found to have higher photosynthesis rates (Fig. 6).

4 Microalgae Consortia in Wastewater Treatment

4.1 Heavy Metal as a Nutrient Source

Exogenous supplementation of various doses of chromium (Cr) considerably delayed the growth and biomass productivity of assorted microalgae monocultures similarly to artificial consortia. To determine the temperature-bearing capacity of cyanophycean, chlorophycean, and microalgae consortia among hexavalent Cr, the biomass of natural habitat species was enriched with four regimes of Cr in several sampling analyses up to a 16-day cultivation time. Regular observance of growth curves showed varied growth behavior in microcosms and metalloid microalgae consortia. The presence of Cr in the culture environs of microalgae greatly distorted their morphology at completely different extents with increasing dosages. Morphological analysis processed the drastic effect within the structure of varied chromium-added microalgae consortia and individual cultures. Within the natural community, niche partitioning of the environment was vital in the acclimation of various microalgae. The branched green algae typically affected aquatic areas: *Lyngbya* was floating on the surface whereas *Oscillatoria* was random within the *Lyngbya cells*. In distinction, unicells of *Chlorella* and coenobia of *Scenedesmus dimorphus* gathered close to the lowest portion. However, with increasing degree of metal toxicity, the niche partitioning and growth rates were altered from the natural community: the stationary phase arrived sooner with synchronization of almost all the cells of *S. dimorphus* at the highest concentration of the metal. In distinction, algae and *Lyngbya* were found to possess sensible stress-tolerating options among all the evaluated organisms; continuous growth observation from the 1st to 16th day showed fascinating changes within the doubling patterns of those microbes. Their growth rates were considerably affected at the various metal concentrations. The growth behavior of those four organisms at 0.5 ppm Cr was not greatly affected ($p \geq 0.05$), and at the end of cultivation about the same quantity of biomass yields was obtained with respect to the natural control. Furthermore, with increasing the Cr doses, *Chlorella* and *Lyngbya* showed good response against Cr up to 1.0 ppm and their growth and biomass productivity were not significantly affected. However, growth of *S. dimorphus* and *Oscillatoria* was highly affected at doses above 0.5 ppm, and between 1.0 and 5.0 ppm their growth rates and biomass yield were reduced significantly ($p \leq 0.01$). Their succession rates were very high at 5.0 ppm, and they arrived at their carrying capacities much earlier with very low cell counts and biomass yields. The structure of those photosynthetic organisms was distorted by increasing the doses of metal. However, at an all-time low dose of 0.5 ppm, slightly swollen cells were seen under the light microscope, which would be the reason behind their increased metabolic attributes and increased yields of biochemicals such as carotenoids and carbohydrates. Moreover, increasing the doses of chromium above 0.5 ppm not only affected the morphology but also the ecophysiology and biomass production of these organisms significantly. A similar trend was recorded in the behavior of the community at different doses of Cr. The natural community of these four organisms produced higher biomass than monocultures and all the communities with these four treatments of Cr. Growth rates, biomass yields, and microscopic examination of various photosynthetic cells verified the following order of tolerance for the tested organisms: *Oscillatoria* sp. $<$ *S. dimorphus* $<$ *Lyngbya* sp. $<$ *Chlorella* $<$ consortia, respectively, in a healthy environment. The monoculture richness or uniformity in the natural community greatly affected community biomass productivity. Within the community, *Chlorella* was the dominant species. Therefore, it contributed significantly in Cr adsorption and biomass production.

Consequently, a notable factor was the improved tolerance behavior and biomass productivity of those microbes at the lowest dose of Cr, indicating their combinatorial positive interactions towards chromium surface assimilation. Interaction of these organisms against Cr tolerance was reduced with regular increase in the doses of Cr, which led to distortion in their morphology and response towards biomass production. As in a natural community, *Chlorella* was found to be dominant in all communities. Compared to a natural community, the 0.5 Cr-treated community produced 1.26% more biomass. With further increase in Cr levels, the morphology of these microbes was greatly distorted, which finally led to regular reduction in biomass yields. With respect to the natural community, communities at 1.0, 3.0, and 5.0 ppm showed 29.4%, 62.0%, and 78.1% of reduction in biomass yields, respectively ($p \leq 0.01$). Also noted at the lowest dose of 0.5 ppm Cr^{+6} was slow augmentation in multiplication rates and altered morphology in monocultures or in consortia. The volumetric appearance of these monocultures and various consortia, respectively, are shown in Fig. 7. *Chlorella* biomass was found to be highest among all the monocultures. Consequently, the maximum cell counts of *Chlorella* that were recorded in the community showed the richness of these algae in a natural community over other individuals. The microalgae consortia could restore water quality by alleviating the Cr levels in water parameters up to 0.5 ppm; beyond this level, these organisms would have a negligible contribution in adsorption of Cr. Consequently, at the end of cultivation, *Chlorella* and *Lyngbya* were found to be dominant species within the community, followed by *S. dimorphus* and *Oscillatoria*, respectively. The total biomass yield of a control community was found to be highest over

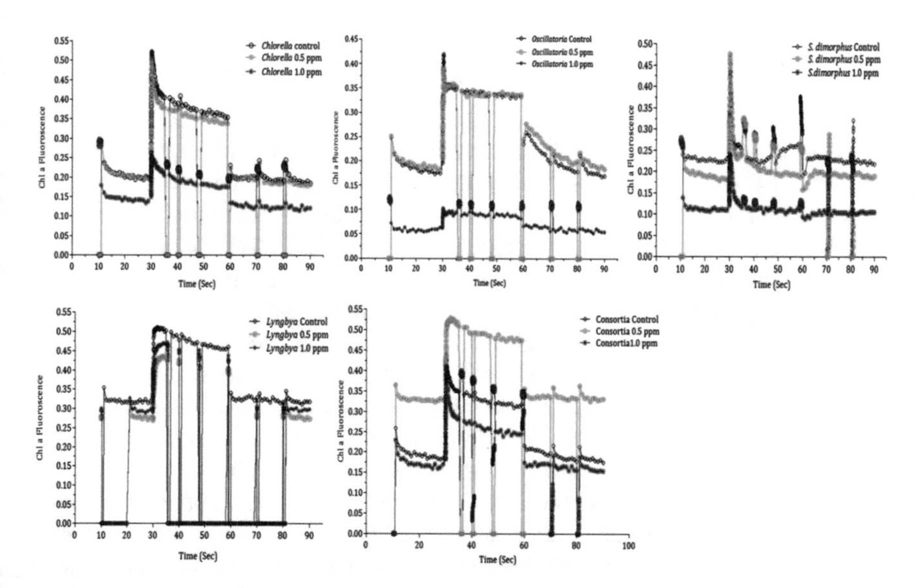

Fig. 7 O First excitation peak, J second excitation peak, K Tertiary excitation signal, I Quartenary peak and P is Stationary peak (OJKIP) curve of different monoculture and consortia showed stress level against heavy metal

monocultures of Cyanobacteria and green algae, providing as a concluding remark that within the community these organisms can have better physiology and biomass yields compared to monocultures. The growth patterns and biomass yields of various organisms and community, respectively, are provided in Fig. 7. The results were further confirmed by microscopic examinations. In various communities, *Chlorella* was the most dominant species, followed by *Lyngbya*, *S. dimorphus*, and *Oscillatoria*, respectively. An interesting characteristic of these microbes was enhanced metabolic rates and other services at the 0.5 ppm dose of Cr. A gradual increase in growth patterns was recorded at that dose until the cultures reached their carrying capacity. However, at higher concentrations (1.0, 3.0, and 5.0 ppm) of Cr, growth and biomass yields were found to be much poorer than the control and 0.5 ppm Cr^{+6} communities. At higher concentrations all the cultures reached their carrying capacity before the control with lower cell counts, decreased biomass yields, and enhanced doubling times. The morphology of cells was highly distorted with increasing Cr^{+6} toxicity levels. Consequently, in the treated community, formation of penetration on the surface of media at the end of cultivation indicated the initiation of their succession. Micrographs recorded at the end showed that during succession *Oscillatoria* sp. was highly affected whereas *Chlorella* dominated in the community. Therefore, in natural habitats affected with severe levels of Cr, similar trends may be expected with some deviations, because in natural water reservoirs there are many different independent factors or variables that could affect the structure and performance of organisms in the community. Therefore, above 1.0 ppm chromium, these microalgae lose their original features (Nath et al. 2017a, b).

4.2 *Enhancing Water Quality by Niche Portioning*

In blue-green algae, a consortium has an increased probability to increase effectual size contrasting to the foremost effective manufacturing respective strains. The principle for the improvement of consortia was measure of joint variability at intervals, the reciprocality effect in consortia, and conjointly the state of strains individually (Cardinale et al. 2006). Explicating compositional and purposeful features in consortia may well be a topic of extraordinary peculiarity that assists us to grasp the interaction of blue-green algae with the environment. Earlier studies have shown many diverse groups of various species drastically utilizing and capturing available restricted environmental resources and with a large amount of biomass production compare to the less different species population. The principle of massive biomass in aggregating abundance production was in response to the phototropic quality of the microalgal population. Korner (in 2004) counselled that algal populations would not intensify the photosynthetic rate exclusively, notwithstanding capture and storage of greenhouse gas from their environment, although this might not correspond to the tangibleness of the biomass by carbon storage attributable to increase in total species population. Through photosynthesis, flora uptake greenhouse gas, produce oxygen, and yield critically quantified biomass. Although various studies have generally acknowledged this, fewer data reveal preliminary activities taking place in photosynthesis in phototropic beings. In this present research we have deliberated that, if the copiousness of strains is inflated at intervals in the quantitative relationships of three, six, or more, then there is a depletion of accessible resources by resource pool methodology, leading to improvement of microalgal consortia biomass stability. The stableness of blue-green algae populations may result in various settings or the best yielding single strain in terms of biomass production, of the most overwhelming and effective strains. Treatment of ethyl methane sulfonate (EMS) with microalgae species with lower doses (0.42 and 1.2 M) significantly ($p < 0.05$) increased Fv/Fm by 2.4%, 7.2%, 32.8%, and 42.2%, respectively, while qP increased by 4.4%, 7.1%, 15.6%, and 30.2%, respectively, and nonphotochemical quenching (NPQ) inclined by −0.73%, −3.65%, −5.85%, and −8.03%, respectively. However, at higher doses (0.72 and 1.2 M) Fv/Fm decreased by 1.3%, −10.3%, and −13.7%, respectively; similarly, photochemical quenching (PQ) also decreased by 3.2%, −10.4%, and −18.1%, respectively, while NPQ increased by 6.57%, 10.95%, and 16.06%, respectively. The NPQ was high in the dark and in consortia that measure the ratio of quenched to remaining fluorescence (Fig. 8). The value of NPQ does suffer from distortion by the underlying protein fluorescence. This potential problem is addressed by fluorescence measurement with an alternate modulated source, which allows excitation of chlorophyll fluorescence without interference from phycobiliprotein emissions. The mutant strains reduce cellular pigment content and exhibited improved photosynthetic activity, thus improved biomass productivity. The isolated mutant strains with photosynthetic alterations pool by better photosynthesis regulation insight. Tillich et al. determined point mutation through chemical mutagenesis raised temperature tolerance about 2 °C in many generations of mutagens in *Synechocystis* PCC6803. Random coupled mutagenesis yields higher fitness for in vitro selection of strains. The quantification of living cells on selection plate rates was equivalent to mutation with nonlethal doses. The developed microalgae consortium has the capability to use enormous amounts of nutrients through the water and improve water quality through trophic factors.

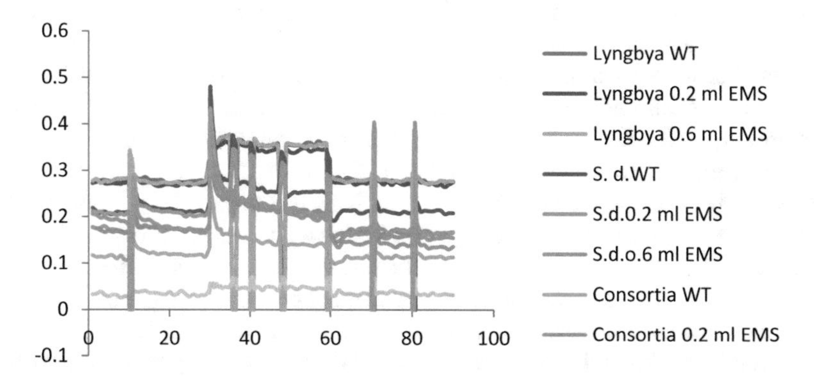

Fig. 8 Chemical mutagenesis by ethyl methane sulfonate (EMS) shows carcinogenesis in monocultures and polycultures

4.3 Improve Water Quality Through Fresh Gross Productivity

The microalgae consortia is only unfeasible for recycling of fresh gross productivity, yet often recycling of trophic factors increased production of biomass. The microalgae consortia, productivity, and tolerance capability with respect to their co-product provides higher amounts in individual species on aqueous-phase recycle time. With the highest tolerance capacity with respect to aqueous-phase co-production, microalgae consortia must have less dilution or a higher concentration during the recycle pathway. The point of interest is ecological interactions in microalgae consortia using the best efficient methods and reducing the waste stream productivity for improving trophic factors substantially harnessed that are significant; otherwise, role eutrophication will ensue (Godwin et al. 2017).

5 Molecular Approach of Microalgae Consortia

5.1 Transcriptomics and Gene Functioning Analysis

The concerned media for growth of organisms were applied to isolation. The partial evaluation of cDNA interpretation systematizes the sequestered as a number of organisms that participated in the microalgae community. These isolates included microalgae orders, families, genera, or species. A number of the isolates that were antagonistic or agonistic within the fermentation medium had properties similar to microorganisms, for example, *Aeromonas*, *Vibrio*, *Acinetobacter*, and *Pseudomonas*, which are directly responsible as causative agents for disease in humans and animals and should be effective in the assessment during negative impact caused by microalgae blooms. The various species inhibited or enhanced the growth of cyanobacteria within community and designer microalgae consortia with respect to media optimization: trophic factors, spatial variation, space, or time could improve it. Such types of consortia had an improving impact on microalgae growth. A large group of isolated microorganisms was applied in the assessment and management of the negative impact of microalgae (Berg et al. 2009). The phytoplankton species interactions and coexistence provided unique insight into the transcriptomic deviations that are acquired without restraints. First, the statistical model was developed with respect to transcriptome-wide investigation of gene expression for associations of genes that were expressed in participating individual cultures. The purpose of RNA spikes to establish detection criteria and similarities–dissimilarities of actual expression intensities provides assistance for future human requirements. The exploration of the constricted roles of shared genes versus uniquely expressed genes for contributing species interfaces, and in precise niche variation and facilitation within microalgae consortia. Second, the gene expression on postmitotic phase point to time during competition; herewith, altered in gene expression in this time would have a role in comparisons of gene expression on behalf of the mid-log growth section, dependence of density, and therefore the complete impact of competitive species. The comparisons of metagenomics basis of huge resources utilized growth versus community-dependent growth of variations versus interspecies microalgae consortia among surroundings would alter gene expression over time. Third, in the thought of the recent analysis during this regard, the illation of gene functions was, by necessity, supporting transitive annotations. The climatic issue is a crucial role example; temporal and abstraction biotic and abiotic complexness of microalgae consortia on gene expression, species interactions, and existence are areas of interest for researchers operating in this area in future. Finally, in this regard to the big plan control notion that visible morphological similarity results in competitive exclusion, the results found that similarity in gene expression among the microalgae community across the transcriptome tends to steer to weaker competition, much possible facilitation, and larger coexistence. The analysis provides support that the expression of the largely preserved normal expressed genes is needed for future survival and fitness in an exceedingly explicit environment; on the opposite hand, niche variations and helpful interactions could also be encoded by simply a number of, or presumably rare, genes. The various varieties of species interactions were investigated directly for their role in deciding known gene functions to lie within the future (Narwani et al. 2017).

5.2 *Proteomics Approach*

Although detectable using a light microscope, the photosynthetic lamellae of the cyanobacteria were first observed with electron microscopy, initially in shadowed metal preparations and later in thin sections. The lamellae are closed discs termed thylakoids. Observance of the lamellae, or the adjacent membranes, varies greatly with the fixative that is employed. After permanganate fixation, the double membrane is seen in the section as three lines in parallel having thickness of 20 Å, 40 Å, and 20 Å, between which lie spaces of 30–35 Å, whereas following osmic acid fixation the membrane apparently appears as two parallel lines each 355 Å to 655 Å thick, separated by a space of about 50 Å. The only chlorophyll that is found in cyanobacteria is chlorophyll *a*. It has been discovered that the chlorophyll from *Phormidium luridum* has phytol. In vivo, the absorption spectrum is shifted to a wavelength around 13 nm greater than the absorption peak in methanol solvent. Representative values of chlorophyll as the percent of algae are 2.2 (667) and 0.2–1.0 (741) [dried weight]. Analysis of the quantities of individual carotenoids found that a large variety of cyanobacteria contains 13-carotene, echinenone, myxoxanthophyll, sometimes zeaxanthin, and most frequently present carotenoids, albeit exceptionally canthaxanthin, caloxanthin, nostoxanthin, oscillaxanthin, and glycoside have been also found to account for almost 10% or more of the total carotenoid present. 13-Carotene is the only one of its kind present in blue-green algae (Fig. 9).

Effect of exogenous bicarbonate-anhydrase addition on total chlorophyll (mg g^{-1} fresh weight), carotenoids (mg g^{-1} fresh weight), total protein (mg g^{-1} fresh weight), and chlorophyll florescence parameters of microalgae and their consortia

Name of organism	Pigments		Biochemical compound		Chlorophyll florescence		
	Chlorophyll-a	Carotenoid	Carbohydrate	Protein	Fv/Fm	qP	NPQ
Chlorella sp.	1.72 ± 0.03e	0.531 ± 0.01f	15.4 ± 0.30ef	10.3 ± 0.18ef	0.73 ± 0.10de	0.758 ± 0.009f	1.37 ± 0.06d
Gleocapsa sp.	1.76 ± 0.12d	0.542 ± 0.01d	15.8 ± 0.21d	10.8 ± 0.21d	0.75 ± 0.07cd	0.791 ± 0.011d	1.36 ± 0.06d
Microcystis aeruginosa	1.81 ± 0.03c	0.548 ± 0.08c	16.3 ± 0.41c	11.2 ± 0.33c	0.78 ± 0.11c	0.812 ± 0.015c	1.32 ± 0.09e
Scytonema sp.	1.87 ± 0.11b	0.553 ± 0.10b	17.2 ± 0.56b	11.8 ± 0.27b	0.80 ± 0.14b	0.876 ± 0.019b	1.29 ± 0.08ef
Scenedesmus abundance	1.92 ± 0.07a	0.559 ± 0.06a	17.9 ± 0.55a	12.4 ± 0.19a	0.82 ± 0.19a	0.987 ± 0.020a	1.26 ± 0.10g
Consortia 1	1.74 ± 0.05de	0.536 ± 0.09e	15.6 ± 0.42de	10.5 ± 0.24de	0.69 ± 0.10f	0.782 ± 0.017e	1.46 ± 0.09c
Consortia 2	1.52 ± 0.04f	0.491 ± 0.07g	14.3 ± 0.32g	9.4 ± 0.33g	0.65 ± 0.06g	0.679 ± 0.016g	1.52 ± 0.14b
Consortia 3	1.48 ± 0.04g	0.472 ± 0.012h	13.8 ± 0.26h	0.61 ± 0.20h	0.61 ± 004h	0.621 ± 0.016h	1.59 = 0.19a

Data are means ± standard error of three replicates. Values with different letters within same column show significant differences at $p < 0.05$ level between treatments according to Duncan's multiple range test

The result was seen as the explanation of the initial rise in the fluorescence signal when CO_2 was suddenly removed from a plan: parts once reduced QA as the reaction center in PSII normally passes electrons, via other electron carriers, to PSI and finally CO_2 via XADP in pathway. As density of treatment increases, it would decrease quantum yield (David Walker)

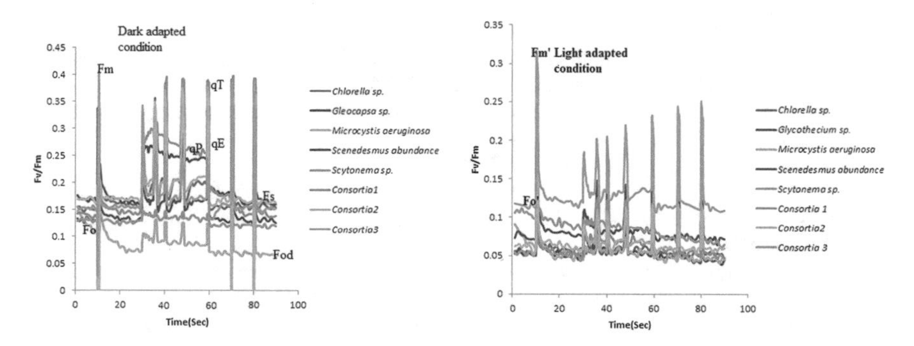

Fig. 9 Light- and dark-adapted analysis in monoculture and consortia for light harvesting complex

6 Conclusion and Future Prospects

Trait-dependent and trait-independent factors, such as level of light, quality of light, water quality, capturing of food, warmth, snow, chemical microorganisms, herbicides, detoxification of metals, pathogen control, and inherent features – all these factors simultaneously affect assimilation in photosynthetic organisms, their health and condition. These factors have a significant role in PS II fluorescence signaling, so by using a pulse amplitude modulator, improvement of biomass production by controlling the foregoing factors was evaluated and easily elucidated the functioning and services of the microbiome. The fluorescence of the PS II reaction center dissipated the energy of photons in the dark-adapted chlorophyll state; the transport of electrons comes back to the maximum oxidized condition after the chain of reduction-oxidation state. Even fluctuations in the microorganisms that affect stoma opening and gas exchange with the atmosphere are suggested by changes in the fluorescence features of an organism. In the present communication, the photochemical quenching (PQ) and nonphotochemical quenching (NPQ) of microalgae and their consortia would be studied in fermentation broth; the Fv/Fm ratio of different individual algal strains and their consortia will be used for determination of maximum quantum yields. The present proposal is expected to develop new technology for sustainable algal biofuels to limit future energy demands and economy-related problems. The designer microalgae consortia interactions would allow for unlimited, less time consuming, and cost-effective technology for bioenergy improvement, by applying the dominant complementarity industrial approach for biofuels.

References

Alpine AE, Cloern JE (1992) Trophic interactions and direct physical control of phytoplankton biomass in an estuary. Limnol Oceanogr 37:946–955

Aneesh CN, Mullalayam F, Manilal VB, Haridas A (2015) Culturing of autoflocculating microalgal consortium in continuous raceway pond reactor. Am J Plant Sci 6:2472–2480. https://doi.org/10.4236/aips.2015.615249

Berg KA, Lyra C, Sivonen K, Paulin L, Suomalainen S (2009) High diversity of cultivable heterotrophic bacteria in association with cyanobacterial water blooms. ISME J 3:314–325. International Society for Microbial Ecology 1751-7362/09

Bruno JF, Boyer KE, Duffy JE, Lee SC, Kertesz JS (2005) Effects of macroalgal species identity and richness on primary production in benthic marine communities. Ecol Lett 8:1165–1174. https://doi.org/10.1111/j.1461-0248.2005.00823.

Bruno JF, Lee SC, Kertesz JS, Carpenter RC, Long ZT, Duffy JE (2006) Partitioning the effects of algal species identity and richness on benthic marine primary production. Oikos 115:170–178. ISSN0030-1299

Burgner T, Hillebrand H (2011) Temperature mean and variance alter phytoplankton biomass and biodiversity in a long term microcosm experiment. Oikos 120:922–933

Cardinale BJ et al (2006) Effects of biodiversity on the functioning of trophic groups and ecosystems. Nature 443:989–992

Cardinale BJ, Wright JP, Cadotte MW, Carroll IT, Hector A, Srivastava DS, Loreau M, Weis JJ (2007) Impacts of plant diversity on biomass production increase through time because of species complementarity. Proc Natl Acad Sci USA 104:18123–18128

Cardinale BJ, Duffy JE, Gonzalez A, Hooper UD, Perrings C, Venil P, Narwani A, Mace MG, Tilman D, Wardle AD, Daily KPA, Gretchen C, Loreau M, Grace BJ, Larigauderie A, Srivastava D, Naeem S (2012) Biodiversity loss and its impact on humanity. Nature 7:459–486

Chinnasamy S, Bhatnagar A, Claxton R, Das KC (2010) Biomass and bioenergy production potential of microalgae consortium in open and closed bioreactors using untreated carpet industry effluent as growth medium. Bioresour Technol 101:6751. https://doi.org/10.1016/j.biortech.2010.03.094

Cho DH, Choi JW, Kang Z, Kim BH, Oh HM, Kim HS, Ramanan R (2017) Microalgal diversity fosters stable biomass productivity in open ponds treating wastewater. Sci Rep 7:1979

Crooks JA (2002) Characterizing ecosystem level consequences of biological invasions: the role of ecosystem engineers. Oikos 97:153–166

Davinson IR (1991) Environmental effects on algal photosynthesis: temperature. J Phycol 27:2–8

Emmerson M, Huxham M (2002) How can marine ecology contribute to the biodiversity-ecosystem functioning debate? In: Loreau M, Naeem S, Inchausti P (eds) Biodiversity and ecosystem functioning: synthesis and perspectives. Oxford University Press, Oxford, pp 139–146

Fox WJ (2004) Effect of algal and herbivore diversity on the portioning of biomass within and among trophic levels. Ecology 85:549–559

Godwin CM, Hietala DC, Lashaway AR, Narwani A, Savage PE, Cardinale BJ (2017) Algal polycultures enhance coproduct recycling from hydrothermal liquefaction. Bioresour Technol 224:630–638

Godwin, William. "Enquiry concerning political justice." The Economics of Population. Routledge, 2018. 37-40

Gross K, Cardinale BJ, Fox WJ, Andrew G, Loreau M, Polly WH, Reich PSB, Ruiwen VJ (2014) Species richness and the temporal stability of biomass production: a new analysis of recent biodiversity experiments. Am Naturalist 183:112

He Z, Piceno Y, Deng Y, Xu M, Lu Z, Santis TD, Andersen G, Hobbie SE, Reich PB, Zhou J (2012) The phylogenetic composition and structure of soil microbial communities shifts in response to elevated carbon dioxide. ISME J 6:259–272. International Society for Microbial Ecology 1751–7362/12

Hena S, Fatimah S, Tabassum S (2015) Cultivation of algae consortium in a dairy farm wastewater for biodiesel production. Water Resour Ind 10:1–14

Hillebrand H, Durselen CD, Kirschtel D, Pollingher U, Zohary T (1999) Biovolume calculation for pelagic and benthic microalgae. J Phycol 35:403–424

Levine JM (2000) Species diversity and biological invasions: relating local process to community pattern. Science 288:852–854

Lima SAC, Raposo MFJ, Castro PML, Morais RM (2003) Biodegradation of *p*-chlorophenol by a microalgae consortium. Water Res 38:97–102

Londt ME, Zeelie B (2013) Outdoor large scale microalgae consortium culture for biofuel production in South Africa: overcoming adverse environmental effects on microalgal growth. J Energy Technol Policy 3(11):39–45. ISSN 2225-0573

Loreau, Michel. "Biodiversity and ecosystem functioning: recent theoretical advances." Oikos 91.1 (2000): 3–17.

Loreau M, Naeem S, Inchausti P, Bengtsson J, Grime JP, Hector A, Hooper DU, Huston MA, Raffaelli D, Schmid B, Tilman D, Wardle DA (2001) Biodiversity and ecosystem functioning: current knowledge and future challenges. Science 294:804–880

Mahapatra DM, Chanakya HN, Ramchandra TV (2014) Bioremediation and lipid synthesis through mixotrophic algal consortia in municipal wastewater. Bioresour Technol 168:142. https://doi.org/10.1016/j.biortech.2014.03.130

Marshall LG, Webb SD, Sepkoski JJ, Raup DM (1982) Mammalian evolution and the Great American interchange. Science 215:1351–1357

May RM (1974) Stability and complexity in model systems, 2nd edn. Princeton University Press, Princeton

Micheli C, Cianchi R, Paperi R, Belmonte A, Pushparaj B (2014) Antarctic cyanobacteria biodiversity based on ITS and TrnL sequencing and ITS ecological implication. Open J Ecol 4:456–467. https://doi.org/10.4236/oje.2014.48039

Munoz R, Guieysse B (2006) Algal-bacterial processes for the treatment of hazardous contaminants: a review. Water Res 40:2799. https://doi.org/10.1016/j.watres.2006.06.011

Narwani A, Bentlage B, Alexandrou MA, Fritschie KJ, Delwiche C, Oakley TH, Cardinale BJ (2017) Ecological interactions and coexistence are predicted by gene expression similarity in freshwater green algae. J Ecol 105:580–591. https://doi.org/10.1111/1365-2745.12759

Nath A, Tiwari PK, Rai AK, Sundaram S (2017a) Microalgal consortia differentially modulate progressive adsorption of hexavalent chromium. Physiol Mol Biol Plants 23:269–280

Nath A, Vajpayee G, Dixit K, Rahman A, Kannaujiya VK, Sundaram S (2017b) Microalgal consortia complexity enhances ecological biomass stability through CO_2 sequestration. J Algal Biomass Utln 8:19–34

Ortiz-Marquez JCF, Nascimento MD, Zehr JP, Curatti L (2013) Genetic engineering of multispecies microbial cell factories as an alternative for bioenergy production. Trends Biotechnol 31(9):521–529. https://doi.org/10.1016/j.tibtech.2013.05.006

Patel VK, Maji D, Singh AK, Suseela MR, Sundaram S, Kalra A (2014) A natural plant growth promoter, calliterpenone, enhances growth and biomass, carbohydrate, and lipid production in cyanobacterium *Synechocystis* PCC 6803. J Appl Phycol 26:279–286

Power DL, Cardinale BJ (2009) Species richness enhances both algal biomass and rates of oxygen production in aquatic microcosms. Oikos 118:1703–1711

Rakowski C, Cardinale BJ (2016) Herbivores control effects of algal species richness on community biomass and stability in a laboratory microcosm experiment. Oikos 125:1627. https://doi.org/10.5061/dryadj1617

Renuka N, Sood A, Ratha SK, Prasanna R, Ahluwalia AS (2013) Evaluation of microalgal consortia for treatment of primary treated sewage effluent and biomass production. J Appl Phycol 25:1529–1537. https://doi.org/10.1007/s10811-013-9982-x

Rippka R, Deruelles J, Waterbury JB, Herdman M, Stanier RY (1979) Genetic assignments, strain histories and properties of pure cultures of cyanobacteria. J Gen Microbiol 111:1–61

Stachowicz JJ, Byrnes JE (2006) Species diversity, invasion success, and ecosystem functioning: disentangling the influence of resource competition, facilitation, and extrinsic factors. Merine Ecol Prog Series 311:251–262

Steneck RS, Dethier MN (1994) A functional group approach to the structure of algal-dominated communities. OIKOS 69:476–498

Sun Z, Dou X, Wu J, He B, Wang Y, Chen YF (2016) Enhanced lipid accumulation of photoautotrophic microalgae by high-dose CO_2 mimics a heterotrophic characterization. World J Microbiol Biotechnol 32:9. https://doi.org/10.1007/s11274-015-1963-6

Tilman D, Wedin D, Knops J (1996) Productivity and sustainability influenced by biodiversity in grassland ecosystem. Nature 379:718–720

Tilman D, Kilham SS, Kilham P (1982) Phytoplankton community ecology: The role of limiting nutrients. Annu Rev Ecol Evol Syst 13:349–372

Tilman D, Lehman C (2001) Biodiversity, composition and ecosystem processes: theory and concepts. In: Kinzing AP, Pacala SW, Tilman D (eds) The functional consequences of biodiversity. Princeton University Press, Princeton, pp 9–41

Vermeij GJ (1991) Anatomy of an invasion: the trans-Arctic interchange. Paleobiology 17:281–307

Viswanathan T, Mani S, Das KC, Chinnasamy S, Bhatnagar A (2011) Drying characteristics of a microalgae consortium developed for biofuel production. Am Soc Agric Biol Eng 54(6):2245–2252

Weis JJ, Madrigal SD, Cardinale BJ (2008) Effects of algal diversity on the production of biomass in homogeneous and heterogeneous nutrient environments: a microcosm experiment. PLoS One 3:e2825

Weyhenmeyer GA, Peter H, Willén E (2013) Shifts in phytoplankton species richness and biomass along a latitudinal gradient – consequences for relationships between biodiversity and ecosystem functioning. Freshw Biol 58:612–623

Yahya L, Nazry M, Zainal A (2016) Screening of microalgae consortium species for enhanced CO_2 fixation under actual flue gas exposure and profiling of its biochemical properties. Int J Adv Sci Eng Tech 4(2):165–169. special issue 2

Zimmerman KE, Cardinale BJ (2014) Is the relationship between algal diversity and biomass in North American lakes consistent with biodiversity experiments? Oikos 123:267–278

Outdoor Microalgae Cultivation for Wastewater Treatment

Djamal Zerrouki and Abdellah Henni

1 Introduction

In algae cultivation, one of the important economic constraints for biomass production is the cost of the nutrient media. Therefore, endeavors are being made to exchange the costlier nutrient media with less expensive supplemental sources. Among current solutions is the utilization of various sorts of wastewater for biomass production. Thus, in recent decades algae research has gained interest mainly oriented to the cultivation of algae in wastewater for cost-effective microalgal biofuel production and waste remediation (Diniz et al. 2016; Salama et al. 2017; Shchegolkova et al. 2018; Wu et al. 2014). The commercial use of algal cultures with application to WWT spans about 70 years. Currently, great interest is being evidenced all over the world including the USA, Australia, Taiwan, Mexico, and Thailand. The utilization of algae in WWT systems for nutrient removal was investigated earlier by Oswald et al. in 1957; wastewaters mainly provide water as well as a large number of necessary nutrients for algae cultivation. Algae are a photosynthetic microorganism with the capacity of converting solar energy into biomass and taking up nutrients such as nitrogen (N) and phosphorus (P) from wastewater (Lam and Lee 2012; Salama et al. 2017). The photosynthetic process produces O_2, which is essential in wastewater to allow aerobic bacteria to break down organic contaminants.

Many studies have shown that nutrient removal from wastewater using microalgae is applicable to various wastewater types including municipal (Álvarez-díaz et al. 2017; Liu and Ruan 2014), domestic (Cabanelas et al. 2013; Dahmani et al. 2016), agro-industrial (Posadas et al. 2014a), piggery (Yuan et al. 2013), and food processing (Lu et al. 2015).

D. Zerrouki (✉) · A. Henni
University of Ouargla, Fac. des sciences appliquées, Lab. dynamique interaction et réactivités des systèmes, Ouargla, Algeria

S. K. Gupta, F. Bux (eds.), *Application of Microalgae in Wastewater Treatment*,
https://doi.org/10.1007/978-3-030-13913-1_5

Coupling WWT with algae cultivation has been reported to be a good economically viable opportunity (Delrue et al. 2016, 2018) for cost reduction for biomass production as well as for water treatment, which provides an opportunity for clean water production in areas of limited water supply, with a very realistic possibility for utilization in the desert environment and hot arid regions (Fig. 1).

To mass develop microalgae with WWT, a variety of issues, mostly technical and economic, must be considered: finding suitable strains of algae, assessing nutrient content to determine readily available forms for algal uptake, determining the presence of biological inhibitories such as fungal and viral infection, solids content and turbidity for wastewater, competition between different microorganism populations, and environmental factors.

Well-balanced formulas of wastewater are necessary to realize maximal growth potential and high removal rate; with most necessary nutrients, municipal wastewaters offer a suitable option as artificial media for algae cultivation. However, inhibitory and toxic substances in the wastewater may smother the development of algal species. Furthermore, not all cultures are suitable for mass cultivation in open systems because of contamination, thus limiting species that can be used for cultivation on a large scale. Finding strains well adapted to cultivation in open ponds through screening of indigenous species or acclimation could provide a solution to this problem. Additional carbon sources (carbon dioxide, for example) from flue gases and different nutrients might be essential for unbalanced wastewater nutrition, especially the N/P ratio.

WWT using various algae strains has been extensively studied indoors at a laboratory scale, needing to be scaled up to outdoors applications to provide an inexpensive process; however, many factors such as illumination, temperature, and seasonal changes can be a challenge (Rawat et al. 2013; Zhu et al. 2016). To the best of our knowledge, only a few pilot plants have been successfully constructed throughout the world, mainly located in high-irradiance areas (Fig. 2).

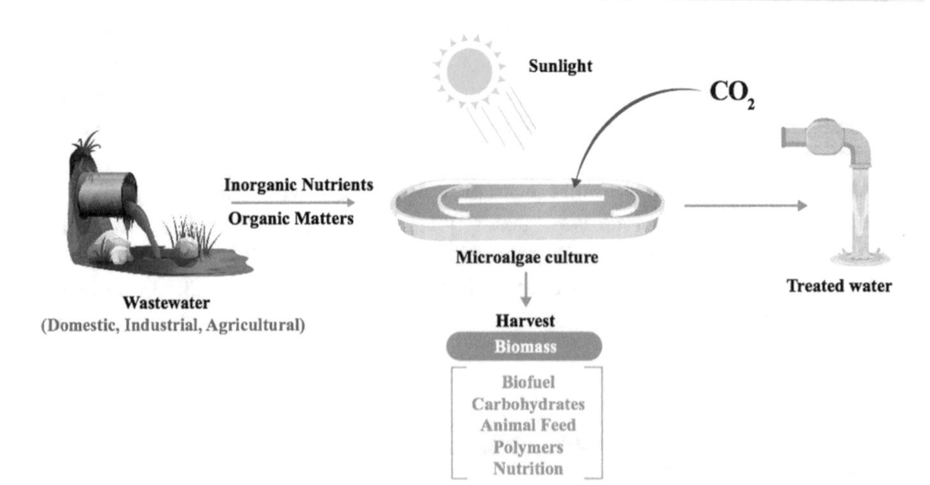

Fig. 1 Principles of microalgae production integration with wastewater treatment (WWT)

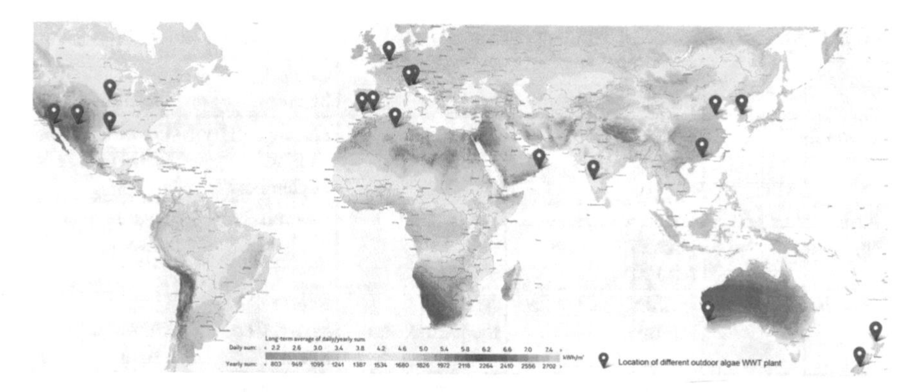

Fig. 2 Location of different outdoor algae wastewater treatment plants and direct normal irradiation (kWh/m^2) (solargis. info)

2 Algae Culture from the Laboratory to Outdoors

Research on microalgae culture in wastewater has increased. The larger part of this exploration is done on a laboratory scale using artificial light, competent technicians, and controlled growth conditions. However, the transfer of results from laboratory research to outdoor production needs further investigation to adjust and optimize the growth kinetics and nutrient uptake parameters.

Biomass productivity and photosynthetic activity are dependent not only on the total amount of solar energy impinging on the culture surface, but also on the various key factors such as pH, nutrients, salinity, and temperature (Cabello et al. 2015; Mahdy 2016). At the laboratory scale, culturing algae for WWT using real and different types of wastewater, many strains have been identified and high nutrient removal efficiency achieved. Even though a strain may grow well in laboratory cultures under particular incubation conditions (e.g., room temperature and relatively low light intensities), there is no warranty that it will grow efficiently in outdoor systems (García et al. 2017), as these are subject to variable seasonal water temperatures and fluctuations in light throughout the day (Tan et al. 2015). In any case, it is difficult to exploit results obtained in the laboratory to determine outdoor operating conditions. Table 1 shows some experiements conducted under outdoor conditions using various media and systems.

3 Challenges in Using Microalgae for WWT

The use of microalgae or microalgae–bacteria symbiosis has been demonstrated to provide a good treated water quality by removing organic matter and nutrients including nitrogen, phosphorus, hazardous contaminants, heavy metals, and also even for the removal of a specific pollutant (Ma et al. 2014; Olguín 2012).

Table 1 Microalgae grown in outdoor condition

Strain	Culture system	Medium	Outdoor location	References
Chlorella	Batch reactor 0.25 m^2 (10 l)	BG11 and F	Florence, Italy (July to September)	Guccione et al. (2014)
Nannochloropsis sp.	Green wall panel photobioreactors (<110 l)	BG11 and F	Livorno, Italy	Rodolfi et al. (2009)
Dunaliella salina	Paddle wheel tanks (20 m^2)	Artificial medium contain	Puerto Santa Maria, Cadiz, Spain	García-González (2013)
Chlorella pyrenoidosa	Photobioreactor (175 l)	Effluent of anaerobically digested activated sludge	Shandong Province, China	Tan et al. (2015)
Chlorella vulgaris	Open photobioreactors (3 l)	Piggery wastewater diluted 10- and 20-fold	Valladolid University, Spain	González-Fernández et al. (2016)
Chlorella zofingiensis	Open plastic pond (100 l)	Dairy wastewater	Foshan City, China	Technology (2016)
Ulothrix sp.	Batch reactors (12 m^3)	Aquacultural wastewater	Inagro, Roeselare, Belgium	Van Den Hende et al. (2014)
Nannochloropsis gaditana	Outdoor bubble columns (100 l)	Real centrate from an urban wastewater	Almería, Spain	Villegas et al. (2017)
Chlorella pyrenoidosa	Closed rectangular tanks (175 l)	Alcohol wastewater	Shandong Province, China	Tan et al. (2017)
Algae–bacteria	(8 m^3)	Domestic wastewater	Hamilton, New Zealand	Park et al. (2011)
Galdieria sulphuraria	Enclosed photobioreactors (700 l)	Urban wastewater	Las Cruces WW Treatment Plant in southern New Mexico	Henkanatte-gedera et al. (2017)
Chlorella pyrenoidosa	Airlift circulation photobioreactor (890 l)	Starch-processing wastewater	Shandong, China	Chu et al. (2015)
Nannochloropsis gaditana	Tubular photobioreactors	Anaerobic digestion of municipal wastewater	Almería, Spain	Ledda et al. (2015)
Microalgal consortium	Fiberglass paddlewheel-driven raceway ponds (1 m^2)	Anaerobic digestate of piggery effluent	Murdoch University, Australia	Ayre et al. (2017)
Indigenous algal consortium	HRAP (100 l)	Municipal wastewater	Daejeon, South Korea	Kim (2014)

(continued)

Table 1 (continued)

Strain	Culture system	Medium	Outdoor location	References
A mixed culture: two *Scenedesmus* sp.	Open circular ponds of 30 l capacity (1 × 1.5 m)	Pulp and paper mill effluent	Karnataka, India	Kim (2014)
Indigenous isolate: *Scenedesmus* sp.	Open pond (4 m × 80 cm) With 38 cm deep,	Produced water	Sultanate, Oman	Winckelmann et al. (2015)
Chlorella, Scenedesmus, Spirulina	Twelve open tanks (2270 l)	Treated wastewater (post-clarifier	Houston, Texas	Bhattacharjee and Siemann (2015)

The main challenges for the application of microalgae for WWT are the variation of cultivation condition and the harvesting of the algae biomass, the result of the settling characteristics and operational conditions. The concentration and composition of the organic material in the influent subjected to daily and seasonal variation must be considered in the design: the hydraulic retention time must be optimized and reduced to face environmental and nutrient fluctuations. Selection of the desired species, and finding and fixing an optimal ratio of algae/bacteria, and micropollutants removal, and a need for external CO_2 present additional obstacles.

Although there are several studies on this topic, additional research is needed to prove the effectiveness of microalgae-based WWT systems at full scale as perspectives for coming years and a promising alternative for the WWT process.

4 Effect of Wastewater Composition

The raw sewage/untreated wastewater composition and characteristics vary depending on the source and the location, and contain in their composition the main nutrients necessary for the growth of microalgae (Table 2).

In recent years, the cultivation of various microalgae strains in various raw wastewaters based on agricultural, municipal, and industrial sources was successful (Álvarez-díaz et al. 2017; Ayre et al. 2017; Zhu et al. 2013) (Fig. 3).

Urban/municipal wastewaters are a mix of a small percentage of industrial with domestic influents, mainly containing small amounts of suspended and dissolved organic and inorganic solids. Among the organic substances present in sewage are carbohydrates, lignin, fats, proteins, soaps, synthetic detergents, and various natural and synthetic organic chemicals from the processing industries. Most of the municipal wastewater is rich in nutrients such as phosphorus, ammonia, nitrogen, and a variety of inorganic substances, for example, calcium, chlorine, sulfur, magnesium, phosphate, and potassium. However, a number of potentially toxic elements such as chromium, silver, mercury, iron, gold, zinc, lead, copper, cadmium, nickel, arsenic, tin, selenium, aluminum, cobalt, manganese, and molybdenum could be present where algae are able to accumulate these toxic elements, when they are present in small concentrations.

Table 2 Different wastewater treatments for outdoor microalgal cultivation

Source of wastewater	Wastewater characteristic (mg/l)		Removal efficiency (%)		Strain	References
Domestic	COD	426	COD	78	*Chlorella pyrenoidosa*	Dahmani et al. (2016)
	TN	1.15	TN	95		
	TP	3.22	TP	81		
	COD	156 ± 79	COD	66	*Stigeoclonium* sp., *Chlorella* sp. *Monoraphidium* sp., *Chlorella* sp. *Stigeoclonium* sp.	Matamoros et al. (2015)
	TN	81 ± 9	TN	99		
	TP	N. R.	TP	N. R.		
	COD	N. R.	COD	N. R.	*Mucidosphaerium pulchellum* (H.C. Wood) C. Bock, Proschold & Krienitz	Sutherland et al. (2014)
	TN	N. R.	TN	79		
	TP	N. R.	TP	49		
	BOD	115.5 ± 71.5	BOD	50	N. R.	Craggs et al. (2012)
	TN	24.2 ± 9.5	TN	65		
	TP	N. R.	TP	19		
	COD	575 ± 84	COD	84	*Scenedesmus* sp.	Posadas et al. (2015a)
	TN	64 ± 15	TN	79		
	TP	9 ± 3	TP	57		
	COD	167 ± 64	COD	89	*Acutodesmus* sp., *Aulacoseira* sp., *Chlorella* sp., *Desmodesmus quadricaudatus*, *Limnothrix redekei*, *Nitzschia* sp., *Planktothrix* cf. *prolifica*, *Pseudanabaena limnetica*, *Synechocystis aquatilis*, *Woronichinia* sp.	Posadas et al. (2014b)
	TN	106 ± 9	TN	92		
	TP	12 ± 3	TP	96		
Urban and municipal	COD	79.62 ± 3.17	COD	N. R.	(TPBR) *Scenedesmus obliquus*	Arbib et al. (2013)
	TN	25.00 ± 1.80	TN	91. 4		
	TP	2.23 ± 0.12	TP	86. 5		
	COD	79.62 ± 3.17	COD	N. R.	(HRAP) *Scenedesmus obliquus*	Arbib et al. (2013)
	TN	25.43 ± 0.54	TN	65. 6		
	TP	2.23 ± 0.12	TP	56. 2		
	COD	N. R.	COD	N. R.	*Actinastrum, Scenedesmus Chlorella, Spirogyra Nitzschia, Golenkinia, Micractinium, Chlorococcum, Closterium, Euglena*	Woertz et al. (2009)
	TN	51	TN	96		
	TP	2.1	TP	99		

(continued)

Table 2 (continued)

Source of wastewater	Wastewater characteristic (mg/l)		Removal efficiency (%)		Strain	References
Agro-industrial	COD	3000 ± 28.1	COD	89	*Scenedesmus* sp.	Usha et al. (2016)
	TN	20.09 ± 1.3	TN	65		
	TP	1.55 ± 0.01	TP	71		
	COD	342 ± 26	COD	56	N. R.	de Godos et al. (2010)
	TN	18.8 ± 9.4	TN	97		
	TP	N. R.	TP	15		
	COD	526 ± 97	COD	76	N. R.	de Godos et al. (2009a)
	TN	59 ± 22	TN	88		
	TP	N. R.	TP	N. R.		
	COD	678 ± 249	COD	77	N. R.	Posadas et al. (2015b)
	TN	31 ± 10	TN	83		
	TP	19 ± 5	TP	94		
	TOC	1247 ± 62	TOC	50	*Chlamydomonas*, *Microspora* *Chlorella*, *Nitzschia*, *Achnanthes* *Protoderma*, *Selenastrum*, *Oocystis*, *Ankistrodesmus*	de Godos et al. (2009b)
	TN	656 ± 37	TN	100		
	TP	117 ± 19	TP	86		

N. R. not reported, *BOD* biochemical oxygen demand, *COD* chemical oxygen demand, *TOC* total organic carbon, *TN* total nitrogen, *TP* total phosphorus, *HRAP* high-rate algal pond

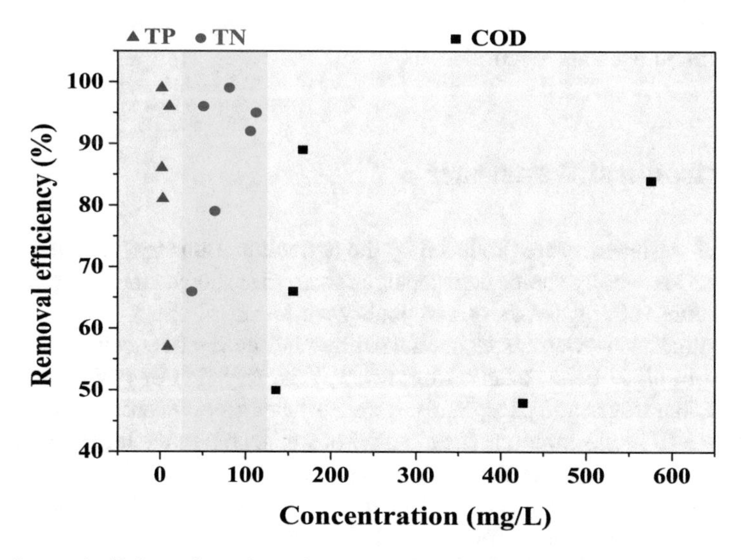

Fig. 3 Removal efficiency from domestic wastewater (outdoor microalgae cultivation based on recent research)

Accumulating processes take place by diverse mechanisms that depend on the species of microalgae and heavy metal ions; diverse factors significantly influence the fixation of heavy metals including the chemical substance concentration, pH, temperature, and accessibility of nutrients in the medium (Kumar et al. 2015; Talebi 2016).

It should be noted that several studies have shown that microalgae biomass has very effective performance for the reduction and elimination of various kinds of pollutants. A range of microalgae strains has been reported to have a potential efficiency of removal of a heavy metal; for example, *Chlorella vulgaris* is able to remove almost 100% of Cu and Fe and more than 60% of Zn, Pb, and Ni when cultured in urban wastewater (Kumar et al. 2015). The elimination of heavy metals in microalgae is mediated by two simultaneous mechanisms: passive and active. Passive mechanisms such as ion exchange, chemisorption, and adsorption occur at the cell surface, are frequently reversible and kinetically fast (Chojnacka et al. 2005). However, active mechanisms for heavy metal removal imply intracellular and extracellular (Ozturk et al. 2014).

Dishwashing liquids and shampoos are mainly constituted of surfactants, discharged into sewage wastewater. A few studies have reported their effect on algae growth (nutrient uptake); inhibition and toxic effect was observed for cationic and anionic surfactants with a concentration exceeding 1 mg/l. However, toxicities of non-ionic surfactant range from 0.003 to 18 mg/l; for example, 0.17–0.69 mg/l of sodium dodecyl sulfate is a minimum inhibitory concentration of *Chlamydomonas reinhardtii* (Groot 1990).

Utilization of microalgae for municipal WWT could be very effectively implemented for small communities where the variations in the effluent concentration could be minimized and controlled.

5 Agricultural Wastewater

Ranges of wastewaters are produced by the agriculture industry. The nutrient concentrations are wholly source dependent; a few studies showed the feasibility of the algal treatment of piggery, dairy, and poultry wastewater.

Agriculture wastewater is characterized by high nutrient concentrations, which may lead to algae being unable to grow. WW is very rich in pollutants such as ammoniacal nitrogen and phosphorus as well as having significant chemical oxygen demand (COD) concentration. The majority of studies report the important need for dilution (50%) for appropriate microalgae growth (Ayre et al. 2017) and a freshwater supply. High ammonia concentrations and high turbidity can be among the major challenges for long-term outdoor microalgae cultivation.

In poultry wastewater, total phosphorus (TP) and total nitrogen (TN) are generally more than 280 mg/l, with a COD exceeding 6000 mg/l; outdoor culture requires diluted poultry media and selection of the right algal strain to avoid photoinhibition. Studies conducted indoors showed that *Arthrospira platensis* cannot survive and grow and was significantly inhibited in 20× and 25× diluted poultry WW; however, *Chlorella vulgaris* displayed better growth in 10× diluted media (Markou et al. 2016).

The outdoor culture of *Scenedesmus* sp. using diluted effluent from a chicken manure biogas plant (25×) has been reported; ammonium and orthophosphate uptake were achieved at 90% but COD and nitrate at more than 50% (Lu et al. 2015).

Increasing use of pesticides, insecticides, antibiotics, and other fertilizers in agriculture will have a significant impact on microalgae growth and the kinetic nutrient uptake process for large-scale application.

6 Industrial Wastewater

Industrial WWT is not simple because its composition fluctuates and may contain poorly biodegradable components and high organic matter content (Dvořák et al. 2014). The characteristics of industrial wastewater and the presence of toxic components vary according to the type of activity that generated the pollution. Wastewater from slaughterhouses includes copper and zinc; petroleum industry wastewater contains benzene, naphthalene, nonylphenol, toluene, and xylene. Plastic plant waste contains octylphenol and chromium; the textile finishing sector produces waters rich in tribulphosphate, naphthalene, xylene, copper, and zinc. The manufacture of biocides or phytosanitary products pollutes the water with arsenic, chromium, copper, and zinc.

Industrial wastewater generally contains many heavy metals compared to municipal wastewater, but less phosphorus and nitrogen. For this purpose, it is necessary to use microalgae species or strains that effectively adsorb heavy metals.

Research studies are devoted to the removal of heavy metals and organic pollutants from microalgae grown in industrial WW. However, the high concentrations of organic toxins and heavy metals present in industrial wastewater tend to limit the possibilities of microalgae cultivation following the inhibition of microalgae growth.

A very few studies have reported outdoor culture using industrial wastewater because of its complex composition. Usha et al. (2016) studied the removal of organic pollution and nutrient rejects from paper and pulp mill effluent in an open outdoor pond diluted with distilled water (60%) using a *Scenedesmus* sp. strain; removal of up to 89% and 75% of COD and biological oxygen demand (BOD), respectively, was achieved. NO_3-N and PO_4-P removal up to 65% and 71.29%, respectively, were observed at the end of 28 days.

The scaleup for outdoor removal in industrial wastewater needs further investigation; indoor cultured studies showed removal efficiency.

7 Systems for Outdoor Microalgae Cultivation

Microalgae cultivation systems are one of the important aspects of biomass productivity. For this, numerous types of algal cultivation systems are developed (Fig. 4); the majority of them are based on closed or open systems. Based on the literature, it is impossible today to compare the performance of different outdoor systems because the reactors are not located in the same places, the mode of operations and measurements are different, and there are several cultivated strains.

7.1 *Open Systems*

A stabilization pond of wastewater is a basin at a designed depth in the ground for the treatment of wastewater. They are used to treat a variety of WW under broad weather conditions and are cost-effective to provide WWT when land is available. Moreover, their operation is very easy it requires minimum maintenance. These ponds are generally used preferably in areas with a warm climate. Raceway ponds are widely used for large-scale algal biomass production.

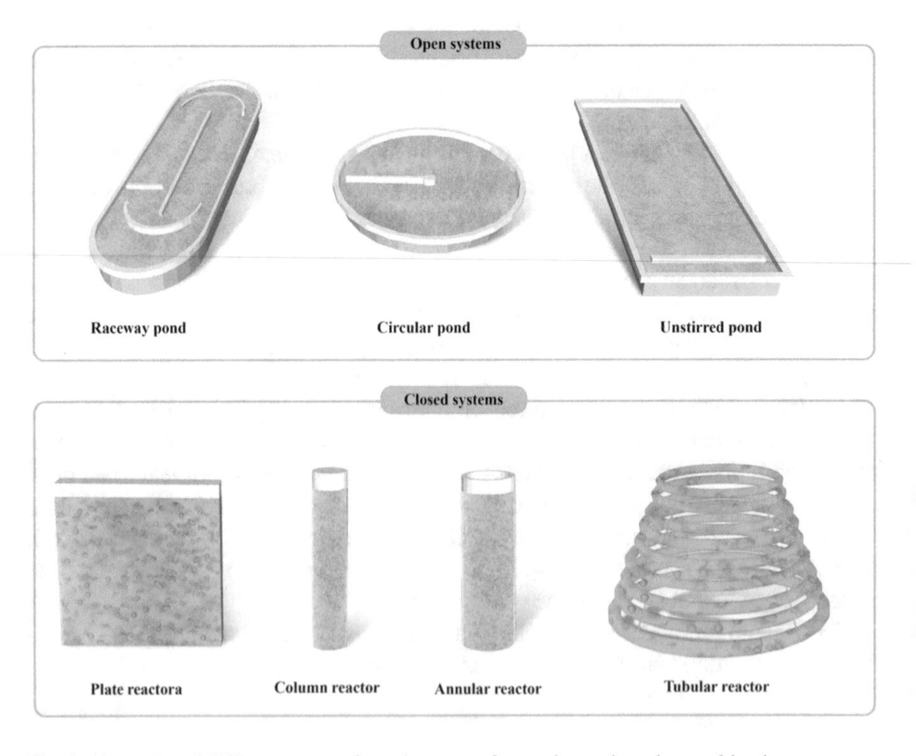

Fig. 4 Examples of different types of pond systems for outdoor microalgae cultivation

The algae ponds used in the WWT plants are constructed in a suitable form for the location; the mixing is generally achieved by a gravity flow. The Werribee WWT plant in Melbourne is one of the largest algal ponds for WWT with a surface area of 11,000 ha. Winckelmann et al. (2015) reported a successful study in an arid desert for cultures of indigenous algae using wastewater in open ponds (38 cm deep, 80 cm wide, 4 m long). However, Ayre et al. (2017) reported outdoor cultivation of microalgae on undiluted anaerobic digested waste using 1 m^2 fiberglass paddlewheel-driven raceway ponds during the winter. Many disadvantages have been observed, such as ammonia lost to the atmosphere or an undesirable contamination of the culture during long-term growth.

Major wastewater pond systems are large shallow ponds, tanks, raceway ponds, and circular ponds (Ugwu et al. 2008). The most popular type for microalgae cultivation and WWT is the raceway pond (de Godos et al. 2014). However, some major limitations can affect biomass productivity in open ponds, such as evaporative losses, easily contaminated cultures, photoinhibition in the summer, light used by the cells, and diffusion of CO_2.

However, the working depth is one of the most important design parameters of a raceway pond. Working with a shallow depth can exposure the algae to high temperature, especially during the summer, but a too great pond depth can prevent efficient light penetration. The optimal depth can be determined according to the quality and quantity of light penetrating, taking into account wastewater turbidity, which induces light-scattering processes and attenuation.

The hydraulic retention time (HRT), defined as the time that a wastewater remains inside the pond before it is evacuated, is an important factor that determines the efficiency and the cost-effectiveness of wastewater treatment. Many authors have suggested keep the HRT long enough to reach a maximum algae growth rate with high nutrient uptake and prevent nutrient limitation (Kim 2014). The HTR should be connected to seasonal variation and WW composition. HTR in the summer must be shorter than in the winter, because for a high nutrient reduction in the summer compared to the winter studies have reported that the removal efficiency of phosphorus and nitrogen with *Scenedesmus obliquus* was 20% higher in the summer (Lu et al. 2016).

7.2 *Closed Systems*

Various closed systems that have been designed for algae culture include tubular, flat-plate, and bag reactors; usually photobioreactors have better light penetration. Most of the closed systems are attractive for cultivation of algae for high-value product. High biomass productivity with minimal contamination has been reported, but a closed system not adapted for algae WWT use because of the high cost of maintenance and the need of more technical knowledge.

The design principles of most bioreactors used for WWT are derived from and similar to the pure culture systems: closed photobioreactors (PBRs) are generally considered to be too complicated to promote to a large scale. Tubular bubble column photobioreactors of 40 l were used for *Chlorella zofingiensis* outdoor culture in artificial wastewater in southern China with temperature varying between 20.6 ° and 30.8 °C; a biomass yield of 17.4 g m^{-2} day^{-1} was achieved (Zemke et al. 2013).

Vertical flat-plate PBR supplemented with municipal wastewater have been reported. Although the experiment was performed indoors under controlled conditions of illumination and temperature, 99% of TN and TP was removed with *Chlorella vulgaris* and *Scenedesmus obliquus* (Salama et al. 2017).

8 Limiting Factors in Outdoor Cultivation

Two competitive processes are responsible for the production of algal biomass: respiration and photosynthesis. However, the outdoor cultivation for nutrient removal from wastewater may depend on various abiotic and biotic factors such as algal species, nutrients, location, season, temperature, and irradiance, amount of rain, and/or wind and turbidity.

To maximize algal productivity, the different growth factors must be maintained. Nutrients can be controlled by adding wastewater as a medium into the culture. However, light intensity, temperature, and evaporation depend on solar irradiation, location, and season, and therefore cannot be controlled during outdoor cultivation.

The following section describes how the different growth factors predict algal productivity in outdoor cultivation.

8.1 Light Limitation and Photoinhibition

Outdoors, depending on the location of the crop and because of seasonal changes, the algae are exposed to irradiation that changes with time, different levels of illumination, and daily fluctuations. The effectiveness of photosynthesis is influenced by the fluctuation in light conditions with regard to intensity, wavelength, and duration. It is essential to evaluate each parameter to determine the optimal light condition that yields maximal productivity.

The intensity and availability of light is the major factor in the growth and productivity of photosynthetic microorganisms. Low light intensity leads to growth limitation whereas light intensity can inhibit the growth process: this phenomenon is known as photoinhibition (Barber 1992; Han 2002). Excess light intensity imposes a serious limitation on photosynthetic efficiency, particularly when coupled with high oxygen level and/or a temperature that is not optimal (Tredici and Zittelli 1998).

In a single day, light intensity changes from zero to saturated or oversaturated light levels. Therefore, the outdoor production of microalgae is light limited at the beginning and end of the day and light saturated in the middle of the day. Furthermore, light intensity also varies throughout the year, leading to complex operational processes in outdoor culture.

Microalgae have developed protective mechanisms to accommodate the changes in light intensity. Photo-acclimation is a process in which the microalgae reduce their photosynthetic pigment content as a way to protect the photosynthetic apparatus against increased irradiance (Sousa et al. 2013). The chlorophyll content per cell varies in relationship to the surrounding light environment (Sousa et al. 2013). Chlorophyll increases in the light-limited phase until the cells become optically dark and decrease under the light-saturated phase, resulting in more transparent cells (Simionato et al. 2011; Sousa et al. 2013). When exposed to strong illumination, microalgae protect their photosynthetic capacity by decreasing chlorophyll content and increasing carotenoid content in their pigmentation. This phenomenon occurs only with a limited number of microalgae, such as *Haematococcus pluvialis* and *Dunaliella salina*.

In most cases, however, chlorophylls and carotenoids decrease when microalgae are exposed to high light intensities, and as a result, the cells gradually turn from green to yellow to orange because the degradation of chlorophylls is usually more rapid than that of carotenoids. The increase in carotenoid content allows algae to dissipate excited chlorophyll energy and remove reactive oxygen species (ROS), thus maintaining the photosystem structure and functions. At very strong illumination, protective mechanisms will not be able to overcome the excess of electrons, ROS accumulation, and singlet oxygen formation, resulting in the death of algal cells (Simionato et al. 2011; Sousa et al. 2013).

In an outdoor culture system, in addition to the total amount of solar energy received on the growing surface that is important for biomass productivity, the amount of energy available at the cell level is necessary for good production (concepts of "light regime" and "light per cell") (Chaumont 1993). However, the decrease in photosynthetic activity is explained by an enzymatic inactivation caused by high temperature, high irradiance, or both. *Scenedesmus obtusiusculus* showed a linear response of photosynthetic activity for irradiances in the range of 8–300 μE m^{-2} s^{-1}, which remains constant until an irradiance of 970 μE m^{-2} s^{-1}. At high irradiances (1600–2360 μE m^{-2} s^{-1}), photoinhibition was observed at temperatures above 35 °C. At high irradiance (1000 μE m^{-2} s^{-1}), photoinhibition has been reported for *Synechococcus, Haematococcus*, *Chlorella*, *Phaedolactinum*, and an *Scenedesmus almeriensis* strain (Revah and Morales 2015).

Photosynthesis of most algal species is saturated at a solar radiation level of 200 μE m^{-2} s^{-1}, which is about 10–17% of summer/winter maximum outdoor light intensity (Park et al. 2011).

The culture density and algal strain affect the light saturation level; the algal productivity from incident solar radiation can be estimated from the maximum efficiency of photosynthetic conversion of algae. For this, it is necessary to study the influence of light on the growth of any particular strain. Algal productivity could be determined from the average solar radiation photoinhibitory process that can result in a loss in biomass productivity and nutrient removal efficiency. This problem can be solved by shading the culture such that the incident photon flux decreases and by avoiding a culture of microorganisms in the heating systems.

8.2 *Temperature Limitation*

In the closed system, especially in the photobioreactor, the temperatures increase with increasing solar radiation and air temperatures. The heat is transferred mainly through radiation, air radiation, direct and diffuse solar radiation; heat is transferred to the medium through natural or forced convection in case of mixing.

In the outdoor culture system, geographic location will define the maximum temperature that algae may be exposed to and be able to grow and survive; in high insolation, temperature can reach 40 °C.

Economically, temperature control for outdoor large-scale ponds is impossible, as temperature varies during the day and with season; thus, the algal strain should be chosen to tolerate a broad range of temperature as well as showing high productivity and nutrient removal during the year in summer when the pond achieves a high temperature in arid areas, especially at night or in the winter when the pond may freeze over.

8.3 *Evaporation*

High evaporation rate is one of the main problems of outdoor algae ponds and is most often seen as a limitation. Evaporation is a surface process, mainly influenced by air temperature, relative humidity, and wind. Evaporation from algae ponds can be estimated from standard evaporation ("Pan A"). The factors affecting evaporation rate are temperature, surface area, humidity of air, and wind speed. However, algae ponds are much shallower and mechanically mixed, and thus are expected to increase in evaporation rates. In most of the areas considered suitable for algae culture the evaporation rate is found to be high, which can affect the "blow-down" ratio (BDR) and induce salinity and nutrient concentration variations. Freshwater evaporation rate in some tropical regions is found to be 0.01 m^3 m^{-2} day^{-1}, or 10 mm per day (Chisti 2012). On the other hand, the composition and nature of the wastewater affect the evaporation rate. The evaporation rate decreases as the solids and the concentrations chemical increase. Under the same environmental condition, the evaporation rate of seawater and wastewater from a pond is generally a little less than the evaporation rate of freshwater.

In reality, a microalgae strain can survive and grow at peak summer temperatures but it develops less well during winter days. For this, the solution is to deploy different strains of microalgae under different climatic variations of the year (different seasonal crops). We can assume the use of alternate microalgal cultivation with seasonal changes for seamless continuity in biomass production over the year (Kenny and Flynn 2017).

9 Conclusion

Much knowledge on the use of microalgae for WWTs has been validated and proved. The removal efficiency of a nutrient or specific pollutant from a different type of wastewater by several microalgae strains has been tested and studied. Although few data on outdoor cultivation performance are available, it appears that selecting the locations for algal cultivation using WWT indeed provides economic and environmental advantages that will, therefore, require careful optimization and assessment. However, further research for performing an algae-based WWT process is still needed to answer to the future challenges such as technical and economic feasibility at an outdoor large scale, optimization of hydraulic retention time, contamination control, and harvesting of algae biomass.

References

Arbib Z, Ruiz J, Álvarez-Díaz P, Garrido-Pérez C, Barragan J, Perales JA (2013) Long term outdoor operation of a tubular airlift pilot photobioreactor and a high rate algal pond as tertiary treatment of urban wastewater. Ecol Eng 52:143–153. https://doi.org/10.1016/j.ecoleng.2012.12.089

Ayre JM, Moheimani NR, Borowitzka MA (2017) Growth of microalgae on undiluted anaerobic digestate of piggery effluent with high ammonium concentrations. Algal Res 24:218–226. https://doi.org/10.1016/j.algal.2017.03.023

Álvarez-díaz PD, Ruiz J, Arbib Z, Barragán J, Garrido-pérez MC, Perales JA (2017) Freshwater microalgae selection for simultaneous wastewater nutrient removal and lipid production. Algal Res. https://doi.org/10.1016/j.algal.2017.02.006

Barber J (1992) Too much of a good thing: light can be bad for photosynthesis. Trends Biochem Sci 17:61–66

Bhattacharjee M, Siemann E (2015) Low algal diversity systems are a promising method for biodiesel production in wastewater fed open reactors. Algae 30:67–79. https://doi.org/10.4490/algae.2015.30.1.067

Cabanelas ITD, Ruiz J, Arbib Z, Chinalia FA, Garrido-Pérez C, Rogalla F, Nascimento IA, Perales JA (2013) Comparing the use of different domestic wastewaters for coupling microalgal production and nutrient removal. Bioresour Technol 131:429–436. https://doi.org/10.1016/j.biortech.2012.12.152

Cabello J, Toledo-Cervantes A, Sanchez L, Revah S, Morales M (2015) Effect of the temperature, pH and irradiance on the photosynthetic activity by Scenedesmus obtusiusculus under nitrogen replete and deplete conditions. Bioresour Technol 181:128–135. https://doi.org/10.1016/j.biortech.2015.01.034

Chaumont D (1993) Biotechnology of algal biomass production: a review of systems for outdoor mass culture. J Appl Phycol 5:593–604

Chojnacka K, Chojnacki A, Górecka H (2005) Biosorption of Cr^{3+}, Cd^{2+} and Cu^{2+} ions by blue-green algae *Spirulina* sp.: kinetics, equilibrium and the mechanism of the process. Chemosphere 59:75–84. https://doi.org/10.1016/j.chemosphere.2004.10.005

Chu H, Tan X, Zhang Y, Yang L, Zhao F, Guo J (2015) Continuous cultivation of *Chlorella pyrenoidosa* using anaerobic digested starch processing wastewater in the outdoors. Bioresour Technol. https://doi.org/10.1016/j.biortech.2015.02.030

Craggs R, Sutherland D, Campbell H (2012) Hectare-scale demonstration of high rate algal ponds for enhanced wastewater treatment and biofuel production. J Appl Phycol 24:329–337. https://doi.org/10.1007/s10811-012-9810-8

Chisti, Y. (2012). Raceways-based production of algal crude oil. In: C. Posten & C. Walter (Eds.), Microalgal biotechnology: Potential and production (pp. 113–146). de Gruyter, Berlin

de Godos I, Blanco S, García-Encina PA, Becares E, Muñoz R (2010) Influence of flue gas sparging on the performance of high rate algae ponds treating agro-industrial wastewaters. J Hazard Mater 179:1049–1054. https://doi.org/10.1016/j.jhazmat.2010.03.112

de Godos I, Blanco S, García-Encina PA, Becares E, Muñoz R (2009a) Long-term operation of high rate algal ponds for the bioremediation of piggery wastewaters at high loading rates. Bioresour Technol 100:4332–4339. https://doi.org/10.1016/j.biortech.2009.04.016

de Godos I, González C, Becares E, García-Encina PA, Muñoz R (2009b) Simultaneous nutrients and carbon removal during pretreated swine slurry degradation in a tubular biofilm photobioreactor. Appl Microbiol Biotechnol 82:187–194. https://doi.org/10.1007/s00253-008-1825-3

de Godos I, Mendoza JL, Acién FG, Molina E, Banks CJ, Heaven S, Rogalla F (2014) Evaluation of carbon dioxide mass transfer in raceway reactors for microalgae culture using flue gases. Bioresour Technol 153:307–314. https://doi.org/10.1016/j.biortech.2013.11.087

Dahmani S, Zerrouki D, Ramanna L, Rawat I, Bux F (2016) Cultivation of *Chlorella pyrenoidosa* in outdoor open raceway pond using domestic wastewater as medium in arid desert region. Bioresour Technol. https://doi.org/10.1016/j.biortech.2016.08.019

Delrue F, Álvarez-Díaz DP, Fon-Sing S, Fleury G, Sassi J-F (2016) The environmental biorefinery: using microalgae to remediate wastewater, a win–win paradigm. Energies 9:132. https://doi.org/10.3390/en9030132

Diniz GS, Silva AF, Araújo OQF, Chaloub RM (2016) The potential of microalgal biomass production for biotechnological purposes using wastewater resources. J Appl Phycol. https://doi.org/10.1007/s10811-016-0976-3

Dvořák L, Lederer T, Jirků V, Masák J, Novák L (2014) Removal of aniline, cyanides and diphenylguanidine from industrial wastewater using a full-scale moving bed biofilm reactor. Process Biochem 49:102–109. https://doi.org/10.1016/j.procbio.2013.10.011

De Francisci D, Su Y, Iital A, Angelidaki I (2018) Evaluation of microalgae production coupled with wastewater treatment. Environ Technol 39:581–592. https://doi.org/10.1080/09593330.2017.1308441

García-gonzález M (2013) Conditions for open-air outdoor culture of *Dunaliella salina* in southern Spain. J Appl Phycol 15:177–184. https://doi.org/10.1023/A:1023892520443

García D, Posadas E, Grajeda C, Blanco S, Martínez-Páramo S (2017) Comparative evaluation of piggery wastewater treatment in algal- bacterial photobioreactors under indoor and outdoor conditions. Bioresour Technol. https://doi.org/10.1016/j.biortech.2017.08.135

González-Fernández C, Mahdy A, Ballesteros I, Ballesteros M (2016) Impact of temperature and photoperiod on anaerobic biodegradability of microalgae grown in urban wastewater. Int Biodeter Biodegr 106:16–23. https://doi.org/10.1016/j.ibiod.2015.09.016

Groot D (1990) Chronic toxicities of surfactants and detergent builders to algae: a review and risk assessment. Ecotoxicol Environ Saf 20:123–140

Guccione A, Biondi N, Sampietro G, Rodolfi L, Bassi N, Tredici MR (2014) *Chlorella* for protein and biofuels: from strain selection to outdoor cultivation in a Green Wall Panel photobioreactor. Biotechnol Biofuels 7:84

Han BP (2002) A mechanistic model of algal photoinhibition induced by photodamage to photosystem. II. J Theor Biol 214:519–527. https://doi.org/10.1006/jtbi.2001.2468

Henkanatte-gedera SM, Selvaratnam T, Karbakhshravari M, Myint M, Nirmalakhandan N, Van Voorhies W, Lammers PJ (2017) Removal of dissolved organic carbon and nutrients from urban wastewaters by *Galdieria sulphuraria*: laboratory to field scale demonstration. Algal Res 4:450–456. https://doi.org/10.1016/j.algal.2016.08.001

Kenny P, Flynn KJ (2017) Physiology limits commercially viable photoautotrophic production of microalgal biofuels. J Appl Phycol 29:2713–2727. https://doi.org/10.1007/s10811-017-1214-3

Kim H (2014) Nutrient removal and biofuel production in high rate algal pond using real municipal wastewater. J Microbiol Biotechnol. https://doi.org/10.4014/jmb.1312.12057

Kumar KS, Dahms H, Won E, Lee J, Shin K (2015) Ecotoxicology and environmental safety microalgae – a promising tool for heavy metal remediation. Ecotoxicol Environ Saf 113:329–352. https://doi.org/10.1016/j.ecoenv.2014.12.019

Lam MK, Lee KT (2012) Potential of using organic fertilizer to cultivate *Chlorella vulgaris* for biodiesel production. Appl Energy 94:303–308. https://doi.org/10.1016/j.apenergy.2012.01.075

Ledda C, Villegas GIR, Adani F, Fernández FGA, Grima EM (2015) Utilization of centrate from wastewater treatment for the outdoor production of *Nannochloropsis gaditana* biomass at pilot-scale. Algal Res 12:17–25. https://doi.org/10.1016/j.algal.2015.08.002

Liu Y, Ruan R (2014) Effect of wastewater-borne bacteria on algal growth and nutrients removal in wastewater-based algae cultivation system. Bioresour Technol 167:8–13. https://doi.org/10.1016/j.biortech.2014.05.087

Lu Q, Zhou W, Min M, Ma X, Chandra C, Doan YTT, Ma Y, Zheng H, Cheng S, Griffith R, Chen P, Chen C, Urriola PE, Shurson GC, Gislerød HR, Ruan R (2015) Bioresource technology: growing *Chlorella* sp. on meat processing wastewater for nutrient removal and biomass production. Bioresour Technol 198:189–197. https://doi.org/10.1016/j.biortech.2015.08.133

Lu Q, Zhou W, Min M, Ma X, Ma Y, Chen P, Zheng H, Doan YTT, Liu H, Chen C, Urriola PE, Shurson GC, Ruan R (2016) Mitigating ammonia nitrogen deficiency in dairy wastewaters for algae cultivation. Bioresour Technol 201:33–40. https://doi.org/10.1016/j.biortech.2015.11.029

Ma X, Zhou W, Fu Z, Cheng Y, Min M, Liu Y, Zhang Y, Chen P, Ruan R (2014) Effect of wastewater-borne bacteria on algal growth and nutrients removal in wastewater-based algae cultivation system. Bioresour Technol 167:8–13. https://doi.org/10.1016/j.biortech.2014.05.087

Mahdy AA (2016) Impact of temperature and photoperiod on anaerobic biodegradability of microalgae grown in urban wastewater. Int Biodeter Biodegr 106:16–23. https://doi.org/10.1016/j.ibiod.2015.09.016

Markou G, Iconomou D, Muylaert K (2016) Applying raw poultry litter leachate for the cultivation of *Arthrospira platensis* and *Chlorella vulgaris*. Algal Res 13:79–84. https://doi.org/10.1016/j.algal.2015.11.018

Matamoros V, Gutiérrez R, Ferrer I, García J, Bayona JM (2015) Capability of microalgae-based wastewater treatment systems to remove emerging organic contaminants: a pilot-scale study. J Hazard Mater 288:34–42. https://doi.org/10.1016/j.jhazmat.2015.02.002

Olguín EJ (2012) Dual purpose microalgae–bacteria-based systems that treat wastewater and produce biodiesel and chemical products within a biorefinery. Biotechnol Adv 30:1031–1046. https://doi.org/10.1016/j.biotechadv.2012.05.001

Oswald WJ, Gotaas HB (1957) Photosynthesis in sewage treatment. Trans Am Soc Civ Eng 122:73–105

Ozturk S, Aslim B, Suludere Z, Tan S (2014) Metal removal of cyanobacterial exopolysaccharides by uronic acid content and monosaccharide composition. Carbohydr Polym 101:265–271. https://doi.org/10.1016/j.carbpol.2013.09.040

Park JB, Craggs RJ, Shilton AN (2011) Wastewater treatment high rate algal ponds for biofuel production. Bioresour Technol 102:35–42. https://doi.org/10.1016/j.biortech.2010.06.158

Posadas E, Bochon S, Coca M, García-González MC, García-Encina PA, Muñoz R (2014a) Microalgae-based agro-industrial wastewater treatment: a preliminary screening of biodegradability. J Appl Phycol 26:2335–2345. https://doi.org/10.1007/s10811-014-0263-0

Posadas E, García-Encina PA, Domínguez A, Díaz I, Becares E, Blanco S, Muñoz R (2014b) Enclosed tubular and open algal-bacterial biofilm photobioreactors for carbon and nutrient removal from domestic wastewater. Ecol Eng 67:156–164. https://doi.org/10.1016/j.ecoleng.2014.03.007

Posadas E, del Mar Morales M, Gomez C, Acién FG, Muñoz R (2015a) Influence of pH and CO_2 source on the performance of microalgae-based secondary domestic wastewater treatment in outdoors pilot raceways. Chem Eng J 265:239–248. https://doi.org/10.1016/j.cej.2014.12.059

Posadas E, Muñoz A, García-González MC, Muñoz R, García-Encina PA (2015b) A case study of a pilot high rate algal pond for the treatment of fish farm and domestic wastewaters. J Chem Technol Biotechnol 90:1094–1101. https://doi.org/10.1002/jctb.4417

Rawat I, Ranjith Kumar R, Mutanda T, Bux F (2013) Biodiesel from microalgae: a critical evaluation from laboratory to large scale production. Appl Energy 103:444–467. https://doi.org/10.1016/j.apenergy.2012.10.004

Revah S, Morales M (2015) Effect of the temperature, pH and irradiance on the photosynthetic activity by *Scenedesmus obtusiusculus* under nitrogen replete and deplete conditions. Bioresour Technol 181:128–135. https://doi.org/10.1016/j.biortech.2015.01.034

Rodolfi L, Zittelli GC, Biondi N, Bonini G, Tredici MR, Padovani G (2009) Microalgae for oil: strain selection, induction of lipid synthesis and outdoor mass cultivation in a low-cost photobioreactor. Biotechnol Bioeng 102:100–112. https://doi.org/10.1002/bit.22033

Salama E, Kurade MB, Abou-shanab RAI, El-dalatony MM (2017) Recent progress in microalgal biomass production coupled with wastewater treatment for biofuel generation. Renew Sustain Energy Rev 79:1189–1211. https://doi.org/10.1016/j.rser.2017.05.091

Shchegolkova N, Shurshin K, Pogosyan S, Voronova E, Matorin D, Karyakin D (2018) Microalgae cultivation for wastewater treatment and biogas production at Moscow wastewater treatment plant. Water Sci Technol. https://doi.org/10.2166/wst.2018.088

Simionato D, Sforza E, Corteggiani Carpinelli E, Bertucco A, Giacometti GM, Morosinotto T (2011) Acclimation of *Nannochloropsis gaditana* to different illumination regimes: effects on lipids accumulation. Bioresour Technol 102:6026–6032. https://doi.org/10.1016/j.biortech.2011.02.100

Sousa C, Compadre A, Vermuë M, Wijffels R (2013) Effect of oxygen at low and high light intensities on the growth of *Neochloris oleoabundans*. Algal Res 2:122–126. https://doi.org/10.1016/j.algal.2013.01.007

Sutherland DL, Turnbull MH, Craggs RJ (2014) Increased pond depth improves algal productivity and nutrient removal in wastewater treatment high rate algal ponds. Water Res 53:271–281. https://doi.org/10.1016/j.watres.2014.01.025

Talebi AF (2016) Potential use of algae for heavy metal bioremediation, a critical review. J Environ Manage. https://doi.org/10.1016/j.jenvman.2016.06.059

Tan X-B, Yang L-B, Zhang Y-L, Zhao F-C, Chu H-Q, Guo J (2015) *Chlorella pyrenoidosa* cultivation in outdoors using the diluted anaerobically digested activated sludge. Bioresour Technol 198:340–350. https://doi.org/10.1016/j.biortech.2015.09.025

Tan X, Zhao X, Zhang Y, Zhou Y, Yang L, Zhang W (2017) Enhanced lipid and biomass production using alcohol wastewater as carbon source for *Chlorella pyrenoidosa* cultivation in anaerobically digested starch wastewater in outdoors. Bioresour Technol. https://doi.org/10.1016/j.biortech.2017.09.152

Tredici MR, Zittelli GC (1998) Efficiency of sunlight utilization: tubular versus flat photobioreactors. Biotechnol Bioeng 57:187–197. https://doi.org/10.1002/(SICI)1097-0290(19980120)57:2<187::AID-BIT7>3.0.CO;2-J

Ugwu CU, Aoyagi H, Uchiyama H (2008) Photobioreactors for mass cultivation of algae. Bioresour Technol 99:4021–4028. https://doi.org/10.1016/j.biortech.2007.01.046

Usha MT, Chandra TS, Sarada R, Chauhan VS (2016) Removal of nutrients and organic pollution load from pulp and paper mill effluent by microalgae in outdoor open pond. Bioresour Technol. https://doi.org/10.1016/j.biortech.2016.04.060

Van Den Hende S, Beelen V, Bore G, Boon N, Vervaeren H (2014) Up-scaling aquaculture wastewater treatment by microalgal bacterial flocs: from lab reactors to an outdoor raceway pond. Bioresour Technol 159:342–354. https://doi.org/10.1016/j.biortech.2014.02.113

Villegas GIR, Fiamengo M, Fernández FGA, Grima EM (2017) Outdoor production of microalgae biomass at pilot-scale in seawater using centrate as the nutrient source. Algal Res 25:538–548. https://doi.org/10.1016/j.algal.2017.06.016

Winckelmann D, Bleeke F, Thomas B, Elle C, Klöck G (2015) Open pond cultures of indigenous algae grown on non-arable land in an arid desert using wastewater. Int Aquat Res 7:221–233. https://doi.org/10.1007/s40071-015-0107-9

Woertz I, Feffer A, Lundquist T, Nelson Y (2009) Algae grown on dairy and municipal wastewater for simultaneous nutrient removal and lipid production for biofuel feedstock. J Environ Eng 135:1115–1122. https://doi.org/10.1061/(ASCE)EE.1943-7870.0000129

Wu Y, Hu H, Yu Y, Zhang T, Zhu S, Zhuang L, Zhang X, Lu Y (2014) Microalgal species for sustainable biomass/lipid production using wastewater as resource: a review. Renew Sustain Energy Rev 33:675–688. https://doi.org/10.1016/j.rser.2014.02.026

Yuan Z, Wang Z, Takala J, Hiltunen E, Qin L, Xu Z, Qin X, Zhu L (2013) Scale-up potential of cultivating *Chlorella zofingiensis* in piggery wastewater for biodiesel production. Bioresour Technol 137:318–325. https://doi.org/10.1016/j.biortech.2013.03.144

Zemke PE, Sommerfeld MR, Hu Q (2013) Assessment of key biological and engineering design parameters for production of *Chlorella zofingiensis* (Chlorophyceae) in outdoor photobioreactors. Appl Microbiol Biotechnol 97:5645–5655. https://doi.org/10.1007/s00253-013-4919-5

Zhu L, Wang Z, Takala J, Hiltunen E, Qin L, Xu X, Qin Z, Yuan Z (2013) Scale-up potential of cultivating *Chlorella zofingiensis* in piggery wastewater for biodiesel production. Bioresource technology, 137:318–325. https://doi.org/10.1016/j.biortech.2013.03.144

Zhu L, Wang Z, Shu Q, Takala J, Hiltunen E, Feng P, Yuan Z (2013) Nutrient removal and biodiesel production by integration of freshwater algae cultivation with piggery wastewater treatment. Water Res 47:4294–4302. https://doi.org/10.1016/j.watres.2013.05.004

Zhu LD, Xu ZB, Qin L, Wang ZM, Hiltunen E, Li ZH, Xu ZB, Qin L, Wang ZM, Hiltunen E, Oil ZHL (2016) Oil production from pilot-scale microalgae cultivation: an economics evaluation. Energy Source Part B Econ Plann Policy 11:11–17. https://doi.org/10.1080/15567249.2015.1052594

Current State of Knowledge on Algae-Mediated Remediation of Endocrine-Disrupting Chemicals (EDCs) from Wastewater

Ritu Singh, Monalisha Behera, Sanjeev Kumar, and Anita Rani

1 Introduction

Industrialization and urbanization have polluted the environment with such contaminants that are not only detrimental to the environment but also to the organisms residing in it including the human beings. The pollutants that have raised a lot of concern recently due to its toxic effects are the endocrine-disrupting chemicals (EDCs). EDCs basically refers to those chemical compounds which have the ability to cause interference in the endocrine system of the organism and create an undesired response leading to various endocrine-related diseases. The main source of EDCs in environmental matrices are the heavy uses of pesticides and insecticides in agricultural sector that leach into the natural water stream, industrial effluents, plastics, personal care products and other daily usable products. Diazinon has been extensively used in agriculture as an insecticide. The United States was found to be one of the countries that tremendously uses diazinon approximately around 6 million kilograms per annum (Li et al. 2015). The World Health Organisation described diazinon as a potential toxic compound that can cause mutagenicity, cytotoxicity and other endocrine disorders in mammals (Wang and Shih 2015). Similarly, another strong endocrine disruptor found in plastics is Bisphenol A (BPA). Several

R. Singh (✉) · M. Behera
Department of Environmental Science, School of Earth Sciences, Central University of Rajasthan, Ajmer, Rajasthan, India
e-mail: ritu_ens@curaj.ac.in

S. Kumar
Centre for Environmental Sciences, School of Natural Resource Management, Central University of Jharkhand, Ranchi, India

A. Rani
Dyal Singh College, University of Delhi, New Delhi, India

S. K. Gupta, F. Bux (eds.), *Application of Microalgae in Wastewater Treatment*,
https://doi.org/10.1007/978-3-030-13913-1_6

products like plastic food packaging, reusable cups, water bottles, baby bottles, tooth coatings, dental sealants, etc. are reported to contain a high degree of BPA in it which can potentially affect each and every one using plastics (Staples et al. 1998, Crain et al. 2007). Despite the lethal effects of BPA, its yearly consumption rate has been observed to increase by 5.5% globally between 2009 and 2012 (Hoepner et al. 2013). Another EDC that has affected most of the water bodies, soil and sediments is phthalic acid esters (PAEs). They are extensively utilized as plasticizers and additives in plastic materials and are reported to cause reproductive and developmental toxicities in mammals (Wang et al. 2008; Matsumoto et al. 2008). The hormones like 17β-estradiol (E2), 17α-ethinylestradiol (EE2) and estrone (E1) are mostly present in sewage waste effluents which are responsible for endocrine disruption that cannot be ignored (Onda et al. 2002). The massive use of disinfectants, fungicides and biocides over the years have accumulated halogenated compounds in the environment which are also reported to cause endocrine-related problems in human and aquatic organisms (Tikoo et al. 1997; Baker et al. 2014).

Among all the remediation technologies available for removal of EDCs, bioremediation has gained significant attention in the past two decades. Microalgae have been used significantly for remediation of EDCs. Minimal growth requirements, broad spectrum of mechanisms and widespread occurrence make microalgae an economic and ecofriendly option for removal of EDCs from wastewater. Since it puts no harm to the environment and removes contaminant efficaciously, it is preferred above all other available removal methods. Moreover, microalgae can accelerate the heterotrophic bacteria to degrade the organic contaminants by providing oxygen to them through photosynthesis (Yan et al. 1995). Microalgae tend to accumulate the pollutants in its body and then biodegrade it efficiently (Newsted 2004; Yang et al. 2002). A variety of pollutants like heavy metals, pesticides and phenols have been reported to get removed by microalgae (Dosnon-Olette et al. 2010). A study revealed that *Chlorella vulgaris* show highest removal capacity for diazinon that is around 94% (Kurade et al. 2016). Jia et al. (2014) demonstrated two freshwater algae, *C. vulgaris* and *C. mexicana*, can be successfully used to remediate BPA from aqueous systems. Another microalgal species, *C. fusca*, was also reported to degrade BPA turning them into intermediates having no estrogenic activity. About 90% of BPA was found to get treated by the algal strain when provided with light and dark conditions of 8:16 h (Hirooka et al. 2005). Gao and Chi (2015) studied that phthalate acid esters can be remediated rapidly by marine microalgal species. He found that extracellular enzymes of algal species are primarily responsible for the removal process. Endocrine-disrupting hormones like 17α-ethynylestradiol and β-estradiol were also found to be successfully removed by microalgal species, *S. capricornutum* and *C. reinhardtii* (Diaz et al. 2015). Since algae can remove the toxic elements efficaciously without putting any extra loads on environment and being economically viable, it is proved as one of the best removal approaches for EDCs.

2 Sources, Occurrence and Fate of EDCs

There are myriads of sources that release EDCs into the environment both directly and indirectly. Table 1 shows variety of EDCs reported in water matrices of different countries. Most of the industrial effluents contaminated with endocrine disruptors are directly dumped into water bodies. The extensive use of pesticides, biocides and fertilizers in agriculture end up delivering huge amount of EDCs into the environment through runoff. Hospital wastes is another major source that adds EDCs into the surface water bodies. Besides, the use of plastics, polybags and different personal care products have made the domestic wastes a serious source of these toxic

Table 1 Sources of endocrine-disrupting chemicals (EDCs)

Endocrine-disrupting chemicals (EDCs)	Sources of EDCs	City/country	References
4-Nonylphenol, BPA, 17β-estradiol and 17α-ethynylestradiol	Wastewater (coastline water)	China	Chiu et al. (2018)
Phthalates, BPA, triclosan, 4-nonylphenol and tris(2-chloroethyl) phosphate	Urban wastewater (i.e. residential, commercial or industrial samples, etc.)	Oakland, CA, United States	Jackson and Sutton (2008)
Endocrine active substances, estrogen and dioxin-responsive elements	Wastewater (urban wastewater)	Brussels region, Belgium	Vandermarken et al. (2018)
Phenolic EDCs, estrogen EDCs, nonylphenol	Freshwater/wastewater	Luoma Lake, China	Liu et al. (2017)
Nonylphenol, diethylhexyl phthalate	Wastewater (coastal region)	Bohai Rim, China	Zhang et al. (2017)
Diltiazem, progesterone, benzyl butyl phthalate, estrone, carbamazepine, acetaminophen	Wastewater (industry influent and effluent)	Ankara, Turkey	Komesli et al. (2012)
Nonylphenol, octylphenol, BPA, estrone, 17α-estradiol, 17β-estradiol, estriol, mestranol, 17α-ethynylestradiol	Water and wastewater (rivers, streams, canals, industrial and municipal wastewater, industrial effluents)	Thessaloniki, Northern Greece	Arditsoglou and Voutsa (2010)
17β-Estradiol, Esteron, ethinylestradiol, bisphenol A, nonylphenol	Water and wastewater (i.e. surface water, municipal water, river water, sewage water, etc.)	Hamadan city, Iran	Jafari et al. (2009)
Nonylphenol, BPA bis(2-ethylhexyl) phthalate, di-n-butyl phthalate, nonylphenol ethoxylates, estriol	Wastewater at different treatment stages, surface water, effluent	Calgary, Alberta, Canada	Chen et al. (2006)
Benzotriazoles, alkylphenols, bisphenol	Municipal wastewater, river water	Glatt River, Switzerland	Voutsa et al. (2006)

contaminants in the environment. These wastes are either directly discharged into the water bodies or they indirectly enter the water bodies through wastewater treatment plants. They have not only contaminated the surface waters but also polluted the groundwater through leaching process. Figure 1 displays the sources, occurrence and fate of EDCs.

There are a number of sources that expose humans to EDCs. They either enter the human body through contaminated food and drinks or through various household products that contain these toxicants or as an accumulant from treatment plants. Occupational exposure is also one of the ways through which humans gets in direct contact with EDCs in manufacturing/production units. The contamination of fishes in the river spread the contamination in the entire food chain including human beings. Some EDCs are banned after knowing their fatal impacts, but since they tend to accumulate in the environmental matrices, their impacts can still be seen on the aquatic organisms and human beings. Nash et al. (2004) reported that PCBs could still be found in drinking water with conc. up to 450 ng/l, after its manufacture being banned by the US government in 1977.

One major source of EDC exposure is the household products. From pharmaceuticals to the colours used in textiles were realized to contain EDCs that make us the direct victim to these poisons. When cosmetics were tested for having endocrine disruptors, it was reported that most sunscreens and fragranced products like air fresheners contain higher concentration of these chemicals (Witorsch and

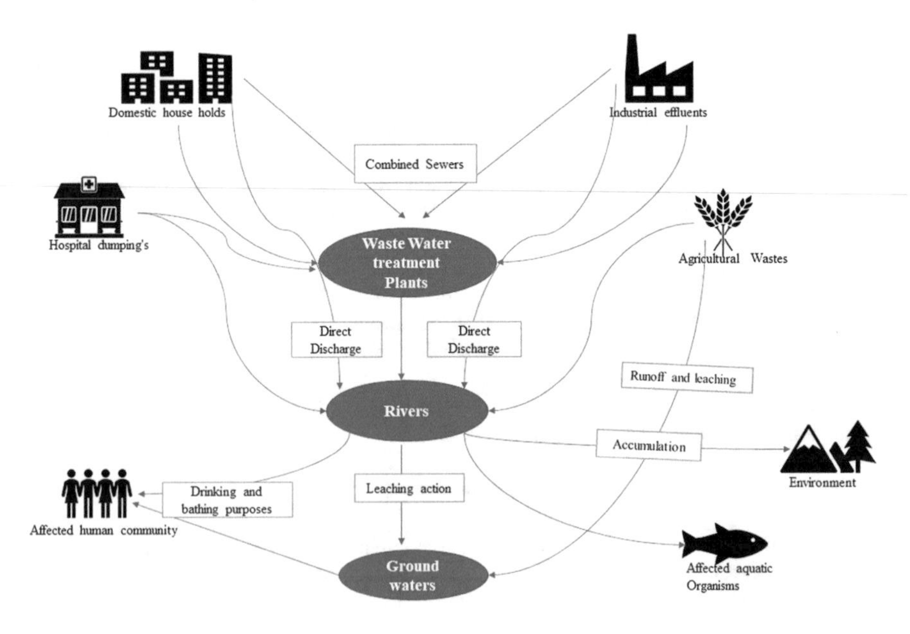

Fig. 1 Source, occurrence and fate of EDCs

Thomas 2010). Another major source of EDCs is the agricultural sector. The extensive use of pesticides, insecticides, herbicides and fertilizers have not only contaminated the crop plants but also the water bodies and soil through runoff and leaching process.

The enormous release of these toxicants from different sources has made their presence prominent in the environment which cannot be ignored. They can be found in several environmental matrices like surface waters, groundwaters, sediments, soil, air, etc. The uncontrolled anthropogenic activities have led to the accumulation of endocrine disruptors in the environment in several ng/L concentrations which is enough to cause endocrine-related diseases to the exposed organisms. The major receptors of these contaminants are groundwater, surface water, wastewater treatment plants and drinking water. Surface waters were first to get contaminated with EDCs since they receive the wastewater from the treatment plants. It is essential to know the levels of endocrine disruptors in surface water as it is used for drinking water production (Balabanič et al. 2017). The steroids and estrogens are found to be present from nanograms to several micrograms making it a matter of concern. BPA and Nonyl Phenols (NPs) are also found in higher concentration in surface water bodies due to the domestic wastewater discharge from urban settlements (Luo et al. 2014). Though wastewater treatment plants (WWTPs) are major receivers of the EDC-contaminated water from various sources, they are found inefficient in removing EDCs from the wastewater through existing technologies. There are several research going on to remove these fatal chemicals before releasing the wastewater to the water bodies, but it's not quite successful yet. So, the EDCs are released to the aquatic environment contaminating increasing their accumulation (Zhang and Zhou 2008). Again, the BPA and phenolic compound are quite high in the wastewater which is the result of high plastic usage and over-the-counter use of different personal care products. Drinking water ultimately gets contaminated with endocrine active chemicals due to the availability of poor treatment methods.

In a wastewater treatment plant, the endocrine disruptors get either absorbed/adsorbed or biodegraded partially or remain as it is which then get released into the water bodies. The conventional remediation methods are not much effective in eliminating the EDCs completely from wastewater. There are some EDCs that get removed easily with the existing treatment methods, while there are some others that hardly get removed from wastewater. When this partially treated wastewater enters the water bodies, they tend to accumulate and affect the organisms exposed to them (Janex-Habibi et al. 2009). Nasu et al. (2001) reported that the biological degradation of E2 hormone increased the concentration of E1 hormone in the treatment plant. The percentage removals of EDCs in anaerobic sludge digestion process were reported to be 12–21% for E1 hormone, 10–15% for EE2 hormone and 44–48% for NP. The endocrine disruptors have a potential to get bioaccumulated and concentrated in the environmental matrices if untreated and released raw to the surface waters (Gomes et al. 2004; Lai et al. 2002).

3 Toxicological Aspects of EDCs

When a group of compounds are named as endocrine-disrupting chemicals, that means there are some serious toxic effects on endocrine systems of organisms exposed to these chemicals. The Endocrine Society of the US Department of Health and Human Services defined EDCs as the exogenous chemicals that disturb/interfere with every aspect of hormonal actions. They mimic the natural hormones and attach to receptor cells producing an undesired response in the victim's body. The compounds are basically reported to affect the metabolic activities and the biosynthesis of hormones which ultimately disturb organism growth, development and reproduction (Hampl et al. 2016). So, they impose a critical risk on the environment by affecting animals living in it, and the victims of these fatal compounds are gradually increasing from aquatic animals to wildlife to humans. The endocrine disruptors are mostly associated with chronic toxicity rather than acute toxicity. The toxic effects of these hazardous compounds like obesity, cancer and developmental disorders are predominantly seen in the long run or in a later stage. An article by the European Commission (EU) reported that developmental abnormalities, retarded reproduction and skeletal deformities are some of the common effects seen in animals having chronic exposure to EDCs. When a population of gastropods *P. lineata* was exposed to BPA, they were observed to have decreased heart rate and lower spawning along with behavioural changes (de de Andrade et al. 2017). When an experiment was conducted to observe the impact of BPA on a group of snails, it was found that the total protein content reduced with decreased sperm production and deformity in eggs (El-Shenawy et al. 2017). The serious human exposure of BPA is through food and drinks as BPA tend to leach from canned food containers, packaged eatables and drinking water and cold drinks bottles. Some of the chronic effects of BPA include brain tumours, asthma and obesity (Rezg et al. 2014). Cadmium being a well-known endocrine disruptor is seen to alter various hormones significantly in male rodents (Lafuente et al. 2004). When exposed to human prostate cancer cell lines, cadmium was observed to mimic different hormones, affect their synthesis and alter the gene expressions (Martin et al. 2002). Phthalates are one of the most common endocrine disruptors found in the environment since it is frequently detected in plastics and is used as plasticizer in most of the plastic manufacturing units (Peter et al. 2007). They are primarily associated with cardiotoxicity, obesity and retarded growth and development. Infants and children are at a greater risk of getting affected than adults due to more use of baby bottles and plastic toys by them (Swan 2008; Sathyanarayana et al. 2008). Besides, it creates hindrance in secretion and proper functioning of testosterone hormone in adults (Helal 2014). Nonylphenols (NPs) are infamous in mimicking natural hormones like 17β-estradiol and bind to estrogen receptors showing various hormonal dysfunctions in the body. They are also related to multiple gene malfunctions (Choi et al. 2011). There are several studies that state that cancer is one of the typical disorders associated with endocrine disruptors. Besides, it is also mentioned that women are more prone to hormone-related cancers than men by EDCs. Women exposed to different endocrine disruptors were most commonly diagnosed with ovarian cancer, hormonal imbalance, damaged oocytes and tumour production in the tissues on surface of the ovary. However, the mechanism of action of these toxic chemicals is still under study and not fully known (Brevini et al. 2005). The properties, uses and toxicological aspects of few EDCs are shown in Table 2.

4 Removal of EDCs from Industrial Effluents

Since industries are the major sources of EDCs, treatment of industrial effluents to remove these compounds is very crucial to save the environment and the ecosystems within it. Industrial wastes have hugely made their way to surface water bodies and wastewater treatment plants. BPA and alkyl phenols are common compounds extensively used in plastic industries, while hormones like ethinylestradiol are widely used in contraceptive pills. Pharmaceutical wastes are also reported to contain paracetamol, caffeine, etc. which are widely taken medication in human population. Activated charcoal was found effective in removing the alkyl phenols and hormones through adsorption process. However, the existence of natural organic matter in real water decreased the efficiency of adsorbent for removing the EDCs (Hemidouche et al. 2017). Di-n-butyl phthalate (DBP), a serious waste of plastic industry effluent, was found to be degraded by UV irradiation of water samples. It could photolyse the DBP efficiently to less toxic intermediates within an hour at a particular pH (Lau et al. 2005). When US surface waters were surveyed for pharmaceutical wastes, significant amount of EDCs were found in water, showing relatively higher concentration of nonylphenol, bisphenol and ethinylestradiol at μg/L concentration. In view of removing the EDCs from water, two techniques, i.e. UV photolysis and advanced oxidation process (AOP), were applied. Though both the technologies show significant degradation of EDCs, AOP was found to be more efficient than UV photolysis when compared (Rosenfeldt and Linden 2004). The extensive use of pesticides and insecticides in agriculture has not only leached the EDCs from agricultural field but also has added the toxicants to the water bodies from pesticide manufacturing industries. Activated carbon filters were used to treat atrazine and its metabolites from water. The filters were found effective in adsorbing the pesticides from contaminated aqueous samples (Faur et al. 2005). Chitosan was also employed successfully as adsorbent in removing bisphenol from water bodies (Dehghani et al. 2016). Low pressure reverse osmosis membrane (LPROM) is also an efficient technology in removing alkyl phenols and steroid hormones from the environment (Razak et al. 2007).

Table 2 Properties and toxicological aspects of selected EDCs

Endocrine-disrupting chemicals (EDCs)	Properties	Application	Concerned hormones	Exposure/ toxicological aspects	References
Phthalates	Usually colourless or slightly yellowish oily and odourless liquid, soluble in solvents	As plasticizers in PVC plastics; floorings, toys, printing inks, perfumes, nail varnishes	Estrogen, androgen, thyroid	Direct contact and uses; indirect contact via leaching and environmental contamination; exposed through inhalation, ingestion and dermal exposure	Mikula et al. (2005), Heudorf et al. (2007), Jackson and Sutton (2008)
Bisphenol A (BPA)	White to light brown flakes or powder; high water solubility; low vapour pressure	Production of polycarbonate plastics, epoxy resins, dental sealants	Thyroid, androgen, estrogen	Acute, short-term and subchronic toxicity; effects on the liver, kidney and body weight; induce mutation	Jackson and Sutton (2008), Cao et al. (2010), Duran and Beiras (2017)
Nonylphenol (NP)	Viscous liquid and soluble in most of the organic solvents, slightly water soluble	Have wide application in industries, consumer products, household laundry detergents, paints, pesticides	Estrogen	Irritation to the eye and respiratory system of human, reduced egg production in female zebrafish	Zoller (2006), Jackson and Sutton (2008), Duran and Beiras (2017)
Benzotriazoles	Heterocyclic compounds, colourless, soluble in organic solvents, less soluble in water	Corrosion inhibitor for copper and copper alloys, used in industry as anticorrosive agents, antifreeze fluids, dishwasher detergents	Thyroid	Metabolic imbalance, carcinogenic activity, hormonal imbalance, acute toxicity in aquatic species	Pillard et al. (2001), Maceira et al. (2018)
Triclosan (TCS)	Polychloro phenoxy phenol, organic compound with white powdered solid, soluble in ethanol, methanol and strongly basic solutions	Antibacterial and antifungal agents in consumer products, i.e. dish soap, toothpaste, liquid hand soap, surgical cleaning treatments; pesticide	Thyroid, estrogen, androgen	Exposure through skin absorption, ingestion, acute and chronic toxicity in aquatic species	Jackson and Sutton (2008), Duran and Beiras (2017)

5 Removal of EDCs from Wastewater Treatment Plants

Another major source of EDCs is the municipal and household wastes. The over-the-counter use of personal care products, plastic packages, polybags and different medications in our daily use have made their accumulation in the environment prominent either through direct discharge of municipal waste into water bodies or through wastewater treatment plants which couldn't efficiently remove the EDCs from the wastewater. A lot of innovative technologies were used for their removal from wastewater treatment plants so there could be a minimal discharge of EDCs into the water bodies. One of them is sequencing batch biofilter granular reactor (SBBGR). When this technology was used in the treatment plant along with activated sludge process for removal of steroid hormones, it was found to perform better than the conventional methods (Balest et al. 2008). Activated oxidation process was used to remove EDCs from urban wastewaters, and it was found highly efficient than other conventional treatment methods as they facilitate partial degradation or complete remediation of the contaminants (Balabanič et al. 2012). This technology was found effective in removing a wide array of organic and inorganic contaminants from wastewater (Cesaro and Belgiorno 2016). Another removal technology, ER/AR competitive ligand binding assay, attaches the steroid hormones to the binding ligands which get removed from the waste treatment plants. It shows higher removal efficiency than other treatment methods (Liu et al. 2009). Sand filtration and ozonation, when analysed for removal of pharmaceutical compounds and personal care products, proved effective in removing EDCs from sewage treatment plant. However, use of only sand filtration was found inefficient than the combined technology (Nakada et al. 2007). When a sequential treatment method was applied to remove estrogens from wastewater treatment plants, it showed greater removal efficiency in eliminating hormones. The sequential method employed a stabilization pond followed by facultative pond and then activated sludge treatment with chlorination followed by a final step of anaerobic sludge blanket reactor (Pessoa et al. 2014). In an advanced treatment plant in Australia, microfiltration and reverse osmosis was found to effectively remove EDCs from recycled wastewater showing around 97% removal. Despite this higher efficiency in removing endocrine disruptors, BPA could still be found in the treated wastewater which is the only disadvantage of using this reverse osmosis method (Al-Rifai et al. 2011). Catalytic photodegradation was also successful in breaking down endocrine active hormones from wastewater treatment plants (Zhang and Zhou 2008).

6 Algal Species

Algal species are reported to remove pollutants since decades. The removals of endocrine disruptors from industrial effluents are reported to be done with various different technologies, but its removal with microorganisms has gained a lot of attention due to its high efficiency and cost-effective nature. Besides, the advanced technologies used are not only on the costlier side but also are not so environment-friendly. Microalgae-mediated remediation is given higher preference as they are more tolerant to the toxicity of endocrine disruptors while degrading them efficaciously in the environment. Algal species break down the contaminants and use them as carbon and nitrogen source for their metabolic activities (Jinqi and Houtian 1992).

Recent studies show that a wide spectrum of harmful compounds could be removed through microalgae. Among all the microalgal species, *Chlorella* species are found to show higher removal of toxic compounds. It is also reported that the algae present in the polluted environment are more resistant to toxicity of contaminants exhibiting their greater removal from environment (Lei et al. 2002; Pinto et al. 2002; Wong and Pak 1992). Microalgae have a potential to convert the contaminants to metabolites that can be used to produce biogas and biodiesel. They produce oxygen during photosynthesis which eliminates the need of external aeration for aerobic biodegradation (Munoz and Guieysse 2006). An algal species *S. capricornutum* completely transformed the pyrene present in the solution efficaciously within 7 days. When microalgae, *Ankistrodesmus braunii* and *Scenedesmus quadricauda*, were used for remediation of phenol-containing wastewater, they biodegraded the phenols within 5 days achieving 70% degradation efficiency. With increased temperature and light intensity, *Chlorella* species remediated the nonylphenol from contaminated water rapidly and effectively (Gao et al. 2011).

Marine microalgal species, when used for the removal of phthalate acid esters, quickly degraded the phthalates with their intra- and extracellular enzymes (Gao and Chi 2015). Fluoranthene and pyrene were seen efficiently degraded by the freshwater microalgae *S. capricornutum* than other species. Results show that removal of toxicant is algae specific (Lei et al. 2007). Algae ponds and duckweed ponds represents an economic alternative for the removal of toxic compounds. When steroid hormones are treated in these ponds, they quickly get sorbed to the algae and duckweeds and then eventually biodegraded by the microorganisms present in the pond systems. The algae present in the ponds also accelerated the removal of estrogens within 6 days (Shi et al. 2010). *Chlorella vulgaris* was also successful in removing p-chlorophenol and p-nitrophenol from wastewaters. A filamentous microalga, *Tribonema minus*, was seen to biotransform the industrial phenol waste into secondary metabolite that can be employed for biofuel generation (Lima et al. 2004; Cheng et al. 2017). When carbamazepine-contaminated water was treated with freshwater microalga for their remediation, *C. mexicana* was found efficient in removing the endocrine active compound from wastewater having a less toxic impact of the harmful compound on its growth and metabolic rate (Xiong et al. 2016). Table 3 displays the EDCs which have been transformed using microalgae as remediation agent.

Table 3 EDCs remediation with variety of microalgal species

S. no.	Categories	EDC	Microalgae	Status	References
1	Personal care products	Bisphenol A	*Chlamydomonas mexicana* and *Chlorella vulgaris*	*C. mexicana* and *C. vulgaris* were found to degrade 24% and 23% of BPA, respectively. *C. mexicana* was observed to be relatively more resistant to BPA indicating that contaminated aqueous systems could be treated with this species efficaciously	Jia et al. (2014)
2			*Chlorella fusca* var. *vacuolata*	When *C. fusca* was provided with 8 to 16 hours of light and dark conditions, it degraded 90% of BPA to intermediate compounds having nonestrogenic activity	Hirooka et al. (2005)
3		Triclosan	*Chlorella pyrenoidosa*	*C. pyrenoidosa* removed 77.2% of triclosan when cultivated with 800 ng/mL of it for 96 h. Dichlorohydroxydiphenyl ether was detected as the major degradation product of triclosan	Wang et al. (2013b)
4	Pesticides/ herbicide	3,4-Dichloroaniline	*Chlorella pyrenoidosa*	*C. pyrenoidosa*, when tested over a 7-day period experiment, showed a removal percentage of 78.4%	Wang et al. (2013a)
5		α-Endosulfan	Green algae, *Chlorococcum* sp. or *Scenedesmus* sp.	Algae successfully transformed α-endosulfan to endosulfan sulphate (major metabolite) and endosulfan ether (minor metabolite) in liquid medium which were reported to degrade further to an aldehyde product	Sethunathan et al. (2004)
6		Diazinon	*Chlorella vulgaris*	*C. vulgaris* effectively degraded diazinon to a lesser toxic product with degradation rate constant ranging between 0.23 and 0.04/day	Kurade et al. (2016)
7	Pharmaceuticals	17α-Estradiol, estrone, 17β-estradiol, estriol	*Scenedesmus dimorphus*	Efficacious removal of steroid hormones (85–95%) was noticed over a period of 8 days	Zhang et al. (2014)

(continued)

Table 3 (continued)

S. no.	Categories	EDC	Microalgae	Status	References
8		Estrone, 17α-ethinylestradiol and 17β-estradiol	*Anabaena cylindrica*, *Scenedesmus quadricauda*, *Chlorella*, *Chlorococcus*, *Spirulina platensis* and *Anabaena* var.	Batch and continuous flow experiments showed that duckweed ponds are more efficient in removing nanogram levels of estrogens from wastewater than algae ponds. Sorption followed by degradation was reported as the chief mechanism of removal	Shi et al. (2010)
9		β-Estradiol (E2) and 17α-ethinylestradiol (EE2)	*Selenastrum capricornutum* and *Chlamydomonas reinhardtii*	Both microalgae strains degraded about 90% of β-estradiol in 24 h. It takes a longer period almost over 7 days to attain high removal of 17α-ethynylestradiol	Diaz et al. (2015)
10		Carbamazepine	*Chlamydomonas mexicana* and *Scenedesmus obliquus*	28% and 35% of carbamazepine degradation was done by *S. obliquus* and *C. mexicana*. *C. mexicana* relatively showed more tolerance towards carbamazepine	Xiong et al. (2016)
11		Progesterone	*Microchaete tenera*	*M. tenera* could convert progesterone into four steroid metabolites with its active 6α- and 6β-hydroxylases components	Safiarian et al. (2011)
12	Plastics	Phthalate acid esters (PAEs)	*Cylindrotheca closterium*, *Dunaliella salina*, *Chaetoceros muelleri*	Both extra- and intracellular enzymes of algae are mainly responsible for biodegradation of DEP and also substantially removed DBP	Gao and Chi (2015)
13	Industrial chemicals	PAHs (fluoranthene and pyrene)	*Chlorella vulgaris*, *Scenedesmus quadricauda*, *Scenedesmus platydiscus* and *Selenastrum capricornutum*	Algae-mediated PAHs remediation found to be species specific and toxicant-dependent. *S. capricornutum* and *C. vulgaris* was reported to be the most efficient and least efficient sp. for PAHs removal, respectively, and *C. vulgaris*	Lei et al. (2007)

S. no.	Categories	EDC	Microalgae	Status	References
14		Nonylphenols (NPs), estradiol (E2), 17α-ethynylestradiol (EE2) and bisphenol A (BPA)	*Navicula incerta*	Among the four tested EDCs, NPs showed lowest degradation rate and highest bioconcentration factor in *N. incerta* indicating its ability to adversely affect the organisms of upper trophic level also	Liu et al. (2010)
15		Polybrominated diphenyl ethers (PBDEs)	*Chlorella* (STCh and SICh), *Parachlorella* (STPa1 and STPa2), *Scenedesmus* (STSc, TPSc1 and TPSc2), *Nitzschia palea* (YLBa) and *Mychonastes* (TPMy)	Among the nine microalgae isolated from the wastewater, four were reported to be highly tolerant towards PBDEs with SICh showing highest PBDEs removal and bioaccumulation	Deng and Tam (2015)
16			*Microcystis aeruginosa*	The degradation ability of *M. aeruginosa* at both lower (0.02–0.5 mg/L) and higher conc. 0 of NPs was found to be quite effective indicating its ability to adapt effectively under varied environmental conditions	Wang and Xie (2007)
17		Nonylphenols (NPs)	*Chlorella vulgaris*, *Chlorella miniata*, *Chlorella* sp. (1uoai) and *Chlorella* sp. (2f5aia)	Temperature and light intensity are found to be most important parameters that increase the biodegradation of nonylphenols which indicates that the removal is directly correlated to the photosynthesis and the metabolism of the species	Gao et al. (2011)
18			*Scenedesmus quadricauda*, *Chlorella vulgaris*, *Ankistrodesmus acicularis*, *Chroococcus minutus*	Among the four species tested for NPs removal, *A. acicularis* and *C. vulgaris* showed highest activity. Measurement of extra- and intracellular concentration of NPs indicated that removal is mainly achieved through degradation	He et al. (2016)

(continued)

Table 3 (continued)

S. no.	Categories	EDC	Microalgae	Status	References
19		4-(1,1,3,3-Tetramethylbutyl) phenol, bisphenol A, 4-n-nonylphenol (4-NP), technical nonylphenol (t-NP)	Microalgae from Chlorophyceae class and cyanobacteria	Aeration assisted the remediation of all the contaminants showing removal greater than 91%, while around 50 to 80% removal was observed in the absence of aeration. However, removal of 4-NP was not much affected with the absence of aeration	Abargues et al. (2013)
20		Nonylphenol (NP) and octylphenol (OP)	*Scenedesmus obliquus*	The algal species show greater than 89% removal of NP and greater than 58% of OP within 5 days. Extra- and intracellular amount of NP and OP showed that the sp. has high bioconcentration factor for both NP and OP and could be utilized as remediation tool efficiently	Zhou et al. (2013)
21		p-Nitrophenol and p-chlorophenol	*Coenochloris pyrenoidosa* and *Chlorella vulgaris*	Complete biodegradation of the contaminants under different photo regimes was reported with the algal sp. in absence of zeolite; however its presence slowed down the degradation rate owing to contaminant adsorption	Lima et al. (2004)
22		Dichlorophenols (DCPs)	*Scenedesmus obliquus*	Depending on the toxicity level of DCP, microalgae channel its energy towards biodegradation and biomass production. For instance, more energy is diverted towards biodegradation in presence of highly toxic DCP, lowering the energy required for biomass production and vice versa	Papazi and Kotzabasis (2013)

7 Advantage and Limitation of Microalgal Remediation

Since nothing in this world is an unmixed blessing, so if a system has advantages, it has its disadvantages too. There are a lot of advantages of using microalgal remediation than other technologies. Microalgae are naturally found in the environment so by maintaining proper temperature and light intensity, we can grow them in large numbers which is possible with minimal expenses. They require low maintenance unlike other technologies. As microalgae are photosynthetic in nature, they use the contaminants to produce food and release oxygen in the process which further facilitates the aerobic degradation of the contaminants. When grown in toxic environment, microalgae acclimatize themselves to the extreme conditions resisting the toxicity of contaminants increasingly and removing them efficiently from water bodies. The contaminant that gets biotransformed with intra- and extracellular enzymes of the microalgae can be used as an excellent feedstock for biofuel production. Besides that, they produce no harm to the environment either living or dead. Sometimes providing nutrient to the microalgae culture can accelerate the degradation rate of the contaminant, removing them faster from the environmental matrices. The whole microalgal system is a cost-effective method; it doesn't require huge amount of capital money to initiate the treatment of contaminated sites.

Canvassing the limitation of microalgal bioremediation, there can be a few disadvantages too that can be listed about the microalgae. Unlike other advanced treatment methods, this bioremediation process is a slow method which requires a long time to remediate the contaminants thoroughly. Secondly, if proper temperature and light are not available, microalgae tend to die or slow down their metabolic activities showing decreased efficiency. When the concentration of the contaminant is increased, the microalgae could not resist the toxicity of contaminant resulting in death of microalgae (Xiong et al. 2016). Since algae and duckweed ponds contain microorganisms also, they may release foul or pungent smell creating aesthetically unpleasant surroundings. Except these limitations, microalgae can be the most suitable and appropriate treatment method for the removal of EDCs from aqueous systems.

References

Abargues MR, Ferrer J, Bouzas A, Seco A (2013) Removal and fate of endocrine disruptors chemicals under lab-scale postreatment stage: removal assessment using light, oxygen and microalgae. Bioresour Technol 149:142–148

Al-Rifai JH, Khabbaz H, Schäfer AI (2011) Removal of pharmaceuticals and endocrine disrupting compounds in a water recycling process using reverse osmosis systems. Sep Purif Technol 77(1):60–67

Arditsoglou A, Voutsa D (2010) Partitioning of endocrine disrupting compounds in inland waters and wastewaters discharged into the coastal area of Thessaloniki, Northern Greece. Environ Sci Pollut Res 17:529–538

Baker BH, Martinovic-Weigelt D, Ferrey M, Barber LB, Writer JH, Rosenberry DO, Schoenfuss HL (2014) Identifying non-point sources of endocrine active compounds and their biological impacts in freshwater lakes. Arch Environ Contam Toxicol 67(3):374–388
Balabanič D, Hermosilla D, Merayo N, Klemenčič AK, Blanco A (2012) Comparison of different wastewater treatments for removal of selected endocrine-disruptors from paper mill wastewaters. J Environ Sci Health A Tox Hazard Subst Environ Eng 47:1350–1363
Balabanič D, Filipič M, Klemenčič AK, Žegura B (2017) Raw and biologically treated paper mill wastewater effluents and the recipient surface waters: cytotoxic and genotoxic activity and the presence of endocrine disrupting compounds. Sci Total Environ 574:78–89
Balest L, Mascolo G, Di Iaconi C, Lopez A (2008) Removal of endocrine disrupter compounds from municipal wastewater by an innovative biological technology. Water Sci Technol 58(4):953–956
Brevini TA, Zanetto SB, Cillo F (2005) Effects of endocrine disruptors on developmental and reproductive functions. Curr Drug Targets Immune Endocr Metabol Disord 5(1):1–10
Cao XL, Corriveau J, Popovic S (2010) Sources of low concentrations of bisphenol A in canned beverage products. J Food Prot 73:1548–1551
Cesaro A, Belgiorno V (2016) Removal of endocrine disruptors from urban wastewater by advanced oxidation processes (AOPs): a review. Open Biotechnol J 10(1):151
Chen M, Ohman K, Metcalfe C, Ikonomou MG, Amatya PL, Wilson J (2006) Pharmaceuticals and endocrine disruptors in wastewater treatment effluents and in the water supply system of Calgary, Alberta, Canada. Water Qual Res J Canada 41(4):c351–c364
Chiu JMY, Po BHK, Degger ND, Tse A, Liu WH, Zheng G, Zhao D, Xu D, Richardson B, Wu RSS (2018) Contamination and risk implications of endocrine disrupting chemicals along the coastline of China: a systematic study using mussels and semipermeable membrane devices. Sci Tot Environ 624:1298–1307
Choi JS, Oh JH, Park HJ, Choi MS, Park SM, Kang SJ, Oh MJ, Kim SJ, Hwang SY, Yoon S (2011) miRNA regulation of cytotoxic effects in mouse Sertoli cells exposed to nonylphenol. Reprod Biol Endocrinol 9:126
Crain DA, Eriksen M, Iguchi T, Jobling S, Laufer H, LeBlanc GA Jr, Guillette LJ (2007) An ecological assessment of bisphenol-A: Evidence from comparative biology. Reprod Toxicol 24:225–239
Cheng T, Zhang W, Zhang W, Yuan G, Wang H, Liu T (2017) An oleaginous filamentous microalgae Tribonema minus exhibits high removing potential of industrial phenol contaminants. Bioresour Technol 238:749–754
de Andrade ALC, Soares PRL, da Silva SCBL, da Silva MCG, Santos TP, Cadena MRS et al (2017) Evaluation of the toxic effect of endocrine disruptor Bisphenol A (BPA) in the acute and chronic toxicity tests with Pomacea lineata gastropod. Comp Biochem Physiol C Toxicol Pharmacol 197:1–7
Dehghani MH, Ghadermazi M, Bhatnagar A, Sadighara P, Jahed-Khaniki G, Heibati B, McKay G (2016) Adsorptive removal of endocrine disrupting bisphenol A from aqueous solution using chitosan. J Environ Chem Eng 4(3):2647–2655
Deng D, Tam NF (2015) Isolation of microalgae tolerant to polybrominated diphenyl ethers (PBDEs) from wastewater treatment plants and their removal ability. Bioresour Technol 177:289–297
Diaz AH, Llorca M, Mozaz SR, Vicent T, Barcel D, Blanquez P (2015) Microalgae cultivation on wastewater digestate: b-estradiol and 17a-ethynylestradiol degradation and transformation products identification. J Environ Manag 155:106–113
Duran I, Beiras R (2017) Acute water quality criteria for polycyclic aromatic hydrocarbons, pesticides, plastic additives, and 4-Nonylphenol in seawater. Environ Pollut 224:384–391
Dosnon-Olette R, Trotel-Aziz P, Couderchet M, Eullaffroy P (2010) Fungicides and herbicide removal in Scenedesmus cell suspensions. Chemosphere 79:117–123
El-Shenawy NS, El Deeb FAA, Mansour SA, Soliman MFM (2017) Biochemical and histopathological changes of Biomphalaria alexandrina snails exposed to bisphenol A. Toxicol Environ Chem 99(3):460–468

Faur C, MEtivier-Pignon H, Le Cloirec P (2005) Multicomponent adsorption of pesticides onto activated carbon fibers. Adsorption 11(5–6):479–490

Gao J, Chi J (2015) Biodegradation of phthalate acid esters by different marine microalgal species. Mar Pollut Bull 99:70–75

Gao QT, Wong YS, Tam NFY (2011) Removal and biodegradation of nonylphenol by different Chlorella species. Mar Pollut Bull 63:445–451

Gomes RL, Deacon HE, Lai KM, Birkett JW, Scrimshaw MD, Lester JN (2004) An assessment of the bioaccumulation of estrone in Daphnia magna. Environ Toxicol Chem 23(1):105–108

Hampl R, Kubátová J, Stárka L (2016) Steroids and endocrine disruptors—History, recent state of art and open questions. Steroid Biochem. Mol Biol 155(2016):217–223

He N, Sun X, Zhong Y, Sun K, Liu W, Duan S (2016) Removal and biodegradation of Nonylphenol by four freshwater microalgae. Int J Environ Res Public Health 13:1239

Helal MA (2014) Celery oil modulates DEHP-induced reproductive toxicity in male rats. Reprod Biol 14(3):182–189

Hemidouche S, Assoumani A, Favier L, Simion AI, Grigoras CG, Wolbert D, Gavrila L (2017) Removal of some endocrine disruptors via adsorption on activated carbon. In: E-Health and Bioengineering Conference (EHB), 2017. IEEE, pp 410–413

Heudorf U, Sundermann VM, Angerer J (2007) Phthalates: Toxicology and exposure. Int J Hyg Environ Health 210(5):623–634

Hirooka T, Nagase H, Uchida K, Hiroshige Y, Ehara Y, Nishikawa JI, Nishihara T, Miyamoto K, Hirata Z (2005) Biodegradation of bisphenol A and disappearance of its estrogenic activity by the green alga Chlorella fusca var. vacuolata. Environ Toxicol Chem 24:1896–1901

Hoepner LA, Whyatt RM, Just AC, Calafat AM, Perera FP, Rundle AG (2013) Urinary concentrations of bisphenol A in an urban minority birth cohort in New York City, prenatal through age 7 years. Environ Res 122:38–44

Jackson J, Sutton R (2008) Sources of endocrine-disrupting chemicals in urban wastewater, Oakland, CA. Sci Total Environ 405:153–160

Jafari AJ, Abasabad RP, Salehzadeh A (2009) Endocrine disrupting contaminants in water resources and sewage in Hamadan city of Iran. Iran J Environ Health Sci Eng 6(2):89–96

Jia MK, Kabra AN, Choi J, Hwang JH, Kim JR, Shanab RAIA, Oh YK, Jeon BH (2014) Biodegradation of bisphenol A by the freshwater microalgae Chlamydomonas mexicana and Chlorella vulgaris. Ecol Eng 73:260–269

Jinqi L, Houtian L (1992) Degradation of azo dyes by algae. Environ Pollut 75(3):273–278

Janex-Habibi ML, Huyard A, Esperanza M, Bruchet A (2009) Reduction of endocrine disruptor emissions in the environment: The benefit of wastewater treatment. Water Res 43(6):1565–1576

Komesli OT, Bakırdere S, Bayören C, Gökçay CF (2012) Simultaneous determination of selected endocrine disrupter compounds in wastewater samples in ultra-trace levels using HPLC-ES-MS/MS. Environ Monit Assess 184:5215–5224

Kurade MB, Kim JR, Govindwar SP, Jeon BH (2016) Insights into microalgae mediated biodegradation of diazinon by Chlorella vulgaris: Microalgal tolerance to xenobiotic pollutants and metabolism. Algal Res 20:126–134

Lafuente A, González-Carracedo A, Romero A, Cano P, Esquifino AI (2004) Cadmium exposure differentially modifies the circadian patterns of norepinephrine at the median eminence and plasma LH, FSH and testosterone levels. Toxicol Lett 146(2):175–182

Lai KM, Scrimshaw MD, Lester JN (2002) Biotransformation and bioconcentration of steroid estrogens by Chlorella vulgaris. Appl Environ Microbiol 68(2):859–864

Lau TK, Chu W, Graham N (2005) The degradation of endocrine disruptor di-n-butyl phthalate by UV irradiation: a photolysis and product study. Chemosphere 60(8):1045–1053

Lei AP, Wong YS, Tam NFY (2002) Removal of pyrene by different microalgal species. Water Sci Technol 46(11–12):195–201

Lei A-P, Hu Z-L, Wong Y-S, Tam NF-Y (2007) Removal of Xuoranthene and pyrene by divergent microalgal species. Bioresour Technol 98:273–280

Lima SAC, Raposo MFJ, Castro PML, Morais RM (2004) Biodegradation of p-chlorophenol by a microalgae consortium. Water Res 38:97–102

Liu ZH, Ito M, Kanjo Y, Yamamoto A (2009) Profile and removal of endocrine disrupting chemicals by using an ER/AR competitive ligand binding assay and chemical analyses. J Environ Sci 21(7):900–906

Liu Y, Guan Y, Gao Q, Tam NF, Zhu W (2010) Cellular responses, biodegradation and bioaccumulation of endocrine disrupting chemicals in marine diatom Navicula incerta. Chemosphere 80:592–599

Liu D, Wu S, Xu H, Zhang Q, Zhang SH, Shi L, Yao C, Liu Y, Cheng J (2017) Distribution and bioaccumulation of endocrine disrupting chemicals in water, sediment and fishes in a shallow Chinese freshwater lake: Implications for ecological and human health risks. Ecotoxicol Environ Saf 140:222–229

Luo Y, Guo W, Ngo HH, Nghiem LD, Hai FI, Zhang J, Liang S, Wang XC (2014) A review on the occurrence of micropollutants in the aquatic environment and their fate and removal during wastewater treatment. Sci Total Environ 473:619–641

Li W, Liu Y, Duan J, van LJ, Saint CP (2015) UV and UV/H2O2 treatment of diazinon and its influence on disinfection byproduct formation following chlorination. Chem Eng J 274:39–49

Maceira A, Marcé RM, Borrull F (2018) Occurrence of benzothiazole, benzotriazole and benzenesulfonamide derivates in outdoor air particulate matter samples and human exposure assessment. Chemosphere 193:557–566

Martin MB, Voeller HJ, Gelmann EP, Lu J, Stoica EG, Hebert EJ, Reiter R, Singh B, Danielsen M, Pentecost E, Stoica A (2002) Role of cadmium in the regulation of AR gene expression and activity. Endocrinology 143(1):263–275

Mikula P, Svobodova Z, Smutna M (2005) Phthalates: toxicology and food safety – a review. Czech J Food Sci 23:217–223

Munoz R, Guieysse B (2006) Algal–bacterial processes for the treatment of hazardous contaminants: a review. Water Res 40(15):2799–2815

Matsumoto M, Hirata-Koizumi M, Ema M (2008) Potential adverse effects of phthalic acid esters on human health: a review of recent studies on reproduction. Regul Toxicol Pharm 50:37–49

Nakada N, Shinohara H, Murata A, Kiri K, Managaki S, Sato N, Takada H (2007) Removal of selected pharmaceuticals and personal care products (PPCPs) and endocrine-disrupting chemicals (EDCs) during sand filtration and ozonation at a municipal sewage treatment plant. Water Res 41(19):4373–4382

Nash JP, Kime DE, Van der Ven LTM, Wester PW, Brion F, Maack G, Stahlschmidt-Allner P, Tyler CR (2004) Long-Term exposure to environmental concentrations of the pharmaceutical ethynylestradiol causes reproductive failure in fish. Environ Health Perspect 112:1725–1733

Nasu M, Oshima Y, Tanaka H (2001) Study on endocrine disrupting chemicals in wastewater treatment plants. Water Sci Technol 43(2):101–108

Newsted JL (2004) Effect of light, temperature, and pH on the accumulation of phenol by Selenastrum capricornutum, a green alga. Ecotoxicol Environ Saf 59(2):237–243

Onda K, Yang SY, Miya A, Tanaka T (2002) Evaluation of estrogen-like activity on sewage treatment processes using recombinant yeast. Water Sci Technol 46:367–373

Papazi A, Kotzabasis K (2013) "Rational" management of Dichlorophenols biodegradation by the microalga Scenedesmus obliquus. PLoS One 8(4)

Pessoa GP, de Souza NC, Vidal CB, Alves JA, Firmino PIM, Nascimento RF, dos Santos AB (2014) Occurrence and removal of estrogens in Brazilian wastewater treatment plants. Sci Total Environ 490:288–295

Peter ML, Friedrich KT, Walter E, Rudolf J, Naresh B, Wolfgang H (2007) Ullmann's (2007) Encyclopedia of industrial chemistry. Wiley Interscience, New York. Phthalic acid and derivatives

Pillard DA, Cornell JS, Dufresne DL, Hernandez MT (2001) Toxicity of benzotriazole and benzotriazole derivatives to three aquatic species. Water Res 35(2):557–560

Pinto G, Pollio A, Previtera L, Temussi F (2002) Biodegradation of phenols by microalgae. Biotechnol Lett 24(24):2047–2051

Razak ARA, Ujang Z, Ozaki H (2007) Removal of endocrine disrupting chemicals (EDCs) using low pressure reverse osmosis membrane (LPROM). Water Sci Technol 56(8):161–168
Rezg R, El-Fazaa S, Gharbi N, Mornagui B (2014) Bisphenol A and human chronic diseases: current evidences, possible mechanisms, and future perspectives. Environ Int 64:83–90
Rosenfeldt EJ, Linden KG (2004) Degradation of endocrine disrupting chemicals bisphenol A, ethinyl estradiol, and estradiol during UV photolysis and advanced oxidation processes. Environ Sci Technol 38(20):5476–5483
Safiarian MS, Faramarzi M, Amini M, Hasan-Beikdashti M, Soltani N, Tabatabaei-Sameni M (2011) Microalgal transformation of progesterone by the terrestrial-isolated cyanobacterium Microchaete tenera. J Appl Phycol 24:777–781
Sathyanarayana S, Karr CJ, Lozano P, Brown E, Calafat AM, Liu F, Swan SH (2008) Baby care products: possible sources of infant phthalate exposure. Pediatrics 121(2):e260–e268
Sethunathan N, Megharaj M, Chen ZL, Williams BD, Lewis G, Naidu R (2004) Algal degradation of a known endocrine disrupting insecticide, α-endosulfan, and its metabolite, endosulfan sulfate, in liquid medium and soil. J Agric Food Chem 52:3030–3035
Shi W, Wang L, Rousseau DPL, Lens PNL (2010) Removal of estrone, 17α-ethinylestradiol, and 17β-estradiol in algae and duckweed-based wastewater treatment systems. Environ Sci Pollut Res 17:824–833
Swan SH (2008) Environmental phthalate exposure in relation to reproductive outcomes and other health endpoints in humans. Environ Res 108(2):177–184
Staples CA, Dorn PB, Klecka GM, O'Block ST, Harris LR (1998) A review of the environmental fate, effects, and exposures of bisphenol-A. Chemosphere 36:2149–2173
Tikoo V, Scragg AH, Shales SW (1997) Degradation of pentachlorophenol by microalgae. J Chem Technol Biotechnol 68(4):425–431
Vandermarken T, Coes K, Langenhove KV, Boonen I, Servais P, Garcia-Armisen T, Brion N, Denison MS, Goeyens L, Elskens M (2018) Endocrine activity in an urban river system and the biodegradation of estrogen-like endocrine disrupting chemicals through a bio-analytical approach using DRE- and ERE-CALUX bioassays. Chemosphere 201:540–549
Voutsa D, Hartmann P, Schaffner C, Giger W (2006) Benzotriazoles, Alkylphenols and Bisphenol A in municipal wastewaters and in the Glatt River, Switzerland. Environ Sci Pollut Res 13(5):333–341
Wang J, Xie P (2007) Antioxidant enzyme activities of Microcystis aeruginosa in response to nonylphenols and degradation of nonylphenols by M. aeruginosa. Environ Geochem Health 29:375–383
Wang S, Poon K, Cai Z (2013a) Biodegradation and removal of 3,4-dichloroaniline by Chlorella pyrenoidosa based on liquid chromatography-electrospray ionization-mass spectrometry. Environ Sci Pollut Res 20:552–557
Wang S, Wang X, Poon K, Wang Y, Li S, Liu H, Lin S, Cai Z (2013b) Removal and reductive dechlorination of triclosan by Chlorella pyrenoidosa. Chemosphere 92:1498–1505
Witorsch RJ, Thomas JA (2010) Personal care products and endocrine disruption: a critical review of the literature. Crit Rev Toxicol 40(sup3):1–30
Wong MH, Pak DC (1992) Removal of Cu and Ni by free and immobilized microalgae. Biomed Environ Sci 5(2):99–108
Wang C, Shih Y (2015) Degradation and detoxification of diazinon by sono-Fenton and sono-Fenton-like processes. Sep Purif Technol 140:6–12
Wang P, Wang SL, Fan CQ (2008) Atmospheric distribution of particulate-and gas-phase phthalic esters (PAEs) in a Metropolitan City, Nanjing, East China. Chemosphere72(10):1567–1572
Xiong J-Q, Kurade MB, Shanab RAIA, Ji MK, Choi J, Kim JO, Jeon BH (2016) Biodegradation of carbamazepine using freshwater microalgae Chlamydomonas mexicana and Scenedesmus obliquus and the determination of its metabolic fate. Bioresour Technol 205:183–190
Yan H, Ye C, Yin, C (1995) Kinetics of phthalate ester biodegradation by Chlorella pyrenoidosa. Environ Toxicol Chem 14(6):931–938
Yang S, Wu RSS, Kong YC (2002) Biodegradation and enzymatic responses in the marine diatom Skeletonema costatum upon exposure to 2,4-dichlorophenol. Aquat Toxicol 59:191–200

Zhang Y, Zhou JL (2008) Occurrence and removal of endocrine disrupting chemicals in wastewater. Chemosphere 73(5):848–853

Zhang Y, Habteselassie MY, Resurreccion EP, Mantripragada V, Peng S, Bauer S, Colosi LM (2014) Evaluating removal of steroid estrogens by a model alga as a possible sustainability benefit of hypothetical integrated algae cultivation and wastewater treatment systems. ACS Sustain Chem Eng 2:2544–2553

Zhang M, Shi Y, Lu YL, Johnson AC, Sarvajayakesavalu S, Liu ZY, Su C, Zhang Y, Juergens MD, Jine XW (2017) The relative risk and its distribution of endocrine disrupting chemicals, pharmaceuticals and personal care products to freshwater organisms in the Bohai Rim, China. Sci Total Environ 590–591:633

Zhou G-J, Peng F-Q, Yang B, Ying G-G (2013) Cellular responses and bioremoval of nonylphenol and octylphenol in the freshwater green microalga Scenedesmus obliquus. Ecotoxicol Environ Saf 87:10–16

Zoller U (2006) Estuarine and coastal zone marine pollution by the nonionic alkylphenol ethoxylates endocrine disrupters: is there a potential ecotoxicological problem? Environ Int 32:269–272

Bioremediation of Municipal Sewage Using Potential Microalgae

Chitralekha Nag Dasgupta, Kiran Toppo, Sanjeeva Nayaka, and Atul K. Singh

1 Introduction

Expansion of city and high population nowadays increase the pollution-associated problems worldwide (Abdel-Raouf et al. 2012). City sewage is carrying the major load of pollution which ultimately is discharged into the nearby rivers. Some countries have made the significant investment in water treatment, but still in many cases, sewage is disposed of directly into the river without treatment (Abdel-Raouf et al. 2012). Even in India economic development and industrialization impose high threats to the quality of water bodies (Kaur et al. 2012). The huge expansion of the water supply networks and sewage removal drainages aggravated the problems. All drainage systems receive a bulk of pollution load contaminated with faecal pathogens, agricultural nutrients, dissolved and suspended solids and other oxygen-demanding materials which ultimately resulted in different health hazards (Hamner et al. 2006; Kaur et al. 2012). Sewage is either treated by 'on-site system' using septic tanks, biofilters or aerobic treatments or transported through pipe lines and pumping stations to a 'centralized system' called 'sewage treatment plant (STP)'. Bharwara Sewage Treatment Plant located in Gomti Nagar, Lucknow, India, is Asia's largest STP (Fig. 1). STPs are designed to remove heavy solids by primary treatment, suspended biological matter by secondary and all the organic ions by tertiary treatment from sewage (Abdel-Raouf et al. 2012). However, existing STPs are now not enough to neutralize the increasing pollution problem (Choksi et al. 2015). Improvement of the existing one and the construction of parallel treatment facilities could help to cope up with the present scenario of pollution (Sivasubramanian 2006).

Several eco-friendly natural methods have emerged for bioremediation of wastewater. In USA, Oswald and Gotaas (1957) launched an innovative idea to treat

C. N. Dasgupta · K. Toppo · S. Nayaka (✉) · A. K. Singh
CSIR-National Botanical Research Institute, Lucknow, U.P., India

S. K. Gupta, F. Bux (eds.), *Application of Microalgae in Wastewater Treatment*,
https://doi.org/10.1007/978-3-030-13913-1_7

Fig. 1 Bharwara Sewage Treatment Plant (STP), Lucknow, India (capacity 345 MLD)

wastewater by algae. Algae can grow naturally and form algal bloom in water contaminated with inorganic and organic pollutants (Grobbelaar 2004). However, most of the bloom-forming algae overconsume oxygen from water and release toxins, which are harmful to the aquatic animals (Hallegraeff 1993). Only some selective nontoxic algal strains could be an attractive solution for the treatment of sewage to improve the water quality (Grobbelaar 2004). During growth, algae can absorb organic and inorganic matters of wastewater as nutrients and reduce pollution (Grobbelaar 2004; Kiran et al. 2008; Sydney et al. 2011; Kshirsagar 2013). Furthermore, harvested algal biomass can be utilized to develop different value-added products such as bioenergy, pharmaceuticals products, nutraceuticals products, genetically engineered products, etc. (Jensen 1993; de la Noue and de Pauw 1988). In addition, unlike the conventional methods, it requires low operational and maintenance cost as well as no hazardous chemicals are used for water purification (Abdel-Raouf et al. 2012).

The algal system can treat human sewage, animal effluent, man-made agricultural wastes and many other wastewaters (Woertz et al. 2009; Wang et al. 2010). Initially, algae-based remediation system has been employed as a tertiary process, but in 1989 Tam and Wong proposed the algal-based system as a secondary treatment system. Since then several laboratories and pilot projects have been constructed, and many STPs started using these systems. In the 'Aquatic Species Program' of the US Department of Energy (National Renewable Energy Laboratory), it has been observed that energy output was twice of the energy input during algal remediation of wastewater (Sheehan et al. 1998). A private corporation 'Sunrise Ridge Algae, Inc.' (Texas USA, 2006) was also engaged to reduce water and air pollution by algae and produce renewable fuel feedstock as well as different animal feeds (www.oilgae.com).

In India, Sivasubramanian (2006) established the first phycoremediation plant at SNAP industry at Ranipet, which helped in pH correction of the effluent. The algal biomass produced was utilized for bio-fertilizer production and sold by the company. They have developed a joint technology with CORE BIOTECH, Colombia, to treat petrochemical wastewater (http://phycoremediation.in/projects.html) by algae. Now the plant has been scaled up to handle 300,000 barrels of effluent water every day. The algae-based remediation technology has also been used in different industrial effluents in Tamilnadu and Ahmedabad. A significant decrease in total hardness was found in all the water samples while using algae-based remediation technology (Sivasubramanian et al. 2012). In Agra, Sengar et al. (2011) collected sewage from drain which opens into the river Yamuna and performed the phycoremediation analysis. A team of scientists from Indian Institute of Science, Bangalore (India), worked on phycoremediation of municipal water of Bellandur Lake, Koramangala region, South of Bangalore (Mahapatra et al. 2014).

2 The Composition of Municipal Sewage

The composition of sewage depends on lifestyle and economic condition of the society (Gray 1989). The quality and quantity of sewage vary from place to place. In India domestic water supply is increased with the rapid expansion of cities. According to the Central Public Health and Environmental Engineering Organisation (CPHEEO), Ministry of Urban Development, Government of India, 70–80% of the total water supplied for domestic purpose gets generated as sewage, which consists of approximately 99% water and 1% mixture of compounds including inorganic, organic and man-made products both suspended and soluble forms (Kaur et al. 2012). The liquid portion of sewage is treated and discharged into the lake or river, and the thick and slurry solid portion of sewage called sludge is dried out on sludge bed.

2.1 Plastic Wastes

Different plastic wastes are nowadays a severe problem of city garbage. Most of the time, polybags, plastic bottles and other garbage are mixed with sewage (Fig. 2). Sometimes massive floating garbage chokes the sewage pipe line (Lazarevic et al. 2010).

Fig. 2 Removal of plastic waste from the sewage inlet of Bharwara STP, Lucknow

Fig. 3 Sludge bed of Bharwara STP, Lucknow

2.2 *Sludge*

Sewage contains very less amount of suspended solids or dissolved solids, which are removed from sewage by sedimentation and grit removal system during sewage treatment (Abdel-Raouf et al. 2012). A thick solid portion of sewage called sludge is dried out on sludge drying beds or by using mechanical devices (Fig. 3). Sludge contains different inorganic and organic materials which are useful for plant growth (Fytili and Zabaniotou 2008). However, due to presence of some toxic chemicals, restricted use of sludge as fertilizer has been observed (Fytili and Zabaniotou 2008). The contribution of total solid in sewage by a person is about 170–220 g/capita/day, and suspended solid is about 70–145 g/capita/day (Arceivala and Asolekar 2010).

2.3 Organic Compounds

A major part of sewage is dissolved organic compounds, which include carbohydrates, amino acids, proteins, fats and acids (Abdel-Raouf et al. 2012). Two criteria are very important to quantify the concentration of organic matters in sewage, one is the ability to oxidize and another is the carbon content (Rogers 1996). Oxidation of organic compounds is responsible for the increase of chemical oxygen demand (COD) of sewage, and carbon content is measured by total organic carbon (TOC) test (Rogers 1996).

2.4 Inorganic Compounds

The inorganic compounds of sewage include salts of sodium, potassium, calcium, magnesium, sulphur, chlorine, bromine, phosphate, bicarbonate, ammonia, nitrate, nitrite and phosphate (Horan 1989). Increase in inorganic compounds in sewage increases the conductivity, salinity and total dissolved solids (TDS) concentration of the sewage (Abdel-Raouf et al. 2012).

2.5 Toxic Elements

Sewage contains some toxic materials and heavy metals such as cadmium, lead, mercury, scandium, chromium, copper, arsenic and cyanide. These materials causes pollution in water and many health problems when present in high concentration (Webber 1972). Sometimes certain toxic pharmaceutical wastes are also mixed with urban wastewater (Gentili and Fick 2017).

2.6 Microorganisms

Sewage provides an ideal environment for the growth of a wide range of microorganisms, and many of them are pathogenic. Mostly the coliform bacteria and parasites cause stomach upset and dysentery (Hamner et al. 2006). Other diseases such as cholera, typhoid and tuberculosis are also waterborne (Rudolfs et al. 1950). The removal of coliform bacteria is not at all easy, and generally, sewage is chlorinated for disinfection and removal of total coliform organisms (Sebastian and Nair 1984). Microbes are also responsible for producing fouling smell by decomposing organic matters. Dissolved oxygen consumed by them resulted in high biological oxygen demand (BOD) of water harmful to aquatic flora and fauna (Wagner and Loy 2002). As per 'Indian National Urban Sanitation Policy', the desirable discharge of faecal coliforms onto land and water is 10^2 CFU/100 ml water, and maximum permissible is 10^3 CFU/100 ml water (Anonymous 2008). Arceivala and Asolekar (2010) described the range of microorganisms contributed by a person per capita per day (Table 1).

3 Physico-Chemical Parameters of Sewage

Physico-chemical parameters such as colour of water, temperature, pH, dissolved oxygen (DO) concentration, conductivity, total dissolved solids (TDS), salinity, biological oxygen demand (BOD), chemical oxygen demand (COD), nitrogen (N) and phosphorus (P) content and presence of heavy metals in water reflect the quality of sewage. There are different guidelines available for standard value of these parameters (Gehm 1945; Patil et al. 2012; Abdel-Raouf et al. 2012; Singh et al. 2015). Singh et al. (2015) have described the overall water quality index (OWQI) for groundwater in Indian context according to the Indian Standards (IS 10500: 1991) and Central Pollution Control Board (CPCB) standards (Table 2).

Physico-chemical parameters of sewage depend on human activities for domestic purposes such as cooking, bathing, laundry and others. Arceivala and Asolekar (2010) estimated the contribution of waste materials in grams per capita per day (Table 3).

3.1 Colour and Odour

Sewage has a fouling smell due to microbial activities. Fresh sewage is cloudy in appearance and becomes dark in colour with time and slightly soapy due to the presence of different salts (Vignesh et al. 2006).

3.2 Temperature

Temperature indicates the solubility of oxygen in sewage (Vignesh et al. 2006). Dissolved oxygen concentration is inversely proportional to the temperature (Barbosa and Sant'Anna Jr 1989). Furthermore, algal growth and treatment system also vary with temperature. The temperature of raw sewage in India depends on seasonal conditions generally about 15–35 °C (Barbosa and Sant'Anna Jr 1989).

Table 1 Contribution of microorganisms by a person per capita per day in sewage (Arceivala and Asolekar 2010)

Microorganisms	Range (CFU/100 ml of sewage/capita/day)
Total bacteria	10^9–10^{10}
Coliforms	10^9–10^{10}
Faecal streptococci	10^5–10^6
Salmonella typhosa	10^1–10^4
Protozoan cysts	Up to 10^3
Helminthic eggs	Up to 10^3
Virus (plaque-forming units)	10^2–10^4

Table 2 Water quality criteria proposed by Indian Standards (IS 10500: 1991) and Central Pollution Control Board (CPCB), India

Sl. no.	Parameters	Water quality				
		Excellent	Good	Fair	Poor	Heavily polluted
1.	Colour (Hazen unit)	10	15	50	175	>175
2.	Turbidity (NTU)	5	10	25	250	>250
3.	pH	6.5–8.5	6.0–9.0	5.5–9.5	<5.5–>9.5	<5.5–>9.5
4.	DO (mg/L)	8	6	4	2	<2
5.	TDS (mg/L)	500	1000	1500	3000	>3000
6.	BOD (mg/L)	2	3	5	7	>7
7.	Total hardness (mg/L)	<300	400	500	600	>600
8.	Nitrate (mg/L)	10	20	50	100	>100
9.	Total phosphate (mg/L)	0.02	0.16	0.4	0.65	>0.65
10.	Sulphate (mg/L)	25	150	250	400	1000
11.	Chloride (mg/L)	200	250	600	800	>800
12.	Fluoride (mg/L)	0.7–1.5	1.6	1.7	2	>2
13.	Iron (mg/L)	0.1	0.3	0.5	1	>1
14.	Arsenic (mg/L)	0.005	0.01	0.05	0.1	0.2
15.	Total coliform (MPN)	50	500	5000	50,000	>50,000

Table 3 Contribution of human wastes in grams per capita per day (Arceivala and Asolekar 2010)

Physico-chemical parameters	Range (g/capita/day)
BOD	45–54
COD	1.6–1.9 times of BOD
TOC	0.6–1.0 times of BOD
Total N	6–12
Organic N	~0.4 of total N
Free ammonia	~0.6 of total N
Nitrate	~0.0–0.5 of total N
Total P	0.6–4.5
Organic P	~0.3 of total P
Inorganic P	~0.7 of total P

3.3 pH

The concentration of hydrogen ion present in sewage is expressed as pH of sewage. It has been observed that pH of the freshly released sewage is slightly higher than the tap water supplied to cities; further decomposition of organic matters lowers the pH of sewage (Vignesh et al. 2006). pH of sewage generally ranges from 5.5 to 8.0, whereas pH 6.5–8.5 is described as the standard value for excellent water quality (Table 2) (Singh et al. 2015). Algae have the ability to increase pH of sewage (Moss 1973).

3.4 Dissolved Oxygen (DO)

Most of the living organisms are depending on oxygen (O_2) to maintain the metabolic processes (Hvitved-Jacobsen 1982). DO is important for oxidization and precipitation of inorganic compounds in water (Vignesh et al. 2006). The value of DO depends on physical, chemical and biological activities carried out in the water (Hvitved-Jacobsen 1982). Under normal atmospheric pressure, the solubility of O_2 in fresh water is about 14.6 mg/L at 0 °C and 7.0 mg/L at 35 °C (Abdel-Raouf et al. 2012). The water with more than 6 mg/L DO is described as good quality water (Table 2) (Singh et al. 2015). Analysis of DO is an important test regarding water pollution control. In wastewaters DO determines whether the biological changes and value of BOD are brought out by aerobic or anaerobic organisms.

3.5 Conductivity

Conductivity is the rapid measurements of the concentration of ionized substances in the water which produce an electric current (Rhoades 1996). In other words it is the rapid estimation of dissolved minerals in water. The value of conductivity is the multiplication of conductance (in mS/cm) by an empirical factor, which may vary from 0.55 to 0.90 depending on the soluble compounds of water (Abdel-Raouf et al. 2012).

3.6 Total Dissolved Solids (TDS)

Total dissolved solids include both filterable and nonfilterable solids. Different types of solids are found in water. According to Indian Standards (IS 10500: 1991) and CPCB norms, more than 3000 mg/L TDS is described as highly polluted water

(Table 2) (Singh et al. 2015). Analysis of total dissolved solids is important to decide the water quality and its treatment processes. Solids affect water quality in many ways (Rhoades 1996).

3.7 Salinity

Salinity is the measurement of dissolved mineral salts such as calcium, magnesium, sodium, potassium, sulphate, chloride, etc. All water supplies naturally contain some salts, but wastewater often contains more. Salts in domestic sewage are contributed by detergents, softeners, cleaning products, bathing soaps and shampoos. Mostly the dissolved salts cannot be effectively removed by conventional treatments. Too much of salts hampers the ecosystem and badly affects the aquatic life. Therefore, it is important to take simple measures to decrease the salinity in wastewater (Rhoades 1996).

3.8 Biological Oxygen Demand (BOD)

BOD is the amount of dissolved oxygen consumed by the aerobic organism for metabolism. Dissolved oxygen is mainly consumed by aerobic bacteria for metabolizing the organic matter. It is also required for the respiratory demand of all aerobes. Excess utilization of dissolved oxygen increases the BOD leading to the death of aquatic life due to anaerobiosis. Thereby the reduction of BOD is one of the major concerns of wastewater treatment (Townsend et al. 1992). The range of BOD observed for raw sewage in average Indian cities is 100 to 400 mg/L (Choksi et al. 2015) which is extremely high according to the standard given by Indian Standards (IS 10500: 1991) and Central Pollution Control Board (CPCB), India (Table 2) (Singh et al. 2015).

3.9 Chemical Oxygen Demand (COD)

The chemical oxygen demand is the measure of chemical oxidation of organic matters. Sewage comprises a large number of organic compounds. The carbon atoms of organic compounds are oxidized to produce carbon dioxide. The COD of raw sewage at various places in India is within the ranges of 200–700 mg/L (Choksi et al. 2015). The COD/BOD ratio for the sewage is observed around 1.7 (Orhon et al. 1994).

3.10 Nitrogen and Phosphorus

Domestic discharge contains the higher amount of nitrogenous compounds such as proteins, amino acids, amines and urea. After metabolic interconversion, ammonia (NH4+) is the major nitrogenous breakdown product and to some extent nitrite and nitrate also (Drizo et al. 1997). Generally in raw sewage, nitrogen (N) content was observed to be in the ranges of 20–50 mg/L (Drizo et al. 1997). Nitrate content of 10 mg/L is described as an excellent water parameter by Indian Standards (IS 10500: 1991) and Central Pollution Control Board (CPCB), India (Table 2) (Singh et al. 2015). Generally ammonia is either adsorbed by soil particles or converted to nitrate and gaseous nitrogen by bacteria (Drizo et al. 1997). However excess non-ionized ammonia is harmful to aquatic organisms (Drizo et al. 1997). Furthermore, nitrogenous compounds in sewage lead to overgrowth of aquatic plants and harmful algal blooms (Hallegraeff 1993), which turn into clogging of water, decomposition of organic matters and reduction in dissolved oxygen concentration. N pollution is a causative factor of a health condition called methemoglobinemia and the source of carcinogenic nitrosamines (Tam and Wong 1989; Berger et al. 1990).

Phosphorus (P) in sewage is coming out of food residues and synthetic detergents. Generally, the concentration of P in sewage is within the range of 5–10 mg/L (Westholm 2006). P concentration more than 0.65 mg/L is considered to be highly polluted water according to Indian Standards (IS 10500: 1991) and Central Pollution Control Board (CPCB), India (Table 2) (Singh et al. 2015). N and P both are the key nutrients for eutrophication (Grobbelaar 2004).

3.11 Heavy Metals

Sewage sometimes contains elevated concentration of different heavy metals, as, for example, zinc (Zn), chromium (Cr), nickel (Ni), cadmium (Cd), lead (Pb), copper (Cu) etc., which causes different health hazards (El-Enany and Issa 2000). They are not degraded by the processes of sewage treatment and present significantly in the sewage sludge. Excess presence of heavy metals is highly toxic to aquatic life (El-Enany and Issa 2000). Algae-based system for the removal of heavy metals is also being developed by many scientists (Mehta and Gaur 2005; Romera et al. 2007).

4 Natural Algal Flora in Sewage

Algae are ubiquitous, occurring in all type of habitats. They are tiny microscopic, single-celled to complex multicellular forms. Algae are the primary producers in all kinds of water bodies. Sewage is also an ideal media for algal growth (Abdel-Raouf et al. 2012). A total of 1090 algal taxa are reported as pollution-tolerant algae which belong to 240 genera in which 60 genera with 80 species are organic pollution-tolerant algae. The most tolerant eight genera are *Oscillatoria*, *Chlamydomonas*, *Chlorella*, *Scenedesmus*, *Navicula*, *Nitzschia*, *Stigeoclonium* and *Euglena* (Palmer 1969). Palmer (1974) has also surveyed for naturally growing microalgae from widely distributed waste stabilization ponds.

In summer season of the year 2017, sewage of Bharwara STP was surveyed to find out the natural algal flora. The analysis of algal diversity data revealed that the chlorophycean are the most dominant flora in sewage followed by bacillariophycean, cyanophycean and euglenophycean algae. A total of 23 algal taxa have been identified from sewage, 10 genera from Chlorophyceae, 6 from Bacillariophyceae, 5 from Cyanophyceae and 2 from Euglenophyceae. They are identified as *Ulothrix*, *Kirchneriella*, *Scenedesmus*, *Cosmarium*, *Gonium*, *Ankistrodesmus*, *Spondylomorum*, *Chlorella*, *Pediastrum* and *Chlamydomonas* from class Chlorophyceae; *Oscillatoria*, *Phormidium*, *Lyngbya*, *Merismopedia* and *Chroocococcus* from Cyanophyceae; *Nitzschia*, *Gomphonema*, *Cyclotella*, *Diploneis*, *Navicula* and *Hantzschia* from Bacillariophyceae; and *Euglena* and *Phycus* from Euglenophyceae (Fig. 4). The similar type of observation was also found in six lagoons in Central Asia where Chlorophyceae was dominant both in diversity and abundance followed by Cyanophyceae, Bacillariophyceae and Euglenophyceae (Ergashev and Tajiev 1986).

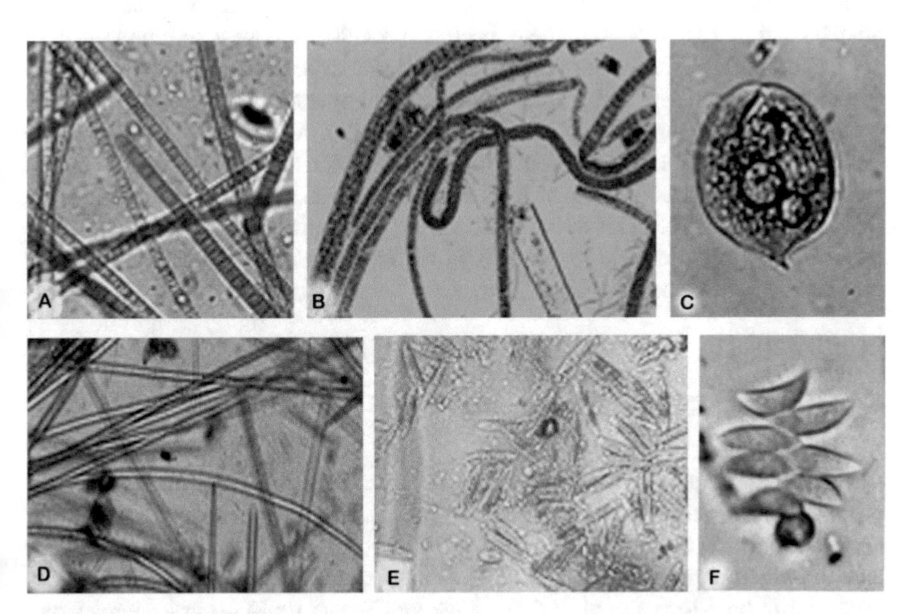

Fig. 4 Naturally occurring algae of Bharwara Sewage Treatment Plant, Lucknow, India. [A. *Oscillatoria* sp., B. *Ulothrix* sp., C. *Phacus* sp., D. *Phormidium* sp., E. *Nitzschia* sp., F. *Scenedesmus* sp.]

5 Potential Microalgae for Sewage Treatment

Biological treatment with microalgae is a fascinating process as they can absorb nutrients (pollutants) from wastewater to produce their own food (de la Noue and De Pauw 1988). This attractive method was launched in 1957 by Oswald and Gotaas (1957). Since then numerous laboratories are involved in pilot studies, and several STPs are using different versions of this process (Shi et al. 2007; Zhu and Liu 2008). Significant interest has been taken by the USA, Mexico, Australia, Thailand and Taiwan to treat wastewater by algae (Borowitzka and Borowitzka 1988; Wang et al. 2008). In India also many scientists are involved in bioremediation of sewage using algae (Sivasubramanian 2006; Sengar et al. 2011; Mahapatra et al. 2014). Microalgae offer a cost-effective, environment-friendly approach to removing pollutants from wastewater (Grobbelaar 2004).

5.1 *Removal of Pathogens*

The disinfection of sewage is measured by the extent of elimination of total coliform organisms from water (Sebastian and Nair 1984). Algae could element the pathogens through competitive growth in sewage. Moawad (1968) observed that the physico-chemical parameters that are favourable for the growth of algae are mostly unfavourable for the survival of bacteria such as *Salmonella* and *Shigella*, viruses, amoeba and protozoa. Among them, bacteria provide the largest community about 10^6 bacteria/ml of wastewater (Horan 1989). Some experimental evidence showed that high-rate ponds are more effective for removal of coliform organisms than conventional treatment system (Parhad and Rao 1976). Malina and Yousef (1964) reported a reduction of 88.8% coliforms in 11.4 days using algae. Many scientists have reported 99% reduction in total coliform counts from sewage using algae (Meron et al. 1965; Shelef et al. 1977; Colak and Kaya 1988). Algae also show disinfecting properties due to increase in pH of the water during photosynthesis process (de la Noue and de Pauw 1988).

5.2 *Improvement in Physico-Chemical Parameters of Water*

Photosynthetic capability of algae help to improve DO concentration, absorb nutrients such as N and P, reduce BOD and COD, and simultaneously convert solar energy into useful biomass (Wang et al. 2010). There is evidence of using *Chlorella* and *Dunaliella* culture for a span of 75 years for mass culturing and wastewater treatment (Abdel-Raouf et al. 2012). Colak and Kaya (1988) observed 68.4% and 67.2% reduction in BOD and COD in domestic wastewater using algal treatment, respectively. They have reported removal of 50.2% N and 85.7% P from industrial wastewater and removal of 97.8% P from domestic wastewater. In India Sengar et al. (2011) collected sewage from the drain from Agra and isolated three most prominent species, *Euglena viridis*, *Gloeocapsa gelatinosa* and *Synedra affinis*, for phycoremediation. Kshirsagar (2013) investigated the efficiency of *C. vulgaris* and *S. quadricauda* on sewage treatment collected from Bopodi of Pune city, India. *C. vulgaris* showed the best result on removal of nitrate and COD while *S. quadricauda* on BOD and phosphate. In Pakistan *Chlorella* and *Scenedesmus* isolated from Kallar Kahar Lake were used for the removal of impurities in wastewater (Ansari et al. 2017). Recently, an effort has been made by the authors to explore the efficiency of two green algae *Scenedesmus* sp. and *Chlorella* sp. as well as two cyanobacteria *Phormidium* sp. and *Oscillatoria* sp. for bioremediation of water of open drains of domestic wastewater disposed to river Gomti, Lucknow (India). Significant improvements are observed in DO, TDS, COD and BOD (Table 4).

Algae could remove N and P within a very short period. Many studies demonstrated the successful removal of nutrients from water rich in N and P (Przytocka-Jusiak et al. 1984; Fierro et al. 2008). A large-scale study was carried out in South Africa for industrial nitrogenous waste removal in high-rate algal ponds followed by harvesting of algal biomass (Bosman and Hendricks 1980). Lau et al. (1996) found *C. vulgaris* has the ability to remove 86% for inorganic N and 70% for inorganic P. González et al. (1997) isolated two green algae *C. vulgaris* and *S. dimorphus*; both strains are highly efficient for ammonia and P removal from agro-industrial wastewater. The microalgae *Chlorella*, *Scenedesmus* and *Arthrospira* are well known for sewage treatment (Lima et al. 2004). Sometimes they have been used in consortia to remove P, N and COD from sewage (Tarlan et al. 2002). Shi et al. (2007) investigated the removal of N and P from wastewater by using a novel method of algal cell immobilization containing *C. vulgaris* and *S. rubescens*. The genus *Phormidium* a cyanobacterium was also found very promising for the removal of nutrients. de la Noue and Basseres (1989) used *Phormidium bohneri*, and Garbisu et al. (1994) used thermophilic *P. laminosum* for the removal of N from the effluents. The thermophilic cyanobacteria can treat the effluent at high temperature, so contamination of other organisms can be avoided (Sawayama et al. 1998).

Though P seems to be safer for human health, removal is required to reduce the eutrophication (Grobbelaar 2004). The uptake of P by cyanobacteria has already been observed in the early 1990s (Garbisu et al. 1993). Li et al. (2011) showed that *Chlorella* sp. grown in 14 days of batch culture could remove ammonia, total N, total P and COD by 93.9%, 89.1%, 80.9% and 90.8%, respectively. In India a team of scientist from Indian Institute of Science, Bangalore, worked on phycoremediation of municipal water of Bellandur Lake, Koramangala region, South of Bangalore. They have fed algae directly the filtered and sterilized municipal wastewater. The nutrient removal efficiencies were found 86%, 90%, 89%, 70% and 76% for TOC, total N, ammonia, total P and orthophosphate, respectively (Mahapatra et al. 2014). Sivasubramanian and his team based on his research work (Vignesh et al. 2006) developed the world's first phycoremediation plant at Ranipet, India, in 2006. Afterwards, many research works have been carried out by this group. They have used *Chlorella vulgaris* as a potential strain for bioremediation (Swetha et al. 2016).

Table 4 Improvement of sewage quality in 10 days of algal cultivation in Gomti River open drainage systems, Lucknow, India

Places		Temp (°C)	pH	DO (mg/L)	Conductivity (µs/cm)	TDS (mg/L)	Salinity (‰)	COD (mg/L)	BOD (mg/L)
Ghaila	On-site	31.6	7.82	1.17	867	425	0.42	72	26
	Blank	23.5	8.51	4.42	748	388	0.37	68	25.8
	Scenedesmus sp.	23.5	8.62	6.38	749	388	0.37	52	20.8
	Chlorella sp.	23.4	8.67	7.37	712	379	0.35	54	21
	Phormidium sp.	23.4	8.73	7.33	725	386	0.36	49	20
	Oscillatoria sp.	23.4	8.68	6.4	751	389	0.37	43	18.2
Bairikala	On-site	32.8	8.24	0.14	2029	1024	1.03	178	57
	Blank	23.5	8.78	6.46	2021	1020	1.03	160	53
	Scenedesmus sp.	23.5	8.84	6.74	1945	980	0.99	114	39.6
	Chlorella sp.	23.5	8.79	6.38	2035	1027	1.04	108	38.4
	Phormidium sp.	23.5	8.74	5.83	2050	1035	1.04	113	40.4
	Oscillatoria sp.	23.4	8.82	6.84	2018	1018	1.03	113	40
Gomti Nagar	On Site	30.7	8.36	5.08	567	275	0.27	37	15
	Blank	23.5	8.24	7.19	435	214	0.21	32	14
	Scenedesmus sp.	23.5	8.4	5.16	499	241	0.24	18	6.8
	Chlorella sp.	23.5	8.55	6.89	433	209	0.21	16	6
	Phormidium sp.	23.4	8.32	5.68	455	220	0.22	20	7.6
	Oscillatoria sp.	23.4	8.73	7.31	495	239	0.24	24	9
Pipraghat	On Site	31.6	8.01	1053	1254	622	0.62	42	17.3
	Blank	23.5	8.56	5.47	1261	626	0.63	37	15.4
	Scenedesmus sp.	23.5	8.98	9.37	1197	593	0.59	23	9.4
	Chlorella sp.	23.5	9.12	9.27	1099	542	0.54	21	6
	Phormidium sp.	23.4	9.05	8.82	1212	600	0.6	26	11
	Oscillatoria sp.	23.4	8.91	8.05	1235	612	0.61	27	11.2

Algae have huge potential to absorb the heavy metals (Afkar et al. 2010; Kumar and Gaur 2011; He and Chen 2014). In living algal cells, these heavy metals are absorbed by active biological transport (Anastopoulos and Kyzas 2015). Bishnoi et al. (2007) reported that *Spirogyra* sp. is capable of accumulating chromium. Among the other algae, the performance of *Scenedesmus* sp. was found far better than the other species because it can uptake and adsorb cadmium and zinc very efficiently (Monteiro et al. 2009, 2011). Studies on isolated two blue-green algae *Nostoc linckia* and *Nostoc spongiaeforme* from salt-affected soils showed high tolerance to chromium (VI) (Kiran et al. 2008). Sydney et al. (2011) observed both *Botryococcus braunii* and *C. vulgaris* are good candidates for nutrient removal from treated sewage, and 79.63% of the N and P was removed after 14 days of culture. Mane and Bhosle (2012) demonstrated the efficiency of metal absorption of two algal species *Spirogyra* sp. and *Spirulina* sp. as well as *C. pyrenoidosa* and *S. obliquus* have proven their efficiency in removing heavy metals (Zhou et al. 2012). Singh et al. (2012) used immobilized *C. minutissima* cells to evaluate the potential of sewage treatment and biosorption of chromium. Cyanobacterial species *Oscillatoria laetevirens* and *Oscillatoria trichoides* are isolated from the polluted environment and determine their chromium removal efficiency from aqueous solutions by Miranda et al. (2012).

6 Factors Affecting the Algal Nutrient Removal

Nutrient uptake by algae not only depends on the availability of minerals; other physico-chemical factors of sewage also contributed to this complex interaction such as pH, temperature, light intensity and other environmental factors (Azov and Shelef 1987; Talbot and de la Noue 1993). For algal growth, optimum range of pH is 7.0–9.0 (Tubea et al. 1981; Munir et al. 2015), optimum temperature is 16 °C to 27 °C. High temperature of water strongly influences growth rates of algae, and higher than 35 °C is lethal for a number of species (Renaud et al. 2002). The optimum light intensity for algal photosynthesis is 3000–5000 lux (Renaud et al. 2002). These factors vary according to the places and in some places it is difficult to grow algae properly. Sometimes, high algal density leads to self-shading and reduction of photosynthetic efficiency (Fogg 1975).

7 Benefits of Microalgal Treatment over Conventional Municipal Sewage Treatment Technology

The conventional sewage treatment process has many bottlenecks which can be overcome by microalgal treatment process (Abdel-Raouf et al. 2012).

- The efficiency of the process varies depending upon the nutrients to be removed. However algae are very efficient to remove all kinds of impurities.

- The conventional process is costly to operate. Algae are naturally growing organism, and treatment process can be more economic.
- Sometimes problem of secondary pollution arises due to the use of chemicals for the conventional treatment, whereas algal treatment is completely environment-friendly as it does not require any hazardous chemical to grow and only utilizes the nutrients present in water.
- During conventional treatment sometimes over-loss of valuable potential nutrients and reduction of DO happened (de la Noüe et al. 1992; Phang 1990). However algal treatment improves the DO concentration by photosynthesis and also uptakes reasonable amount of nutrients from the water.

There are few bottlenecks of microalgal treatment of wastewater which needs to be addressed for implementation of this process universally.

- Sometimes it is difficult to introduce microalgal treatment as a universal method for wastewater treatment because it needs to meet local conditions of that area.
- There is a huge land requirement to construct a high rate algal pond. That's why efforts have been made to develop hyper-concentrated algal culture (de Pauw and Van Vaerenbergh 1983).
- The depth of the pond is also important to get sufficient exposure to sunlight to reduce self-shading.
- Sometimes it is difficult to produce sufficient pure algal inoculum, and growth of the algae is very much area specific as it depends on environmental conditions.
- Harvesting of algal biomass is another major bottleneck of this process (Mohn 1988).

8 Recovery of Algal Biomass from Sewage

The most difficult part is harvesting of algal biomass (Mohn 1980). Many different techniques have been adopted to recover the algal biomass. One of them is a sedimentation technique where the solid and liquid part is separated by gravitational force. Others are filtration method and cell immobilization (Danquah et al. 2009).

8.1 Sedimentation

Different types of settling tank are used for sedimentation. Algal biomass is carried from cultivation tank to settling tank with a pump. Lamella separator settling plates in the settling tank are arranged in such a way that algal biomass is settled by movement of slanted plates (Danquah et al. 2009). Ultimately the biomass is concentrated in a funnel-shaped bottom of the cylindrical tank and easily collected by outflow of water. Sedimentation rate and cell recovery depends on many factors (Danquah et al. 2009; Hattab et al. 2015);

- Algal size and density of the culture.
- The optimum temperature for sedimentation.
- Algal cell age significantly affected the settling rate, increased in stationary growth phase (10–12 days) and low in log phase (4–10 days) (Danquah et al. 2009).
- Time required for settling depends on the density of algae.
- Flocculating agents can be used for settling which add cost to the process.
- External energy is required for pumping.

8.2 *Filtration*

In the filtration method, the liquid part is removed by the filter to accumulate the algal biomass. Membrane filter of different pore sizes is used depending on the microalgae dimension such as macro-filtration (>10 μm), microfiltration (0.1–10 μm), ultrafiltration (0.02–0.20 μm) and reverse osmosis (<0.001 μm) (Ras et al. 2011). Some external pressure across the membrane is also required for liquid to pass through which can be driven by pressure, vacuum or gravity (Hattab et al. 2015).

8.3 *Immobilized Cell System*

Immobilization of microalgae into a matrix improves the harvesting problem to a great extent (Chevalier and de la Noue 1985). This technology is flexible and easy to operate (Mallick and Rai 1993). Many scientists reported higher nutrient removal ability of the same algal species by immobilized algae than the freely suspended cells (de la Noue and Proulx 1988; Hameed 2002).

9 Possible Uses of Harvested Algal Biomass

Microalgae biomass worldwide is utilized for production of numerous value-added products which include bioenergy, pharmaceutical products and nutraceutical products.

Algae grown in wastewater potentially provide a cost-effective and sustainable way of biomass production combining with wastewater treatment (Pittman et al. 2011). Different types of bioenergies can be obtained from algae such as biodiesel, bioethanol, biogases and biohydrogen (Demirbas 2011). They can be grown in sewage or other wastewater, harvested and dried for lipid extraction (Fig. 5). Algal lipid contains a high amount of palmitic (16:0), steric (18:0), oleic (18:1) and linolenic acids (Dasgupta et al. 2015). These fatty acids can be used for several product

Fig. 5 (**a**) Harvested algal biomass, (**b**) sun-drying of algal biomass, (**c**) Soxhlet extraction of lipid, (**d**) unpurified lipid, (**e**) purified lipid

developments majorly in petrochemical industries. Steric and oleic acids can be used for the production of lubricants and fuel additives (Dasgupta et al. 2015). Saturated fatty acids are useful for the production of soap and cosmetics. Some species of marine algae contain a high amount of carbohydrate useful for bioethanol production (Demirbas 2011). Sydney et al. (2011) used treated domestic sewage for algal growth for production of biodiesel. In Virginia (USA, 2008), researchers at Old Dominion University produced biodiesel feedstock by growing algae at STP (www.oilgae.com). Kingsburgh Sewage Project in Durban (South Africa, 2008) focused on growing algae in semi-purified sewage and then converting it into liquid fuel (www.oilgae.com). Besides nutrient removal *Chlorella* sp. has shown high potential for FAME content and biomass productivity (Li et al. 2011). Microalgae are well known as an important source of bioactive compounds and toxins which can be used for pharmaceutical product development (Schwartz et al. 1990). Algae are also well known as a food supplement. Some algae contain a high amount of protein such as *Spirulina* sp. They are rich in pigments and vitamins. They are widely used as fish feed and animal feed (Habib et al. 2008). Linolenic acid is the precursor of highly nutritious omega-3 fatty acids (Bishop and Zubeck 2012). Algae are source of different value-added pigments including chlorophylls, carotenoids and phycobiliproteins which are used for biotechnological applications including food colourant, nutraceuticals, pharmaceuticals and cosmetic products (Wang et al. 2015; Dasgupta 2015).

10 Future Prospective

The recent advances in scientific researches on algae could open a new avenue for environment-friendly wastewater treatment. Conventional sewage treatment processes have many disadvantages which can be overcome by the algal treatment process. Though it has been found that algae are efficient organism for absorption of domestic waste, it has few bottlenecks. The removal of pathogens and pollutants can be improved by the addition of other organisms, helpful bacteria and aquatic plants. The consortia of different organisms might help to resolve the drawbacks of the individuals. In the future genetically modified algae can be introduced to improve the bioremediation of waste and better production of algae. These futuristic options can overcome the obstacle associated with the implementation of the algal treatment of domestic wastewater worldwide.

Acknowledgements We thank the Director of CSIR-NBRI for providing laboratory facilities to Indo-US Science and Technology Forum, New Delhi, for the financial assistance under i-CRAFT project; to CSIR-Scientist's Pool Scheme; to authorities of 345 MLD Bharwara Sewage Treatment Plant, Lucknow; and to Mr. Gurubachan and other members of Algology Laboratory for their cooperation during the study.

References

Abdel-Raouf N, Al-Homaidan AA, Ibraheem IBM (2012) Microalgae and wastewater treatment. Saudi J Biol Sci 19(3):257–275

Afkar E, Ababna H, Fathi AA (2010) Toxicological response of the green alga *Chlorella vulgaris*, to some heavy metals. Am J Environ Sci 6(3):230–237

Anastopoulos I, Kyzas GZ (2015) Progress in batch biosorption of heavy metals onto algae. J Mol Liq 209:77–86

Anonymous (2008) MOEF, Government of India. National Urban Sanitation Policy

Ansari AA, Khoja AH, Nawar A et al (2017) Wastewater treatment by local microalgae strains for CO_2 sequestration and biofuel production. Appl Water Sci 7(7):4151–4158

Arceivala SJ, Asolekar SR (2010) Wastewater treatment for pollution control and reuse. Tata McGraw Hill Education Pvt Ltd, New Delhi

Azov Y, Shelef G (1987) The effect of pH on the performance of high-rate oxidation ponds. Water Sci Technol 19(12):381–383

Barbosa RA, Sant'Anna GL Jr (1989) Treatment of raw domestic sewage in an UASB reactor. Water Res 23(12):1483–1490

Berger MR, Schmähl D, Edler L (1990) Implications of the carcinogenic hazard of low doses of three hepatocarcinogenic N- Nitrosamines. Cancer Sci 81(6–7):598–606

Bishnoi NR, Kumar R, Kumar S, Rani S (2007) Biosorption of Cr (III) from aqueous solution using algal biomass *Spirogyra* spp. J Hazard Mater 145(1–2):142–147

Bishop WM, Zubeck HM (2012) Evaluation of microalgae for use as nutraceuticals and nutritional supplements. J Nutr Food Sci 2(5):1–6

Borowitzka MA, Borowitzka LJ (1988) Micro-algal biotechnology. Cambridge University Press, Cambridge

Bosman J, Hendricks F (1980) The development of an algal pond system for the removal of nitrogen from an inorganic industrial; effluent. In: Proceedings of international symposium on aquaculture in wastewater NIWP. CSIR, Pretoria, pp 26–35
Chevalier P, de la Noue J (1985) Wastewater nutrient removal with microalgae immobilized in carrageenan. Enzym Microb Technol 7(12):621–624
Choksi KN, Sheth MA, Mehta D (2015) To evaluate the performance of sewage treatment plant: a case study of Surat city. Int J Eng Technol 2(8):1076–1080
Colak O, Kaya Z (1988) A study on the possibilities of biological wastewater treatment using algae. Doga Biyoloji Serisi 12:18–29
Danquah MK, Gladman B, Moheimani N, Forde GM (2009) Microalgal growth characteristics and subsequent influence on dewatering efficiency. Chem Eng J 151(1–3):73–78
Dasgupta CN (2015) Algae as a source of phycocyanin and other industrially important pigments. In: Algal Biorefinery: An Integrated Approach. Springer, Cham, pp 253–276
Dasgupta CN, Suseela MR, Mandotra SK et al (2015) Dual uses of microalgal biomass: an integrative approach for biohydrogen and biodiesel production. Appl Energy 146:202–208
de la Noue J, Basseres A (1989) Biotreatment of anaerobically digested swine manure with microalgae. Biol Wastes 29(1):17–31
de la Noue J, de Pauw N (1988) The potential of microalgal biotechnology: a review of production and uses of microalgae. Biotechnol Adv 6(4):725–770
de la Noue J, Proulx D (1988) Biological tertiary treatment of urban wastewaters with chitosan-immobilized *Phormidium*. Appl Microbiol Biotechnol 29(2–3):292–297
de la Noüe J, Laliberté G, Proulx D (1992) Algae and waste water. J Appl Phycol 4(3):247–254
de Pauw N, Van Vaerenbergh E (1983) Microalgal wastewater treatment systems: potentials and limits. In: Internation convention on Phytodepurization and the use of the produced biomass, Parma, pp 211–287
Demirbas MF (2011) Biofuels from algae for sustainable development. Appl Energy 88(10):3473–3480
Drizo AFCA, Frost CA, Smith KA et al (1997) Phosphate and ammonium removal by constructed wetlands with horizontal subsurface flow, using shale as a substrate. Water Sci Technol 35(5):95–102
El-Enany AE, Issa AA (2000) Cyanobacteria as a biosorbent of heavy metals in sewage water. Environ Toxicol Pharmacol 8(2):95–101
Ergashev AE, Tajiev SH (1986) Seasonal variations of phytoplankton in a series of waste treatment lagoons (Chimkent, Central Asia) Part 2: Distribution of phytoplankton numbers and biomass. CLEAN–Soil, Air, Water 14(6):613–625
Fierro S, del Pilar Sánchez-Saavedra M, Copalcua C (2008) Nitrate and phosphate removal by chitosan immobilized *Scenedesmus*. Bioresour Technol 99(5): 1274–1279
Fogg GE (1975) Primary productivity. Chem Oceanogr 2:385–453
Fytili D, Zabaniotou A (2008) Utilization of sewage sludge in EU application of old and new methods - a review. Renew Sustain Energy Rev 12(1):116–140
Garbisu C, Hall DO, Serra JL (1993) Removal of phosphate by foam immobilized *Phormidium laminosum* in batch and continuous flow bioreactors. J Chem Technol Biotechnol 57(2):181–189
Garbisu C, Hall DO, Llama MJ et al (1994) Inorganic nitrogen and phosphate removal from water by free-living and polyvinyl-immobilized *Phormidium laminosum* in batch and continuous-flow bioreactors. Enzym Microb Technol 16(5):395–401
Gehm HW (1945) Characteristics of Sewage. Sewage Work J 17(5):984–987
Gentili FG, Fick J (2017) Algal cultivation in urban wastewater: an efficient way to reduce pharmaceutical pollutants. J Appl Phycol 29(1):255–262
González LE, Cañizares RO, Baena S (1997) Efficiency of ammonia and phosphorus removal from a Colombian agroindustrial wastewater by the microalgae *Chlorella vulgaris* and *Scenedesmus dimorphus*. Bioresour Technol 60(3):259–262
Gray BE (1989) A primer on California water transfer law. Ariz L Rev 31:745–781

Grobbelaar JU (2004) Algal nutrition - mineral nutrition. In: Richmond A (ed) Handbook of microalgal culture: biotechnology and applied phycology. Blackwell Publishing Ltd, Ames, pp 95–115
Habib MAB, Parvin M, Huntington TC et al (2008) A review on culture, production and use of *Spirulina* as food for humans and feeds for domestic animals and fish. Food and agriculture organization of the United Nations. J Clean Prod 91:1–11
Hallegraeff GM (1993) A review of harmful algal blooms and their apparent global increase. Phycologia 32(2):79–99
Hameed MSA (2002) Effect of immobilization on growth and photosynthesis of the green alga *Chlorella vulgaris* and its efficiency in heavy metals removal. Bull Fac Sci Assiut Univ 31(1-D):233–240
Hamner S, Tripathi A, Mishra RK et al (2006) The role of water use patterns and sewage pollution in incidence of water-borne enteric diseases along the Ganges River in Varanasi, India. Int J Environ Health Res 16(2):113–132
Hattab MA, Ghaly A, Hammoud A et al (2015) Microalgae harvesting methods for industrial production of biodiesel: critical review and comparative analysis. J Fundam Renew Energy Appl 5(154):10–4172
He J, Chen JP (2014) A comprehensive review on biosorption of heavy metals by algal biomass: materials, performances, chemistry, and modeling simulation tools. Bioresour Technol 160:67–78
Horan NJ (1989) Biological wastewater treatment systems: theory and operation. Wiley, New York
Hvitved-Jacobsen T (1982) The impact of combined sewer overflows on the dissolved oxygen concentration of a river. Water Res 16(7):1099–1105
Jensen A (1993) Present and future needs for algae and algal products. In: Fourteenth International Seaweed Symposium. Springer, Dordrecht, pp 15–23
Kaur R, Wani SP, Singh AK et al (2012) Wastewater production, treatment and use in India. In National Report presented at the 2nd regional workshop on Safe Use of Wastewater in Agriculture
Kiran B, Rani N, Kaushik A (2008) Chromium (VI) tolerance in two halotolerant strains of *Nostoc*. J Environ Biol 29(2):155–158
Kshirsagar AD (2013) Bioremediation of wastewater by using microalgae: an experimental study. Int J Life Sci Biotechnol Pharm 2(3):339–346
Kumar D, Gaur JP (2011) Chemical reaction and particle diffusion based kinetic modeling of metal biosorption by a *Phormidium* sp. dominated cyanobacterial mat. Bioresour Technol 102(2):633–640
Lau PS, Tam NFY, Wong YS (1996) Wastewater nutrients removal by *Chlorella vulgaris*: optimization through acclimation. Environ Technol 17(2):183–189
Lazarevic D, Aoustin E, Buclet N et al (2010) Plastic waste management in the context of a European recycling society: comparing results and uncertainties in a life cycle perspective. Resour Conserv Recycl 55(2):246–259
Li Y, Chen YF, Chen P et al (2011) Characterization of a microalga *Chlorella* sp. well adapted to highly concentrated municipal wastewater for nutrient removal and biodiesel production. Bioresour Technol 102(8):5138–5144
Lima SA, Raposo MFJ, Castro PM et al (2004) Biodegradation of p-chlorophenol by a microalgae consortium. Water Res 38(1):97–102
Mahapatra DM, Chanakya HN, Ramachandra TV (2014) Bioremediation and lipid synthesis through mixotrophic algal consortia in municipal wastewater. Bioresour Technol 168:142–150
Malina JF, Yousef YA (1964) The fate of coliform organisms in waste stabilization ponds. J Water Pollut Control Fed 36:1432–1442
Mallick N, Rai LC (1993) Influence of culture density, pH, organic acids and divalent cations on the removal of nutrients and metals by immobilized *Anabaena doliolum* and *Chlorella vulgaris*. World J Microbiol Biotechnol 9(2):196–201

Mane PC, Bhosle AB (2012) Bioremoval of some metals by living algae *Spirogyra* sp. and *Spirullina* sp. from aqueous solution. Int J Environ Res 6(2):571–576
Mehta SK, Gaur JP (2005) Use of algae for removing heavy metal ions from wastewater: progress and prospects. Crit Rev Biotechnol 25(3):113–152
Meron A, Rebhun M, Sless B (1965) Quality changes as a function of detention time in wastewater stabilization ponds. J Water Pollut Control 37(12):1657–1670
Miranda J, Krishnakumar G, Gonsalves R (2012) Cr 6+ bioremediation efficiency of *Oscillatoria laetevirens* (Crouan & Crouan) Gomont and *Oscillatoria trichoides* Szafer: kinetics and equilibrium study. J Appl Phycol 24(6):1439–1454
Moawad SK (1968) Inhibition of coliform bacteria by algal population in microoxidation ponds. Environ Health 10(2):106–112
Mohn FH (1980) Experiences and strategies in the recovery of biomass from mass cultures of microalgae biomass; production and use. In: Shelef G, Soeder CJ (eds) National Council for Research and Development, Israel and the Gesellschaft fur Strahlen-und Umweltforschung (GSF), Munich, Germany
Mohn FH (1988) Harvesting of micro-algal biomass. In: Borowitzka MA, Borowitzka LJ (eds) Micro-algal biotechnology. Cambridge U.P., Cambridge, UK, pp 395–414
Monteiro CM, Castro PM, Malcata FX (2009) Use of the microalga *Scenedesmus obliquus* to remove cadmium cations from aqueous solutions. World J Microbiol Biotechnol 25(9):1573–1578
Monteiro CM, Fonseca SC, Castro PM et al (2011) Toxicity of cadmium and zinc on two microalgae, *Scenedesmus obliquus* and *Desmodesmus pleiomorphus*, from Northern Portugal. J Appl Phycol 23(1):97–103
Moss B (1973) The influence of environmental factors on the distribution of freshwater algae: an experimental study: II. The role of pH and the carbon dioxide-bicarbonate system. J Ecol 61(1):157–177
Munir N, Imtiaz A, Sharif N, Naz S (2015) Optimization of growth conditions of different algal strains and determination of their lipid contents. J Anim Plant Sci 25(2):546–553
Orhon D, Artan N, Ateş E (1994) A description of three methods for the determination of the initial inert particulate chemical oxygen demand of wastewater. J Chem Technol Biotechnol 61(1):73–80
Oswald WJ, Gotaas HB (1957) Photosynthesis in sewage treatment. Trans Am Soc Civ Eng 122(1):73–105
Palmer CM (1969) A composite rating of algae tolerating organic pollution. J Phycol 5(1):78–82
Palmer CM (1974) Algae in American sewage stabilization's ponds. Rev Microbiol 5:75–80
Parhad NM, Rao NU (1976) Decrease of bacterial content in different types of stabilization ponds. Indian J Environ Health 18(1):33–46
Patil PN, Sawant DV, Deshmukh RN (2012) Physico-chemical parameters for testing of water – A review. Int J Environ Sci 3(3):1194–1207
Phang SM (1990) Algal production from agro-industrial and agricultural wastes in Malaysia. Ambio 19:415–418
Pittman JK, Dean AP, Osundeko O (2011) The potential of sustainable algal biofuel production using wastewater resources. Bioresour Technol 102(1):17–25
Przytocka-Jusiak M, Duszota M, Matusiak K et al (1984) Intensive culture of *Chlorella vulgaris* as the second stage of biological purification of nitrogen industry wastewaters. Water Res 18(1):1–7
Ras M, Lardon L, Bruno S et al (2011) Experimental study on a coupled process of production and anaerobic digestion of *Chlorella vulgaris*. Bioresour Technol 102:200–206
Renaud SM, Thinh LV, Lambrinidis G et al (2002) Effect of temperature on growth, chemical composition and fatty acid composition of tropical Australian microalgae grown in batch cultures. Aquaculture 211(1–4):195–214
Rhoades JD (1996) Salinity: electrical conductivity and total dissolved solids. In: Methods of soil analysis part 3, chemical methods. Soil Science Society of America and American Society of Agronomy, USA, pp 417–435

Rogers HR (1996) Sources, behavior and fate of organic contaminants during sewage treatment and in sewage sledges. Sci Total Environ 185(1–3):3–26

Romera E, González F, Ballester A et al (2007) Comparative study of biosorption of heavy metals using different types of algae. Bioresour Technol 98(17):3344–3353

Rudolfs W, Falk L, Ragotzkie RA (1950) Literature review on the occurrence and survival of enteric, pathogenic, and relative organisms in soil, water, sewage, and sludges, and on vegetation: I. Bacterial and virus diseases. Sewage and Industrial Wastes, Water Environment Federation, New Jersey, pp 1261–1281

Sawayama S, Rao KK, Hall DO (1998) Nitrate and phosphate ion removal from water by *Phormidium laminosum* immobilized on hollow fibres in a photobioreactor. Appl Microbiol Biotechnol 49(4):463–468

Schwartz RE, Hirsch CF, Sesin DF et al (1990) Pharmaceuticals from cultured algae. J Ind Microbiol 5(2–3):113–123

Sebastian S, Nair KVK (1984) Total removal of coliforms and *E. coli* from domestic sewage by high-rate pond mass culture of *Scenedesmus obliquus*. Environ Pollut A - Ecol Biol 34(3):197–206

Sengar RMS, Singh KK, Singh S (2011) Application of phycoremediation technology in the treatment of sewage water to reduce pollution load. Ind J Sci Res 2(4):33–39

Sheehan J, Dunahay T, Benemann J et al (1998) Look back at the US department of energy's aquatic species program: biodiesel from algae; close-out report (No. NREL/TP-580-24190). National Renewable Energy Lab., Golden, CO. (US)

Shelef G, Moraine R, Meydan A, Sandbank E (1977) Combined algae production-wastewater treatment and reclamation systems. In: Microbial energy conversion. American Society for Microbiology, USA, pp 427–442

Shi J, Podola B, Melkonian M (2007) Removal of nitrogen and phosphorus from wastewater using microalgae immobilized on twin layers: an experimental study. J Appl Phycol 19(5):417–423

Singh SK, Bansal A, Jha MK et al (2012) An integrated approach to remove Cr (VI) using immobilized *Chlorella minutissima* grown in nutrient rich sewage wastewater. Bioresour Technol 104:257–265

Singh S, Ghosh NC, Krishan G et al (2015) Development of an overall water quality index (OWQI) for surface water in Indian context. Curr World Environ 10(3):813–822

Sivasubramanian V (2006) Phycoremediation of industrial effluents. http://phycoremediation.in/projects.html

Sivasubramanian V, Subramanian VV, Muthukumaran M et al (2012) Algal technology for effective reduction of total hardness in wastewater and industrial effluents. Phykos 42(1):51–58

Swetha C, Sirisha K, Swaminathan D, Sivasubramanian V (2016) Study on the treatment of dairy effluent using *Chlorella vulgaris* and production of biofuel (Algal treatment of dairy effluent). Biotechnol Ind J 21(1):12–17

Sydney ED, Da Silva TE, Tokarski A et al (2011) Screening of microalgae with potential for biodiesel production and nutrient removal from treated domestic sewage. Appl Energy 88(10):3291–3294

Talbot P, de la Noue J (1993) Tertiary treatment of wastewater with *Phormidium bohneri* (Schmidle) under various light and temperature conditions. Water Res 27(1):153–159

Tam NFY, Wong YS (1989) Wastewater nutrient removal by *Chlorella pyrenoidosa* and *Scenedesmus* sp. Environ Pollut 58(1):19–34

Tarlan E, Dilek FB, Yetis U (2002) Effectiveness of algae in the treatment of a wood-based pulp and paper industry wastewater. Bioresour Technol 84(1):1–5

Townsend SA, Boland KT, Wrigley TJ (1992) Factors contributing to a fish kill in the Australian wet/dry tropics. Water Res 26(8):1039–1044

Tubea B, Hawxby K, Mehta R (1981) The effects of nutrient, pH and herbicide levels on algal growth. Hydrobiol 79(3):221–227

Vignesh MS, Shivsankar R, Rao PR et al (2006) Phycoremediation of effluent from tannery and pharmaceutical industries – a lab study. Ind Hydrobiol 9(1):51–60

Wagner M, Loy A (2002) Bacterial community composition and function in sewage treatment systems. Curr Opin Biotechnol 13(3):218–227
Wang B, Li Y, Wu N, Lan CQ (2008) CO_2 bio-mitigation using microalgae. Appl Microbiol Biotechnol 79(5):707–718
Wang L, Min M, Li Y et al (2010) Cultivation of green algae *Chlorella* sp. in different wastewaters from municipal wastewater treatment plant. Appl Biochem Biotechnol 162(4):1174–1186
Wang HMD, Chen CC, Huynh P et al (2015) Exploring the potential of using algae in cosmetics. Bioresour Technol 184:355–362
Webber J (1972) Effects of toxic metals in sewage on crops. Water Pollut Control 71(4):404–413
Westholm LJ (2006) Substrates for phosphorus removal – potential benefits for on-site wastewater treatment. Water Res 40(1):23–36
Woertz I, Feffer A, Lundquist T et al (2009) Algae grown on dairy and municipal wastewater for simultaneous nutrient removal and lipid production for biofuel feedstock. J Environ Eng 135(11):1115–1122
Zhou GJ, Peng FQ, Zhang LJ et al (2012) Biosorption of zinc and copper from aqueous solutions by two freshwater green microalgae *Chlorella pyrenoidosa* and *Scenedesmus obliquus*. Environ Sci Pollut Res 19(7):2918–2929
Zhu Z, Liu J (2008) Remote monitoring system of urban sewage treatment based on Internet. In: Automation and logistics, 2008. ICAL 2008. IEEE International Conference on IEEE, pp 1151–1155

Phycoremediation of Petroleum Hydrocarbon-Polluted Sites: Application, Challenges, and Future Prospects

Pankaj Kumar Gupta, Shashi Ranjan, and Sanjay Kumar Gupta

1 Introduction

The expanded generation, transport, utilization, and transfer of mixed petrochemicals have made these one of the main contaminants in nature (Margesin 2000). Aromatic mixes and their quality in the soil and groundwater framework is of great concern because these contaminants exist as discrete stages that cause danger in the long term to downstream receptors. The most extreme permissible levels for these mixes in drinking water are 1 ppm for the benzene, toluene, ethylbenzene, and xylene (BTEX pollutants (USEPA 2006). Although hydrocarbon pollutants do not form a homogeneous mixture when mixed with water, their solubility is a few levels in extent greater than the admissible limit for drinking water (Voudrias and Yeh 1994). Some of the traditional remediation methods such as pump-and-treat, air sparging, booming and skimming used for the removal of these mono-aromatic pollutants are not only typically costly but also destroy the native biota capable of natural bioremediation (Basu et al. 2015; Gupta et al. 2013). Moreover, these techniques might result in incomplete mass removal or toxicity and are often not feasible in remote locations (Olson and Sale 2015; Van Stempvoort and Biggar 2008). The other promising technique is bioremediation in which microbes degrade contaminants into harmless products. This natural bioremediation is safer and less disruptive than some of the conventional technologies; however, it takes quite a long time to restore the polluted site under prevailing environmental conditions (Abhishek et al. 2018a, b). Therefore, engineered bioremediation techniques such as biostimulation, and phyto/phycoremediation, are gaining popularity because of their faster

P. K. Gupta, PhD (✉) · S. Ranjan
Indian Institute of Technology Roorkee, Department of Hydrology,
Roorkee, Uttarakhand, India

S. K. Gupta
Environmental Engineering, Department of Civil Engineering,
Indian Institute of Technology – Delhi, New Delhi, Delhi, India

S. K. Gupta, F. Bux (eds.), *Application of Microalgae in Wastewater Treatment*,
https://doi.org/10.1007/978-3-030-13913-1_8

rates of remediation. In designed bioremediation, the microbial populations and their encompassing ecological conditions are in fact altered to hasten the procedure of biodegradation.

Phycoremediation offers cost-effective, nonintrusive, and safe cleanup technology in which the potential macro- or microalgae are used to treat a large group of pollutants. These photoautotrophic species act as ecological biotransformers to pollutants originating from wastewater discharge. Algae contain chlorophyll, which transforms sunlight and CO_2 into chemical energy for its growth. Most algae grow comparatively faster than other plants, resulting in faster removal/biotransformation of pollutants by easier uptake of water and nutrients/pollutants. Algae can survive better under nutrient stress and limited conditions by fixing atmospheric nitrogen-N. Overall, the potential treatment of polluted sites using algae is more sustainable for natural resource management.

2 Petroleum Hydrocarbons: Source, Types, Toxicity

Hydrocarbons are natural mixes containing hydrogen and carbon, which are ordered as aliphatic (straight chain) and aromatic (cyclic). Crude oil produces the hydrocarbon products, and leakage during production may cause subsurface contamination (Gupta and Yadav, 2017a,b; Gupta et al. 2017; Gupta et al. 2018b; Gupta et al. 2019). General subsurface hydrocarbon pollutants include BTEX, naphthalene and fluorine, and other constituents of petroleum products (Kumar and Kumari 2015). Refinery and industry wastewater is the main source of hydrocarbons in the subsurface environment (Fuentes et al. 2014). The negative impact of hydrocarbons on the environment results from the impeding effect of the oil layer to the flow of soil moisture, nutrients, and oxygen to the subsurface system (Fuentes et al. 2014). Soil contamination with these pollutants is a global concern because of the heavy pollution load to groundwater (drinking water), which causes ecological toxicity (Hentati et al. 2013; Wang et al. 2008). Moreover, such pollutants are slowly degraded and thus remain as long-term residual chemicals in the subsurface soil layers (Margesin et al. 2000; Kumar and Kumari 2015). The exposure of humans to hydrocarbons occurs in three main ways (Tormoehlen et al. 2014): (1) accidental ingestion, (2) dermal or work-related/industrial exposures, and (3) drinking water and food. Chemicals like hydrocarbons can be toxic if exposed/inhaled at more than the limit amounts because these alter metabolism (Tormoehlen et al. 2014). Figure 1 present the petrochemical spill in subsurface from underground storage tank.

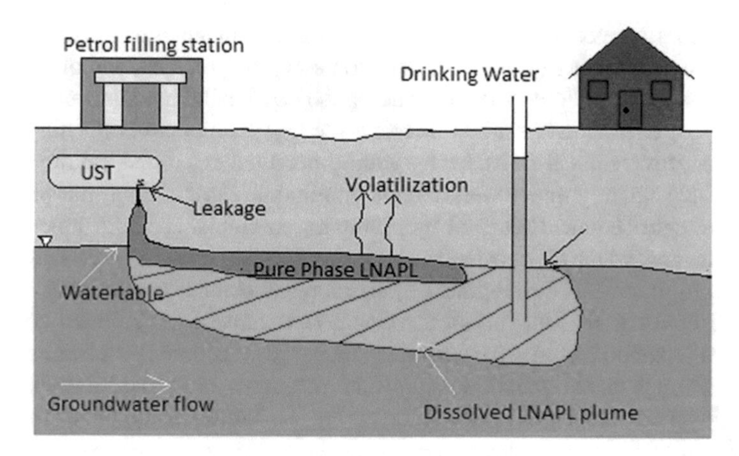

Fig. 1 Schematic diagram of light nonaqueous-phase liquids (LNAPL) leakage from underground storage tanks (UTS) and its migration in the subsurface

3 Engineered Bioremediation Techniques: Comparative Accounting

Bioremediation is a developing innovation that holds extraordinary promise for the practical removal of a wide assortment of natural contaminants. Fruitful uses of bioremediation have been recorded for some locales debased by dangerous oil hydrocarbons. Bioremediation offers some favorable circumstances and refinements in contrast with conventional locale remediation methodologies, for example, pump-and-treat or soil removal followed by other physicochemical remediation procedures. Two types of bioremediation are practiced, generally referred to as in situ and ex situ. The methods by which polluted sites are remediated in place are known as in situ, whereas ex situ practices include the elimination of the contaminated natural resources at a distance away from the site. In situ bioremediation of aquifers contaminated by petrochemicals has been in use for more than four decades and is mostly reliant on native microbes to reduce the pollutants. In this technique, no further extraction of natural resources is needed to make it more effective; therefore, it is distinguished by non-disruption of soil morphology. Preferably, in situ methods should to be less costly compared to ex situ techniques, as no supplementary cost is essential for mine practices. However, the cost of the plan and the on-site setting up of some advanced tools to increase bacterial growth is of key concern. The strength of in situ bioremediation can be enhanced by biostimulation and phytoremediation, while other methods do not require any form of enrichment. The native microbial population is specialized to degrade the pollutant visibly for a long time, but cleanup takes a significantly longer time.

Engineered bioremediation has emerged to accelerate bioremediation by modifying the environmental conditions and native microflora. This advanced bioremediation influences the microbial activities and their neighboring environmental conditions by accelerating biostimulation and bioaugmentation. Biostimulation is enhanced by the adding of nutrients, electron acceptors, oxygen, and other relevant compounds to the polluted sites, enhancing the co-metabolic actions of the microflora. Bioaugmentation is a microorganism-seeding practice for cultivating the volume of a petrochemical degrader by adding potential microbial cultures that are grown independently under well-defined conditions. Furthermore, the plants also accelerate petrochemical removal by promoting microbial reestablishment in polluted soils and water by constantly delivering oxygen by root-zone aeration and nutrients for microbial development by fixation and exudation. Similarly, the technique of constructed wetlands is a concurrent treatment for polluted soil–water resources (Azubuike et al. 2016). In this chapter, the art pertaining to bioremediation of petrochemical polluted soil–water resources is presented with special emphasis on engineered bioremediation strategies. Table 1 listed the different bioremediation techniques for petrochemicals in subsurface.

Plant-assisted bioremediation refers to the use of selected plant species for the targeted pollutants to mitigate the toxic effects and removal of pollutant mass from the subsurface. This technique uses the plant–geochemical interaction to modify the polluted site and also supply micronutrients, oxygen, etc. to the subsurface level for better performance of petrochemical degraders on targeted pollutants (Susarla et al. 2002). Petrochemicals are mostly removed by rhizoremediation mechanisms of such plants as *Canna generalis* (Basu et al. 2015). The plant–geochemical interaction enhances the (1) physicochemical properties of polluted sites, (2) nutrient supply (releasing root exudates) (Shimp et al. 1993), (3) aeration by transfer of oxygen (Burken and Schnoor 1998), (4) the movements of chemicals, and (5) plant enzymatic transformation that is resistant to the migration of contaminants (Narayanan et al. 1998). Similarly, the plant–microbe interactions increased (1) mineralization

Table 1 Summary of different integrated bioremediation strategies and associated dominant bio-agents for remediation of hydrocarbon-polluted sites

Integrated strategy	Dominant bio-agents	References
Bioaugmentation	*Dehalococcoides* sp., *Azoarcus* sp., *Corynebacterium variabilis*, *Pseudomonas aeruginosa*, *Pseudomonas putida*, *Rhodococcus* sp., *Mycobacterium* sp., *Pseudomonas* sp., *Pseudoxanthomonas spadix*, *Cladophialophora* sp.	Karamalidis et al. (2010)
Phytoremediation	*Juncus subsecundus*, *Canna generalis*, *Scirpus grossus Polygonum aviculare*, *Mirabilis jalapa*	Yadav et al. (2013)
Biostimulation with microbial seeding	*Bacillus* sp., *Pseudomonas* sp., *Bacillus* sp., *Trichoderma* sp., *Candida catenulata*, *Dehalococcoldes* sp., *Pseudomonas* sp., *Achromobacter* sp., *Xanthomonas* sp., *Alcaligenes* sp., *Acinetobacter* sp., *Baumannii* sp., *Bacillus* sp.	Sarkar et al. (2005)
Plant-enhanced biostimulation	*Galega orientalis*, *Rhizobium* sp., *Scorzonera mongolica*, *Atriplex centralasiatica*, *Limonium bicolor*, *Lolium perenne*, *Typha domingensis*, *Vetiver* (grass), *Achromobacter xylosoxidans*	Shehzadi et al. (2014)

in the rhizosphere and (2) the numbers of degraders and a shorter lag phase until disappearance of the compound. Selection of plants having deep root systems can improve subsurface aeration, which maintains the oxygen level in the deep vadose zone. The root exudates, dead root hairs, and fine roots serve as important sources of carbon for microbial growth (Shimp et al. 1993). The root exudates also accelerate the enzyme synthesis of microbial metabolisms (Dzantor 2007). Overall, the plants have a crucial role in removal of petrochemical mass, but investigation of the many issues related to the application of plants to petrochemical-polluted sites are required before the implementation of such techniques. Further, the impacts of static and dynamic environmental variables on pollutant removal, and the combination of other bioremediation techniques with plant-assisted bioremediation, need full investigation. A constructed wetlands is a precisely engineered structure using selected plant species for water quality improvement. A wetlands system offers a low-cost alternative technology for remediation of industrially polluted sites (Cottin and Merlin 2008). In the plant-based framework, complex physical, synthetic, and natural procedures may happen at the same time, including volatilization, sorption and sedimentation, phytodegradation, plant uptake and gathering, and microbial debasement (Matamoros et al. 2005).

4 Phycoremediation of Petroleum Hydrocarbons

Natural treatment of these natural defiled assets is attracting expanding interest and, where appropriate, can fill in as a financially useful remediation elective. In this direction, phycoremediation also getting more attention for decontamination of petrochemically polluted sites. Ongoing investigations have hence demonstrated that when appropriate strategies for algal determination and development are utilized, it is conceivable to deliver the oxygen required by acclimatized microorganisms to biodegrade oil hydrocarbons. Furthermore, many studies show effective performance of selected potential algal species including diatoms and phytoplankton in removal of petroleum hydrocarbons. Thus, the integral action of potential algal–bacterial consortia has a significant role in the degradation of such hazardous pollutants (Jacques and McMartin 2009).

Two possible mechanisms, an anaerobic and the oxygenic pathway, have been reported in the literature as involved in the breakdown of petroleum hydrocarbons. The oxygen photosynthetically produced by algae in the presence of sunlight and CO_2 is used by heterotrophic bacteria to oxidize the petroleum hydrocarbons in situ, producing in return the CO_2 needed for algae photosynthesis. In the oxygenic process, architecturally diverse petrochemical contaminants are first transformed into intermediates through a number of peripheral pathways such as *orto* and *meta* cleavage, which are then further directed via some central pathways to the cellular metabolism. In algae–bacterial consortia, electron flow is mediated and transferred to electron carriers such as nicotinamide adenine dinucleotide (NADH). The electron gained is further transferred to an extracellular electron acceptor, being referred to as respiration or reoxidation of NADH. The energy gained in this process can be stored as adenosine triphosphate (ATP) together with NADH, in which NADH is reinvested to maintain the cell growth (Naas et al. 2014). Degradation of petroleum hydrocarbons includes the two enzymatic systems, dioxygenase and monooxygenase, under terminal, β-, and ω-oxidation pathways. For example, during biodegradation of benzene and toluene, the main intermediate products are catechol and 3-methyl catechol, respectively. Subsequently, these intermediate products are mineralized by either the enzyme catechol 1,2-dioxygenase under *ortho*-cleavage or by β-ketoadipate or the enzyme catechol 2,3-dioxygenase under *meta*-cleavage. Finally, the ring is opened, producing the lower molecular weight compounds such as pyruvate and acetaldehyde, which can be further broken down by the tricarboxylic adic (TCA) cycle (Naas et al. 2014; Tsao et al. 1998).

Algae-supported biodegradation of petrochemical contaminants has been scarcely investigated, and the catabolic pathways of biodegradation of these compounds in algae are still largely unknown (Jacques and McMartin 2009). There are only a few studies comparatively evaluating the performance of algae species and algal–bacterial systems for the handling of petroleum hydrocarbon-polluted sites (Semple et al. 1999; Hammed et al. 2016). Initially, Walker et al. (1975a, b) implemented experimentation with the achlorophyllous alga *Prototheca zopfii*, which was established to degrade petroleum hydrocarbons found in Louisiana crude oils. Jacobson and Alexander (1981), having grown *Chlamydomonas* sp. with or without light on acetate, reported a significant degradation of hydrocarbons. Cerniglia et al. (1979, 1980a, b) reported that both Cyanobacteria (blue-green algae) and eukaryotic microalgae were proficient of biodegradation of naphthalene to nontoxic prod-

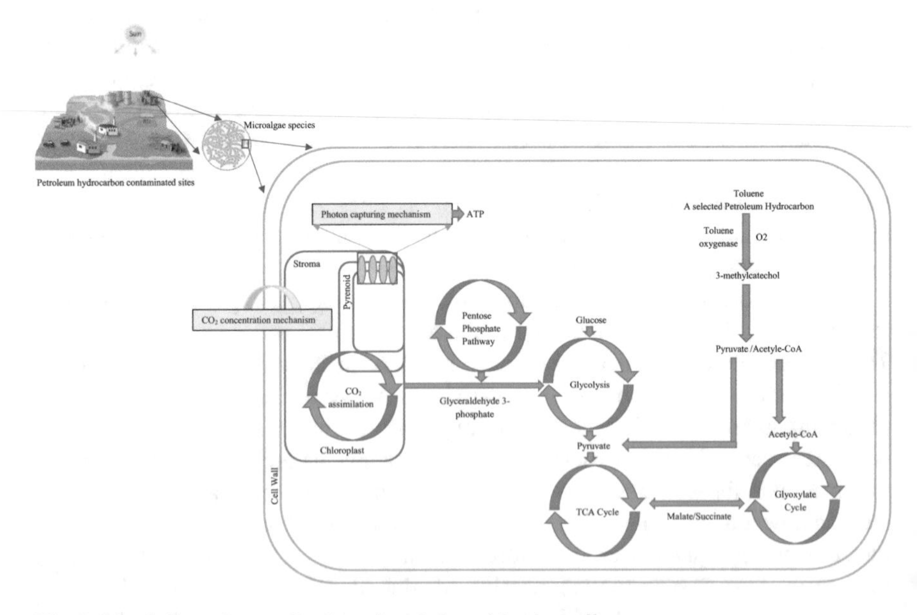

Fig. 2 Metabolic pathway of hydrocarbon (toluene) in alga cell

Table 2 List of potential algae and their role in remediation of hydrocarbon-polluted sites

Algae	Role	References
Selenastrum capricornutum	Bioaccumulation/biotransformation of BTEX, chlorobenzene, 1,2-dichlorobenzene, nitrobenzene, etc.	Semple et al. (1999)
Prototheca zopfii	Biodegradation of total petroleum hydrocarbons (TPH)	Walker et al. (1975a, b)
Chlamydomonas sp.	Biodegradation of petroleum hydrocarbons	Jacobson and Alexander (1981)
Agmenellum quadruplicatum	Biotransformation of naphthalene	Cerniglia et al. (1979, 1980a, b)
Chlamydomonas reinhardtii	Removal of iso-octane-extracted PAHs	Liebe and Fock (1992)
Scenedesmus obliquus	Biotransformation of naphthalene	Luther and Soeder (1987), Luther (1990)
Sphingomonas sp.	Biodegradation of 4,4P- and 2,4-dihalodiphenyl ethers	Schmidt et al. (1993)
Chlorococcum sp.	Removal of diclofop-methyl	Wolfaardt et al. (1994)
Ochromonas danica	Biodegradation of phenolics	Semple and Cain (1996)
Selenastrum capricornutum	Metabolism of benzo[*a*]pyrene	Warshawsky et al. (1988)
Prototheca zopfii	Highlighted as most popular hydrocarbon-degrading microalga	Suzuki and Yamaya (2005), Ueno et al. (2006, 2007), de-Bashan and Bashan (2010)
Scenedesmus obliquus *Nitzschia linearis*	Isolate from Nile River capable of degradation of hydrocarbon pollutants	Ibrahim and Gamila (2004)
Nitzschia sp. *Skeletonema costatum*	Biodegradation of fluoranthene and phenanthrene	Hong et al. (2008)
Scenedesmus obliquus	Microalgal–bacterial consortium for high degradation of hydrocarbons	Tang et al. (2010)
Chlorella sorokiniana	Removal of phenanthrene	Muñoz et al. (2003)
Parachlorella kessleri	Biodegradation of BTEX compounds	Takacova et al. (2015)
Enteromorpha sp. *Cladophora glomerata*	Removal of benzo[*a*]pyrene	Kirso and Irha (1998)

ucts. Liebe and Fock (1992) found that *Chlamydomonas reinhardtii* is capable of removing some of the iso-octane-extracted polyaromatic hydrocarbons (PAHs) from diesel particulate exhaust. Wolfaardt et al. (1994) showed a significant role of a *Chlorococcum* sp. present in an algal–bacterial consortium in removing the diclofop-methyl. Later on, Semple et al. (1999) reviewed and identified several potential microalgae utilizing hydrocarbons as the carbon source. Suzuki and Yamaya (2005), Ueno et al. (2006, 2007), and de-Bashan and Bashan (2010) identified *Prototheca zopfii* as the most popular hydrocarbon-degrading microalga. Ibrahim and Gamila (2004) isolated seven microalgae from the Nile River capable of the degradation of hydrocarbon pollutants based on their affinity. They showed that *Scenedesmus obliquus* had greater affinity toward degradation of hydrocarbons, whereas *Nitzschia linearis* demonstrated better removal of n-alkanes. The biodegradation of fluoranthene and phenanthrene, typical PAHs, was found to be accelerated by *Nitzschia* sp. and *Skeletonema costatum* more than natural attenuation (Hong et al. 2008; Tang et al. 2010). Tang et al. (2010) showed that a microalga–bacterial consortium containing *Scenedesmus obliquus* removed an effective amount of the aromatic hydrocarbons of crude oil. Similarly, Muñoz et al. (2003) investigated a potential consortium of *Chlorella sorokiniana* and *Pseudomonas migulae* to remove phananthrene and observed its effective elimination under photosynthetic conditions without external supply of oxygen. Hammed et al. (2016) reviewed the literature related to the phycoremediation capabilities of potential microalgae and well-identified growth regimes, metabolic pathways, and factors affecting its performance.

Based on the available literature, a biodegradation pathway of toluene, a selected petroleum hydrocarbon, is shown in Fig. 2. Autotrophic microalgae were used to degrade toluene from polluted sites. Figure 2 demonstrates that microalgal pigments, that is, chlorophylls, absorb photons from sunlight and generate electrons by photon-capturing mechanisms (Kruse et al. 2005). ATP is generated by this mechanism and used in the dark reaction to drive biochemical CO_2 digestion through the Calvin Benson cycle (Wang et al. 2014). Toluene is converted into a 3-methylcatechol intermediate by toluene oxygenase and O_2 and finally into pyruvate/acetyl-CoA via the *orto-* and *meta*-cleavage pathways. Later, the pyruvate/acetyl-CoA participates in the TCA and glyoxylate cycle (Hammed et al. 2016) (Table 2).

5 Recent Advances in Phycoremediation Techniques

Implementation of phycoremediation techniques is attracting more attention in recent years, not only for decontaminating polluted sites but also for sequestering atmospheric CO_2 and production of biofuels to achieve environmental and economic sustainability. Algal techniques are advanced by (1) applying molecular or genetic modifications, (2) establishing potential microalgae–bacterial consortia, (3) algae-based biofilm, and (4) designing algal photo-bioreactors.

5.1 Applying Molecular or Genetic Modifications

Genetic engineering approaches are applied to manipulate genes of interest of microalgae to improve its growth and actions. In this regard, reverse and forward types of genetic processes are used to modify algal applications. Reverse genetics refers to identification and manipulation of the gene, whereas forward genetics refers to physical or chemical mutagenesis. In physical mutagenesis, UV/gamma/heavy ion beams are applied to develop a target strain having specified features (de Jaeger et al. 2014). Najafi et al. (2011) also demonstrated mutagenesis using gamma rays. RNA interference (RNAi) or artificial RNA (amiRNA) are recent developments of this process (Hlavová et al. 2015). Radakovits et al. (2010) reported genetic modification of several species including green (Chlorophyta), red (Rhodophyta), and brown (Phaeophyta) algae; diatoms; euglenids; and dinoflagellates. Most of the studies show such genetic transformation of algae species for biofuel production. Genetic modification of potential algae species will also be helpful for remediation of polluted sites.

5.2 *Establishing Potential Microalgae–Bacterial Consortia*

The peer-reviewed literature reported that algae–bacterial consortia are more effective at remediating petroleum hydrocarbons than a single culture of algae, because algae transfer more oxygen to the polluted domain, which helps to accelerate bacterial growth. At the same time, bacteria produce CO_2, which effectively moderates algal growth. Cerniglia et al. (1979) isolated and observed nine Cyanobacteria, five green algae, one red alga, one brown alga, and two diatoms that could oxidize naphthalene. Wolfaardt et al. (1994) showed a significant role of a *Chlorococcum* sp. present in an algal–bacterial consortium to remove diclofop-methyl. Al-Turki AI (2009) reported that consortia of bacteria–algae are capable of adding more oxygen in the aerobic pathway of hydrocarbon degradation. Tang et al. (2010) investigated a microalgal-bacterial consortium of *Scenedesmus obliquus*, which effectively removes aromatic hydrocarbons of crude oil. Recently, many studies have reported microalgae as hosts of distinct bacterial species with mutualistic interactions (Jasti et al. 2005; Sapp et al. 2007; Allgaier et al. 2008). In such consortia, microalgae provide more favorable environmental conditions (Allgaier et al. 2008), nutrients (Amin et al. 2009), iron (Hopkinson et al. 2008), haem (Mouget et al. 1995), and vitamins (Croft et al. 2005) to consume oxygen. Kneip et al. (2007) reported symbiotic Cyanobacteria supplying nitrogen and acting as N_2 fixing for algae growth. These recent advances pooled significant knowledge, but better understanding of biogeochemical interactions of potential microalgae–bacterial consortia is still required for effective remediation practices.

5.3 *Algal Photo-Bioreactors*

Algal photo-bioreactors are a designed open/closed tank system, in which selected algae are produced to remove target pollutants and/or for production of biofuels (Richmond 2004; Carlsson et al. 2007). Generally, algae photo-bioreactors are designed to achieve high efficiency with more biomass concentration in a shorter timeframe (Wang et al. 2008). This advance is also the result of an improved control of farm settings, yielding greater productivity and reproducibility, with less pollution risk, and permitting better collection of potential algal species. These photo-bioreactors are of different shape and size, depending on the interest of cultivation (Ugwu et al. 2008). Generally, however, tube photo-bioreactors, horizontal/serpentine tubular airlift, flat-plated, double/multiple column, and internally illuminated photo-bioreactors are developed for phycoremediation studies. Ugwu et al. (2008) intensively reviewed the different types of algae photo-bioreactors, highlighting associated prospects and limitations. Fernandez et al. (2004) designed a pilot-scale vertical column 0.19 m in diameter and 2 m long with the capacity of 60 l to cultivate *Phaeodactylum tricornutum*. Most of the studies are limited to use of algae production for biofuels; use of such algae photo-bioreactors for in situ phycoremediation is still awaited. Furthermore, optimization of reactor dimension, flow rate, light requirement, culture condition, algae species, reproducibility, and economic value to decontaminate large diverse pollutants are major challenges associated with designing such algae photo-bioreactors.

5.4 Algae-Based Biofilms

Algae biofilms refer to colonized algal communities on a solid surface having a matrix of extracellular substances. Recent renewed interest in algal biofilms has been driven by the need for polluted site remediation strategies, biofuel feedstocks, and effective low-cost biomass harvesting techniques. In general, algal biofilm productivity values from bench- to pilot-scale operations ranged from 0.6 to 31 $g/m^2/day$ with municipal wastewater (Kesaano and Sims 2014). Many studies have been conducted to develop algae-based biofilms to treat wastewater and for removal of nutrients (Roeselers et al. 2008; Mata et al. 2010; Babu 2011; Pittman et al. 2011). Only a few studies are reported as specific for petroleum hydrocarbon treatment. Future research on algal systems needs to focus on (1) identifying the appropriate strains and materials for optimal biofilm formation and growth to remediate petrochemical pollutants, (2) exploring the value of specialty for diverse pollutants, (3) long-duration pilot- and demonstration-scale studies, and (4) the economics and sustainability of algal biofilm systems.

6 Research Challenges and Future Prospects

As each of the disciplines advances and as new cleanup needs arise, opportunities for new phycoremediation techniques will emerge. Until now, the following limitations have restricted the use of phycoremediation to clean up contaminated sites:

6.1 Identification and Characterization of Polluted Sites

The focus has now shifted in favor of using green and sustainable approaches to clean contaminated sites. Thus, the following prospectives are important for identification and characterization of polluted sites.

- Estimation of pollution types, load, and extent will help to establish effective bioremediation techniques in field conditions. For example, Gupta and Yadav (2017) classified applicability of bioremediation techniques based on level and scale of pollution.
- The emphasis must be on practical applications of emerging technologies like nanometerial based treament of the polluted sites (Ranjan et al. 2018), to select site-specific remediation practices (Caliman et al. 2011).
- Hydrogeochemical/geophysical investigation techniques will be introduced to identify and manage effectively the data associated with problematic sites (Ayolabi et al. 2013; Ojo et al. 2014).
- Similarly, application of satellite-based investigations (Kumari et al. 2019), monitoring, and assessment will be more effective to identify petroleum hydrocarbon-polluted sites (Rodell and Famiglietti 2002; Jasmin and Mallikarjuna 2011).
- Application of artificial intelligence and optimization tools will help frame optimal phycoremediation design for petroleum hydrocarbon-polluted sites (Fijani et al. 2013).

6.2 Fate and Transport Behavior of Petroleum Hydrocarbons

Fate and transport are governing processes that control the extend and load of pollution in soil–water systems. Thus, better understanding of the fate and transport of petrochemicals in the subsurface environment will enhance existing knowledge and aid effective implementation of bio-/phycoremediation under varying environmental conditions. The following research challenges are highlighted in the literature:

- Fate (adsorption, volatilization, biodegradation) and transport (advective, diffusive and dispersive flux) studies are required to forecast future pollution load and the time/cost of remediation (Oostrom et al. 2007).
- Most of the fate and transport studies are conducted in the numerical domain (Gupta et al. 2018b; Gupta and Yadav, 2019); thus, demonstration of practical/field experiments will add a new aspect in the near future (Essaid et al. 2015).
- Incorporation of a heterogeneous domain is very rare; thus, consideration of the heterogeneous domain will scale up current results and understanding (Essaid et al. 2015).
- Similarly, consideration of the prevailing site conditions will effectively improve outcomes and be most significant in the selection of proper bioremediation tech-

niques. For example, only a few studies have been reported on petroleum hydrocarbon transport under fluctuating groundwater conditions (Dobson et al. 2007).
- Impact of hydrogeological and meteorological variables are significant for effective implementation of phycoremediation techniques for polluted sites (Gupta and Joshi, 2017; Gupta and Sharma, 2018; Gupta and Yadav 2017).

6.3 Response of Algae/Microbial Community to Polluted Sites

- To maintain optimal biotransformation, a better understanding of the response of algae–microbial communities to polluted site conditions is required, especially climate change conditions (Zhou et al. 2015).
- As novel biotransformation become better understood at the ecological, biochemical, and genetic levels, new bio-/phycoremediation strategies (Mustapha et al. 2018; Gupta et al. 2018c) will become available for various polluted sites (Arnot et al. 2010).
- Identification of potential algae and/or microbial communities may also improve bioremediation effectiveness in meeting cleanup standards (Hammed et al. 2016).
- Effective design of biodegradation systems by supplying nutrients and electron acceptors will expand the capabilities of potential algae–microbe consortia (Yadav and Hassanizadeh 2011; Gupta and Yadav 2017).
- Algae/microbial response to polluted sites is important to design advanced bioremediation applications such as biofilms (Zhou et al. 2015).

6.4 Upgrading Existing Tools and Techniques

- Photo-bioreactors, such as high-rate ponds, are generally used in algae-based remediation, with comparatively low removal of petrochemical pollutants. Thus, it is important to improve these structures for better performance under varying environmental conditions.
- Optimization of longevity, remediation time, and cost using various hybrid simulation-optimization tools will be helpful to design an effective phycoremediation approach to decontaminate polluted sites.

7 Summary

Petrochemicals such as BTEX compounds are anthropogenically released into soil and water resources by accidental spillages and leaks from tankers, pipelines, and storage tanks. Petrochemicals are very mobile in nature and are highly soluble in soil–water systems, creating major subsurface pollution problems. Further, the provision of safe drinking water to the second largest population of the globe is a challenging task for policy makers. Of the physical, chemical, and biological methods applied for the removal of these contaminants, the most environmentally benign option is through bioremediation using microbes and plants/algae. To frame such an engineered bioremediation strategy for these contaminants, it is important to compare all the bioremediation techniques and their effectiveness. In this chapter, the detailed literature on petrochemical sources, toxicity, and fate in the subsurface environment is presented first. Thereafter, a comparative account of different bio-/phytoremediation techniques of polluted sites is discussed. Phycoremediation processes involved in biodegradation of petrochemicals are elaborated next, presenting recent developments. Finally, research challenges and recommendations based on the detailed literature survey are listed for future studies. This chapter may help to implement effective bioremediation techniques in the field to decontaminate petrochemical-polluted sites.

References

Abhishek GPK, Yadav BK, Amandeep TAS, Kataria S, Kumar S (2018a) Phytoremediation of toluene polluted groundwater under nutrient loading using constructed wetlands. poster presentation (B33G-2766) in AGU Fall Meeting 2018 held in Washington DC, USA during 10–14 Dec 2018

Abhishek YBK, Gupta PK (2018b) Morphological variations in unsaturated porous media due to LNAPL contamination. Poster in Japan Geoscience Union (JpGU) Chiba-city, Japan, May 20–24 2018

Allgaier M, Riebesell U, Vogt M, Thyrhaug R, Grossart HP (2008) Coupling of heterotrophic bacteria to phytoplankton bloom development at different pCO_2 levels: a mesocosm study. Biogeosciences 5:1007–1022

Al-Turki AI (2009) Microbial polycyclic aromatic hydrocarbons degradation in soil. Res J Environ Toxicol 3:1–8

Amin SA, Green DH, Hart MC, Küpper FC, Sunda WG, Carrano CJ (2009) Photolysis of iron-siderophore chelates promotes bacterial-algal mutualism. Proc Natl Acad Sci USA 106:17071–17076

Arnot JA, Mackay D, Parkerton TF, Zaleski RT, Warren CS (2010) Multimedia modeling of human exposure to chemical substances: the roles of food web biomagnification and biotransformation. Environ Toxicol Chem 29(1):45–55

Ayolabi EA, Folorunso AF, Kayode OT (2013) Integrated geophysical and geochemical methods for environmental assessment of municipal dumpsite system. Int J Geosci 4:850–862

Azubuike CC, Chikere CB, Okpokwasili GC (2016) Bioremediation techniques–classification based on site of application: principles, advantages, limitations and prospects. World J Microbiol Biotechnol 32:180. https://doi.org/10.1007/s11274-016-2137-x

Babu M (2011) Effect of algal biofilm and operational conditions on nitrogen removal in wastewater stabilization ponds. PhD Dissertation, Wageningen University, Netherlands

Basu S, Yadav BK, Mathur S (2015) Enhanced bioremediation of BTEX contaminated groundwater in pot-scale wetlands. Environ Sci Pollut Res 22(24):20041–20049

Burken JG, Schnoor JL (1998) Predictive relationships for uptake of organic contaminants by hybrid poplar trees. Environ Sci Technol 32(21):3379–3385. https://doi.org/10.1021/es9706817

Caliman FA, Robu BM, Smaranda C, Pavel VL, Gavrilescu M (2011) Soil and groundwater cleanup: benefits and limits of emerging technologies. Clean Techn Environ Policy 13(2):241–268

Carlsson AS, van Beilen JB, Möller R, Clayton D (2007) Micro- and macro-algae: utility for industrial applications. In: Bowles D (ed) Outputs from the EPOBIO: Realising the economic potential of sustainable resources – bioproducts from non-food crops project. CNAP, University of York

Cerniglia CE, Gibson DT, van Baalen C (1979) Algal oxidation of aromatic hydrocarbons: formation of 1-naphthol from naphthalene by *Agmenellum quadruplicatum*, strain PR-6. Biochem Biophys Res Commun 88:50–58

Cerniglia CE, Gibson DT, van Baalen C (1980a) Oxidation of naphthalene by cyanobacteria and microalgae. J Gen Microbiol 116:495–500

Cerniglia CE, van Baalen C, Gibson DT (1980b) Metabolism of naphthalene by the cyanobacterium *Oscillatoria* sp., strain JCM. J Gen Microbiol 116:485–494

Cottin N, Merlin G (2008) Removal of PAHs from laboratory columns simulating the humus upper layer of vertical flow constructed wetlands. Chemosphere 73:711–716

Croft MT, Lawrence AD, Raux-Deery E, Warren MJ, Smith AG (2005) Algae acquire vitamin B12 through a symbiotic relationship with bacteria. Nature 438:90–93

de Jaeger L, Verbeek REM, Draaisma RB (2014) Superior triacylglycerol (TAG) accumulation in starchless mutants of *Scenedesmus obliquus*: (I) mutant generation and characterization. Biotechnol Biofuels 7:69. https://doi.org/10.1186/1754-6834-7-69

de-Bashan LE, Bashan Y (2010) Immobilized microalgae for removing pollutants: review of practical aspects. Bioresour Technol 101:1611–1627

Dobson R, Schroth MH, Zeyer J (2007) Effect of water-table fluctuation on dissolution on and biodegradation of a multi-component, light nonaqueous-phase liquid. J Contam Hydrol 94:235–248

Dzantor EK (2007) Phytoremediation: the state of rhizosphere "engineering" for accelerated rhizodegradation of xenobiotic contaminants. J Chem Technol Biotechnol 82(3):228–232. https://doi.org/10.1002/jctb.1662

Essaid HI, Bekins BA, Cozzarelli IM (2015) Organic contaminant transport and fate in the subsurface: evolution of knowledge and understanding. Water Resour Res 51(7):4861–4902

Fernandez SJ, Ceron GM, Sanchez MA (2004) Pilot-plant-scale outdoor mixotrophic cultures of *Phaeodactylum tricornutum* using glycerol in vertical bubble column and airlift photobioreactors: studies in fed-batch mode. Biotechnol Prog 20:728–736

Fijani E, Nadiri AA, Moghaddam AA, Tsai FTC, Dixon B (2013) Optimization of DRASTIC method by supervised committee machine artificial intelligence to assess groundwater vulnerability for Maragheh–Bonab plain aquifer, Iran. J Hydrol 503:89–100

Fuentes S, Méndez V, Aguila P, Seeger M (2014) Bioremediation of petroleum hydrocarbons: catabolic genes, microbial communities, and applications. Appl Microbiol Biotechnol 98:4781–4794

Gupta PK, Joshi P (2017) Assessing groundwater resource vulnerability by coupling GIS based DRASTIC and solute transport model in Ajmer District, Rajasthan. J Geol Soc India (Springer). https://doi.org/10.1007/s12594-018-0958-y

Gupta PK, Yadav BK, Hassanizadeh SM (2017) Engineered bioremediation of LNAPL polluted soil-water resources under changing climatic conditions. Proceedings of international conference on modeling of environmental and water resources systems (ICMEWRS-2017), HBTU Kanpur, 24–26th March, 2017 (ISBN 978-93-85926-53-2)

Gupta PK, Yadav BK (2017a) Role of climatic variability on fate and transport of LNAPL pollutants in subsurface. Session H060: groundwater response to climate change and variability, AGU fall meeting 2017, New Orleans, USA. (Abstract ID: 220494)

Gupta PK, Yadav BK (2017b) Chapter 8: Effects of climatic variation on dissolution of LNAPL pollutants in subsurface environment. In: Climate change resource conservation and sustainability strategies. DBH Publishers and Distributors, New Delhi. isbn:9789384871086

Gupta PK, Sharma D (2018) Assessments of hydrological and hydro-chemical vulnerability of groundwater in semi-arid regions of Rajasthan, India. Sustain Water Resour Manag:1–15. https://doi.org/10.1007/s40899-018-0260-6

Gupta PK, Abhishek YBK (2018a) Chapter 5: Impact of hydrocarbon pollutants on partially saturated soil media in batch system: morphological analysis using SEM techniques. In: Water quality management. Water science and technology library, vol 79. Springer. isbn:978-981-10-5794-6

Gupta PK, Ranjan S, Kumar D (2018b) Chapter 2: Groundwater pollution by emerging industrial pollutants and its remediation techniques. In: Recent advances in environmental management, vol 1. CRC Press, Taylor & Francis Group. isbn:9780815383147

Gupta PK, Yadav B, Yadav BK (2018c) Transport of LNAPL and biofilm growth in subsurface under dynamic groundwater conditions. C001723-Oral presentation in Japan Geoscience Union (JpGU) Chiba-city, Japan, May 20–24 2018

Gupta PK, Yadav B, Yadav BK (2019) Assessment of LNAPL in subsurface under fluctuating groundwater table using 2D sand tank experiments. ASCE J Environ Eng. https://doi.org/10.1061/(ASCE)EE.1943-7870.0001560

Gupta PK, Yadav BK (2019) Subsurface processes controlling reuse potential of treated wastewater under climate change conditions. In: Water conservation, recycling and reuse: issues and challenges. Springer, Singapore, pp 147–170

Gupta PK, Yadav BK (2017) Bioremediation of non-aqueous phase liquids (NAPLs) polluted soil-water resources Chapter 8. In: Bharagava RN (ed) Environmental pollutants and their bioremediation approaches. CRC Press, Boca Raton

Gupta PK, Shashi R, Yadav BK, (2013) BTEX biodegradation in soil-water system having different substrate concentrations. Int J Eng 2(12):1765–1772

Hammed SK, Prajapati S, Simsek H, Simsek S (2016) Growth regime and environmental remediation of microalgae. Algae 31:189–204. https://doi.org/10.4490/algae.2016.31.8.28

Hentati O, Lachhab R, Ayadi M, Ksibi M (2013) Toxicity assessment for petroleum-contaminated soil using terrestrial invertebrates and plant bioassays. Environ Monit Assess 185:2989–2998

Hlavová M, Turóczy Z, Bišová K (2015) Improving microalgae for biotechnology-from genetics to synthetic biology. Biotechnol Adv. https://doi.org/10.1016/j.biotechadv.2015.01.009

Hong Y-W, Yuan D-X, Lin Q-M, Yang T-L (2008) Accumulation and biodegradation of phenanthrene and fluoranthene by the algae enriched from a mangrove aquatic ecosystem. Mar Pollut Bull 56:1400–1405

Hopkinson BM, Roe K, Barbeau KA (2008) Heme uptake by *Microscilla marina* and evidence for heme uptake systems in the genomes of diverse marine bacteria. Appl Environ Microbiol 74:6263–6270

Ibrahim MBM, Gamila HA (2004) Algal bioassay for evaluating the role of algae in bioremediation of crude oil: II. Freshwater phytoplankton assemblages. Bull Environ Contam Toxicol 73:971–978

Jacobson SN, Alexander M (1981) Enhancement of the microbial dehalogenation of a model chlorinated compound. Appl Environ Microbiol 42:1062–1066

Jacques NR, McMartin DW (2009) Evaluation of algal phytoremediation of light extractable petroleum hydrocarbons in subarctic climates. Remediat J 20:119–132

Jasmin I, Mallikarjuna P (2011) Satellite-based remote sensing and geographic information systems and their application in the assessment of groundwater potential, with particular reference to India. Hydrogeol J 19(4):729–740

Jasti S, Sieracki ME, Poulton NJ, Giewat MW, Rooney-Varga JN (2005) Phylogenetic diversity and specificity of bacteria closely associated with *Alexandrium* spp. and other phytoplankton. Appl Environ Microbiol 71:3483–3494

Karamalidis AK, Evangelou AC, Karabika E, Koukkou AI, Drainas C, Voudrias EA (2010) Laboratory scale bioremediation of petroleum-contaminated soil by indigenous microorganisms and added *Pseudomonas aeruginosa* strain Spet. Bioresour Technol 101(16):6545–6552. https://doi.org/10.1016/j.biortech.2010.03.055

Kesaano M, Sims R (2014) Algal biofilm based technology for wastewater treatment. Algal Res 5:231–240
Kirso U, Irha N (1998) Role of algae in fate of carcinogenic polycyclic aromatic hydrocarbons in the aquatic environment. Environ Toxicol Environ Saf 41:83–89
Kneip C, Lockhart P, Voß CMaier U-G (2007) Nitrogen fixation in eukaryotes: new models for symbiosis. BMC Evol Biol 7:55
Kruse O, Rupprecht J, Mussgnug JH, Dismukes GC, Hankamer B (2005) Photosynthesis: a blueprint for solar energy capture and biohydrogen production technologies. Photochem Photobiol Sci 4:957–970
Kumar H, Kumari JP (2015) Heavy metal lead influative toxicity and its assessment in phytoremediating plants: a review. Water Air Soil Pollut 226:324
Kumari B, Gupta PK, Kumar D (2019) In-situ observation and Nitrate-N load assessment in Madhubani District, Bihar, India. J Geol Soc India (Springer) 93(1):113–118. https://doi.org/10.1007/s12594-019-1130-z
Liebe B, Fock HP (1992) Growth and adaption of the green alga *Chlamydomonas reinhardtii* on diesel exhaust particle extracts. J Gen Microbiol 138:973–978
Luther M (1990) Degradation of di¡erent substituted aromatic compounds as nutrient sources by the green alga *Scenedesmus obliquus*. Dechema Biotechnol Conf 4:613–615
Luther M, Soeder CJ (1987) Some naphthalene sulphonic acids as sulphur sources for the green microalga, *Scenedesmus obliquus*. Chemosphere 16:1565–1578
Margesin R (2000) Potential of cold-adapted microorganisms for bioremediation of oil-polluted alpine soils. Int Biodeterior Biodegradation 46:3–10
Margesin R, Walder G, Schinner F (2000) The impact of hydrocarbon remediation (diesel oil and polycyclic aromatic hydrocarbons) on enzyme activities and microbial properties of soil. Acta Biotechnol 20:313–333
Mata T, Martins A, Caetanao N (2010) Microalgae for biodiesel production and other applications: a review. Renew Sust Energ Rev 14:217–232
Matamoros V, García J, Bayona JM (2005) Behavior of selected pharmaceuticals in subsurface flow constructed wetlands: a pilot-scale study. Environ Sci Technol 39:5449–5454
Mouget JL, Dakhama A, Lavoie MC, Delanoue J (1995) Algal growth enhancement by bacteria – is consumption of photosynthetic oxygen involved. FEMS Microbiol Ecol 18:35–43
Muñoz R, Guieysse B, Mattiasson B (2003) Phenanthrene biodegradation by an algal-bacterial consortium in two-phase partitioning bioreactors. Appl Microbiol Biotechnol 61:261–267
Mustapha IH, Gupta PK, Yadav BK, van Bruggen JJA, Lens PNL (2018) Performance evaluation of duplex constructed wetlands for the treatment of diesel contaminated wastewater. Chemosphere. https://doi.org/10.1016/j.chemosphere.2018.04.036
Naas MH, Acio JA, El Telib AE (2014) Aerobic biodegradation of BTEX: progresses and prospects. J Environ Chem Eng 2(2):1104–1122. https://doi.org/10.1016/j.jece.2014.04.009
Najafi N, Hosseini R, Ahmadi A (2011) Impact of gamma rays on the *Phaffia rhodozyma* genome revealed by RAPD-PCR. Iran J Microbiol 3:216
Narayanan M, Tracy JC, Davis LC, Erickson LE (1998) Modeling the fate of toluene in a chamber with alfalfa plants 1. Theory and modelling concepts. J Hazard Subst Res 1:1–30
Ojo AO, Oyelami CA, Adereti AO (2014) Hydro-geochemical and geophysical study of groundwater in the suburb of Osogbo, South Western Nigeria. J Earth Sci Clim Change 5:205. https://doi.org/10.4172/2157-7617.1000205
Olson MR, Sale TC (2015) Implications of soil mixing for NAPL source zone remediation: column studies and modeling of field-scale systems. J Contam Hydrol 177-178:206–219. https://doi.org/10.1016/j.jconhyd.2015.04.008
Oostrom M, Dane JH, Wietsma TW (2007) A review of multidimensional, multifluid, intermediate-scale experiments: flow behavior, saturation imaging, and tracer detection and quantification. Vadose Zone J 6(3):610–637
Pittman J, Dean A, Osundeko O (2011) The potential of sustainable algal biofuel production using wastewater resources. Bioresour Technol 102:17–25

Radakovits R, Jinkerson RE, Darzins A, Posewitz MC (2010) Genetic engineering of algae for enhanced biofuel production. Eukaryot Cell 9:486–501
Ranjan S, Gupta PK, Yadav BK (2018) Chapter 6: Application of nano-materials in subsurface remediation techniques – challenges and future prospects. In: Recent advances in environmental management, vol 1. CRC Press Taylor & Francis Group. isbn:9780815383147
Richmond A (2004) Principles for attaining maximal microalgal productivity in photobioreactors: an overview. In: Asian Pacific phycology in the 21st century: prospects and challenges. Springer, Dordrecht, pp 33–37
Rodell M, Famiglietti JS (2002) The potential for satellite-based monitoring of groundwater storage changes using GRACE: the high plains aquifer, Central US. J Hydrol 263(1-4):245–256
Roeselers G, van Loosdrecht MCM, Muyzer G (2008) Phototrophic biofilms and their potential applications. J Appl Phycol 20:227–235
Sapp M, Schwaderer AS, Wiltshire KH, Hoppe HG, Gerdts G, Wichels A (2007) Species-specific bacterial communities in the phycosphere of microalgae? Microb Ecol 53:683–699
Sarkar D, Ferguson M, Datta R, Birnbaum S (2005) Bioremediation of petroleum hydrocarbons in contaminated soils: comparison of biosolids addition, carbon supplementation, and monitored natural attenuation. Environ Pollut 136(1):187–195. https://doi.org/10.1016/j.envpol.2004.09.025
Schmidt S, Fortnagel P, Wittich RM (1993) Biodegradation and transformation of 4,4P- and 2,4-dihalodiphenyl ethers by *Sphingomonas* sp. strain SS33. Appl Environ Microbiol 59:3931–3933
Semple KT, Cain RB (1996) Biodegradation of phenolics by *Ochromonas danica*. Appl Environ Microbiol 62:1265–1273
Semple KT, Cain RB, Schmidt S (1999) Biodegradation of aromatic compounds by microalgae. FEMS Microbiol Lett 170:291–300
Shehzadi M, Afzal M, Khan MU, Islam E, Mobin A, Anwar S, Khan QM (2014) Enhanced degradation of textile effluent in constructed wetland system using *Typha domingensis* and textile effluent-degrading endophytic bacteria. Water Res 58:152–159. https://doi.org/10.1016/j.watres.2014.03.064
Shimp JF, Tracy JC, Davis LC, Lee E, Huang W, Erickson LE, Schnoor JL (1993) Beneficial effects of plants in the remediation of soil and groundwater contaminated with organic materials. Crit Rev Environ Sci Technol 23(1):41–77
Susarla S, Medina VF, McCutcheon SC (2002) Phytoremediation: an ecological solution to organic chemical contamination. Ecol Eng 18(5):647–658. https://doi.org/10.1016/S0925-8574(02)00026-5
Suzuki T, Yamaya S (2005) Removal of hydrocarbons in a rotating biological contactor with biodrum. Process Biochem 40:3429–3433
Takacova A, Smolinská M, Semerád M, Matúš P (2015) Degradation of BTEX by microalgae *Parachlorella kessleri*. Pet Coal 57(2):101–107
Tang X, He LY, Tao XQ, Dang Z, Guo CL, Lu GN, Yi XY (2010) Construction of an artificial microalgal bacterial consortium that efficiently degrades crude oil. J Hazard Mater 181:1158–1162
Tormoehlen LM, Tekulve KJ, Agas KA (2014) Hydrocarbon toxicity: a review. Clin Toxicol 52:479–489
Tsao C, Song H, Bartha R (1998) Metabolism of benzene, toluene, and xylene hydrocarbons in soil. Appl Environ Microbiol 64(12):4924
Ueno R, Wada S, Urano N (2006) Synergetic effects of cell immobilization in polyurethane foam and use of thermotolerant strain on degradation of mixed hydrocarbon substrate by *Prototheca zopfii*. Fish Sci 72:1027–1033
Ueno R, Wada S, Urano N (2007) Repeated batch cultivation of the hydrocarbon-degrading, microalgal strain *Prototheca zopfii* RND16 immobilized in polyurethane foam. Can J Microbiol 54:66–70
Ugwu CU, Aoyagi H, Uchiyama H (2008) Photobioreactors for mass cultivation of algae. Bioresour Technol 99:4021–4028

USEPA (2006) In situ and ex situ biodegradation technologies for remediation of contaminated sites, EPA/625/R-06/015

Van Stempvoort D, Biggar K (2008) Potential for bioremediation of petroleum hydrocarbons in groundwater under cold climate conditions: a review. Cold Reg Sci Technol 53:16–41. https://doi.org/10.1016/j.coldregions.2007.06.009

Voudrias EA, Yeh MF (1994) Dissolution of a toluene pool under constant and variable hydraulic gradients with implications for aquifer remediation. Groundwater 32(2):305–311

Walker JD, Colwell RR, Petrakis L (1975a) Degradation of petroleum by an alga, *Prototheca zop¢i*. Appl Microbiol 30:79–81

Walker JD, Colwell RR, Vaituzis Z, Meyer SA (1975b) Petroleum degrading achlorophyllous alga *Prototheca zopfii*. Nature 254:423–424

Wang D, Liu Y, Lin Z, Yang Z, Hao C (2008) Isolation and identification of surfactin producing *Bacillus subtilis* strain and its effect of surfactin on crude oil. Wei Sheng Wu Xue Bao 48(3):304–311

Wang S-K, Stiles AR, Guo C, Liu CZ (2014) Microalgae cultivation in photobioreactors: an overview of light characteristics. Eng Life Sci 14:550–559

Warshawsky D, Radike M, Jayasimhulu K, Cody T (1988) Metabolism of benzo(a)pyrene by a dioxygenase system of the fresh green alga *Selenastrum capricornutum*. Biochem Biophys Res Commun 152:540–544

Wolfaardt GM, Lawrence JR, Robarts RD, Caldwell DE (1994) The role of interactions, sessile growth, and nutrient amendments on the degradative efficiency of a microbial consortium. Can J Microbiol 40:331–340

Yadav BK, Hassanizadeh SM (2011) An overview of biodegradation of LNAPLs in coastal (semi)-arid environment. Water Air Soil Pollut 220:225–239

Yadav BK, Ansari FA, Basu S, Mathur A (2013) Remediation of LNAPL contaminated groundwater using plant-assisted biostimulation and bioaugmentation methods. Water Air Soil Pollut 225(1):1793. https://doi.org/10.1007/s11270-013-1793-9

Zhou AX, Zhang YL, Dong TZ, Lin XY, Su XS (2015) Response of the microbial community to seasonal groundwater level fluctuations in petroleum hydrocarbon-contaminated groundwater. Environ Sci Pollut Res 22(13):10094–10106

Genetic Technologies and Enhancement of Algal Utilization in Wastewater Treatment and Bioremediation

Mohamed A. El-Esawi

1 Introduction

Bioremediation is the process through which the environmental pollution is controlled, and the transformation or degradation of toxic chemicals could be enhanced to become less harmful environmental forms using biological systems (Prabha 2012). Nowadays, this process to an increasing extent has become one of the most popular biological processes due to its cost effective and efficient method of purification (Prabha 2012). The evacuation of liquid wastes – effluents – from the industrial or residential areas is highly regarded as a main source of water pollution (Prabha 2012). The effluents from the industrial areas, for instance, have been drained into open outlets from where the rivers are jointly met (Kumari et al. 2006; Prabha 2012). The wastewater discharges of industries are considered as the main issues of water pollution which regrettably lead to incrementing the promotion of an unstable toxic aquatic ecosystem and lessening the desideratum for nutrient loading and oxygen of water bodies (Morrison et al. 2001; Prabha 2012). Moreover, it has been reported that the values of low or high pH in the rivers have a high impact on the aquatic life which causes a substantial change in the level of pollution toxicity in one or another form (DWAF 1996; Prabha 2012). Furthermore, a reduction in pH values enhances the solubility of some environmental elements such as Cd, Fe, Al, B, Hg, Mn, and Cu and decrements the solubility of other main elements like selenium (Prabha 2012).

Good quality water availability is essential for alleviating diseases and enhancing life quality (Oluduro and Adewoye 2007; Prabha 2012). Natural water comprises

M. A. El-Esawi (✉)
Botany Department, Faculty of Science, Tanta University, Tanta, Egypt

The Sainsbury Laboratory, University of Cambridge, Cambridge, UK
e-mail: mohamed.elesawi@science.tanta.edu.eg

S. K. Gupta, F. Bux (eds.), *Application of Microalgae in Wastewater Treatment*,
https://doi.org/10.1007/978-3-030-13913-1_9

some types of debasements, for instance, minerals are brought into the ocean framework through the weathering of the rock, soil purification or filtering, and the dissolution of the vaporized particles from the atmosphere along with other several forms of human activities (Asaolu et al. 1997; Prabha 2012). After entering water, minerals can then be absorbed by plants and ultimately gathered in marine life forms that then get consumed by humans (Asaolu 1998; Prabha 2012).

Recent technologies have been introduced to perform the treatment of wastewater and bioremediation. Besides their use for biofuel production, algae represent an essential component among these technologies used for wastewater treatment and bioremediation to minimize phosphorus and nitrogen contents in sewage and some agricultural residues as well as to remove toxic metals from water-related industrial wastes (Hallmann 2007) (Fig. 1). Algae used in wastewater must tolerate the variable media conditions (e.g., salinity). The macroalgae *Monostroma* spp. and *Ulva* spp. were successfully used in reducing phosphorus and nitrogen compounds (Hallmann 2007). Through their comprehensive study for more than 40 years, Ryther et al. (1972), Romero-Gonzalez et al. (2001), and Kuyucak and Volesky (1988) have also studied the bioremediation of algae and elaborated their main processes. Cyanobacteria exhibit a significant potential in industrial effluents and wastewater treatment, bioremediation, bio-fertilizers, and food and chemical industries (Cairns Jr. and Dickson 1971; Prabha 2012). *Spirulina* sp. is a rapidly growing cyanobacterium that has a pronounced level of lead and mercury under contaminated conditions, suggesting its main role in absorbing toxic metals from the environment (Prabha 2012). *Synechococcus* sp., *Plectonema terebrans*, *Oscillatoria salina*, and *Aphanocapsa* sp. are some common cyanobacterial species which have been successfully utilized in the bioremediation of oil spills worldwide (Cohen 2002; Prabha 2012). Microalgae have also demonstrated their main role in removing main heavy metals, including lead, mercury, nickel, or cadmium (Chen et al. 1998) from the effluents. *Spirogyra* sp. induced the accumulation of heavy metals, such as chromium, copper, and zinc (Prabha 2012). Moreover, marine algae (e.g., *Sargassum* and *Ascophyllum*) induce pollutant biosorption (Yu et al. 1999). Too much industrial wastes are accumulated in rivers, resulting in impure water (Prabha 2012). Therefore, developing transgenic algae for wastewater bioremediation is of utmost need.

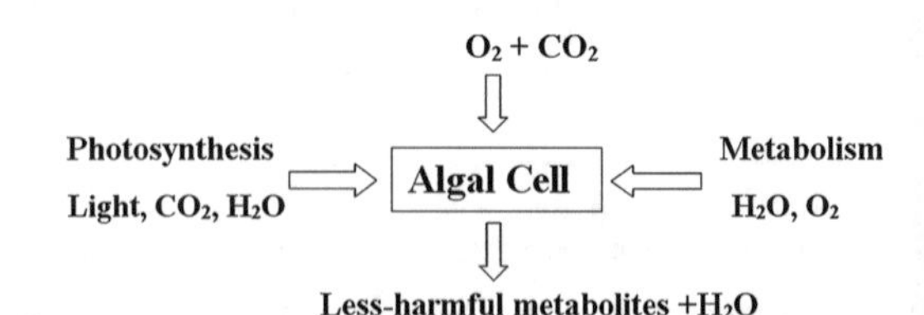

Fig. 1 Algae bioremediation cycle

2 Algal Transgenic Basis

The algal-related genome research is considered as a basic foundation that is highly required for another effective level in the utilization of gene technologies and the application of biotechnology to algae and their main products (Hallmann 2007). Fortunately, the genomic information that is retrieved from algae, as well as from living things, have increased exponentially in the last recent years (Hallmann 2007). The most significant advances in the microalgal genome projects are those related to the red alga such as *Ostreococcus tauri*, *Cyanidioschyzon merolae*, *Volvox carteri*, and *Chlamydomonas reinhardtii* (Hallmann 2007). In the last decade, several studies revealed the completion of such genomes; for instance, Barbier et al. (2005) and Matsuzaki et al. (2004) demonstrated that the annotation and the sequencing of the genome of 16.5 Mb *Cyanidioschyzon merolae* have been done (Hallmann 2007). Grossman et al. (2003) and Shrager et al. (2003) showed likewise that the sequencing of the genome of ~120 Mb *Chlamydomonas reinhardtii* was completed (Hallmann 2007). Further, the genome of the ~140 Mb-sized *Volvox carteri* was sequenced. Other studies such as Derelle et al. (2002, 2006) established the completion and the sequencing of the 11.5 Mb-sized genome of *Ostreococcus tauri*. In addition to these findings, Armbrust et al. (2004) demonstrated the completion of the annotation and sequencing processes of the genome of 34 Mb *Thalassiosira pseudonana*. Extra projects of algal genome have seen extensive advances. This could be found, for instance, in ~40 Mb genome such as *Chlorella vulgaris*; ~32 Mb genomes like *Aureococcus anophagefferens*, *Amphidinium operculatum*, and *Alexandrium tamarense*; ~15 Mb genomes like *Heterocapsa triquetra*, *Lotharella amoeboformis*, *Guillardia theta*, *Micromonas pusilla*, *Isochrysis galbana*, and *Karenia brevis* (Hallmann 2007); also in ~130 Mb genomes such as those in *Dunaliella salina* and *Cyanophora paradoxa*; ~30 Mb genomes as those in *Phaeodactylum tricornutum* and *Pavlova lutheri* in addition to *Porphyra yezoensis* and *Porphyra purpurea* (Hallmann 2007); further, *Ostreococcus lucimarinus* and *Ochromonas danica* with ~12 Mb genome type; *Emiliania huxleyi* (~220 Mb); *Ectocarpus siliculosus* (~214 Mb); and finally ~12 Mb type of genomes in *Galdieria sulphuraria* and *Euglena gracilis* (Hallmann 2007). At last, because of their little measured genomes, completion of genome sequencing from ~30 cyanobacteria is accessible in various databases such as those in *Gloeobacter violaceus* (~4.6 Mb), *Synechocystis* sp. (~3.6 Mb), *Anabaena* sp. (6.4 Mb), *Synechococcus elongatus* (~2.7 Mb), *Prochlorococcus marinus* (1.7–2.4 Mb; measure relies upon ecotype), and in addition *Thermosynechococcus elongatus* (~2.6 Mb) (Hallmann 2007). The list above isn't finished since there are (less broad) genomic sequences from numerous algal species existed in GenBank as well as different databases. In addition, new projects of genome ceaselessly come along (Hallmann 2007).

Similarly likewise with genome projects, there has as of late been a significant increment in the sequenced expressed sequence tags in algae (ESTs). Broad EST information originates from different types of algae such as those of *Phaeodactylum tricornutum* and the diatom *Thalassiosira pseudonana* (Scala et al. 2002; Hallmann

2007); green algae like *Ostreococcus tauri* and *Acetabularia acetabulum* (Henry et al. 2004), and *Chlamydomonas reinhardtii* (Shrager et al. 2003); also, from brown algae such as *Laminaria digitata* (Crepineau et al. 2000) and *Pavlova lutheri* (Pereira et al. 2004); and the haptophytes *Emiliania huxleyi* (Wahlund et al. 2004). In addition, several studies have also shown that these sorts of data can likewise be retrieved from red algae, namely, as haptophytes *Emiliania huxleyi* (Wahlund et al. 2004), *Gracilaria gracilis* (Lluisma and Ragan 1997), *Galdieria sulphuraria* (Weber et al. 2004), and the *Porphyra yezoensis* (Nikaido et al. 2000; Hallmann 2007), as well as from chlorarachniophyte *Bigelowiella natans* (Archibald et al. 2003) and dinoflagellates *Amphidinium carterae*, *Karlodinium micrum* (Bachvaroff et al. 2004), *Lingulodinium polyedrum* (Bachvaroff et al. 2004), and *Alexandrium tamarense* (Hackett et al. 2004). EST information from numerous other algal species are accessible at the TBestDB, Taxonomically Broad EST Database, available at http://amoebidia.bcm.umontreal.ca/pepdb/looks/welcome.php) and further available at the EST sequencing databases of the National Center for Biotechnology Information (NCBI) (http://www.ncbi.nlm.nih.gov/ventures/dbEST/).

The sequences of chloroplast and mitochondrial related genomes have been fulfilled with significantly more algal species than the genome sequencing projects or the EST because of plastid genomes which have a considerably littler size (Hallmann 2007). These types of sequences are accessible at the GOBASE, organelle genome database (http://www.bch.umontreal.ca/gobase/gobase.html), and also is found at NCBI organelle database (http://www.ncbi.nlm.nih.gov/genomes/ORGANELLES/organelles.html). Several studies reported that the eukaryotic algae are known to be contaminated by infections; a few genomes of such infections, all the more absolutely dsDNA infections, have been sequenced (van Etten et al. 2002; Hallmann 2007). Furthermore, Eukaryotic cells can forever gain chloroplasts by engulfing an alga. Much of the time, little stays of the engulfed alga separated from its chloroplast; however in two gatherings, the chlorarachniophytes and the cryptomonads, in addition to a little remnant nucleus of the engulfed alga, are as yet present. These minor nuclei, called nucleomorphs, have been sequenced in the chlorarachniophyte *Bigelowiella natans* and the cryptomonad alga *Guillardia theta* (Gilson and McFadden 2002).

3 Application Constraints of Genetic Technologies of Algae for Bioremediation

Cellular and molecular genetic techniques have been successfully used in plants and organisms for various purposes (El-Esawi and Sammour 2014; El-Esawi 2016a, b, c; El-Esawi 2017a, b; Consentino et al. 2015; El-Esawi et al. 2015; Jourdan et al. 2015; El-Esawi et al. 2016a, b; Arthaut et al. 2017; Elansary et al. 2017, 2018; El-Esawi et al. 2017a, b, c, d; Vwioko et al. 2017; El-Esawi et al. 2018a, b; Sherrard et al. 2018). Additionally, the genetic engineering of algal species has recently been applied to enhance pollutant bioremediation (Hallmann 2007) (Table 1). Most of these transformed algae were performed using nuclear transformation. For instance, the genetic transformation of *Chlamydomonas reinhardtii* (Kindle et al. 1989), *Ulva lactuca* (Huang et al. 1996), and *Volvox carteri* (Schiedlmeier et al. 1994) has been demonstrated. Furthermore, red algae species were genetically transformed (Cheney et al. 2001; Gan et al. 2003; Minoda et al. 2004; Hallmann 2007). *Laminaria japonica* and *Undaria pinnatifida* (Qin et al. 1999; Qin et al. 2003) were also stably transformed. Additionally, cyanobacterial species, such as *Anabaena*, *Spirulina*, and *Synechocystis*, have been transformed using conjugation and electroporation methods (Koksharova and Wolk 2002; Hallmann 2007).

Previous algal transformation studies demonstrated that the transformation efficiency and producible transformant number depend on the species transformed (Hallmann 2007). For instance, in *Cyanidioschyzon merolae* approximately 200 of transformants per μg of plasmid-DNA were produced when $3–4 \times 10^8$ cells were distributed on an agar plate (Minoda et al. 2004). Moreover, transformation efficiency in *Chlamydomonas reinhardtii* ranges between 10^{-4} and 10^{-5}, and 8×10^6 cells could be distributed on a plate (Kindle 1990; Hallmann 2007). Transformation efficiency of *Volvox carteri* is also 2.5×10^{-5} (Schiedlmeier et al. 1994).

The green alga *Chlamydomonas reinhardtii* tolerates cadmium; however the genetically transformed *Chlamydomonas*, expressing the moth bean *P5CS* gene, can grow under much higher concentrations of heavy metals (Hallmann 2007). *P5CS* gene expression, involved in proline biosynthesis, in the genetically transformed cells causes a fourfold increase in cadmium-binding capacity and 80% higher proline level as compared to wild-type cells (Hallmann 2007). Furthermore, *P5CS* gene expression causes a rapid growth at deadly cadmium concentrations (Siripornadulsil et al. 2002; Hallmann 2007). This might be attributed to that proline mitigated heavy metal stress via detoxifying free radicals generated. Generation of this transformed *Chlamydomonas* is essential for bioremediation of contaminated areas and water (Hallmann 2007).

Table 1 Transformable algal species

Species	Transformation	References
Chlamydomonas reinhardtii	Stable	Kindle et al. (1989)
Ulva lactuca	Transient	Huang et al. (1996)
Volvox carteri	Stable	Schiedlmeier et al. (1994)
Porphyra yezoensis	Stable	Cheney et al. (2001)
Gracilaria changii	Transient	Gan et al. (2003)
Cyanidioschyzon merolae	Stable	Minoda et al. (2004)
Laminaria japonica	Stable	Qin et al. (1999)
Undaria pinnatifida	Stable	Qin et al. (2003)

4 Selectable Marker Genes

As a limited number of treated organisms are successfully transformed, selectable marker gene utilization is needed for the experiments which aim at producing stable transgenic algae (Hallmann 2007). These markers are frequently antibiotic resistance genes that are highly regarded as prevailing markers as they give another pivotal trait to some other transformed targets of a species, regardless of the particular genotype (Hallmann 2007). The most noteworthy number of these selectable markers has been demonstrated for the *Chlamydomonas reinhardtii*: the R100.1 plasmid/bacteriophage T4/synthetic aminoglycoside adenyltransferase gene aadA provides resistance to streptomycin and spectinomycin (Cerutti et al. 1997; Hallmann 2007), the mutated *Chlamydomonas reinhardtii* protoporphyrinogen oxidase gene PPX1 confers resistance to the N-phenyl heterocyclic herbicide S-23142 (Randolph-Anderson et al. 1998; Hallmann 2007), the *Streptoalloteichus hindustanus* ble gene confers resistance to phleomycin and zeomycin (Stevens et al. 1996), the *Streptomyces rimosus* aminoglycoside phosphotransferase aphVIII (aphH) gene confers resistance to paromomycin (Sizova et al. 2001; Hallmann 2007), the *Streptomyces hygroscopicus* aminoglycoside phosphotransferase aph7″ gene confers resistance to hygromycin B (Berthold et al. 2002), the mutated *Chlamydomonas reinhardtii* acetolactate synthase gene ALS confers resistance to sulfonylurea herbicides (Kovar et al. 2002; Hallmann 2007), and the mutated version of the *Chlamydomonas reinhardtii* ribosomal protein gene S14 (CRY1) confers resistance to emetine and cryptopleurine (Nelson et al. 1994). Correspondingly, in *Volvox carteri* multicellular algae, *Streptoalloteichus hindustanus* ble gene provides a high level of resistance to both phleomycin and zeomycin (Hallmann 2007). The *Streptomyces rimosus* aminoglycoside phosphotransferase aphVIII (aphH) gene confers resistance to paromomycin (Hallmann and Wodniok 2006). The modified *Haematococcus pluvialis* gene pdsMod4.1 confers enhanced astaxanthin biosynthesis and resistance to the bleaching herbicide norflurazon (Steinbrenner and Sandmann 2006). The *Streptomyces hygroscopicus* aminoglycoside phosphotransferase gene in *Chlorella vulgaris* was expressed under the control of the cauliflower mosaic virus promoter (CaMV35S) for selection with hygromycin (Chow and Tung 1999; Hallmann 2007). A mutant of the gene encoding acetohydroxyacid synthase [AHAS (W492S)] served as a selectable marker for transformation of chloroplast in *Porphyridium* (Lapidot et al. 2002). SV40 promoter-hygromycin phosphotransferase chimeric gene in *Laminaria japonica* provides resistance to hygromycin (Qin et al. 1999). The *Streptoalloteichus hindustanus* ble gene in *Phaeodactylum tricornutum* diatom was shown to provide resistance to zeomycin (Falciatore et al. 1999), sat-1 and nat genes confer resistance to nourseothricin (Zaslavskaia et al. 2000), and the neomycin phosphotransferase II (nptII) gene resists the aminoglycoside antibiotic G418 (Zaslavskaia et al. 2000; Hallmann 2007). The endogenous calcium-binding glycoprotein α-frustulin fruα3 promoter in *Cylindrotheca fusiformis* diatom was utilized for the expression of *Streptoalloteichus hindustanus* ble gene and provides resistance to zeomycin (Fischer et al. 1999; Hallmann 2007). Additionally, nptII gene provides resistance to the antibiotic G418 in *Navicula saprophila* and *Cyclotella cryptica* diatoms (Dunahay et al. 1995). The R100.1 plasmid/bacteriophage T4/synthetic aminoglycoside adenyltransferase gene aadA served as a selectable marker for transformation of chloroplast in *Euglena gracilis* and provides resistance to spectinomycin (Doetsch et al. 2001; Hallmann 2007). *Symbiodinium microadriaticum* and *Amphidinium* sp. transformation was performed using the hygromycin B phosphotransferase gene (hpt) fused to *Agrobacterium* p1'2′ promoter or the nptII gene mediated by the *Agrobacterium* nos promoter (Hallmann 2007).

There are also many specific recessive markers besides the dominant ones for algal systems (Hallmann 2007). These markers have the colossal favorable position that an entire gene is normally utilized with its own particular promoter, in spite of the fact that recessive markers need mutants with changes in the comparing homogenous and the relating flawless genes (Hallmann 2007). Hence, in this way as opposed to those numerous dominant markers, the expression and capacity of the selectable marker in the organism are quite certain beforehand. Nitrate reductase gene (nit) is a common recessive marker used for functional complementation of nitrate reductase defective mutants of *Volvox carteri* (Schiedlmeier et al. 1994; Hallmann 2007), *Chlamydomonas reinhardtii* (Kindle et al. 1989), *Chlorella sorokiniana* (Dawson et al. 1997), *Dunaliella viridis* (Sun et al. 2006), and *Ulva lactuca* (Huang et al. 1996). Nitrate reductase reduces chlorate and nitrate, and the produced chlorite is toxic (Hallmann 2007). Chlorate may be utilized to recheck putative nit^+ transformants. It could also be used to get the needed auxotrophic target organisms for the experiments of nit transformation (Hallmann 2007). Chlorate may also be utilized as a negative selectable marker to recognize nit mutants (Hallmann 2007). Mutations in argininosuccinate lyase defective *Chlamydomonas reinhardtii* mutants could be complemented by *Chlamydomonas reinhardtii* argininosuccinate lyase gene ASL (Debuchy et al. 1989; Hallmann 2007).

5 Methods Used for DNA Introduction into Algal Cells

There are two main transformation methods which assist in viable transformant recovery in algae (Hallmann 2007). Microparticle bombardment represents the most common method. This method is also termed micro-projectile bombardment, gene gun transformation, particle gun transformation, or biolistics (Hallmann 2007). It utilizes DNA-coated heavy metal micro-projectiles and performs the transformation of organelles or cells, regardless of the rigidity or thickness of the cell wall (Hallmann 2007). This method has been successfully performed in *Volvox carteri* (Schiedlmeier et al. 1994), *Chlamydomonas reinhardtii* (Kindle et al. 1989), *Dunaliella salina* (Tan et al. 2005), *Laminaria japonica* (Jiang et al. 2003),

Gracilaria changii (Gan et al. 2003), *Phaeodactylum tricornutum* (Apt et al. 1996), *Cyclotella cryptica* (Dunahay et al. 1995), *Navicula saprophila* (Dunahay et al. 1995), *Euglena gracilis* (Doetsch et al. 2001), *Cylindrotheca fusiformis* (Fischer et al. 1999), *Porphyridium* sp. (Lapidot et al. 2002), *Chlorella sorokiniana* (Dawson et al. 1997), and *Haematococcus pluvialis* (Steinbrenner and Sandmann 2006).

Another transformation technique which is cheaper includes the preparation of algal suspension which is then agitated in the presence of polyethylene glycol, micro- or macro-particles and DNA (Hallmann 2007). Numerous studies have utilized silicon carbide (SiC) bristles, which are roughly 0.3 to 0.6 μm in thickness and about 5 to 15 μm long as microscale particles (Hallmann 2007). SiC is a ceramic compound of both carbon and silicon. These microparticles assisted in the transformation of some algae, such as *Chlamydomonas reinhardtii* (Dunahay 1993) and *Amphidinium* sp. (ten Lohuis and Miller 1998). Agitation was used to transform the cell wall-reduced mutants in *Chlamydomonas reinhardtii* in using large glass beads, DNA, and polyethylene glycol (Kindle 1990; Hallmann 2007). This cheap technique is applied in *Chlamydomonas* transformation. The green algal protoplast could be transformed in the absence of particles (Jarvis and Brown 1991) but in the presence of DNA and polyethylene glycol. Cell wall-reduced mutants (Hallmann 2007), protoplasts, and naked cells could also be transformed via electroporation in which the large electronic pulse disturbs cell membrane phospholipid bilayer, which allows DNA and other molecules to pass. Cells of *Cyanidioschyzon merolae* (Minoda et al. 2004; Hallmann 2007), *Chlamydomonas reinhardtii* (Brown et al. 1991), *Chlorella vulgaris* (Chow and Tung 1999), and *Dunaliella salina* (Geng et al. 2003) were transformed using this procedure. *Agrobacterium tumefaciens*-mediated transformation via tumor-inducing (Ti) plasmids has been used to genetically modify two algal species (Hallmann 2007). *Agrobacterium*-mediated transformation has been applied in the unicellular green alga *Chlamydomonas reinhardtii* (Kumar et al. 2004; Hallmann 2007) and the multicellular red alga *Porphyra yezoensis* (Cheney et al. 2001).

6 Conclusions

Bioremediation allows the control of the environmental pollution and enhances the transformation or degradation of toxic chemicals to become less harmful environmental forms using biological systems. Algae could be utilized in bioremediation and wastewater treatment to decrease nitrogen and phosphorus contents in agricultural wastes. Genetic technologies and transgenesis and their potential roles in enhancing the potential utilization of algae in wastewater treatment and bioremediation have been reviewed and discussed in this chapter. Selectable marker genes and methods of DNA introduction into algal cells have also show their potential.

References

Apt KE, Kroth-Pancic PG, Grossman AR (1996) Stable nuclear transformation of the diatom *Phaeodactylum tricornutum*. Mol Gen Genet 252:572–579

Archibald JM, Rogers MB, Toop M, Ishida K, Keeling PJ (2003) Lateral gene transfer and the evolution of plastid-targeted proteins in the secondary plastid-containing alga *Bigelowiella natans*. Proc Natl Acad Sci U S A 100:7678–7683

Armbrust EV, Berges JA, Bowler C, Green BR, Martinez D, Putnam NH, Zhou S, Allen AE, Apt KE, Bechner M, Brzezinski MA, Chaal BK, Chiovitti A, Davis AK, Demarest MS, Detter JC, Glavina T, Goodstein D, Hadi MZ, Hellsten U, Hildebrand M, Jenkins BD, Jurka J, Kapitonov VV, Kröger N, Lau WW, Lane TW, Larimer FW, Lippmeier JC, Lucas S, Medina M, Montsant A, Obornik M, Parker MS, Palenik B, Pazour GJ, Richardson PM, Rynearson TA, Saito MA, Schwartz DC, Thamatrakoln K, Valentin K, Vardi A, Wilkerson FP, Rokhsar DS (2004) The genome of the diatom *Thalassiosira pseudonana*: ecology, evolution, and metabolism. Science 306:79–86

Arthaut L, Jourdan N, Mteyrek A, Procopio M, El-Esawi M, d'Harlingue A, Bouchet P, Witczak J, Ritz T, Klarsfeld A, Birman S, Hoecker U, Martino C, Ahmad M (2017) Blue-light induced accumulation of Reactive Oxygen Species is a consequence of the Drosophila cryptochrome photocycle. PLoS One 12(3):e0171836

Asaolu SS (1998) Chemical pollution studies of coastal water of OndoState. Ph.D Thesis, Fed. Univ. Technol

Asaolu SS, Ipinmoroti KO, Adeyinowo CE, Olaofe O (1997) Interrelationship 0f heavy metals concentration in water, sediment as fish samples from Ondo State coastal Area. Nig Afr J Sci 1:55–61

Bachvaroff TR, Concepcion GT, Rogers CR, Herman EM, Delwiche CF (2004) Dinoflagellate expressed sequence tag data indicate massive transfer of chloroplast genes to the nuclear genome. Protist 155:65–78

Barbier G, Oesterhelt C, Larson MD, Halgren RG, Wilkerson C, Garavito RM, Benning C, Weber AP (2005) Comparative genomics of two closely related unicellular thermo-acidophilic red algae, *Galdieria sulphuraria* and *Cyanidioschyzon merolae*, reveals the molecular basis of the metabolic flexibility of *Galdieria sulphuraria* and significant differences in carbohydrate metabolism of both algae. Plant Physiol 137:460–474

Berthold P, Schmitt R, Mages W (2002) An engineered *Streptomyces hygroscopicus aph 7"* gene mediates dominant resistance against hygromycin B in *Chlamydomonas reinhardtii*. Protist 153:401–412

Brown LE, Sprecher SL, Keller LR (1991) Introduction of exogenous DNA into *Chlamydomonas reinhardtii* by electroporation. Mol Cell Biol 11:2328–2332

Cairns J Jr, Dickson KL (1971) A simple method for the biological assessment of the effects of waste discharge on aquatic bottom dwelling organisms. J Water Pollut Control Fed 43:722–725

Cerutti H, Johnson AM, Gillham NW, Boynton JE (1997) A eubacterial gene conferring spectinomycin resistance on *Chlamydomonas reinhardtii*: integration into the nuclear genome and gene expression. Genetics 145:97–110

Chen B, Huang Q, Lin X, Shi Q, Wu S (1998) Accumulation of Ag, Cd, Co, Cu, Hg, Ni and Pb in *Pavlova viridis* Tseng (Haptophyceae). J Appl Phycol 10:371–376

Cheney D, Metz B, Stiller J (2001) *Agrobacterium*-mediated genetic transformation in the macroscopic marine red alga *Porphyra yezoensis*. J Phycol 37:11

Chow K-C, Tung WL (1999) Electrotransformation of *Chlorella vulgaris*. Plant Cell Rep 18:778–780

Cohen Y (2002) Bioremediation of oil by marine microbial mats. Int Microbiol 5:189–193

Consentino L, Lambert S, Martino C, Jourdan N, Bouchet PE, Witczak J, Castello P, El-Esawi M, Corbineau F, d'Harlingue A, Ahmad M (2015) Blue-light dependent reactive oxygen species formation by *Arabidopsis* cryptochrome may defi ne a novel evolutionarily conserved signalling mechanism. New Phytol 206:1450–1462

Crepineau F, Roscoe T, Kaas R, Kloareg B, Boyen C (2000) Characterisation of complementary DNAs from the expressed sequence tag analysis of life cycle stages of *Laminaria digitata* (Phaeophyceae). Plant Mol Biol 43:503–513

Dawson HN, Burlingame R, Cannons AC (1997) Stable transformation of *Chlorella*: rescue of nitrate reductase-deficient mutants with the nitrate reductase gene. Curr Microbiol 35:356–362

Debuchy R, Purton S, Rochaix JD (1989) The argininosuccinate lyase gene of *Chlamydomonas reinhardtii*: an important tool for nuclear transformation and for correlating the genetic and molecular maps of the ARG7 locus. EMBO J 8:2803–2809

Derelle E, Ferraz C, Lagoda P, Eychenié S, Cooke R, Regad F, Sabau X, Courties C, Delseny M, Demaille J, Picard A, Moreau H (2002) DNA libraries for sequencing the genome of *Ostreococcus tauri* (Chlorophytae; Prasinophyceae): the smallest free-living eukaryotic cell. J Phycol 38:1150–1156

Derelle E, Ferraz C, Rombauts S, Rouze P, Worden AZ, Robbens S, Partensky F, Degroeve S, Echeynie S, Cooke R, Saeys Y, Wuyts J, Jabbari K, Bowler C, Panaud O, Piegu B, Ball SG, Ral JP, Bouget FY, Piganeau G, de Baets B, Picard A, Delseny M, Demaille J, van de Peer Y, Moreau H (2006) Genome analysis of the smallest free-living eukaryote *Ostreococcus tauri* unveils many unique features. Proc Natl Acad Sci U S A 103:11647–11652

Doetsch NA, Favreau MR, Kuscuoglu N, Thompson MD, Hallick RB (2001) Chloroplast transformation in *Euglena gracilis*: splicing of a group III twintron transcribed from a transgenic *psbK* operon. Curr Genet 39:49–60

Dunahay TG (1993) Transformation of *Chlamydomonas reinhardtii* with silicon carbide whiskers. BioTechniques 15:452–455

Dunahay TG, Jarvis EE, Roessler PG (1995) Genetic transformation of the diatoms *Cyclotella cryptica* and *Navicula saprophila*. J Phycol 31:1004–1012

DWAF (1996) South Africa water quality guidelines. 7: aquatic ecosystems, 1st edn. Department of Water Affairs and Forestry, Pretoria

Elansary HO, Yessoufou K, Abdel-Hamid AME, El-Esawi MA, Ali HM, Elshikh MS (2017) Seaweed extracts enhance Salam Turfgrass performance during prolonged irrigation intervals and saline shock. Front Plant Sci 8:830

Elansary HO, Szopa A, Kubica P, Ekiert H, Ali HM, Elshikh MS et al (2018) Bioactivities of traditional medicinal plants in Alexandria. Evid Based Complement Alternat Med 2018:1463579

El-Esawi MA (2016a) Micropropagation technology and its applications for crop improvement. In: Anis M, Ahmad N (eds) Plant tissue culture: propagation, conservation and crop improvement. Springer, Singapore, pp 523–545

El-Esawi MA (2016b) Nonzygotic embryogenesis for plant development. In: Anis M, Ahmad N (eds) Plant tissue culture: propagation, conservation and crop improvement. Springer, Singapore, pp 583–598

El-Esawi MA (2016c) Somatic hybridization and microspore culture in Brassica improvement. In: Anis M, Ahmad N (eds) Plant tissue culture: propagation, conservation and crop improvement. Springer, Singapore, pp 599–609

El-Esawi MA (2017a) Genetic diversity and evolution of *Brassica* genetic resources: from morphology to novel genomic technologies – a review. Plant Genet Resour C 15:388–399

El-Esawi MA (2017b) SSR analysis of genetic diversity and structure of the germplasm of faba bean (*Vicia faba* L.). C R Biol 340:474–480

El-Esawi MA, Sammour R (2014) Karyological and phylogenetic studies in the genus *Lactuca* L. (Asteraceae). Cytologia 79:269–275

El-Esawi M, Glascoe A, Engle D, Ritz T, Link J, Ahmad M (2015) Cellular metabolites modulate *in vivo* signaling of *Arabidopsis* cryptochrome-1. Plant Signal Behav 10(9):e1063758

El-Esawi MA, Germaine K, Bourke P, Malone R (2016a) Genetic diversity and population structure of *Brassica oleracea* germplasm in Ireland using SSR markers. C R Biol 339:133–140

El-Esawi MA, Germaine K, Bourke P, Malone R (2016b) AFLP analysis of genetic diversity and phylogenetic relationships of *Brassica oleracea* in Ireland. C R Biol 133:163–170

El-Esawi M, Arthaut L, Jourdan N, d'Harlingue A, Martino C, Ahmad M (2017a) Blue-light induced biosynthesis of ROS contributes to the signaling mechanism of Arabidopsis cryptochrome. Sci Rep 7:13875
El-Esawi MA, Elansary HO, El-Shanhorey NA, Abdel-Hamid AME, Ali HM, Elshikh MS (2017b) Salicylic acid-regulated antioxidant mechanisms and gene expression enhance rosemary performance under saline conditions. Front Physiol 8:716
El-Esawi MA, Elkelish A, Elansary HO et al (2017c) Genetic transformation and hairy root induction enhance the antioxidant potential of *Lactuca serriola* L. Oxid Med Cell Long 2017:5604746, 8 pages
El-Esawi MA, Mustafa A, Badr S, Sammour R (2017d) Isozyme analysis of genetic variability and population structure of *Lactuca* L. germplasm. Biochem Syst Ecol 70:73–79
El-Esawi MA, Alaraidh IA, Alsahli AA, Alamri SA et al (2018a) *Bacillus firmus* (SW5) augments salt tolerance in soybean (*Glycine max* L.) by modulating root system architecture, antioxidant defense systems and stress-responsive genes expression. Plant Physiol Biochem 132:375–384
El-Esawi MA, Al-Ghamdi AA, Ali HM, Alayafi AA, Witczak J, Ahmad M (2018b) Analysis of genetic variation and enhancement of salt tolerance in French pea (*Pisum sativum* L.). Int J Mol Sci 19(8)
Falciatore A, Casotti R, Leblanc C, Abrescia C, Bowler C (1999) Transformation of nonselectable reporter genes in marine diatoms. Mar Biotechnol 1:239–251
Fischer H, Robl I, Sumper M, Kröger N (1999) Targeting and covalent modification of cell wall and membrane proteins heterologously expressed in the diatom *Cylindrotheca fusiformis*. J Phycol 35:113–120
Gan S-Y, Qin S, Othman RY, Yu D, Phang S-M (2003) Transient expression of *lacZ* in particle bombarded *Gracilaria changii* (Gracilariales, Rhodophyta). J Appl Phycol 15:345–349
Geng D, Wang Y, Wang P, Li W, Sun Y (2003) Stable expression of hepatitis B surface antigen gene in *Dunaliella salina* (Chlorophyta). J Appl Phycol 15:451–456
Gilson PR, McFadden GI (2002) Jam packed genomes – a preliminary, comparative analysis of nucleomorphs. Genetica 115:13–28
Grossman AR, Harris EE, Hauser C, Lefebvre PA, Martinez D, Rokhsar D, Shrager J, Silflow CD, Stern D, Vallon O, Zhang Z (2003) *Chlamydomonas reinhardtii* at the crossroads of genomics. Eukaryot Cell 2:1137–1150
Hackett JD, Yoon HS, Soares MB, Bonaldo MF, Casavant TL, Scheetz TE, Nosenko T, Bhattacharya D (2004) Migration of the plastid genome to the nucleus in a peridinin dinoflagellate. Curr Biol 14:213–218
Hallmann A (2007) Algal transgenics and biotechnology. Transgenic Plant J 1(1):81–98
Hallmann A, Wodniok S (2006) Swapped green algal promoters: aphVIIIbased gene constructs with *Chlamydomonas* flanking sequences work as dominant selectable markers in *Volvox* and vice versa. Plant Cell Rep 25:582–591
Henry IM, Wilkinson MD, Hernandez JM, Schwarz-Sommer Z, Grotewold E, Mandoli DF (2004) Comparison of ESTs from juvenile and adult phases of the giant unicellular green alga *Acetabularia acetabulum*. BMC Plant Biol 4:3
Huang X, Weber JC, Hinson TK, Mathieson AC, Minocha SC (1996) Transient expression of the GUS reporter gene in the protoplasts and partially digested cells of *Ulva lactuca*. Bot Mar 39:467–474
Jarvis EE, Brown LM (1991) Transient expression of firefly luciferase in protoplasts of the green alga *Chlorella ellipsoidea*. Curr Genet 19:317–321
Jourdan N, Martino C, El-Esawi M, Witczak J, Bouchet PE, d'Harlingue A, Ahmad M (2015) Bluelight dependent ROS formation by Arabidopsis Cryptochrome-2 may contribute towards its signaling role. Plant Signal Behav 10(8):e1042647
Kindle KL (1990) High-frequency nuclear transformation of *Chlamydomonas reinhardtii*. PNAS 87:1228–1232

Kindle KL, Schnell RA, Fernandez E, Lefebvre PA (1989) Stable nuclear transformation of *Chlamydomonas* using the *Chlamydomonas* gene for nitrate reductase. J Cell Biol 109:2589–2601

Koksharova OA, Wolk CP (2002) Genetic tools for cyanobacteria. Appl Microbiol Biotechnol 58:123–137

Kovar JL, Zhang J, Funke RP, Weeks DP (2002) Molecular analysis of the acetolactate synthase gene of *Chlamydomonas reinhardtii* and development of a genetically engineered gene as a dominant selectable marker for genetic transformation. Plant J 29:109–117

Kumar SV, Misquitta RW, Reddy VS, Rao BJ, Rajam MV (2004) Genetic transformation of the green alga *Chlamydomonas reinhardtii* by *Agrobacterium tumefaciens*. Plant Sci (Shannon, Ireland) 166:731–738

Kumari SB, Kirubavathy AK, Thirumalnesan R (2006) Stability and water quality criteria of open drainage municipal sewage water at Coimbatore used for irrigation. J Environ Biol 27(4):709–712

Kuyucak N, Volesky B (1988) Biosorbents for recovery of metals from industrial solutions. Biotechnol Lett 10:137–142

Lapidot M, Raveh D, Sivan A, Arad SM, Shapira M (2002) Stable chloroplast transformation of the unicellular red alga *Porphyridium* species. Plant Physiol 129:7–12

Lluisma AO, Ragan MA (1997) Expressed sequence tags (ESTs) from the marine red alga *Gracilaria gracilis*. J Appl Phycol 9:287–293

Matsuzaki M, Misumi O, Shin IT, Maruyama S, Takahara M, Miyagishima SY, Mori T, Nishida K, Yagisawa F, Nishida K, Yoshida Y, Nishimura Y, Nakao S, Kobayashi T, Momoyama Y, Higashiyama T, Minoda A, Sano M, Nomoto H, Oishi K, Hayashi H, Ohta F, Nishizaka S, Haga S, Miura S, Morishita T, Kabeya Y, Terasawa K, Suzuki Y, Ishii Y, Asakawa S, Takano H, Ohta N, Kuroiwa H, Tanaka K, Shimizu N, Sugano S, Sato N, Nozaki H, Ogasawara N, Kohara Y, Kuroiwa T (2004) Genome sequence of the ultrasmall unicellular red alga *Cyanidioschyzon merolae* 10D. Nature 428:653–657

Minoda A, Sakagami R, Yagisawa F, Kuroiwa T, Tanaka K (2004) Improvement of culture conditions and evidence for nuclear transformation by homologous recombination in a red alga, *Cyanidioschyzon merolae* 10D. Plant Cell Physiol 45:667–671

Morrison GO, Fatoki OS, Ekberg A (2001) Assessment of the impact of point source pollution from the Keiskammahoek sewage treatment plant on the Keiskamma River. Water SA 27:475–480

Nelson JA, Savereide PB, Lefebvre PA (1994) The *CRY1* gene in *Chlamydomonas reinhardtii*: structure and use as a dominant selectable marker for nuclear transformation. Mol Cell Biol 14:4011–4019

Nikaido I, Asamizu E, Nakajima M, Nakamura Y, Saga N, Tabata S (2000) Generation of 10,154 expressed sequence tags from a leafy gametophyte of a marine red alga, *Porphyra yezoensis*. DNA Res 7:223–227

Oluduro AO, Adewoye BI (2007) Efficiency of *moringa Oleifera* seed extract on the microflora of surface and ground water. J Plant Sci 6:453–438

Pereira SL, Leonard AE, Huang YS, Chuang LT, Mukerji P (2004) Identification of two novel microalgal enzymes involved in the conversion of the ω3-fatty acid, eicosapentaenoic acid, into docosahexaenoic acid. Biochem J 384:357–366

Prabha Y (2012) Potential of algae in bioremediation of waste water. PhD thesis, Deemed University

Qin S, Sun G-Q, Jiang P, Zou L-H, Wu Y, Tseng C-K (1999) Review of genetic engineering of *Laminaria japonica* (Laminariales, Phaeophyta) in China. Hydrobiologia:398–399

Qin S, Yu D-Z, Jiang P, Teng C-Y, Zeng C-K (2003) Stable expression of *lacZ* reporter gene in seaweed *Undaria pinnatifida*. High Technol Lett 13:87–89

Randolph-Anderson BL, Sato R, Johnson AM, Harris EH, Hauser CR, Oeda K, Ishige F, Nishio S, Gillham NW, Boynton JE (1998) Isolation and characterization of a mutant protoporphyrinogen oxidase gene from *Chlamydomonas reinhardtii* conferring resistance to porphyric herbicides. Plant Mol Biol 38:839–859

Romero-Gonzalez ME, Williams CJ, Gardiner PHE (2001) Study of the mechanisms of cadmium biosorption by dealginated seaweed waste. Environ Sci Technol 35:3025–3030
Ryther JH, Tenore KR, Dunstan WM, Huguenin JE (1972) Controlled eutrophication increasing food production from sea by recycling human wastes. Bioscience 22:144
Scala S, Carels N, Falciatore A, Chiusano ML, Bowler C (2002) Genome properties of the diatom *Phaeodactylum tricornutum*. Plant Physiol 129:993–1002
Schiedlmeier B, Schmitt R, Müller W, Kirk MM, Gruber H, Mages W, Kirk DL (1994) Nuclear transformation of *Volvox carteri*. PNAS 91:5080–5084
Sherrard RM, Morellini N, Jourdan N, El-Esawi M, Arthaut L-D, Neissner C, Rouyer F, Klarsfeld A, Doulazmi M, Witczak J, d'Harlingue A, Mariani J, Mclure I, Martino CF, Ahmad M (2018) Low-intensity electromagnetic fields stimulate human cryptochrome to modulate intracellular reactive oxygen species. PLoS Biol 6(10):e2006229
Shrager J, Hauser C, Chang CW, Harris EH, Davies J, McDermott J, Tamse R, Zhang Z, Grossman AR (2003) *Chlamydomonas reinhardtii* genome project. A guide to the generation and use of the cDNA information. Plant Physiol 131:401–408
Siripornadulsil S, Traina S, Verma DP, Sayre RT (2002) Molecular mechanisms of proline-mediated tolerance to toxic heavy metals in transgenic microalgae. Plant Cell 14:2837–2847
Sizova I, Fuhrmann M, Hegemann P (2001) A *Streptomyces rimosus* aphVIII gene coding for a new type phosphotransferase provides stable antibiotic resistance to *Chlamydomonas reinhardtii*. Gene 277:221–229
Steinbrenner J, Sandmann G (2006) Transformation of the green alga *Haematococcus pluvialis* with a phytoene desaturase for accelerated astaxanthin biosynthesis. Appl Environ Microbiol 72:7477–7484
Stevens DR, Rochaix JD, Purton S (1996) The bacterial phleomycin resistance gene *ble* as a dominant selectable marker in *Chlamydomonas*. Mol Gen Genet 251:23–30
Sun Y, Gao X, Li Q, Zhang Q, Xu Z (2006) Functional complementation of a nitrate reductase defective mutant of a green alga *Dunaliella viridis* by introducing the nitrate reductase gene. Gene 377:140–149
ten Lohuis MR, Miller DJ (1998) Genetic transformation of dinoflagellates (*Amphidinium* and *Symbiodinium*): expression of GUS in microalgae using heterologous promoter constructs. Plant J 13:427–435
Tan C, Qin S, Zhang Q, Jiang P, Zhao F (2005) Establishment of a micro-particle bombardment transformation system for *Dunaliella salina*. J Microbiol 43:361–365
van Etten JL, Graves MV, Müller DG, Boland W, Delaroque N (2002) Phycodnaviridae - large DNA algal viruses. Arch Virol 147:1479–1516
Vwioko E, Adinkwu O, El-Esawi MA (2017) Comparative physiological, biochemical and genetic responses to prolonged waterlogging stress in okra and maize given exogenous ethylene priming. Front Physiol 8:632
Wahlund TM, Hadaegh AR, Clark R, Nguyen B, Fanelli M, Read BA (2004) Analysis of expressed sequence tags from calcifying cells of marine coccolithophorid (*Emiliania huxleyi*). Mar Biotechnol 6:278–290
Weber AP, Oesterhelt C, Gross W, Bräutigam A, Imboden LA, Krassovskaya I, Linka N, Truchina J, Schneidereit J, Voll H, Voll LM, Zimmermann M, Jamai A, Riekhof WR, Yu B, Garavito RM, Benning C (2004) EST-analysis of the thermo-acidophilic red microalga *Galdieria sulphuraria* reveals potential for lipid A biosynthesis and unveils the pathway of carbon export from rhodoplasts. Plant Mol Biol 55:17–32
Yu Q, Matheickal JT, Yin P, Kaewsar P (1999) Heavy metals uptake capacities of common microalgal biomass. Water Res 33:1534–1537
Zaslavskaia LA, Lippmeier JC, Kroth PG, Grossman AR, Apt KE (2000) Transformation of the diatom *Phaeodactylum tricornutum* (Bacillariophyceae) with a variety of selectable marker and reporter genes. J Phycol 36:379–386

Potential and Feasibility of the Microalgal System in Removal of Pharmaceutical Compounds from Wastewater

Mayuri Chabukdhara, Manashjit Gogoi, and Sanjay Kumar Gupta

1 Introduction

The revolutionized development of resources and technologies has produced a lot of chemicals which possess potential threats to the living systems (Bolong et al. 2009). Environment pollution due to presence of pharmaceuticals in wastewater streams is recognized as a major threat to aquatic environment globally as they affect the terrestrial and aquatic organisms (Khetan and Collins 2007). Pharmaceuticals are defined as therapeutic products used to prevent or treat human or animal diseases. These products are consumed by human and animals and then released into sewage streams as parent compounds or their metabolites during excretion as urine or feces, etc. (Al Aukidy et al. 2014; Daughton 2001). Apart from this, pharmaceutical manufacturing processes like chemical synthesis and fermentation processes are responsible for generation of large volume of wastewater containing very high levels of spent solvents, recalcitrant organics, pharmaceutical residues, as well as salts (Chen et al. 2008). Since conventional wastewater treatment plants are designed to remove these products, they are available in different concentrations in different natural water bodies (Ternes et al. 2004). These pharmaceuticals products finally enter into surface water after partial removal by wastewater treatment plants. Pharmaceuticals and personal care products, surfactants, surfactant residues, plasticizers and various industrial additives, and a large group of chemicals are collectively known as

M. Chabukdhara (✉)
Department of Environmental Biology and Wildlife Sciences, Cotton University, Guwahati, Assam, India

M. Gogoi
Department of Biomedical Engineering, North Eastern Hill University, Shillong, Meghalaya, India

S. K. Gupta
Environmental Engineering, Department of Civil Engineering, Indian Institute of Technology – Delhi, New Delhi, Delhi, India

S. K. Gupta, F. Bux (eds.), *Application of Microalgae in Wastewater Treatment*,
https://doi.org/10.1007/978-3-030-13913-1_10

endocrine disruptors, which are not metabolized and released into wastewater treatment plants (Boxall et al. 2012). Currently these pharmaceutical products are extensively available in different aquatic streams around the world. Their omnipresence possesses a huge problem to the terrestrial ecosystems as well as animal kingdom.

In the recent years, algae-based technologies in wastewater treatment have been drawing huge attention from researchers due to their cost-effectiveness (Abinandan and Shanthakumar 2015; Wang et al. 2016; Wu et al. 2012). Algal biomass assimilates nutrients from wastewater, and valuable products such as biofertilizer (Cai et al. 2013), biofuel, proteins, carbohydrates, pigments, and vitamins can be produced by algae during treating wastewater (Cuellar-Bermudez et al. 2017; da Silva et al. 2014; Úbeda et al. 2017). Moreover, algae capture CO_2 via photosynthesis process and hence reduce greenhouse gas (Razzak et al. 2013; Subashchandrabose et al. 2011). Apart from these, algae-based technologies are capable of removal of heavy metals and other hazardous materials from wastewater through surface sorption, bioaccumulation, and precipitation (Zeraatkar et al. 2016). Therefore, algae-based technologies are proved to be more sustainable compared to the other conventional wastewater treatment technologies. Research on removal of pharmaceuticals and other emerging contaminants using algal-based technologies have increased in the recent decades.

In this chapter, discussion is made on the presence of pharmaceutical compounds in different wastewater as well as aquatic streams, their side effect on terrestrial ecosystems, as well as animal kingdom and algal-based methods for their removal. Role of microalgae in CO_2 mitigation and microalgal biomass for biofuels production are also briefly discussed in this chapter.

2 Pharmaceutical Compounds in Wastewater and Associated Risks

All over the world, surface water and ground water are the main sources of drinking water. Presence of chemicals such as pharmaceutical and personal care products or their derivatives in drinking water may cause detrimental effects on our health. A large number of studies demonstrated the presence of chemical compounds related to pharmaceutical and personal care products in surface water streams globally. Surface water such as streams and rivers are reported to be contaminated with anti-inflammatory drugs (diclofenac, ibuprofen, indometacine, naproxen, and phenazone), lipid regulators (bezafibrate, gemfibrozil, clofibric acid, fenofibric acid), β-blockers (metoprolol, propranolol), antiepileptic (carbamazepine), antimicrobial, cytostatic agents and 17β-estradiol from ng/L range, up to μg/L (Bendz et al. 2005; Bruchet et al. 2005; Calamari et al. 2003; Delgado et al. 2012; Miao et al. 2002; Moldovan 2006; Thacker 2005). India being one of the top producers of pharmaceutical products, its annual turnover is expected to rise to USD45 billion by 2020, and drug export is growing at 30% annually (KPMG International 2006). Though India produces large number of pharmaceutical product, the sewage treatment capacity is far below than the generated sewage (Subedi et al. 2015). Presence of emerging contaminants in wastewater and surface water streams in the UK is shown in Table 1.

It is observed that more than 200 pharmaceuticals are found in the river waters globally, and the maximum reported concentration is 6.5 mg/l for the antibiotic ciprofloxacin (Hughes et al. 2013) (Fig. 1).

Consumption of such contaminated water may have detrimental effects on human and animal health. Very little is known about the ecological effects of these compounds in contrast to their pharmacological and toxicological effects at high concentrations (Santos et al. 2010; Boxall et al. 2012). Ecological effects of different classes of pharmaceutical agents are discussed in the following section:

Steroidal hormones, analgesics, and nonsteroidal anti-inflammatory drugs These chemicals can affect human, animals, and plant life. Exposure to estrogens at pollutant levels have been linked with breast cancer in women and prostate cancer in men. Estrogens are reported to affect reproductive development in both domestic and wild animals. In plant, root and shoot development, flowering, and germination processes were reported to be affected when they undergone treatment with steroid estrogen hormones or their precursors. However, estrogens help in ameliorating the effects of other environmental stresses on the plant (Adeel et al. 2017).

Nonsteroidal anti-inflammatory drugs (NSAID) such as ibuprofen, naproxen, and diclofenac and some of their metabolites (e.g., hydroxyl-ibuprofen and carboxy-ibuprofen) are very often detected in sewage and surface water. Among the NSAID group of compounds, diclofenac seems to have highest acute toxicity toward algae and invertebrates (Fent et al. 2006; Webb 2001; Cleuvers 2003).

Blood lipid regulators Acute toxicity of blood lipid regulators such as bezafibrate, clofibric acid, gemfibrozil, and fenofibric acid was investigated in different aquatic organisms like *Vibrio fischeri*, *Daphnia magna*, and *Anabaena*. Results showed varying degree of toxicity of these compounds on different organisms. Fenofibric acid was found to be most toxic for *V. fischeri* with EC(50) of 1.72 mg/l for. The wastewater was found to be very toxic to *Anabaena* CPB4337 as it inhibited 84% of its bioluminescence (Rosal et al. 2010). In another study, plasma testosterone of goldfish was reported to reduce on exposure to gemfibrozil in aqueous conditions (Mimeault et al. 2005).

Antibiotics Presence of large amount of antibiotics in environment as well as in water streams leads to development of antibiotic-resistant species of bacteria which is a cause of great concern (Daughton and Ternes 1999). The reported ciprofloxacin concentration in the outlet of WTP at Okhla, Delhi (Mutiyar and Mittal 2014), is higher than the outlet concentration of WTPs at Australia (Al-Rifai et al. 2007) and Italy (Verlicchi et al. 2012) by 2.5 and 5 times, respectively. Ash et al. (2002) reported presence of antibiotic resistance bacteria in US rivers. Presence of antibiotics may adversely affect the microorganism population present in soil and farmland

Table 1 Emerging contaminant occurrence information for wastewaters and surface waters in the UK (all data reported as mean concentrations and aqueous phases unless otherwise stated) (Petrie et al. 2015)

Emerging contaminant	Family/use	Known metabolites	Influent (ng/l)	Removal (%)	Sludge (ng/kg)	Final effluent (ng/l)	Surfaces water (ng/l)	Surfaces water max (ng/l)
Estrone	Steroid estrogen	Sulfate and glucuronide conjugate	49[c]	L-H[c,m]	–	4.3–12	–	–
17β-estradiol	Steroid estrogen	Sulfate and glucuronide conjugates	20[c]	M-H[c,m]	–	0.4–1.3[b,m]	–	–
17α-ethinylestradiol	Synthetic estrogen	Sulfate and glucuronide conjugates	1.0[c]	L-H[c,m]	–	0.20–0.47[b,m]	–	–
Propranolol	Beta-blocker	4-Hydroxypropranolol (active), glucuronide conjugates (20%)[f]	60–638[c,e,g,j]	L-H[c,g,j]	170[c]	93–388[b,e,g,j,k]	<0.5–107[d–g,j,k]	215[k]
Metoprolol	Beta-blocker	No active metabolites[f]	75–110[e,g]	L-M[g]	–	41–69[e,g]	<0.5–10[d–g]	12[f]
Salbutamol	Beta-blocker	Sulfate conjugate	0.1–130[e,g]	M-H[g]	–	63–66[e,g]	<0.5–2[e–g]	8[f]
Atenolol	Beta-blocker	Hydroxylated metabolite (3%)[f]	12,913–14,223[e,g]	M-H[g]	–	2123–2,870[e,g]	<1–487[d–8]	560[f]
Carbamazepine	Antiepileptic	Hydroxylated (10,11-epoxide) (active), conjugated metabolites[f]	950–2,593[e,g]	L[g]		826–3,117[e,g]	<0.5–251[d–g]	684[f]
Gabapentin	Antiepileptic	–	15,034–18,474[e,g]	L-H[g]		2592–21,417[e,g]	<0.6–1,879[d–g]	1,887[f]
Acetaminophen	NSAID	Sulfate conjugate (30%), paracetamol cysteinate, mercapturate (5%)[f]	6924–492,340[e,g,j]	H[g,j]		<20–11,733[e,g,j]	<1.5–1,388[d–g]	2,382[f]
Diclofenac	NSAID	Glucuronide, sulfate conjugates[f]	69–1,500[c,e,g,j]	L-H[c,g,j]	70[c]	58–599[b,e,g,j,k]	<0.5–154[e–g,k]	568[k]
Ibuprofen	NSAID	(+)-2-42)-′-Hydroxy-2-methylpropyl)-phenylpropionic acid (25%) and (+)-2-40-(2-carboxypropyl)-phenylpropionic acid (37%), conjugated ibuprofen (14%)[f]	1681–33,764[c,e,g,j]	H[c,g]	380[c]	143–4,239[b,e,g,j,k]	1–2,370[e–g,j,k]	5,044[k]

Naproxen	NSAID	6-o-Desmethylnaproxen (<1%), conjugates (66–92%)[f]	838–1,173[g]	M[g]	–	170–370[g]	1–59[f,g]	146[f]
Ketoprofen	NSAID	2-(3-benzoylphenyl)-propanoic acid and glucuronides	28–102[e,g]	M[g]	–	16–23[e,g]	1–4[e–g]	14[f]
Clofibric acid	Metabolite	–	1–651[e,g,j]	L-H[g,j]	–	6–44[e,g,j]	<0.3–101[e–g]	164[f]
Salicylic acid	Metabolite	–	5866–52,000[c,e,g]	L-H[c,g]	–	75–209[e]	4–62[e–g]	302[f]
Ranitidine	H_2 receptor agonist	N-oxide (3-6%), S-oxide (1-2%) Desmethyl Ranitidine (1–2%)[f]	<12–5,060[e,g]	L-H[g]	–	<9–425[e,g]	<3–32[d–g]	73[f]
Cimetidine	H_2 receptor agonist	Cimetidine N-glucuronide (24%), Cimetidine sulfoxide (7–14%), hydroxymethyl cimetidine (4%)[f]	2219–3,452[e,g]	L-M[g]	–	462–2,605[e,g]	<0.5–105[e–g]	220[f]
Furosemide	Diuretic	Mainly glucuronides[f]	1476–2,789[e,g]	L-M[g]	–	629–1,161[e,g]	<6–129[e–g]	630[f]
Bezafibrate	Lipid regulator	Glucuronides (20%)[f]	420–971[e,g]	L-M[g]	–	177–418[e,g]	<10–60[e–g]	90[g]
Simvastatin	Lipid regulator	β-Hydroxy acid metabolite[f]	<7–115[e,g]	–	–	<3–5[e,g]	<0.6[d,e,g]	–
Fluoxetine	Antidepressant	Norfluoxetine	14–86[c,h,i,l]	L-H[c,i]	170[c]	16–29[b,h,i]	5.8–14[h,i]	14[h,i]
Norfluoxetine	Metabolite	–	3.3–63[h,i,l]	M[i]	–	5.8–13[h,i]	1.3–2.8[h,i]	3.5[i]
Venlafaxine	Antidepressant	Desmethylvenlafaxine	120–249[h,i,l]	L[i]	–	95–188[h,i]	1.1–35[h,i]	85[i]
Dosulepin	Antidepressant	Desmethyldosulepin	21–228[h,i,l]	M–H[i]	–	57[h]	0.5–25[h,i]	32[h,i]
Amitriptyline	Antidepressant	Nortriptyline, 10-hydroxyamitriptyline (active), 10-Hydroxynortriptyline (active)[f]	106–2,092[e,g,h,i,l]	M-H[g,i]	–	66–207[e–i]	<0.5–30[d–i]	72[h]
Nortriptyline	Antidepressant	Hydroxynortriptyline	5.1–114[h,i,l]	M[i]	–	7.6–33[h,i]	0.8–6.8[h,i]	19[h,i]
Valsartan	Hypertension	Valeryl 4-hydroxy valsartan[f]	342–1,734[e,g]	L-H[g]	–	192–344[e,g]	<1–55[d–g]	144[f]
Diltiazem	Calcium channel blocker	Desacetyldiltiazem, N-monodemethyldiltiazem (active)[f]	770–1,559[e,g]	L-M[g]	–	95–357[e,g]	<1–17[d–g]	65[f]

(continued)

Table 1 (continued)

Emerging contaminant	Family/use	Known metabolites	Influent (ng/l)	Removal (%)	Sludge (ng/kg)	Final effluent (ng/l)	Surfaces water (ng/l)	Surfaces water max (ng/l)
Theophylline	Bronchodilator	Caffeine, 3-methylxanthine	9467–20,400[h,i,l]	H[i]	–	1220–3,169[h,i]	76–558[h,i]	1,439[i]
Tramadol	Analgesic	Desmethyltramadol (active)[f]	733–48,488[e-i,l]	L[g,i]	–	739–59,046[e,g,h,i]	<30–5,970[d–f,h,i]	7,731[f]
Nortramadol	Metabolite	–	226–2,457[h,i,l]	M-H[i]	–	145–433[h,i]	11–181[h,i]	410[i]
Codeine	Various	Codeine-6-glucuronide (main), free/conjugated morphine (10–15%), norcodeine (10–20%)[f]	1088–10,321[e-i,l]	L-H[g,i]	–	372–5,271[e,g–i]	<1.5–347[d–i]	815[f]
Norcodeine	Metabolite	–	30–112[h,i,l]	M[i]	–	24–33[h,i]	2.1–9[h,i]	20[h]
Oxycodone	Analgesic	Noroxycodone, oxymorphone						
Oxymorphone	Analgesic	Noroxycodone, Noroxymorphone	11–20[h,i,l]	–	–	<1.7–8.4[h,i]	<0.1–2.3[h,i]	3.5[i]
Morphine	Analgesic	Morphine-3-glucuronide, morphine-6-glucuronide, normorphine	340–481[h,i,l]	H[i]	–	59–131[h,i]	1.6–36[h,i]	36[h]
Normorphine	Metabolite	–	51–203[h,i,l]	H[i]	–	20–62[h,i]	<5–5.7[h,i]	5.7[i]
Dihydrocodeine	Analgesic	Dihydromorphine (active)	227–386[h,i,l]	L[i]	–	118–146[h,i]	2.9–36[h,i]	97[i]
Buprenorphine	Various	Norbuprenorphine and Glucuronides	33–47[h,i]	–	–	14[h]	<0.5–15[h,i]	14[i]
Norbuprenorphine	Metabolite	–	<1–19[h,i]	–	–	<0.7–7.5[h,i]	<0.5–12.2[h,i]	3.4[i]
Methadone	Analgesic	EDDP, EMDP	52–88[h,i,l]	L[i]		42–50[h,i]	0.6–12[h,i]	24[i]
EDDP	Metabolite	–	71–193[h,i,l]	L-M[i]	–	32–89[h,i]	1.2–19[h,i]	38[h]
EMDP	Metabolite	–	1.8–5.7[h,i]	–	–	1.3–1.7[h,i]	0.6–1.0[h,i]	1.1[h]

Fentanyl	Analgesic	Norfentanyl, Despropionylfentanyl	1.3–1.7[h,i]	–	–	<0.1–0.5[h,i]	<0.1[h]	–
Norfentanyl	Metabolite	–	6.9[h]	–	–	1.1[h]	<0.1[h]	–
Propoxyphene	Analgesic	Norproxyphene	8.2–11[h,i]	–	–	7.1[h]	<0.1[h]	–
Norpropoxyphene	Metabolite	–	<4.2–184[h,i,l]	L[i]	–	91–106[h,i]	6.5–31[h,i]	80[i]
Temazepam	Hypnotic	Oxazepam	85–208[h,i,l]	L[i]	–	135–179[h,i]	3.2–34[h,i]	78[i]
Diazepam	Hypnotic	Nordiazepam, temazepam, oxazepam	<0.9–7.6[h,i]	–	–	1.6–5.1[h,i]	0.6–0.9[h,i]	1.1[h,i]
Nordiazepam	Metabolite	–	12–25[h,i,l]	M[i]	–	5.8–9.9[h,i]	0.7–3.2[h,i]	6.8[i]
7-Aminonitrazepam	Metabolite	–	<1.9–205[h,i]	–	–	<1.2[h]	<0.5[h]	–
Oxazepam	Hypnotic	Glucuronide metabolite	22–50[h,i,l]	L[i]	–	33–58[h,i]	2.4–11[h,i]	21[i]
Ketamine	Anesthetic	Norketamine	52–235[h,i,l]	L[i]	–	83–130[h,i]	0.6–27[h,i]	54[i]
Norketamine	Metabolite	–	11–85[h,i,l]	L[i]	–	14–28[h,i]	1.7–5.8[h,i]	14[h]
Sildenafil	Erectile dysfunction	N-desmethyl sildenafil	8.3–25[h,i,l]	L[i]	–	7.0–9.7[h,i]	<1–2.2[h,i]	2.9[i]
Ephedrine/ pseudoephedrine	Various	Cathine	476–966[h,i,l]	H[i]	–	35–70[h,i]	7.7–15[h,i]	17–29[h,i]
Norephedrine	Various	–	40–86[h,i]	H[i]	–	59[h]	<5[h]	–
Amoxicillin	Antibacterial	Amoxicilloic acid	<87[e]	–	–	31[e]	<2.5–245[d,e,f]	622[f]
Erythromycin	Antibacterial	Erythromycin-H_2O[f]	71–2,530[c,e,g,j]	L-H[c,g,j]	50[c]	109–1,385[b,e,g,j,k]	<0.5–159[d–g,j,k]	1,022[k]
Metronidazole	Antibacterial	1-(b-hydroxymethyl-5-nitroimidazole, 2-methyl-5-nitroimidazole-1yl-acetic acid[f]	569–2,608[e,g]	L[g]	–	265–373[e,g]	<1.5–12[d–g]	24[f]
Ofloxacin	Antibacterial	Desmethyl, N-oxide Metabolites	180[c]	M-H[c]	210[c]	10[b]	–	–

(continued)

Table 1 (continued)

Emerging contaminant	Family/use	Known metabolites	Influent (ng/l)	Removal (%)	Sludge (ng/kg)	Final effluent (ng/l)	Surfaces water (ng/l)	Surfaces water max (ng/l)
Chloramphenicol	Antibacterial	Glucuronide conjugates[f]	<4–248[e,g]	H[g]	–	<6–21[e,g]	<10[d,f,g]	40[f]
Sulfamethoxazole	Antibacterial	N_4 –acetylated Metabolite	<3–115[e,g]	L-M[g]	–	10–19[e,g]	<0.5–2[d–g]	8[g]
Sulfapyridine	Antibacterial	Hydroxyl, acetyl Metabolites	914–4,971[g]	L-H[g]	–	277–455[g]	<2–28[f,g]	142[f]
Sulfasalazine	Chronic bowel disorders	5-Aminosalicylic acid (active), sulfapyridine (active)[f]	0.2–116[e,g]	L[g]	–	0.3–484[e,g]	<1.5–76[e–g]	168[f]
Trimethoprim	Antibacterial	1,3-oxides, 3′ 4-hydroxy derivatives[f]	213–2,925[e,g,j]	L-M[j,g]	–	128–1,152[e,g,j,k]	<1.5–108[d–g,j,k]	183[f]
Oxytetracycline	Antibacterial	N-desmethyl Oxytetracycline	3600	H[c]	6710	170[b]	–	–
Tamoxifen	Anticancer	Hydroxytamoxifen (active)	143–215[j]	L[j]	–	<10–369[j,k]	<10–212[j,k]	–
Illicit drugs and licit stimulants								
MDMA	Hallucinogen	MDA, HMMA, HMA	10–231[h,i,l]	L[i]	–	13–38[h,i]	0.5–8.7[h,i]	25[h,i]
MDEA	Hallucinogen	MDA	<0.3–1.4[h,i]	–	–	<0.5[h]	<0.1[h]	–
MDA	Hallucinogen/ metabolite	–	10–18[h,i,l]	–	–	11–15[h,i]	<0.1–1.7[h,i]	1.7[i]
Amphetamine	Stimulant	25% Phenylacetone, benzoic acid, hippuric acid, <10% 4-hydroxy-norephedrine, norephedrine[f]	77–5,236[e–i,l]	H[g,i]	–	2–201[e,g–i]	<1–9[d–i]	4.3[h]
Methamphetamine	Stimulant	Amphetamine, 4-hydroxymethamphetamine	2–40[h,i,l]	M-H[i]	–	0.8–1[h,i]	<0.1–0.3[h,i]	0.3[i]

Cocaine	Stimulant	35–54% benzoylecgonine, 32–49% ecgonine[f]methyl ester (main)	57–526[e,g–i,l]	L-H[i]	–	<1–149[e,g–i]	<0.3–6[d–i]	17[i]
Benzoylecgonine	Metabolite	–	196–1,544[e,g–i,l]	L-H[g]	–	13–1,597[e,g–i]	<1–92[d–i]	72[i]
Norbenzoylecgonine	Metabolite	–	6–54[h,i,l]	–	–	3.6–7[h,i]	0.3–2[h,i]	3.1[i]
Norcocaine	Metabolite	–	1[h]	–	–	<0.2[h]	0.1[h]	0.1[h]
Cocaethylene	Metabolite	–	3–45[h,i,l]	L-M[i]	–	1.7–3.8[h,i]	0.1–0.6[h,i]	1.4[h]
Benzylpiperazine	Stimulant	Hydroxylated metabolite	22–25[h,i]	L-H[i]	–	31–44[h,i]	1.1–26[h,i]	65[h]
Trifluoromethyl phenylpiperazine	Stimulant	Hydroxylated metabolite	2.4–5.1[h,i,l]	–	–	4.6–6.6[h,i]	0.2–1.2[h,I]	6.1[i]
6-acetylmorphine	Metabolite	–	3–22[h,i,l]	–	–	<0.3–2.2[h,i]	<0.1[h]	–
Caffeine	Human indicator	Acids and 1-methylxanthine	9902–25,138[h,i,l]	H[i]	–	1744–2,048[h,i]	163–743[h,i]	1,716[i]
Nicotine	Human indicator	–	3919–9,684[h,i,l]	H[i]	–	52[h]	12–86[h,i]	148[i]
Personal care products								
Triclosan	Antibacterial	–	70–2,500[c,e,g]	M-H[c,g]	5,140[c]	25–200[b,e,g]	5–48[e–g]	–
Bisphenol A	Plasticizer	–	416–2,050[c,e,g]	M-H[c,g]	320[c]	35–86[b,e,g]	<6–34[e–g]	–
1-benzophenone	Sunscreen agent	–	134–306[e,g]	H[g]	–	12–32[e,g]	<0.3–9[e,f]	–
2-benzophenone	Sunscreen agent	–	25–194[e,g]	H[g]	–	1–4[e,g]	<0.5–18[e–g]	–
3-benzophenone	Sunscreen agent	–	638–1,195[e,g]	M-H[g]	–	22–231[e,g]	<15–36[e–g]	–
4-benzophenone	Sunscreen agent	–	3597–5,790[e,g]	L[g]	–	2701–4,309[e,g]	<3–227[e–g]	–
Methylparaben	Preservative	–	2642–11,601[e,g]	H[g]	–	<3–50[e,g]	<0.3–68[e–g]	–

(continued)

Table 1 (continued)

Emerging contaminant	Family/use	Known metabolites	Influent (ng/l)	Removal (%)	Sludge (ng/kg)	Final effluent (ng/l)	Surfaces water (ng/l)	Surfaces water max (ng/l)
Ethylparaben	Preservative	–	589–2,002[e,g]	H[g]	–	4–50[e,g]	1–13[e–g]	–
Propylparaben	Preservative	–	598–3,090[e,g]	H[g]	–	26–63[e,g]	<0.2–7[e–g]	–
Butylparaben	Preservative	–	50–723[e,g]	H[g]	–	<1[e,g]	<0.3–6[e–g]	–

Key: NSAID, nonsteroidal anti-inflammatory drug; L, secondary removal is <50%; M, secondary removal is within the range 50–80%; H secondary removal >80%

[a]Taken from NHS Prescription Cost Analysis information for England (National Health Service (NHS) 2012)

[b]Median data reported from 162 WwTWs (Gardner et al. 2012)

[c]Data obtained from 14 WwTWs (Inc. activated sludge, trickling filters, membrane bioreactors, and biological nutrient removal plants (Gardner et al. 2013)

[d]Kasprzyk-Hordern et al. (2007)

[e]Kasprzyk-Hordern et al. (2008a)

[f]Kasprzyk-Hordern et al. (2008b)

[g]Two WwTWs monitored (activated sludge and trickling filter) (Kasprzyk-Hordern et al. 2009)

[h]Baker and Kasprzyk-Hordern (2011)

[i]Seven WwTWs (activated sludge and trickling filters), median data reported (Baker and Kasprzyk-Hordern (2013)

[j]WwTWs is a train of processes (trickling filter, activated sludge, and UV treatment) (Roberts and Thomas 2006)

[k]Five WwTWs (oxidation ditch, activated sludge and trickling filters) (Ashton et al. 2004)

[l]Baker et al. (2014)

[m]Includes particulate concentrations (Koh et al. 2009)

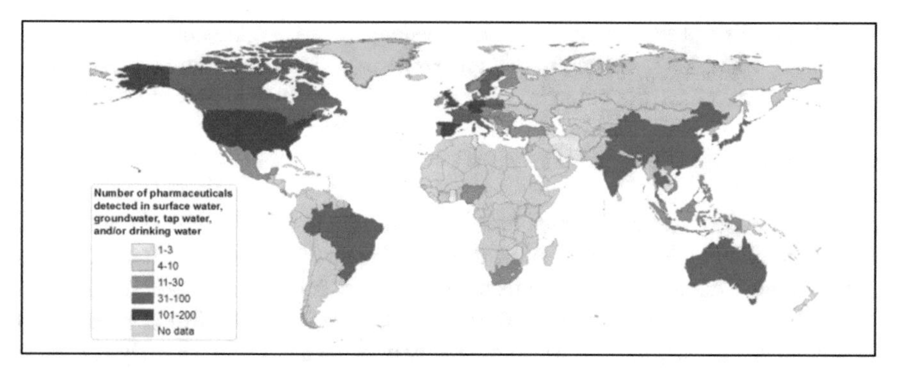

Fig. 1 Country survey on the number of pharmaceutical substances detected in surface water, groundwater, or tap/drinking water. (With permission from (aus der Beek et al. 2016))

that help in nitrogen fixation and other nutrient fluxes required for agriculture processes (Ash et al. 2002).

Beta blockers Beta blockers are among the most widely used pharmaceutical products used for treatment of cardiovascular disorders such as high blood pressure, ischemic heart disease, anxiety therapies, and heart rhythm disturbances (British Pharmacopoeia Commission 2005; United States Pharmacopeia 2005). These compounds are available in the aquatic and terrestrial environments in the range of ngL^{-1} even up to μgL^{-1} due to different anthropological activities. Beta blockers such as propranolol, metoprolol, and nadolol were reported to be highly stabile in aqueous conditions, and their estimated half-lives are more than 1 year (Maszkowska et al. 2014a). These chemicals may have deleterious effects on different aquatic organisms such as fish (Japanese medaka, rainbow trout), invertebrates (*Daphnia magna*, *Hyalella azteca*, *Daphnia lumholtzi*, *Ceriodaphnia dubia*), and green algae (*Pseudokirchneriella subcapitata*) (Maszkowska et al. 2014b; Santos et al. 2010). Since beta blockers are endocrine disruptive compounds (ECDs), it has been proven that they affect both free and total testosterone levels in male organisms (Rosen et al. 1988; el-Sayed et al. 1998).

Neuroactive compounds (antiepileptics, antidepressants) Human antiepileptic drug carbamazepine (CBZ) was reported to significantly decrease the siphoning behavior (filtration rates), superoxide dismutase (SOD), and glutathione reductase activities of freshwater clams *Corbicula fluminea, whereas the* catalase activity and malondialdehyde (MDA) content were increased in the gills and digestive gland, suggesting that CBZ induced an oxidative effect. In short, exposure to CBZ at environmentally relevant concentration exerts a negative effect on *C. fluminea* tissue at the molecular and protein level (Chen et al. 2014). Antidepressants are human pharmaceuticals extensively detected in the aquatic environment, and they act by modulating the neurotransmitters serotonin, dopamine, and norepinephrine. Presence of these antidepressants at environmentally relevant concentrations was reviewed and reported to disrupt the normal biological systems of two highly abundant and ecologically important invertebrate groups, i.e., molluscs and crustaceans. Antidepressants have the potential to affect multiple biological processes including reproduction, growth, metabolism, immunity, feeding, locomotion, color physiology, and behavior (Fong and Ford 2014).

Antineoplastics Antineoplastics or anticancer agents are highly toxic by nature. Different antineoplastic agents show different levels of toxicity to different organisms. Acute toxicity of four cytostatic drugs, 5-fluorouracil (5-FU), cisplatin (CDDP), etoposide (ET), and imatinib mesylate (IM) in zebrafish (*Danio rerio*) embryos, in adult fish and subchronic toxicity of 5-FU and IM in the early-life stage were reported to be low acute and subchronic toxicity, which indicates low susceptibility of fish toward these drugs (Kovacs et al. 2016). However, an earlier study reported that the chronic two-generation exposure of zebrafish to 5-FU at environmentally relevant concentrations (10 ng L^{-1}) caused histopathological changes in the liver and kidney, impaired their DNA integrity, and induced massive whole-transcriptome changes (Kovacs et al. 2015). In another study, it was reported that the fertility of higher plants was adversely affected by anticancer drugs such as 5-FU, CDDP, and ET. These drugs increased the frequencies of abortive grains with the lowest effective doses between 1 and 10 mg/kg of dry soil (Misik et al. 2016). In higher plants such as *Tradescantia* and *Allium*, cytotoxic drugs induced genotoxic effects including DNA damage (Misik et al. 2014).

Various other compounds Apart from these compounds, a large number of pharmaceutical compounds such as impotence drugs, retinoids, diagnostic contrast media, tranquilizers, bronchodilators, pharmaceutical belonging to the group of endocrine-disrupting chemicals (EDCs), etc. have been detected in sewage and surface water (Daughton and Ternes 1999; Heberer 2002; Fent et al. 2006). In US water streams, the antacid cimetidine and ranitidine were detected in the concentrations of 0.58 and 0.01 μg/L, respectively (Kolpin et al. 2002). Contrast media used in X-ray (Iopamidol) and MRI (gadolinium-based complex) were reported at various concentrations in different water streams (Kümmerer and Helmers 2000; Putschew et al. 2000). EDCs are drawing huge attention in last decade as they can adversely affect human and animal health. Exposure to EDCs is responsible for decrease in male sperm count, increase in cases of testicular, prostate, ovarian, and breast cancers, and reproductive malfunctions (Joffe 2001). Fetuses and newborn babies are the most vulnerable to EDCs (Sharpe and Irvine 2004).

As discussed presence of pharmaceutical compounds in different aquatic streams has large number of negative impacts on human, animals, and aquatic ecology; it is necessary to remove the pharmaceuticals from aquatic streams. The different microalgal-based treatment processes are being discussed in the following section.

3 Different Wastewater Treatment Processes

Conventional water and wastewater treatment plants have been set up using the best designs for treatment and maximum removal of contaminants and eutrophicating pollution loads which are specified in the existing regulations. However, the conventional treatment processes are not capable of handling the new and "unregulated" micro-contaminants such as EDCs, pharmaceutical, and personal care products effectively. Due to presence of wide range of contaminants, treatment of wastewater is much more complicated than treatment of water. Advanced treatment processes are required for treating such micro-contaminants (Bolong et al. 2009). Discharge of wastewater effluent into water bodies results in significant reduction in the concentration of these compounds due to their degradation or binding with natural organic matter or soil along the riverbank, resulting in lower concentrations in downstream (Verstraeten et al. 2003; Bowman et al. 2002). Still, these contaminants are available in the range of ηg/L to μg/L in drinking water. A bench-scale water treatment study was conducted using water from a treatment plant model and natural waters containing 30 pharmaceuticals to 80 different EDCs. Results (shown in Table 2) showed that there is no significant removal except using activated carbon (powder) and oxidation via chlorination and ozonation. As per Westerhoff et al. (2005), conventional treatment processes such as coagulation, flocculation, or lime softening were reported to be ineffective in removing EDCs and PPCPs (pharmaceuticals and personal care products). Similarly, in few other studies also, it was demonstrated that conventional treatment processes were not able to remove pharmaceutical compounds effectively (Adams et al. 2002; Petrovic et al. 2003; Vieno et al. 2006).

Advanced technologies like oxidation, photodegradation and photocatalytic degradation, membrane filtration, and hybrid processing using multiple different technologies have been investigated for removal of PPCP from wastewater. Oxidation processes using O_3, H_2O_2, and Fenton (Fe^{2+}/H_2O_2) have been reported to be highly efficient in removing PPCPs (Ghatak 2014; Esplugas et al. 2007). These processes generate hydroxyl radical to breakdown PPCPs oxidatively. In photodegradation and photocatalytic degradation processes, PPCPs are degraded using direct photolysis and indirect photolysis (Kanakaraju et al. 2014; Yang et al. 2014). Other technologies such as membrane filtration, activated carbon adsorption, and hybrid processes have also been found to be efficient in removing PPCPs (Rodriguez et al. 2016; Yang et al. 2011; Ahmed et al. 2017). Although these technologies are demonstrating promising results in treating PPCPs, they are very expensive, and large-scale application is not economic and sustainable at present level. Therefore, search for more cost-effective and sustainable solutions are going on (Wang et al. 2017). Biological treatment using algal-based technologies such as constructed wetlands, i.e., shallow pond, beds, or trenches, provides sustainable and cost-effective solution in treating wastewater as they require low energy, less operational requirements (Wu et al. 2015b). We are discussing the different microalgal-based treatment processes for treatment of wastewater.

Table 2 Removal performance of EDCs by selected treatment processes (Westerhoff et al. 2005; Bolong et al. 2009)

Treatment process	Removal performance
Coagulation by alum or ferric sulphate	<20% of compound removed, specially associated with particulate matter. Presence of hydrophobic dissolved organic carbon enhances removal and provides partitioning
Lime softening	<20% of compound concentration was removed at pH 9–11
Powder activated carbon (Battin et al. 2016)	>90% of many EDCs removed (at 5 mg/l dose PAC of 4 hour contact time). Yet some EDCs (ibuprofen, sulfamethoxazole, meprobamate) had lower removals (40–60%) Hydrophobic compounds (octanol-water partition coefficient, log K ow >5) have better removal than polar compounds
Biofilm	Removal depends on biodegradability of compounds, but removal rate is unclear
Chlorination	Able to remove >90% for more reactive compounds containing aromatic structures with hydroxide functional groups Not suitable because it produces chlorine by-product (react with EDCs) and should be avoided
Ozonation	Oxidized similar to chlorination but at slightly higher removal rates Addition of hydrogen peroxide during ozone addition slightly increased the EDC removal

4 Mechanism Involved in Removal of Pharmaceuticals Using Microalgal-Based Treatment Processes

Algal-based methods can remove pharmaceuticals and other personal care products from wastewater using mechanisms like sorption, biodegradation, photodegradation, and volatilization as shown in Fig. 2.

4.1 *Sorption and Volatilization*

Sorption is considered as redistribution of a substance from liquid phase to solid phase. Adsorption is adhesion of substances on solid surface, while absorption is transfer of substance into a sorbent. Maes et al. (2014) reported efficient removal and biotransformation of EE2 by *Desmodesmus subspicatus*. Aliphatic amine groups containing ionizable pharmaceuticals have been accumulated in algae cells through ion-trapping effect (Neuwoehner and Escher 2011), and positively charged pharmaceuticals are adsorbed onto negatively charged surface due to electrostatic attractions (Stevens-Garmon et al. 2011). However, sorption of acidic pharmaceuticals is found to be relatively weak at elevated pH (Duan et al. 2013). PPCPs like fluoroquinolones and tetracyclines have been adsorbed strongly on soil through surface complexation (Carrasquillo et al. 2008; Vasudevan et al. 2009) with the help of metal ions precipitated in algae biofilms. Algae or algal-bacterial consortia bioaccumulate PPCPs from wastewater streams instead of breaking them down via

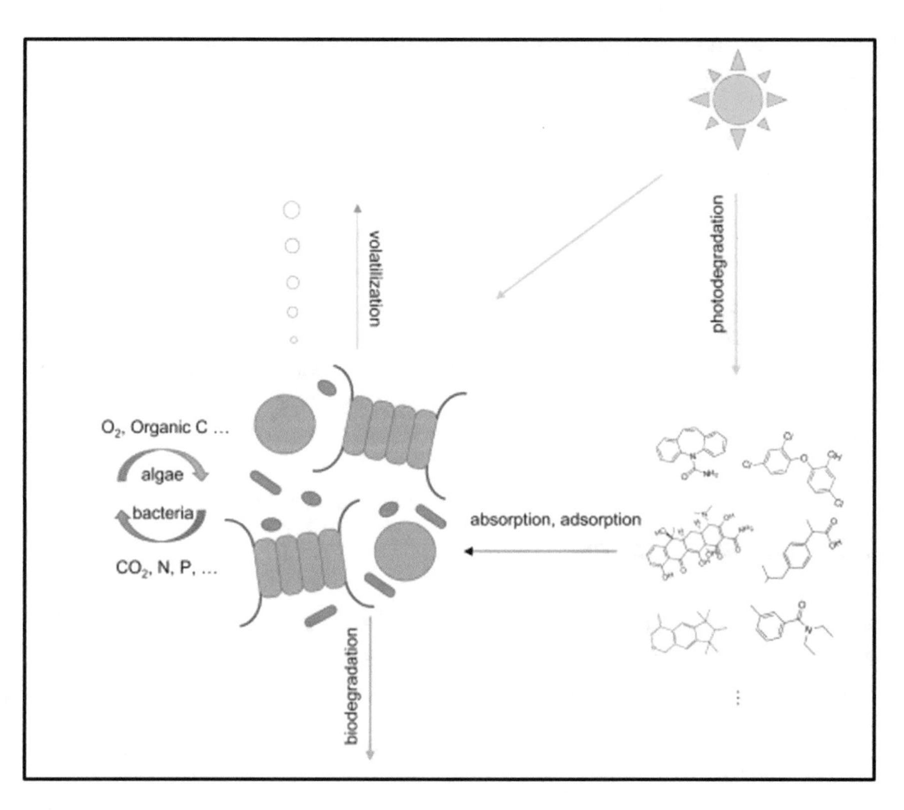

Fig. 2 Processes involved in PPCP removal using algae-based technologies (With Permission from Wang et al. (2017))

sorption processes. Therefore, PPCP containing biomass generated from algae-based treatment systems must be disposed properly to avoid release of the sorbed PPCPs, and this contamination should be considered when biomasses are used for the production of other valuable products (Wang et al. 2017).

Volatilization is one of the methods for removal of PPCPs in CAS systems (Suárez et al. 2008). Volatilization helps in removal of volatile and semi-volatile PPCPs from open algae-based treatment system. However, volatilization transfers pollutants from the water bodies to the atmosphere and don't break them down.

4.2 *Biodegradation*

Biodegradation is a process where organic chemicals are broken down with the help of enzymes produced by microorganisms. In algae-based technologies, algae can play an important role in biodegradation of organic contaminants. Enzymes present in algae metabolize variety of xenobiotics with a process divided into three phases

(Torres et al. 2008). In phase I, hydrophobic xenobiotics are oxidized, reduced, or hydrolyzed to transform into more hydrophilic compounds, and these facilitate their excretion. Cytochrome P450, a group of microsomal heme-thiolate proteins anchored in algal membrane, plays an important role in this phase (Zangar et al. 2004). In phase II hydrophilic moieties are conjugated to xenobiotics to facilitate their excretion. Xenobiotics containing COOH, OH, or NH_2 groups and the metabolites from phase I are catalyzed by glutathione S-transferases (GSTs) or glucosyltransferases to conjugate glutathione (GSH) or glucuronic acid (Nakajima et al. 2007; Pflugmacher et al. 1999; Yang et al. 2002). In phase III, xenobiotics are packed in vacuoles or by cell wall fractions (Dietz and Schnoor 2001; Petroutsos et al. 2008). Detoxification process of xenobiotics from environment by algae is similar to mammalian liver, and, hence, algae are considered as “green livers” (Torres et al. 2008).

4.3 *Photodegradation*

Photodegradation is one of the major mechanisms to remove pharmaceuticals from wastewater streams, and many of them are reported to be photodegradable under sunlight irradiation (Hanamoto et al. 2014; Lam et al. 2004; Lin et al. 2006). Target compounds with aromatic rings, conjugated π systems, heteroatoms, and other functional groups are degraded due to strong absorption of UV light from solar radiation (Challis et al. 2014). In addition, algae can generate free radicals such as hydroxyl radicals (OH-), peroxyl radicals (ROO-), and singlet oxygen ($^1O_2$) under illuminating sunlight in presence of photosensitizers and certain metal ions (Boreen et al. 2003). Then these free radicals facilitate biodegradation of PPCPs as demonstrated by Eqs. (1) and (2) (Collén et al. 1995; Liu et al. 2004; Peng et al. 2006).

$$O_2 + \text{cell organelles of algae} + hv \rightarrow {}^1O_2 / O_2^{\cdot-} \quad (1)$$

$$O_2 + \text{cell secretion of algae} + hv \rightarrow \cdot OH \quad (2)$$

Rate of photodegradation depends upon the latitude and varies seasonally due to angle and duration of the solar irradiance, and estimated half-life ($t_{1/2}$) periods for photodegradation of pharmaceuticals under sunlight vary from a few hours to hundreds of days (Andreozzi et al. 2003; Wu et al. 2015a; Yamamoto et al. 2009). In algae-based treatment systems, sunlight is the preferred for the photosynthesis is one of the major mechanisms to r of algae as well as photodegradation of pharmaceuticals to reduce cost of extra energy.

5 Removal of Pharmaceuticals in Microalgal-Based Treatment Systems

Research on use of microalgae for the removal of pharmaceuticals has grown since the last decade. Investigation in the treatment of urban wastewater with pharmaceutical pollutants using algal cultivation with an initial inoculation with *Tetradesmus dimorphus* showed that removal efficiencies were very high (>90%), moderate (50–90%), low (10–50%), and very low or nonquantifiable (<10%) for 9, 14, 11, and 18 pharmaceuticals, respectively, over a 7-day period (Gentili and Fick 2017). Among all pharmaceuticals tested, high removal rates were found for atenolol, bisoprolol, metoprolol, clarithromycine, bupropion, atracurium, diltiazem, and terbutaline. Further the most frequent genus in the batches we examined was the green alga *Dictyosphaerium* (Gentili and Fick 2017). Study using freshwater microalgae such as *Chlamydomonas mexicana* and *Scenedesmus obliquus* showed a maximum of 35% and 28% biodegradation of carbamazepine (CBZ), respectively, and further demonstrated that *C. mexicana* was more tolerant to CBZ and could be used for treatment of CBZ contaminated wastewater (Xiong et al. 2016). In a study by Matamoros et al. (2016), *Chlorella* sp. and *Scenedesmus* sp. were inoculated in urban or synthetic wastewater containing caffeine, ibuprofen, galaxolide, tributyl phosphate, 4-octylphenol, tris(2-chloroethyl) phosphate, and carbamazepine and were incubated for 10 days in aerated reactors. Results showed that 99% of the micro-contaminants (4-octylphenol, galaxolide, and tributyl phosphate) were removed by volatilization due to the effect of air stripping and 95% and 99% of ibuprofen and caffeine were removed, respectively, by biodegradation (Matamoros et al. 2016).

In another study, removal of six spiked (100–350 μg/l) pharmaceuticals (diclofenac, ibuprofen, paracetamol, metoprolol, carbamazepine, and trimethoprim) were investigated through batch experiments with the microalgae *Chlorella sorokiniana* grown on urine, anaerobically treated black water, and synthetic urine (Wilt et al. 2016). 60–100% removal of diclofenac, ibuprofen, paracetamol, and metoprolol was by biodegradation and photolysis, and carbamazepine and trimethoprim were recalcitrant toward both biodegradation and photolysis, and removal did not exceed 30% and 60%, respectively, while sorption to algal biomass accounted for less than 20% of the micropollutant removal (Wilt et al. 2016).

Chlorella sorokiniana is reported as a robust strain for the bioremediation of paracetamol and salicylic acid concentrated wastewaters (Escapa et al. 2017). At two different concentrations of these pharmaceuticals (I, 25 mg/l; II, 250 mg/l), *C. sorokiniana* showed removal efficiencies above 41% and 69% for PCI and PCII, respectively, and above 93% and 98% for SaCI and SaCII, respectively (Escapa et al. 2017). In similar study, removal kinetics using *Chlorella sorokiniana* were 2.3 times greater for the salicylic acid than paracetamol, reaching volumetric efficiencies above 93% for salicylic acid in the semicontinuous culture, and removal of nutrients were >70% for nitrates and >89% for phosphates (Escapa et al. 2015).

In another study, three different microalgae strains, namely, *Chlorella sorokiniana*, *Chlorella vulgaris*, and *Scenedesmus obliquus*, were used to see the removal efficiency of diclofenac and to verify the effect of diclofenac on nutrient removal capacity. *S. obliquus* showed the highest efficiency in the removal of diclofenac (>79%) and nutrients (>87% nitrates, >99% phosphates) per liter and per gram of biomass, while *C. sorokiniana* was the strain showing the largest increase of growth rate and microalgae density, which were above 25% and 31%, respectively (Escapa et al. 2016).

Shallow high rate algal ponds (HRAPs) showed 69 ± 1% tetracycline removal with influent concentration of 2 mg/l, and removal was mainly caused by photodegradation and biosorption (de Godos et al. 2012). Four microalgal species was screened (*Scenedesmus obliquus*, *Chlamydomonas mexicana*, *Chlorella vulgaris*, *and Chlamydomonas pitschmannii*) to remove diazinon from the aqueous phase (Kurade et al. 2016). *Chlorella vulgaris* showed the highest removal capacity (94%) at 20 mg/l; however, the growth of *C. vulgaris* was significantly affected above 40 mg/l of diazinon, showing >30% growth inhibition after 12 days of cultivation (Kurade et al. 2016). Naproxen is one of the most prevalent pharmaceuticals, and biodegradation study of naproxen using freshwater algae *Cymbella sp.* and *Scenedesmus quadricauda* showed higher removal efficiency for *Cymbella* sp., and hydroxylation, decarboxylation, demethylation, tyrosine conjunction, and glucuronidation contributed to naproxen transformation in algal cells (Ding et al. 2017). Xiong et al. (2017) reported that after 11 days of cultivation of *Chlamydomonas Mexicana*, 13 ± 1% of ciprofloxacin (2 mg/l) was removed, and its removal was enhanced by >threefold (56 ± 1.8%) after addition of electron donor (Sodium acetate, 4 g/l). Gao et al. (2011) reported that *Chlorella vulgaris* and three local isolates, *Chlorella* sp., *Chlorella* sp., and *Chlorella miniata*, had a rapid and high ability to remove nonylphenol. Further, highest nonylphenol (more than 80%) was degraded by *C. vulgaris* after 168 hours of cultivation followed by *Chlorella miniata*.

6 Additional Potential Role of Microalgal-Based Treatment Processes

6.1 Biofuel Production from Microalgae

Microalgal biomass which is produced during wastewater is not safe to be used as human food or animal feed. However, phycoremediation-based algal biomass along with biodegradable organic municipal waste has a strong potential for the sustainable production of biofuels (Demirbas 2009; Gibbons and Hughes 2009). The oil content of algae in relation to their dry weight made them ideal for the production of biofuel or biogas, through different conversion processes such as pyrolysis, direct combustion or thermochemical liquefaction, fermentation, and biochemical processes (Briens et al. 2008; Rizwan et al. 2018), and sources of renewable energy. Some specific biofuels that can be produced from algal and residual biomass are biodiesel, bioethanol, biogas, and biohydrogen production (Sivaramakrishnan and Incharoensakdi 2018; Zhang et al. 2018; Zhang et al. 2016). Bioethanol from microalgal biomass can be produced through fermentation and gasification (Pittman et al. 2011), and proficient microalgal species for bioethanol production via starch fermentation are *Chlorella vulgaris* and *Chlamydomonas* sp. (Saharan et al. 2013).

Although microalgae offer a good potential for biogas and bioethanol production, research on production of biogas and bioethanol from microalgal biomass is still in infancy and not yet commercialized (Harun et al. 2010). Biogas is considered to be an extremely versatile and eco-friendly fuel as it releases significantly lower amounts of greenhouse gases and other particulate matter relative to conventional and nonrenewable fuels. Methane that can be produced by anaerobic digestion of residual biomass after lipid extraction can be used to generate the electrical power necessary for running the microalgal biomass facility, reducing the cost of making microalgal biodiesel (Chisti 2007). Liquefaction of microalgae has resulted in production of between 30% and 65% dry weight of oil depending on species used (Mutanda et al. 2011; Brennan and Owende 2010; Ross et al. 2010), and the major benefit of thermochemical liquefaction is the ability to use wet biomass. Pyrolysis of algal biomass has given promising results and has been shown to produce higher-quality bio-oil than lignocellulosic compounds (Brennan and Owende 2010), and lipid containing biomass has been shown to produce higher heat balances and bio-oil yields (Ross et al. 2010). The potential of microalgae consortia used in dairy wastewater treatment combined with microalgae biodiesel feedstock production was evaluated by comparing the nutrient removal of dairy wastewater, the growth of cells, and the lipid content and composition of biomass between monoalgae and microalgae consortia cultivation system (Qin et al. 2016).

6.2 Microalgal-Based CO_2 Fixation

Microalgae are capable of fixing CO_2 photosynthetically and have greater efficiency than terrestrial plants in converting solar energy to chemical energy (Khan et al. 2009; Li et al. 2008; Rosenberg et al. 2011). In addition to CO_2 fixation, Chlorophyta showed 10–50 times better solar energy absorbing efficiency than terrestrial plants (Wang et al. 2008). Microalgae utilized CO_2 from atmosphere as carbon source to grow and reproduce, and efficient microalgal growth and metabolism require enhanced CO_2 levels typically well above its atmospheric level and currently are major contributors to the overall cost of microalgal cultivation (Kumar et al. 2010). Currently, research is also focused to identify suitable microalgae strains that can grow under high concentration of CO_2 while producing lipid for subsequent biodiesel production.

Microalgae can tolerate high CO_2 concentration in feeding air streams and help in biofuel production (Chang and Yang 2003), and it also allows an efficient capturing of CO_2 content (5–15%) from the flue and flaring gasses (Hsueh et al. 2007). *Chlorella sp*. eliminated 10–50% CO_2 level from flue gas, and the efficiency of absorbing CO_2 was decreased by more injection of flue gas into microalgae culture (Doucha et al. 2005). Supplying high concentration of CO_2 to microalgae could enhance the accumulation of polyunsaturated fatty acid in the microalgae cells (Tang et al. 2011).

A marine alga *Chlorococcum littorale* displayed a notable tolerance to high CO_2 content of up to 40% (Iwasaki et al. 1998), and *Chlorella* strains from hot springs are tolerant to high CO_2 concentration (40%) and a temperature of 42 °C (Sakai et al. 1995).

Chlorella sp., *Scenedesmus* sp., and *Botryococcus braunii* are among the microalgae strains that have shown promising result to bio-mitigate CO_2 emission with typical CO_2 consumption rate of 200–1300 mg/L/day (Rosenberg et al. 2011; Sydney et al. 2010; Yoo et al. 2010; Zhao et al. 2011). *Scenedesmus obliquus* was able to tolerate high concentration of CO_2 up to 12% (v/v) with optimal removal efficiency of 67% (Li et al. 2011). Algae *Monoraphidium minutum* could produce a significant amount of biomass by efficiently utilizing flue gas comprising high content of CO_2 along with sulfur and nitrogen oxides (Zeiler et al. 1995). Consequently, microalgae has attracted a great deal of attention as to serve the multiple purpose of biofuels production or other valuable by-products, together with a potential greenhouse gases (GHGs) mitigation and wastewater treatment (Rizwan et al. 2018; Kumar et al. 2010).

7 Challenges and Future Prospects

Removal of pollutants including pharmaceuticals by algal-based technologies has multiple advantages, but it still faces several challenges. The treatment efficiency can be affected by wastewater strength and flow rate; toxicity of target pollutants; availability of nutrients; physicochemical parameters such as pH, temperature, light intensity, and CO_2 concentration; and microalgal species used and production and operation cost (Lam and Lee 2012; Rawat et al. 2013; Liang 2013; Chen et al. 2015). Water loss due to wastewater evaporation, as well as a greater contamination risk, may lead to higher costs and energy consumption associated with replenishment of lost water and selection of risk-tolerance species (Rogers et al. 2014). Further, many good technologies have been developed for microalgal-based biofuel conversion, land availability for commercial-scale microalgal cultivation is a constrain, and scalability remains a challenge before they are economically feasible (Ribeiro and da Silva 2013; Chen et al. 2015). Significant improvement must be done on the current economics of algae biofuel production in order to permit a good competition with fossil fuel before the ultimate displacement (Rastogi et al. 2018).

More research is needed to optimize the natural method of microalgal cultivation for multiple roles of microalgae in pharmaceutical removal, energy production, and CO_2 mitigation. Genetic engineering may play important role to improve the microalgal strain for integrated approach. The combination of the three roles of microalgae – CO_2 fixation, wastewater treatment, and biofuel production – has the potential to maximize the impact of microalgal biofuel production systems (Guzzon et al. 2008; Kumar et al. 2010). An integrated process considering the various biotransformations and biomass biorefineries is the best solution to maximize economic and environmental benefits while minimizing waste and pollution (Briens et al. 2008; Singh and Gu 2010).

8 Conclusions

Pharmaceuticals as emerging contaminants will remain a great environmental challenge in the years to come due to increasing and extensive application and associated potential environmental risk. Wastewater-based microalgal cultivation is an ideal platform to simultaneously remove pharmaceuticals from wastewater. Algae-based technologies show several advantages over conventional wastewater treatment technologies including reduction in energy consumption, biological sequestration of CO_2, and biofuel production. It is only by continuous research and development that the possibility of the cost-effectiveness and global availability of microalgae for treatment of emerging contaminants such as pharmaceuticals along with cost-effective biofuel production and CO_2 reduction from flue gas could be achieved during the next decade. Research should also be focused on degradation pathways and potential risks of the degradation products.

References

Abinandan S, Shanthakumar S (2015) Challenges and opportunities in application of microalgae (Chlorophyta) for wastewater treatment: a review. Renew Sust Energ Rev 52:123–132. https://doi.org/10.1016/j.rser.2015.07.086

Adams C, Wang Y, Loftin K, Meyer M (2002) Removal of antibiotics from surface and distilled water in conventional water treatment processes. J Environ Eng 128(3):253–260. https://doi.org/10.1061/(ASCE)0733-9372(2002)128:3(253)

Adeel M, Song X, Wang Y, Francis D, Yang Y (2017) Environmental impact of estrogens on human, animal and plant life: a critical review. Environ Int 99:107–119. https://doi.org/10.1016/j.envint.2016.12.010

Ahmed MB, Zhou JL, Ngo HH, Guo W, Thomaidis NS, Xu J (2017) Progress in the biological and chemical treatment technologies for emerging contaminant removal from wastewater: a critical review. J Hazard Mater 323:274–298. https://doi.org/10.1016/j.jhazmat.2016.04.045

Al Aukidy M, Verlicchi P, Voulvoulis N (2014) A framework for the assessment of the environmental risk posed by pharmaceuticals originating from hospital effluents. Sci Total Environ 493:54–64. https://doi.org/10.1016/j.scitotenv.2014.05.128

Al-Rifai JH, Gabelish CL, Schafer AI (2007) Occurrence of pharmaceutically active and non-steroidal estrogenic compounds in three different wastewater recycling schemes in Australia. Chemosphere 69(5):803–815. https://doi.org/10.1016/j.chemosphere.2007.04.069

Andreozzi R, Raffaele M, Nicklas P (2003) Pharmaceuticals in STP effluents and their solar photodegradation in aquatic environment. Chemosphere 50(10):1319–1330

Ash RJ, Mauck B, Morgan M (2002) Antibiotic resistance of gram-negative bacteria in rivers, United States. Emerg Infect Dis 8(7):713–716. https://doi.org/10.3201/eid0807.010264

Ashton D, Hilton M, Thomas KV (2004) Investigating the environmental transport of human pharmaceuticals to streams in the United Kingdom. Sci Total Environ 333(1):167–184. https://doi.org/10.1016/j.scitotenv.2004.04.062

aus der Beek T, Weber FA, Bergmann A, Hickmann S, Ebert I, Hein A, Kuster A (2016) Pharmaceuticals in the environment – Global occurrences and perspectives. Environ Toxicol Chem 35(4):823–835. https://doi.org/10.1002/etc.3339

Baker DR, Kasprzyk-Hordern B (2011) Multi-residue analysis of drugs of abuse in wastewater and surface water by solid-phase extraction and liquid chromatography–positive electrospray ionisation tandem mass spectrometry. J Chromatogr A 1218(12):1620–1631. https://doi.org/10.1016/j.chroma.2011.01.060

Baker DR, Kasprzyk-Hordern B (2013) Spatial and temporal occurrence of pharmaceuticals and illicit drugs in the aqueous environment and during wastewater treatment: new developments. Sci Total Environ 454-455:442–456. https://doi.org/10.1016/j.scitotenv.2013.03.043

Baker DR, Barron L, Kasprzyk-Hordern B (2014) Illicit and pharmaceutical drug consumption estimated via wastewater analysis. Part A: chemical analysis and drug use estimates. Sci Total Environ 487:629–641. https://doi.org/10.1016/j.scitotenv.2013.11.107

Battin TJ, Besemer K, Bengtsson MM, Romani AM, Packmann AI (2016) The ecology and biogeochemistry of stream biofilms. Nat Rev Microbiol 14:251. https://doi.org/10.1038/nrmicro.2016.15

Bendz D, Paxéus NA, Ginn TR, Loge FJ (2005) Occurrence and fate of pharmaceutically active compounds in the environment, a case study: Höje River in Sweden. J Hazard Mater 122(3):195–204. https://doi.org/10.1016/j.jhazmat.2005.03.012

Bolong N, Ismail AF, Salim MR, Matsuura T (2009) A review of the effects of emerging contaminants in wastewater and options for their removal. Desalination 239(1):229–246. https://doi.org/10.1016/j.desal.2008.03.020

Boreen AL, Arnold WA, McNeill K (2003) Photodegradation of pharmaceuticals in the aquatic environment: a review. Aquat Sci 65(4):320–341. https://doi.org/10.1007/s00027-003-0672-7

Bowman JC, Zhou JL, Readman JW (2002) Sediment–water interactions of natural oestrogens under estuarine conditions. Mar Chem 77(4):263–276. https://doi.org/10.1016/S0304-4203(02)00006-3

Boxall AB, Rudd MA, Brooks BW, Caldwell DJ, Choi K, Hickmann S, Innes E, Ostapyk K, Staveley JP, Verslycke T, Ankley GT, Beazley KF, Belanger SE, Berninger JP, Carriquiriborde P, Coors A, Deleo PC, Dyer SD, Ericson JF, Gagne F, Giesy JP, Gouin T, Hallstrom L, Karlsson MV, Larsson DG, Lazorchak JM, Mastrocco F, McLaughlin A, McMaster ME, Meyerhoff RD, Moore R, Parrott JL, Snape JR, Murray-Smith R, Servos MR, Sibley PK, Straub JO, Szabo ND, Topp E, Tetreault GR, Trudeau VL, Van Der Kraak G (2012) Pharmaceuticals and personal care products in the environment: what are the big questions? Environ Health Perspect 120(9):1221–1229. https://doi.org/10.1289/ehp.1104477

Brennan L, Owende P (2010) Biofuels from microalgae—a review of technologies for production, processing, and extractions of biofuels and co-products. Renew Sust Energ Rev 14(2):557–577. https://doi.org/10.1016/j.rser.2009.10.009

Briens C, Piskorz J, Berruti F (2008) Biomass valorization for fuel and chemicals production – a review. Int J Chem React Eng 6. https://doi.org/10.2202/1542-6580.1674

British Pharmacopoeia Commission Uhbgcibie (2005) British pharmacopoeia 2005. Stationery Office

Bruchet A, Hochereau C, Picard C, Decottignies V, Rodrigues JM, Janex-Habibi ML (2005) Analysis of drugs and personal care products in French source and drinking waters: the analytical challenge and examples of application. Water Sci Technol J Int Assoc Water Pollut Res 52(8):53–61

Cai T, Park SY, Li Y (2013) Nutrient recovery from wastewater streams by microalgae: status and prospects. Renew Sust Energ Rev 19:360–369. https://doi.org/10.1016/j.rser.2012.11.030
Calamari D, Zuccato E, Castiglioni S, Bagnati R, Fanelli R (2003) Strategic survey of therapeutic drugs in the Rivers Po and Lambro in Northern Italy. Environ Sci Technol 37(7):1241–1248. https://doi.org/10.1021/es020158e
Carrasquillo AJ, Bruland GL, MacKay AA, Vasudevan D (2008) Sorption of ciprofloxacin and Oxytetracycline zwitterions to soils and soil minerals: influence of compound structure. Environ Sci Technol 42(20):7634–7642. https://doi.org/10.1021/es801277y
Challis JK, Hanson ML, Friesen KJ, Wong CS (2014) A critical assessment of the photodegradation of pharmaceuticals in aquatic environments: defining our current understanding and identifying knowledge gaps. Environ Sci: Processes Impacts 16(4):672–696. https://doi.org/10.1039/c3em00615h
Chang E-H, Yang S-S (2003) Some characteristics of microalgae isolated in Taiwan for biofixation of carbon dioxide. Bot Bull Acad Sin 44:43–52
Chen Z, Ren N, Wang A, Zhang Z-P, Shi Y (2008) A novel application of TPAD–MBR system to the pilot treatment of chemical synthesis-based pharmaceutical wastewater. Water Res 42(13):3385–3392. https://doi.org/10.1016/j.watres.2008.04.020
Chen H, Zha J, Liang X, Li J, Wang Z (2014) Effects of the human antiepileptic drug carbamazepine on the behavior, biomarkers, and heat shock proteins in the Asian clam *Corbicula fluminea*. Aquat Toxicol 155:1–8. https://doi.org/10.1016/j.aquatox.2014.06.001
Chen G, Zhao L, Qi Y (2015) Enhancing the productivity of microalgae cultivated in wastewater toward biofuel production: a critical review. Appl Energy 137:282–291
Chisti Y (2007) Biodiesel from microalgae. Biotechnol Adv 25:294–306
Cleuvers M (2003) Aquatic ecotoxicity of pharmaceuticals including the assessment of combination effects. Toxicol Lett 142(3):185–194
Collén J, Del Rio MJ, Gea G-R (1995) Photosynthetic production of hydrogen peroxide by *Ulva rigida* C. Ag. (Chlorophyta). Planta 196(2):225–230. https://doi.org/10.1007/BF00201378
Cuellar-Bermudez SP, Aleman-Nava GS, Chandra R, Garcia-Perez JS, Contreras-Angulo JR, Markou G, Muylaert K, Rittmann BE, Parra-Saldivar R (2017) Nutrients utilization and contaminants removal. A review of two approaches of algae and cyanobacteria in wastewater. Algal Res 24:438–449. https://doi.org/10.1016/j.algal.2016.08.018
da Silva TL, Gouveia L, Reis A (2014) Integrated microbial processes for biofuels and high value-added products: the way to improve the cost effectiveness of biofuel production. Appl Microbiol Biotechnol 98(3):1043–1053. https://doi.org/10.1007/s00253-013-5389-5
Daughton CG (2001) Pharmaceuticals and personal care products in the environment: overarching issues and overview. In: Pharmaceuticals and care products in the environment, vol 791. ACS Symposium Series, vol 791. American Chemical Society, pp 2–38. https://doi.org/10.1021/bk-2001-0791.ch00110.1021/bk-2001-0791.ch001
Daughton CG, Ternes TA (1999) Pharmaceuticals and personal care products in the environment: agents of subtle change? Environ Health Perspect 107(Suppl 6):907–938
de Godos I, Munoz R, Guieysse B (2012) Tetracycline removal during wastewater treatment in high-rate algal ponds. J Hazard Mater 229–230:446–449. https://doi.org/10.1016/j.jhazmat.2012.05.106
Delgado LF, Charles P, Glucina K, Morlay C (2012) The removal of endocrine disrupting compounds, pharmaceutically activated compounds and cyanobacterial toxins during drinking water preparation using activated carbon--a review. Sci Total Environ 435–436:509–525. https://doi.org/10.1016/j.scitotenv.2012.07.046
Demirbas A (2009) Biorefineries: current activities and future developments. Energy Convers Manag 50(11):2782–2801. https://doi.org/10.1016/j.enconman.2009.06.035
Dietz AC, Schnoor JL (2001) Advances in phytoremediation. Environ Health Perspect 109 (Suppl 1):163–168
Ding T, Lin K, Yang B, Li J, Li W, Gan J (2017) Biodegradation of naproxen by freshwater algae Cymbella sp. and Scenedesmus quadricauda and the comparative toxicity. Bioresour Technol 238:164–173

Doucha J, Straka F, Lívanský K (2005) Utilization of flue gas for cultivation of microalgae (Chlorella sp.) in an outdoor open thin-layer photobioreactor. J Appl Phycol 17:403–412

Duan YP, Meng XZ, Wen ZH, Chen L (2013) Acidic pharmaceuticals in domestic wastewater and receiving water from hyper-urbanization city of China (Shanghai): environmental release and ecological risk. Environ Sci Pollut Res Int 20(1):108–116. https://doi.org/10.1007/s11356-012-1000-3

el-Sayed MG, el-Sayed MT, Elazab Abd el S, Hafeiz MH, el-Komy AA, Hassan E (1998) Effects of some beta-adrenergic blockers on male fertility parameters in rats. Dtsch Tierarztl Wochenschr 105(1):10–12

Escapa C, Coimbra RN, Paniagua S, García AI, Otero M (2015) Nutrients and pharmaceuticals removal from wastewater by culture and harvesting of *Chlorella sorokiniana*. Bioresour Technol 185:276–284. https://doi.org/10.1016/j.biortech.2015.03.004

Escapa C, Coimbra RN, Paniagua S, García AI, Otero M (2016) Comparative assessment of diclofenac removal from water by different microalgae strains. Algal Res 18:127–134. https://doi.org/10.1016/j.algal.2016.06.008

Escapa C, Coimbra RN, Paniagua S, García AI, Otero M (2017) Paracetamol and salicylic acid removal from contaminated water by microalgae. J Environ Manag 203:799–806. https://doi.org/10.1016/j.jenvman.2016.06.051

Esplugas S, Bila DM, Krause LG, Dezotti M (2007) Ozonation and advanced oxidation technologies to remove endocrine disrupting chemicals (EDCs) and pharmaceuticals and personal care products (PPCPs) in water effluents. J Hazard Mater 149(3):631–642. https://doi.org/10.1016/j.jhazmat.2007.07.073

Fent K, Weston AA, Caminada D (2006) Ecotoxicology of human pharmaceuticals. Aquat Toxicol 76(2):122–159. https://doi.org/10.1016/j.aquatox.2005.09.009

Fong PP, Ford AT (2014) The biological effects of antidepressants on the molluscs and crustaceans: a review. Aquat Toxicol 151:4–13. https://doi.org/10.1016/j.aquatox.2013.12.003

Gao QT, Wong YS, Tam NFY (2011) Removal and biodegradation of nonylphenol by different Chlorella species. Mar Pollut Bull 63(5):445–451. https://doi.org/10.1016/j.marpolbul.2011.03.030

Gardner M, Comber S, Scrimshaw MD, Cartmell E, Lester J, Ellor B (2012) The significance of hazardous chemicals in wastewater treatment works effluents. Sci Total Environ 437:363–372. https://doi.org/10.1016/j.scitotenv.2012.07.086

Gardner M, Jones V, Comber S, Scrimshaw MD, Coello - Garcia T, Cartmell E, Lester J, Ellor B (2013) Performance of UK wastewater treatment works with respect to trace contaminants. Sci Total Environ 456-457:359–369. https://doi.org/10.1016/j.scitotenv.2013.03.088

Gentili FG, Fick J (2017) Algal cultivation in urban wastewater: an efficient way to reduce pharmaceutical pollutants. J Appl Phycol 29(1):255–262. https://doi.org/10.1007/s10811-016-0950-0

Ghatak HR (2014) Advanced oxidation processes for the treatment of biorecalcitrant organics in wastewater. Crit Rev Environ Sci Technol 44(11):1167–1219. https://doi.org/10.1080/10643389.2013.763581

Gibbons WR, Hughes SR (2009) Integrated biorefineries with engineered microbes and high-value co-products for profitable biofuels production. In Vitro Cell Dev Biol Plant 45(3):218–228. https://doi.org/10.1007/s11627-009-9202-1

Guzzon A, Bohn A, Diociaiuti M, Albertano P (2008) Cultured phototrophic biofilms for phosphorus removal in wastewater treatment. Water Res 42:4357–4367

Hanamoto S, Kawakami T, Nakada N, Yamashita N, Tanaka H (2014) Evaluation of the photolysis of pharmaceuticals within a river by 2 year field observations and toxicity changes by sunlight. Environ Sci: Processes Impacts 16(12):2796–2803. https://doi.org/10.1039/c4em00448e

Harun R, Singh M, Forde GM, Danquah MK (2010) Bioprocess engineering of microalgae to produce a variety of consumer products. Renew Sust Energ Rev 14(3):1037–1047. https://doi.org/10.1016/j.rser.2009.11.004

Heberer T (2002) Occurrence, fate, and removal of pharmaceutical residues in the aquatic environment: a review of recent research data. Toxicol Lett 131(1):5–17. https://doi.org/10.1016/S0378-4274(02)00041-3

Hsueh HT, Chu H, Yu ST (2007) A batch study on the bio-fixation of carbon dioxide in the absorbed solution from a chemical wet scrubber by hot spring and marine algae. Chemosphere 66(5):878–886. https://doi.org/10.1016/j.chemosphere.2006.06.022

Hughes SR, Kay P, Brown LE (2013) Global synthesis and critical evaluation of pharmaceutical data sets collected from river systems. Environ Sci Technol 47(2):661–677. https://doi.org/10.1021/es3030148

Iwasaki I, Hu Q, Kurano N, Miyachi S (1998) Effect of extremely high-CO_2 stress on energy distribution between photosystem I and photosystem II in a 'high-CO_2' tolerant green alga, Chlorococcum littorale and the intolerant green alga *Stichococcus bacillaris*. J Photochem Photobiol B Biol 44(3):184–190. https://doi.org/10.1016/S1011-1344(98)00140-7

Joffe M (2001) Are problems with male reproductive health caused by endocrine disruption? Occup Environ Med 58(4):281–288. https://doi.org/10.1136/oem.58.4.281

Kanakaraju D, Glass BD, Oelgemöller M (2014) Titanium dioxide photocatalysis for pharmaceutical wastewater treatment. Environ Chem Lett 12(1):27–47. https://doi.org/10.1007/s10311-013-0428-0

Kasprzyk-Hordern B, Dinsdale RM, Guwy AJ (2007) Multi-residue method for the determination of basic/neutral pharmaceuticals and illicit drugs in surface water by solid-phase extraction and ultra performance liquid chromatography–positive electrospray ionisation tandem mass spectrometry. J Chromatogr A 1161(1):132–145. https://doi.org/10.1016/j.chroma.2007.05.074

Kasprzyk-Hordern B, Dinsdale RM, Guwy AJ (2008a) Multiresidue methods for the analysis of pharmaceuticals, personal care products and illicit drugs in surface water and wastewater by solid-phase extraction and ultra performance liquid chromatography-electrospray tandem mass spectrometry. Anal Bioanal Chem 391(4):1293–1308. https://doi.org/10.1007/s00216-008-1854-x

Kasprzyk-Hordern B, Dinsdale RM, Guwy AJ (2008b) The occurrence of pharmaceuticals, personal care products, endocrine disruptors and illicit drugs in surface water in South Wales, UK. Water Res 42(13):3498–3518. https://doi.org/10.1016/j.watres.2008.04.026

Kasprzyk-Hordern B, Dinsdale RM, Guwy AJ (2009) The removal of pharmaceuticals, personal care products, endocrine disruptors and illicit drugs during wastewater treatment and its impact on the quality of receiving waters. Water Res 43(2):363–380. https://doi.org/10.1016/j.watres.2008.10.047

Khan SA, Rashmi HMZ, Prasad S, Banerjee UC (2009) Prospects of biodiesel production from microalgae in India. Renew Sust Energ Rev 13(9):2361–2372. https://doi.org/10.1016/j.rser.2009.04.005

Khetan SK, Collins TJ (2007) Human pharmaceuticals in the aquatic environment: a challenge to Green Chemistry. Chem Rev 107(6):2319–2364. https://doi.org/10.1021/cr020441w

Koh YKK, Chiu TY, Boobis AR, Scrimshaw MD, Bagnall JP, Soares A, Pollard S, Cartmell E, Lester JN (2009) Influence of operating parameters on the biodegradation of steroid estrogens and nonylphenolic compounds during biological wastewater treatment processes. Environ Sci Technol 43:6646–6654. https://doi.org/10.1021/es901612v

Kolpin DW, Furlong ET, Meyer MT, Thurman EM, Zaugg SD, Barber LB, Buxton HT (2002) Pharmaceuticals, hormones, and other organic wastewater contaminants in U.S. streams, 1999-2000: a national reconnaissance. Environ Sci Technol 36(6):1202–1211

Kovacs R, Csenki Z, Bakos K, Urbanyi B, Horvath A, Garaj-Vrhovac V, Gajski G, Geric M, Negreira N, Lopez de Alda M, Barcelo D, Heath E, Kosjek T, Zegura B, Novak M, Zajc I, Baebler S, Rotter A, Ramsak Z, Filipic M (2015) Assessment of toxicity and genotoxicity of low doses of 5-fluorouracil in zebrafish (*Danio rerio*) two-generation study. Water Res 77:201–212. https://doi.org/10.1016/j.watres.2015.03.025

Kovacs R, Bakos K, Urbanyi B, Kovesi J, Gazsi G, Csepeli A, Appl AJ, Bencsik D, Csenki Z, Horvath A (2016) Acute and sub-chronic toxicity of four cytostatic drugs in zebrafish. Environ Sci Pollut Res Int 23(15):14718–14729. https://doi.org/10.1007/s11356-015-5036-z

KPMG International (2006) The Indian pharmaceutical industry: collaboration for growth. pp 2–42

Kumar A, Ergas S, Yuan X, Sahu A, Zhang Q, Dewulf J, Malcata FX, van Langenhove H (2010) Enhanced CO(2) fixation and biofuel production via microalgae: recent developments and future directions. Trends Biotechnol 28(7):371–380. https://doi.org/10.1016/j.tibtech.2010.04.004

Kümmerer K, Helmers E (2000) Hospital effluents as a source of gadolinium in the aquatic environment. Environ Sci Technol 34(4):573–577. https://doi.org/10.1021/es990633h

Kurade MB, Kim JR, Govindwar SP, Jeon B-H (2016) Insights into microalgae mediated biodegradation of diazinon by *Chlorella vulgaris*: microalgal tolerance to xenobiotic pollutants and metabolism. Algal Res 20:126–134. https://doi.org/10.1016/j.algal.2016.10.003

Lam MK, Lee KT (2012) Microalgae biofuels: a critical review of issues, problems and the way forward. Biotechnol Adv 30:673–690

Lam MW, Young CJ, Brain RA, Johnson DJ, Hanson MA, Wilson CJ, Richards SM, Solomon KR, Mabury SA (2004) Aquatic persistence of eight pharmaceuticals in a microcosm study. Environ Toxicol Chem 23(6):1431–1440

Li Y, Horsman M, Wu N, Lan CQ, Dubois-Calero N (2008) Biofuels from microalgae. Biotechnol Prog 24(4):815–820. https://doi.org/10.1021/bp070371k

Li F-F, Yang ZH, Zeng R, Yang G, Chang X, Yan JB, Hou YL (2011) Microalgae capture of CO_2 from actual flue gas discharged from a combustion chamber. Ind Eng Chem Res 50(10):6496–6502

Liang Y (2013) Producing liquid transportation fuels from heterotrophic microalgae. Appl Energy 104:860–868

Lin AY, Plumlee MH, Reinhard M (2006) Natural attenuation of pharmaceuticals and alkylphenol polyethoxylate metabolites during river transport: photochemical and biological transformation. Environ Toxicol Chem 25(6):1458–1464

Liu X, Wu F, Deng N (2004) Photoproduction of hydroxyl radicals in aqueous solution with algae under high-pressure mercury lamp. Environ Sci Technol 38(1):296–299

Maes HM, Maletz SX, Ratte HT, Hollender J, Schaeffer A (2014) Uptake, elimination, and biotransformation of 17α- ethinylestradiol by the freshwater alga Desmodesmus subspicatus. Environ Sci Technol 48(20):12354–12361. https://doi.org/10.1021/es503574z

Maszkowska J, Stolte S, Kumirska J, Lukaszewicz P, Mioduszewska K, Puckowski A, Caban M, Wagil M, Stepnowski P, Bialk-Bielinska A (2014a) Beta-blockers in the environment: part I. Mobility and hydrolysis study. Sci Total Environ 493:1112–1121. https://doi.org/10.1016/j.scitotenv.2014.06.023

Maszkowska J, Stolte S, Kumirska J, Lukaszewicz P, Mioduszewska K, Puckowski A, Caban M, Wagil M, Stepnowski P, Bialk-Bielinska A (2014b) Beta-blockers in the environment: part II. Ecotoxicity study. Sci Total Environ 493:1122–1126. https://doi.org/10.1016/j.scitotenv.2014.06.039

Matamoros V, Uggetti E, García J, Bayona JM (2016) Assessment of the mechanisms involved in the removal of emerging contaminants by microalgae from wastewater: a laboratory scale study. J Hazard Mater 301:197–205. https://doi.org/10.1016/j.jhazmat.2015.08.050

Miao X-S, Koenig BG, Metcalfe CD (2002) Analysis of acidic drugs in the effluents of sewage treatment plants using liquid chromatography–electrospray ionization tandem mass spectrometry. J Chromatogr A 952(1):139–147. https://doi.org/10.1016/S0021-9673(02)00088-2

Mimeault C, Woodhouse AJ, Miao XS, Metcalfe CD, Moon TW, Trudeau VL (2005) The human lipid regulator, gemfibrozil bioconcentrates and reduces testosterone in the goldfish, *Carassius auratus*. Aquat Toxicol 73(1):44–54. https://doi.org/10.1016/j.aquatox.2005.01.009

Misik M, Pichler C, Rainer B, Filipic M, Nersesyan A, Knasmueller S (2014) Acute toxic and genotoxic activities of widely used cytostatic drugs in higher plants: possible impact on the environment. Environ Res 135:196–203. https://doi.org/10.1016/j.envres.2014.09.012

Misik M, Kundi M, Pichler C, Filipic M, Rainer B, Misikova K, Nersesyan A, Knasmueller S (2016) Impact of common cytostatic drugs on pollen fertility in higher plants. Environ Sci Pollut Res Int 23(15):14730–14738. https://doi.org/10.1007/s11356-015-4301-5

Moldovan Z (2006) Occurrences of pharmaceutical and personal care products as micropollutants in rivers from Romania. Chemosphere 64(11):1808–1817. https://doi.org/10.1016/j.chemosphere.2006.02.003

Mutanda T, Ramesh D, Karthikeyan S, Kumari S, Anandraj A, Bux F (2011) Bioprospecting for hyper-lipid producing microalgal strains for sustainable biofuel production. Bioresour Technol 102(1):57–70. https://doi.org/10.1016/j.biortech.2010.06.077

Mutiyar PK, Mittal AK (2014) Occurrences and fate of selected human antibiotics in influents and effluents of sewage treatment plant and effluent-receiving river Yamuna in Delhi (India). Environ Monit Assess 186(1):541–557. https://doi.org/10.1007/s10661-013-3398-6

Nakajima N, Teramoto T, Kasai F, Sano T, Tamaoki M, Aono M, Kubo A, Kamada H, Azumi Y, Saji H (2007) Glycosylation of bisphenol A by freshwater microalgae. Chemosphere 69(6):934–941. https://doi.org/10.1016/j.chemosphere.2007.05.088

National Health Service (NHS) (2012). http://www.nhsbsa.nhs.uk/PrescriptionServices/3494.aspx Accessed 18 May 2018

Neuwoehner J, Escher BI (2011) The pH-dependent toxicity of basic pharmaceuticals in the green algae Scenedesmus vacuolatus can be explained with a toxicokinetic ion-trapping model. Aquat Toxicol 101(1):266–275. https://doi.org/10.1016/j.aquatox.2010.10.008

Peng Z, Wu F, Deng N (2006) Photodegradation of bisphenol A in simulated lake water containing algae, humic acid and ferric ions. Environ Pollut 144(3):840–846. https://doi.org/10.1016/j.envpol.2006.02.006

Petrie B, Barden R, Kasprzyk-Hordern B (2015) A review on emerging contaminants in waste-waters and the environment: current knowledge, understudied areas and recommendations for future monitoring. Water Res 72:3–27. https://doi.org/10.1016/j.watres.2014.08.053

Petroutsos D, Katapodis P, Samiotaki M, Panayotou G, Kekos D (2008) Detoxification of 2,4-dichlorophenol by the marine microalga Tetraselmis marina. Phytochemistry 69(3):707–714. https://doi.org/10.1016/j.phytochem.2007.09.002

Petrovic M, Diaz A, Ventura F, Barceló D (2003) Occurrence and removal of estrogenic short-chain Ethoxy Nonylphenolic compounds and their halogenated derivatives during drinking water production. Environ Sci Technol 37(19):4442–4448. https://doi.org/10.1021/es034139w

Pflugmacher S, Wiencke C, Sandermann H (1999) Activity of phase I and phase II detoxication enzymes in Antarctic and Arctic macroalgae. Mar Environ Res 48(1):23–36. https://doi.org/10.1016/S0141-1136(99)00030-6

Pittman JK, Dean AP, Osundeko O (2011) The potential of sustainable algal biofuel production using wastewater resources. Bioresour Technol 102(1):17–25. https://doi.org/10.1016/j.biortech.2010.06.035

Putschew A, Wischnack S, Jekel M (2000) Occurrence of triiodinated X-ray contrast agents in the aquatic environment. Sci Total Environ 255(1–3):129–134

Qin L, Wang Z, Sun Y, Shu Q, Feng P, Zhu L, Xu J, Yuan Z (2016) Microalgae consortia cultivation in dairy wastewater to improve the potential of nutrient removal and biodiesel feedstock production. Environ Sci Pollut Res Int 23(9):8379–8387. https://doi.org/10.1007/s11356-015-6004-3

Rastogi RP, Pandey A, Larroche C, Madamwar D (2018) Algal green energy – R&D and technological perspectives for biodiesel production. Renew Sustain Energy Rev 82:2946–2969

Rawat I, Ranjith Kumar R, Mutanda T, Bux F (2013) Biodiesel from microalgae: a critical evaluation from laboratory to large scale production. Appl Energy 103:444–467

Razzak SA, Hossain MM, Lucky RA, Bassi AS, de Lasa H (2013) Integrated CO_2 capture, wastewater treatment and biofuel production by microalgae culturing—a review. Renew Sust Energ Rev 27:622–653. https://doi.org/10.1016/j.rser.2013.05.063

Ribeiro LA, da Silva PP (2013) Surveying techno-economic indicators of microalgae biofuel technologies. Renew Sustain Energy Rev 25:89–96

Rizwan M, Mujtaba G, Memon SA, Lee K, Rashid N (2018) Exploring the potential of microalgae for new biotechnology applications and beyond: a review. Renew Sust Energ Rev 92:394–404. https://doi.org/10.1016/j.rser.2018.04.034

Roberts PH, Thomas KV (2006) The occurrence of selected pharmaceuticals in wastewater effluent and surface waters of the lower Tyne catchment. Sci Total Environ 356(1):143–153. https://doi.org/10.1016/j.scitotenv.2005.04.031

Rodriguez E, Campinas M, Acero JL, João Rosa MJ (2016) Investigating PPCP removal from wastewater by powdered activated carbon/ultrafiltration. Water Air Soil Pollut 227(6):177. https://doi.org/10.1007/s11270-016-2870-7

Rogers JN, Rosenberg JN, Guzman BJ, Oh VH, Mimbela LE, Ghassemi A et al (2014) A critical analysis of paddlewheel-driven raceway ponds for algal biofuel production at commercial scales. Algal Res 4:76–88

Rosal R, Rodea-Palomares I, Boltes K, Fernandez-Pinas F, Leganes F, Gonzalo S, Petre A (2010) Ecotoxicity assessment of lipid regulators in water and biologically treated wastewater using three aquatic organisms. Environ Sci Pollut Res Int 17(1):135–144. https://doi.org/10.1007/s11356-009-0137-1

Rosen RC, Kostis JB, Jekelis AW (1988) Beta-blocker effects on sexual function in normal males. Arch Sex Behav 17(3):241–255

Rosenberg JN, Mathias A, Korth K, Betenbaugh MJ, Oyler GA (2011) Microalgal biomass production and carbon dioxide sequestration from an integrated ethanol biorefinery in Iowa: a technical appraisal and economic feasibility evaluation. Biomass Bioenergy 35(9):3865–3876. https://doi.org/10.1016/j.biombioe.2011.05.014

Ross AB, Biller P, Kubacki ML, Li H, Lea-Langton A, Jones JM (2010) Hydrothermal processing of microalgae using alkali and organic acids. Fuel 89(9):2234–2243. https://doi.org/10.1016/j.fuel.2010.01.025

Saharan BS, Sharma D, Sahu R, Sahin O, Warren A (2013) Towards algal biofuel production: a concept of green bio energy development. Innov Rom Food Biotechnol 12:1–21

Sakai N, Sakamoto Y, Kishimoto N, Chihara M, Karube I (1995) Chlorella strains from hot springs tolerant to high temperature and high CO_2. Energ Conver Manage 36(6):693–696. https://doi.org/10.1016/0196-8904(95)00100-R

Santos LH, Araujo AN, Fachini A, Pena A, Delerue-Matos C, Montenegro MC (2010) Ecotoxicological aspects related to the presence of pharmaceuticals in the aquatic environment. J Hazard Mater 175(1–3):45–95. https://doi.org/10.1016/j.jhazmat.2009.10.100

Sharpe RM, Irvine DS (2004) How strong is the evidence of a link between environmental chemicals and adverse effects on human reproductive health? BMJ. (Clinical research ed) 328(7437):447–451. https://doi.org/10.1136/bmj.328.7437.447

Singh J, Gu S (2010) Commercialization potential of microalgae for biofuels production. Renew Sustain Energy Rev 14:2596–2610

Sivaramakrishnan R, Incharoensakdi A (2018) Utilization of microalgae feedstock for concomitant production of bioethanol and biodiesel. Fuel 217:458–466. https://doi.org/10.1016/j.fuel.2017.12.119

Stevens-Garmon J, Drewes JE, Khan SJ, McDonald JA, Dickenson ERV (2011) Sorption of emerging trace organic compounds onto wastewater sludge solids. Water Res 45(11):3417–3426. https://doi.org/10.1016/j.watres.2011.03.056

Suárez S, Carballa M, Fea O (2008) How are pharmaceutical and personal care products (PPCPs) removed from urban wastewaters? Rev Environ Sci Biotechnol 7(2):125–138. https://doi.org/10.1007/s11157-008-9130-2

Subashchandrabose SR, Ramakrishnan B, Megharaj M, Venkateswarlu K, Naidu R (2011) Consortia of cyanobacteria/microalgae and bacteria: biotechnological potential. Biotechnol Adv 29(6):896–907. https://doi.org/10.1016/j.biotechadv.2011.07.009

Subedi B, Balakrishna K, Sinha RK, Yamashita N, Balasubramanian VG, Kannan K (2015) Mass loading and removal of pharmaceuticals and personal care products, including psychoactive and illicit drugs and artificial sweeteners, in five sewage treatment plants in India. J Environ Chem Eng 3(4, Part A):2882–2891. https://doi.org/10.1016/j.jece.2015.09.031

Sydney EB, Sturm W, de Carvalho JC, Thomaz-Soccol V, Larroche C, Pandey A, Soccol CR (2010) Potential carbon dioxide fixation by industrially important microalgae. Bioresour Technol 101(15):5892–5896. https://doi.org/10.1016/j.biortech.2010.02.088

Tang D, Han W, Li P, Miao X, Zhong J (2011) CO_2 biofixation and fatty acid composition of Scenedesmus obliquus and *Chlorella pyrenoidosa* in response to different CO_2 levels. Bioresour Technol 102(3):3071–3076. https://doi.org/10.1016/j.biortech.2010.10.047

Ternes TA, Joss A, Siegrist H (2004) Scrutinizing pharmaceuticals and personal care products in wastewater treatment. Environ Sci Technol 38(20):392a–399a

Thacker PD (2005) Pharmaceutical data elude researchers. Environ Sci Technol 39(9):193a–194a

Torres MA, Barros MP, Campos SCG, Pinto E, Rajamani S, Sayre RT, Colepicolo P (2008) Biochemical biomarkers in algae and marine pollution: a review. Ecotoxicol Environ Saf 71(1):1–15. https://doi.org/10.1016/j.ecoenv.2008.05.009

Úbeda B, Gálvez JÁ, Michel M, Bartual A (2017) Microalgae cultivation in urban wastewater: Coelastrum cf. pseudomicroporum as a novel carotenoid source and a potential microalgae harvesting tool. Bioresour Technol 228:210–217. https://doi.org/10.1016/j.biortech.2016.12.095

United States Pharmacopeial C (2005) The United States Pharmacopeia: USP 29: the National Formulary: NF 24: by authority of the United States Pharmacopeial convention, Inc., meeting at Washington, D.C., March 9–13, 2005. United States Pharmacopeial Convention, Rockville

Vasudevan D, Bruland GL, Torrance BS, Upchurch VG, MacKay AA (2009) pH-dependent ciprofloxacin sorption to soils: Interaction mechanisms and soil factors influencing sorption. Geoderma 151(3):68–76. https://doi.org/10.1016/j.geoderma.2009.03.007

Verlicchi P, Al Aukidy M, Galletti A, Petrovic M, Barceló D (2012) Hospital effluent: investigation of the concentrations and distribution of pharmaceuticals and environmental risk assessment. Sci Total Environ 430:109–118. https://doi.org/10.1016/j.scitotenv.2012.04.055

Verstraeten IM, Heberer T, Vogel JR, Speth T, Zuehlke S, Duennbier U (2003) Occurrence of endocrine-disrupting and other wastewater compounds during water treatment with case studies from Lincoln, Nebraska and Berlin, Germany. Pract Period Hazard Toxic Radioact Waste Manage 7(4):253–263. https://doi.org/10.1061/(ASCE)1090-025X(2003)7:4(253)

Vieno N, Tuhkanen T, Kronberg L (2006) Removal of pharmaceuticals in drinking water treatment: effect of chemical coagulation. Environ Technol 27(2):183–192. https://doi.org/10.1080/09593332708618632

Wang B, Li Y, Wu N, Lan CQ (2008) CO(2) bio-mitigation using microalgae. Appl Microbiol Biotechnol 79(5):707–718. https://doi.org/10.1007/s00253-008-1518-y

Wang Y, Ho S-H, Cheng C-L, Guo W-Q, Nagarajan D, Ren N-Q, Lee D-J, Chang J-S (2016) Perspectives on the feasibility of using microalgae for industrial wastewater treatment. Bioresour Technol 222:485–497. https://doi.org/10.1016/j.biortech.2016.09.106

Wang Y, Liu J, Kang D, Wu C, Wu Y (2017) Removal of pharmaceuticals and personal care products from wastewater using algae-based technologies: a review. Rev Environ Sci Biotechnol 16(4):717–735. https://doi.org/10.1007/s11157-017-9446-x

Webb SF (2001) A data based perspective on the environmental risk assessment of human pharmaceuticals II — aquatic risk characterisation. In: Kümmerer K (ed) Pharmaceuticals in the environment: sources, fate, effects and risks. Springer Berlin Heidelberg, Berlin/Heidelberg, pp 203–219. https://doi.org/10.1007/978-3-662-04634-0_16. isbn:978-3-662-04634-0

Westerhoff P, Yoon Y, Snyder S, Wert E (2005) Fate of endocrine-disruptor, pharmaceutical, and personal care product chemicals during simulated drinking water treatment processes. Environ Sci Technol 39(17):6649–6663. https://doi.org/10.1021/es0484799

Wilt A, Butkovskyi A, Tuantet K, Leal LH, Fernandes TV, Langenhoff A, Zeeman G (2016) Micropollutant removal in an algal treatment system fed with source separated wastewater streams. J Hazard Mater 304:84–92. https://doi.org/10.1016/j.jhazmat.2015.10.033

Wu Y, Li T, Yang L (2012) Mechanisms of removing pollutants from aqueous solutions by microorganisms and their aggregates: a review. Bioresour Technol 107:10–18. https://doi.org/10.1016/j.biortech.2011.12.088

Wu C, Huang X, Lin J, Liu J (2015a) Occurrence and fate of selected endocrine-disrupting chemicals in water and sediment from an urban lake. Arch Environ Contam Toxicol 68(2):225–236. https://doi.org/10.1007/s00244-014-0087-6

Wu H, Zhang J, Ngo HH, Guo W, Hu Z, Liang S, Fan J, Liu H (2015b) A review on the sustainability of constructed wetlands for wastewater treatment: design and operation. Bioresour Technol 175:594–601. https://doi.org/10.1016/j.biortech.2014.10.068

Xiong JQ, Kurade MB, Abou-Shanab RA, Ji MK, Choi J, Kim JO, Jeon BH (2016) Biodegradation of carbamazepine using freshwater microalgae Chlamydomonas mexicana and Scenedesmus obliquus and the determination of its metabolic fate. Bioresour Technol 205:183–190. https://doi.org/10.1016/j.biortech.2016.01.038

Xiong J-Q, Kurade MB, Patil DV, Jang M, Paeng K-J, Jeon B-H (2017) Biodegradation and metabolic fate of levofloxacin via a freshwater green alga, *Scenedesmus obliquus* in synthetic saline wastewater. Algal Res 25:54–61. https://doi.org/10.1016/j.algal.2017.04.012

Yamamoto H, Nakamura Y, Moriguchi S, Nakamura Y, Honda Y, Tamura I, Hirata Y, Hayashi A, Sekizawa J (2009) Persistence and partitioning of eight selected pharmaceuticals in the aquatic environment: laboratory photolysis, biodegradation, and sorption experiments. Water Res 43(2):351–362. https://doi.org/10.1016/j.watres.2008.10.039

Yang S, Wu RS, Kong RY (2002) Biodegradation and enzymatic responses in the marine diatom *Skeletonema costatum* upon exposure to 2,4-dichlorophenol. Aquat Toxicol 59(3–4):191–200

Yang X, Flowers RC, Weinberg HS, Singer PC (2011) Occurrence and removal of pharmaceuticals and personal care products (PPCPs) in an advanced wastewater reclamation plant. Water Res 45(16):5218–5228. https://doi.org/10.1016/j.watres.2011.07.026

Yang W, Zhou H, Cicek N (2014) Treatment of organic micropollutants in water and wastewater by UV-based processes: a literature review. Crit Rev Environ Sci Technol 44(13):1443–1476. https://doi.org/10.1080/10643389.2013.790745

Yoo C, Jun S-Y, Lee J-Y, Ahn C-Y, Oh H-M (2010) Selection of microalgae for lipid production under high levels carbon dioxide. Bioresour Technol 101(1 Suppl):S71–S74. https://doi.org/10.1016/j.biortech.2009.03.030

Zangar RC, Davydov DR, Verma S (2004) Mechanisms that regulate production of reactive oxygen species by cytochrome P450. Toxicol Appl Pharmacol 199(3):316–331. https://doi.org/10.1016/j.taap.2004.01.018

Zeiler KG, Heacox DA, Toon ST, Kadam KL, Brown LM (1995) The use of microalgae for assimilation and utilization of carbon dioxide from fossil fuel-fired power plant flue gas. Energ Conver Manage 36(6):707–712. https://doi.org/10.1016/0196-8904(95)00103-K

Zeraatkar AK, Ahmadzadeh H, Talebi AF, Moheimani NR, McHenry MP (2016) Potential use of algae for heavy metal bioremediation, a critical review. J Environ Manag 181:817–831. https://doi.org/10.1016/j.jenvman.2016.06.059

Zhang X, Yan S, Tyagi RD, Surampalli RY, Valéro JR (2016) Energy balance of biofuel production from biological conversion of crude glycerol. J Environ Manag 170:169–176. https://doi.org/10.1016/j.jenvman.2015.09.031

Zhang L, Cheng J, Pei H, Pan J, Jiang L, Hou Q, Han F (2018) Cultivation of microalgae using anaerobically digested effluent from kitchen waste as a nutrient source for biodiesel production. Renew Energy 115:276–287. https://doi.org/10.1016/j.renene.2017.08.034

Zhao B, Zhang Y, Xiong K, Zhang Z, Hao X, Liu T (2011) Effect of cultivation mode on microalgal growth and CO_2 fixation. Chem Eng Res Des 89(9):1758–1762. https://doi.org/10.1016/j.cherd.2011.02.018

Phycoremediation of Persistent Organic Pollutants from Wastewater: Retrospect and Prospects

Ashutosh Pandey, Manish Pratap Singh, Sanjay Kumar, and Sameer Srivastava

Abbreviations

2,4-DNP	2,4-dinitrophenol
ARISA	Automated ribosomal intergenic spacer analysis
B[k]F	Benzo[k] fluoranthene
BaA	Benzo[a] anthracene
BaF	Benzo[b] fluoranthene
BGA	Blue-green algae
BghiP	Benzo[g,h,i] erylene
BPA	Bisphenol A
DDT	Dichlorodiphenyltrichloroethane
DGEG	Denaturing gradient gel electrophoresis
DNA	Deoxyribonucleic acid
EDCs	Endocrine disruptors
EE2	17-α- Ethinylestradiol
FISH	Fluorescent in situ hybridization
Fla.	Fluoranthene
HAB	Harmful algal bloom
HMW-PAH	High molecular weight polycyclic aromatic hydrocarbon
LH-PCR	Length Heterogeneity Polymerase Chain Reaction
LMW-PAH	Low molecular weight polycyclic aromatic hydrocarbon

A. Pandey · M. P. Singh · S. Srivastava (✉)
Department of Biotechnology, Motilal Nehru National Institute of Technology Allahabad, Prayagraj 211004, Uttar Pradesh, India
e-mail: sameers@mnnit.ac.in

S. Kumar
School of Biochemical Engineering, Indian Institute of Technology (BHU) Varanasi, Varanasi 221005, Uttar Pradesh, India

S. K. Gupta, F. Bux (eds.), *Application of Microalgae in Wastewater Treatment*,
https://doi.org/10.1007/978-3-030-13913-1_11

LTRR	Long tandemly repeated repetitive
OC	Organochlorine
OP	4-Octylphenol
PAH	Polycyclic aromatic hydrocarbon
PCB	Polychlorinated biphenyls
PCR	Polymerase chain reaction
PHC	Petroleum hydrocarbon
Phe	Phenanthrene
PNP	Poly-nitrophenol
POP	Persistent organic pollutants
PS I, II	Photosystem I, II
Pyr	Pyrene
RAPD	Random Amplification of Polymorphic DNA
RFLP	Restriction Fragment Length Polymorphism
RNA	Ribonucleic acid
SSCP	Single-strand conformation polymorphism
STRR	Short tandemly repeated repetitive

1 Introduction

Growing population and its selective demand, industrialization and other anthropogenic activity around the world are destined to grow at present rate in future. This growth is concomitant with a significant increase in contamination of water streams with an extensive range of organic pollutants. Persistent organic pollutants (POPs) are carbon-based heterogenous set of toxic compounds that adversely affect human health and the environment. POPs can accumulate in the environment and can pass between species to species through the food chain. It can also be transported by wind, water, and migratory species across international boundaries (Fig. 1). United Nations Environment Programme, 2006, in its report suggested and noted that POPs generated in one country may also affect the people and wildlife far from where they are produced. Although insoluble in water, POPs are readily absorbed in fatty tissues and can accumulate in animals high up the trophic pyramids. Cancer, allergies, damage to the central and peripheral nervous systems, reproductive disorders, and disruption of the immune system are some specific damages to animals which could be attributed to POP toxicity (http://chm.pops.int). Under Stockholm convention, a list of new POPs are regularly updated. Presently, there are 26 POPs listed which pose threat in various forms (For more detail visit at http://www.chem.pop.int).

Contamination of aquatic system by POPs, for instance, polyaromatic hydrocarbons (PAHs), polychlorinated biphenyls (PCBs), pesticides, insecticides, phenolics, petroleum hydrocarbons (PHCs), and antibiotics poses significant public and environmental threat (Table 1). Considerable efforts are made to develop methods to eliminate PAHs and other organic pollutants from wastewater. Methods for removal of complex organic pollutants include activated oxidation using ozone, UV-radiations, hydrogen peroxides, and membrane filtration (Zaini et al. 2010;

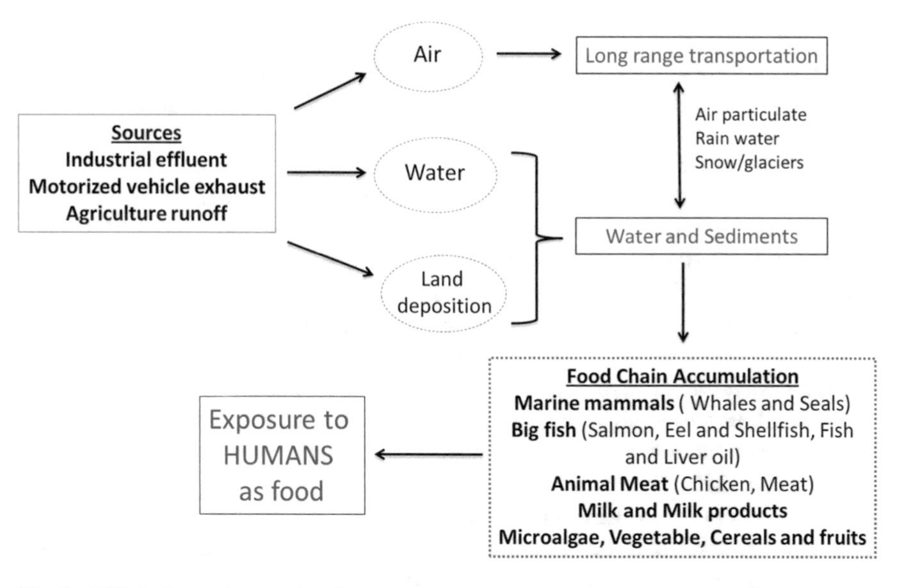

Fig. 1 POPs in the environment and exposure route

Table 1 POPs and its harmful effect on human health

Organic pollutant	Disorders	References
Polyaromatic hydrocarbons	Carcinogenic, mutagenic, and teratogenic	Hussein and Mansour (2016)
Pesticides	Neurotoxicity, carcinogenic, reproductive defect, birth defect, and fatal effect	Sanborn et al. (2007); Jurewicz and Hanke (2008)
Polychlorobiphenyl	Dermal and ocular lesions, irregular menstrual cycles, compromised immune responses, liver damage, poor cognitive development in children, mutagenic, carcinogenic, estradiol inhibition, hypothyroidism	Aoki (2001); Winneke (2011); Crinnion (2011); www.foxriverwatch.com
Phenolics	Endocrine disruptor	Rubin (2011)
Petroleum hydrocarbons	Nervous system disorder, immunogenic, respiratory disorders, carcinogenic	Bahadar et al. (2014); Greenberg (1997); Kraut et al. (1988).

Naturevardsverket 2008). The suggested methods are fairly energy and cost-intensive (MistraPharma 2011). Therefore, the need of the hour is for a sustainable remediation method that can remove such contaminants efficiently even if they are present at low concentrations.

Bioremediation has the potential to provide an economical alternative to conventional treatment methods (Montero- Rodriguez et al. 2015). Bioremediation can be defined as a collective name for cleaning processes utilizing biological material such as microbes (bacteria, fungi) and plants. On the other hand, phytoremediation can also be pitched as a sustainable reclamation strategy that uses sunlight for remediating contaminated sites (Abou-shanab 2011). Last few decades have witnessed enormous amount of research utilizing photoautotrophic aquatic organisms (such as algae) for successful phytoremediation of organic and inorganic pollutants (Ji et al. 2011; Dosnon-Olette et al. 2010). Alternatively, microalgae have also been used indicating the potential to bioremediate pollutants from wastewater (Safonova et al. 2005; Yang et al. 2002). Microalgae are relatively sensitive to chemicals but play a very important role in maintaining the equilibrium of aquatic ecosystem. Microalgae are primary producers residing at the base trophic level of the ecological pyramid. Besides this alga has a ubiquitous distribution. They also have heterotrophic abilities and play a central role in carbon fixation and turnover (Semple and Cain 1995). Many studies have reported microalgae as an effective bioremediating agent for polyaromatic hydrocarbons and many phenolic derivatives (Dosnon Olette et al. 2010). However, algae are known to remove organic pollutants via bioabsorption, bioaccumulation, and biotransformation (Priyadarshani et al. 2011; Kobayashi and Rittman 1982).

This chapter tries to stress and focus on the research covering the remediation of POPs by algae. We have discussed bioremediating methodologies including whole cell heterotrophic studies to detailed remediation mechanism of common POPs found in the aquatic system. Limitations and advance approaches used to enhance phycoremediation of POPs are also discussed briefly.

2 Microalgae and Blue-Green Algae

2.1 *Biodiversity*

Cyanobacteria are reported from ancient time based on a phylogenetic relationship with existing living organism having the specialized ability to nitrogen fixation in special cells (heterocyst) and symbiotic association with several eukaryotes (plant, fungi, and bryophytes). Cyanobacteria exhibit an extremely diverse association with other planktons and have an extensive range of ecological adaptation. They are predominantly reported in the vast range of ecosystems including cold and tropical, fresh, saline, marine, and terrestrial water. This environmental adaptability is attributed to the presence of chloroplast and mitochondria, as a part of photosynthetic machinery and energy metabolism (Bergman et al. 1996). Photosynthetic mechanism is extremely dependent upon PSII reaction center that can easily harbor electrons from the aquatic environment, so this is the main reason that can help in ecological adaptation of algae in various water sources. In summer they tend to form algal bloom in temperate eutrophic lakes. These blooms are unique characteristics of gas vacuolated algal species including *Anabaena*, *Microcystis*, *Anacystis*, *Aphanizomenon*, and *Planktothrix*. Usually, these blooms are dangerous for plankton and humans and add severe impact on economy globally. Harmful algal bloom (HAB) species are the producer of highly toxic compounds that inhibit the growth of another organism present in the respective ecological zone. Another mechanism suggests that toxin produced by them may protect them from various predators including bacteria, virus, and fungi. One possible mechanism is that the toxin produced by these HABs promote the competitive growth of algae and diminish the competitor's growth (Jonsson et al. 2009; Turner 2014).

Cyanobacteria were conventionally grouped as blue-green algae (BGA) based on their morphological characteristics distribution among algal species. The study of these microscopic bodies at an advanced (biochemical and genotypic) level proposed that these should be classified as cyanobacteria instead of cyanophyceae (Stanier et al. 1978). Cyanobacteria can be classified into several groups based on their genotypic characteristics, but till date, there is no clear consensus for establishing the unit of biological diversity within the species. Recently a polyphasic approach has been put forward for classification of cyanobacteria. It relies upon the combination of several methods and resemblance of unique features constituting the identification of morphological and genetic properties organism in disparate environmental conditions. The genetic features configure the foundation of classification that will eventually combine with other secondary feature like morphology, physiological conditions, and ecological survey.

Currently, there are two most frequent classification systems to classify the cyanobacteria: bacteriological approach and botanical approach. Komárek (2016) incorporated modern molecular marker methods to improve classification of cyanobacteria based on systemic of bacteria. Botanical nomenclature code grouped the cyanobacteria in BGA group under eukaryotic algae on the basis of morphology. Further, cyanobacteria are divided into four orders (Nostocales, Stigonematales, Chroococcales, and Oscillatoriales), families, subfamilies, genera, and species (Stanier et al. 1978). Hoffman and others proposed modern classification criteria established on the grounds of ultrastructural, phenotypic data and genetic information of the cyanobacteria (Hoffmann et al. 2005).

Molecular Methods Applied in Cyanobacterial Diversity

Taxonomical classification is a very promising method for exploring significant aspects of organismal diversification. Furthermore, it is very intricated in prokaryotic cyanobacteria, which exist in various cellular forms ranging from unicellular to multicellular having the remarkable appearance of the thallus. They exist in several ecological zones and adapted to that specialized ecosystem and pose unique adaptation and expression pattern in their genome. So it is recommended by various taxonomist that the introduction of modern molecular methods with the old cytological, morphological methods in cyanobacterial diversity may provide a reliable outcome (Komárek 2016). Nowadays scientist uses DNA sequencing (16 s DNA), metagenomics, DNA fingerprinting, and polyphasic and few non-PCR-based techniques to classify cyanobacteria. A probe-based detection system for 16 s rDNA sequence has been developed, for classifying 19 species of cyanobacterial groups from a mixed culture (Castiglioni et al. 2004). In several cases cyanobacteria exert similar morphological characteristics having different toxigenic potentials; therefore identification of these strains becomes very difficult. However, molecular methods provide an accurate distinction between closely related strains. Baker et al. (2002) reported a PCR-based method to discriminate between toxic and nontoxic *Microcystis aeruginosa* in an algal bloom. Phycocyanin intergenic spacers operons are routinely used in PCR-based amplification of alpha and beta subunit, a rpoC1 gene in a straight forward analysis of environmental samples (Baker et al. 2002). Molecular-based identification methods utilize multiple techniques such as real-time PCR, microarray, FISH, RFLP, and more updated next-generation sequencing of given species in a rapid and robust manner to classify cyanobacteria (Castiglioni et al. 2004; Humbert et al. 2010). DNA fingerprinting of cyanobacteria in mixed culture is used to study diversity and resemblance between the given sample. Commonly used techniques in DNA fingerprinting are a different version of PCR amplified products, namely, RAPD, RFLP LH-PCR, ARISA, DGEG/TGEG, and SSCP.

Valério et al. (2009) studied 118 cyanobacterial isolates from freshwater using STRR and LTRR PCR fingerprinting profiles along with 16S rRNA gene and rpoC1 gene and established a very congruent consortium suitable for taxonomic affiliation (Valério et al. 2009). Cyanobacteria can be studied using several non-PCR-based techniques without culture and very specific conditions. FISH is one of the major technique that permeates through the cell wall and specifically binds to r RNA. *Microcystis aeruginosa* is directly identified using ring FISH in algal bloom (potential harm to humans). The RING FISH targets the microcystein synthase D gene to identify toxin-producing strain from a mixed culture of nontoxic cyanobacteria (Dziallas and Grossart 2011).

2.2 *Mode of Cultivation*

Photoautotrophy

Algae are able to exploit light energy and atmospheric CO_2 into energy for normal cellular processes known as the phototrophic mode of cultivation. Cyanobacteria are the only prokaryotes that have two reaction centers, Fe-S type (PS-I) and pheophytin-quinone type (PSII), to carry out photosynthesis. The microalgae are considered as better photosynthesizer than that of higher terrestrial plants, and their growth is very fast (Wang et al. 2008). Till date it has been recognized that only phototrophic method is theoretically and economically viable to culture microalgae on an industrial scale. An added advantage of phototrophic culture is that it can capture carbon dioxide from flue gases, making it a superior carbon sink. Temperate countries with low available sunlight throughout the year cannot rely on the phototrophic method of cultivation (Lam and Lee 2012). It has also been estimated that microalgae can use up to 9% of the total sunlight received during the year and can sequester up to 513 tons of carbon dioxide and produce up to 280 tons of dry biomass ha^{-1} $year^{-1}$ (Bilanovic et al. 2009; Chisti 2007). Besides light, the availability of carbon source and temperature has a major effect on the algal growth rate and productivity (Juneja et al. 2013). Carbon is a very important factor for both autotrophs and heterotrophs.

Heterotrophy

Heterotrophs do not depend on light energy and utilize organic substrate (e.g., glucose, acetate, glycerol) as both carbon and energy source. For decades the heterotrophic growth of many algal strains have been studied (Droop 1974). The heterotrophic growth of microalgae is always found to be better over the photoautotrophic in mass production (Borowitzka 1999). A number of microalgal strains such as *Chlorella protothecoides* (Cheng et al. 2009; Xiong et al. 2008), *Chlorella vulgaris* (Liang et al. 2009), *Schizochytrium limacinum* (Johnson and Wen 2009), and *Crypthecodinium cohnii* (Couto et al. 2010) have been demonstrated to grow in dark and accumulate good amount of lipids. In another report, *Rhodobacter sphaeroides* and *Chlorella sorokiniana* were shown to efficiently remove nutrient from wastewater under heterotrophic conditions (Ogbonna et al. 2000). This method has few advantages over other methods of wastewater remediation; however, this area of research is not explored to a great extent. Approximately 30% of research work on microalgae have reported using wastewater as nutrients source whereas; the remaining 70% of published work reported the use of chemical fertilizers which are easily available in the market (Lam and Lee 2012).

Mixotrophy

The mixotrophic culture regime is an alternate to heterotrophic culture. In this, mode of cultivation algae utilizes both dissolved inorganic through photosynthesis (a light-dependent reaction) and organic carbon sources (glucose, acetate, glycerol, molasses etc.) (Andrate and Costa 2007). Wang et al. (2014) reviewed and reported number of algal strains such as *Nannochloropsis oculata*, *Dunaliella salina*, *Botryococcus braunii*, *Chlorella* sp., *Scenedesmus* sp., *Phaeodactylum tricornutum*, *Chlamydomonas globosa*, *Spirulina platensis*, *Nostoc flagelliforme*, *Pleurochrysis carterae*, etc. can be efficiently grown mixotrophically and produce more biomass and other photosynthetic metabolites as compared to auto- and heterotrophic mode of cultivation. Lin and Wu (2015) reported that light is not a limiting factor for mixotrophic cultivation, and it can reduce the photoinhibition under high and low illumination condition. Mixotrophic cultivation of microalgae not only enhances the specific growth rate but also enhances the biomass yield many folds as compared to autotrophic cultivation. Subashchandrabose et al. (2013) reviewed that the light and nutrient limitation in a polluted environment encourage the mixotrophic cultivation in microalgae. The organic chemical substances such as acetate, ethanol, and glycerol (a by-product of various industrial processes) can support the mixotrophic growth of microalgae during the periods of low nutrient concentration. Phycoremediation of POPs by using algal strains able to grow mixotrophically would be a good strategy for industrial wastewater treatment.

2.3 Molecular Mechanisms of Phycoremediation

The phycoremediation is based on the concept of uptake of a complex pollutant from the environment and utilizes or transforms them to nontoxic form (Quintana et al. 2011). The algae can remove POPs mainly by "biodegradation" and "bioaccumulation." On one hand, biodegradation involves decomposition of the complex toxic organic polymer into a simpler nontoxic organic compound such as carbon dioxide and water. On the other hand, algae are also able to bind pollutant on their outer surface, due to the presence of various surface receptors such as lipid, proteins, and polysaccharides (Priyadarshani et al. 2011). Pollutants trapped by algal surface binds passively, and the phenomenon is known as "bioabsorption." Uptake of pollutants inside the cell is generally done via "active transport." Both the passive absorption and active uptake of the pollutant by the algal cell is termed as "bioaccumulation." A number of algal species capable to remediate POPs are widely studied (Table 2), but the information on the bioremediation mechanism by algae is by enlarging less traveled road. Semple (1998) reviewed the interaction between algae and aromatic pollutants and highlighted the biodegradability capacity of algae for a range of complex aromatic compounds (monocyclic to complex polycyclic pollutants). One of the comprehensive reports on biodegradation and bioaccumulation of POPs was compiled by Kobayashi and Rittman (1982). They studied the interaction between eukaryotic algae with various forms of POPs (Table 2). The bioremediation of various forms of organic pollutant by algae is discussed in detail in Sect. 3.

Algae are omnipresent in nature and are present widely in the photic zone of the aquatic ecosystem. They may be considered as a major sink for the transformation of PAHs. Cerniglia et al. (1980) have demonstrated that cyanobacteria as well as microalgae under photoautotrophic conditions oxidize naphthalene into four major metabolites. Predominantly they transform naphthene to 1-naphthol along with a small quantity of cis-1,2-dihydroxy-1,2-dihydronaphthalene, trans-naphthalene dihydrodiol, and 4-hydroxy-1-tetralone (Figs. 2 and 3a). Narro et al. (1992a) first time reported the mechanism of PAH oxidation in the marine cyanobacterium *Oscillatoria* sp. strain JCM. They demonstrated the formation of 1-naphthol via an arene oxide intermediate that further isomerizes nonenzymatically, similar to that of monooxygenase catalyzed reaction reported for fungal and mammalian enzymes (Fig. 3b). In the same study, Narro et al. (1992b) also found that *Agmenellum quadruplicatum* PR-6 metabolizes phenanthrene (PHE) and transforms it into trans-9,10-dihydroxy-9,10-dihydrophenanthrene and 1-methoxyphenanthrene as major metabolites. Studies by different groups have shown that the green alga, *Selenastrum capricornutum*, metabolizes benzo[a]pyrene into cis-11,12-dihydroxy-11,12-

Table 2 POPs of industrial origin and its remediation by algae

Algal species	Persistent organic pollutants (POPs)	References
A. *Biodegradation/biotransformation of persistent organic pollutants*		
Selanastrum capricornutum	Atrazine	Friesen-Pankratz et al. (2003)
Selanastrum capricornutum, Scenedesmus acutus	Benzopyrene	Semple et al. (1999); Kobayashi and Rittman (1982); de Llasera et al. (2016)
Selanastrum capricornutum	Benzo[a]pyrene	de Llasera et al. (2016); Warshawsky et al. (1990)
Navicula pelliculosa	Naphthenic acid	Mahdavi et al. (2015)
Cymbella sp., *Scenedesmus quadricauda*	Naproxen	Ding et al. (2017)
Chlamydomonas sp., *Chlorella* sp.	Linden	Semple et al. (1999); Kobayashi and Rittman (1982)
Dunaliella sp., *Euglena gracilis, Chlamydomonas* sp., *Scenedesmus obliquus*	Naphthalene	Semple et al. (1999); Kobayashi and Rittman (1982)
Euglena gracilis, Chlamydomonas sp.	Phenol	Semple et al. (1999); Kobayashi and Rittman (1982)
Chlamydomonas reinhardtii	Prometryne	Jin et al. (2012)
Nitzschia sp.	Fluoranthene, Phenanthrene	Hong et al. (2008)
Desmodesmus sp.	Bisphenol-A	Wang et al. (2017a, 2017b)
Chlorella sp.	Chlordimeform	Semple et al. (1999); Kobayashi and Rittman (1982)
Chlorella vulgaris	Diazinon	Kurade et al. (2016)
Chlorococcum sp. *and Scenedesmus* sp.	Alpha-endosulfan	Sethunathan et al. (2004)
B. *Bioaccumulation of persistent organic pollutants*		
Scenedesmus quadricauda	Dimethomorph	Olette et al. (2010)
Monoraphidium braunii	Bisphenol-A	Gattullo et al. (2012)
Scenedesmus quadricauda	Pyrimethanil	Olette et al. (2010)
Selanastrum capricornutum	Chlorobenzene and 1,2-dichlorobenzene	Semple et al. (1999); Kobayashi and Rittman (1982)
Chlorella sp.	Toxaphene	Semple et al. (1999); Kobayashi and Rittman (1982)
Selanastrum capricornutum	Benzene	Semple et al. (1999); Kobayashi and Rittman (1982)

(continued)

Table 2 (continued)

Algal species	Persistent organic pollutants (POPs)	References
Selanastrum capricornutum	2,6-Dinitrotoluene	Semple et al. (1999); Kobayashi and Rittman (1982)
Cylindrotheca sp., *Euglena gracilis, Scenedesmus obliquus*	DDT	Semple et al. (1999); Kobayashi and Rittman (1982)
Chlorella sp.	Methoxychlor	Semple et al. (1999); Kobayashi and Rittman (1982)
Chlamydomonas sp., *Chlorococcum* sp., *Dunaliella* sp.	Mirex	Semple et al. (1999); Kobayashi and Rittman (1982)
Selanastrum capricornutum	Naphthalene	Semple et al. (1999); Kobayashi and Rittman (1982)
Selanastrum capricornutum	Nitrobenzene	Semple et al. (1999); Kobayashi and Rittman (1982)
Euglena gracilis and Scenedesmus obliquus	Parathion	Semple et al. (1999); Kobayashi and Rittman (1982)
Selanastrum capricornutum	Pyrene	Semple et al. (1999); Kobayashi and Rittman (1982)
Ulva lactuca	Chloramphenicol	Leston et al. (2013)
Ulva lactuca, Ulva sp., *and Cystophora* sp.	DDT	Qiu et al. (2017)
Ulva lactuca	Poly aromatic hydrocarbons	Net et al. (2015)
Ulva lactuca	Polybrominated diphenyl ethers	Qiu et al. (2017)
Ulva lactuca, Fucus vesiculosus, Fucus virsoides, Cystoseira barbata, Gracilaria gracilis	Polychlorobiphenyl	Lauze and Hable (2017); Pavoni et al. (2003); Net et al. (2015)
Lessonia nigrescens, Macrocystis integrifolia	Phenol	Navarro et al. (2008)

dihydrobenzo[a]pyrene and other minor cis-dihydrodiols which eventually are converted to sulfate esters and α and β-glucoside conjugates. Synthesis of cis-dihydrodiols suggests dioxygenase-catalyzed reactions, similar to that found in bacteria (Warshawsky et al. 1990; Lindquist and Warshawsky 1985). In another preliminary work, Cerniglia et al. (1982) demonstrated the degradation of naphthalene in diatoms. Semple and Cain (1996) analyzed the degradation of phenol and its derivatives by eukaryotic alga *Ochromonas danica* (CCAP 933/2B) and elucidated that hydroxylation of phenol to catechol was catalyzed by whole algal cell and not by the cell-free extract. This was possible because of an enzyme phenol monooxy-

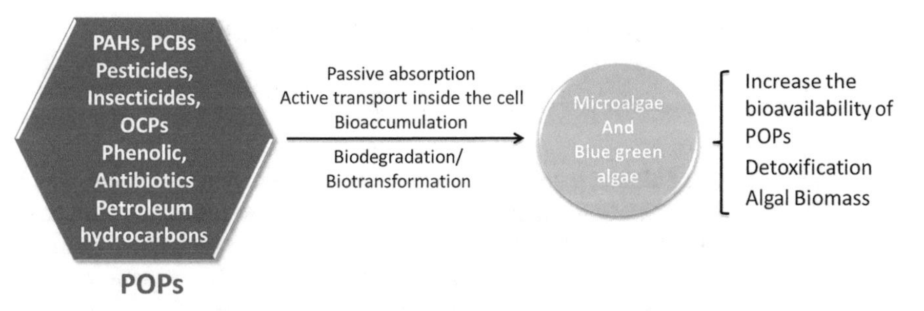

Fig. 2 An overview of POPs remediation by microalgae and blue-green algae

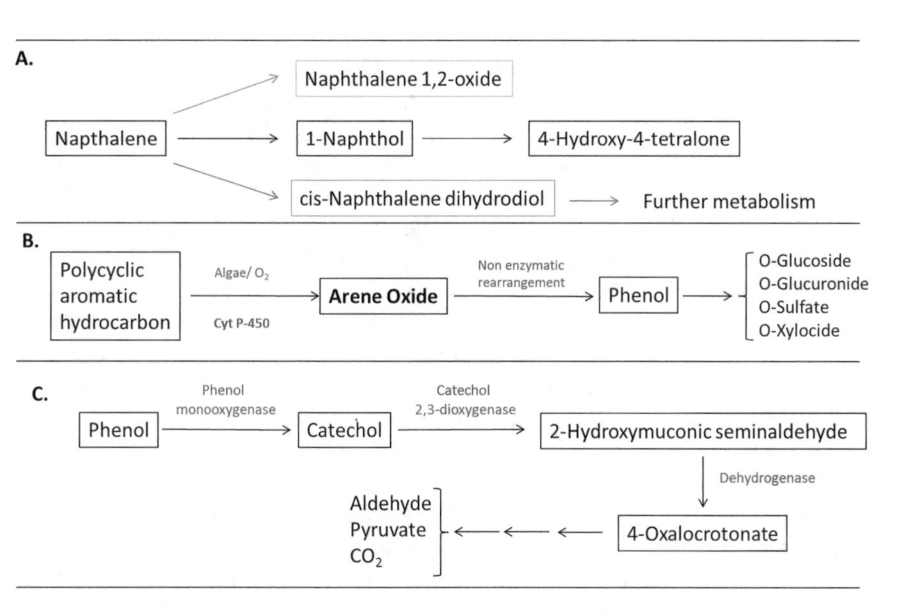

Fig. 3 The proposed pathways of (**a**) biotransformation of naphthalene by algae. (Cerniglia et al. 1980), (**b**) degradation/oxidation of PAHs. (Narro et al. 1992a) (**c**) metabolism of phenol via the meta-cleavage pathway. (Semple and Cain 1996)

genase that has high affinity toward phenol and its methylated derivatives (Fig. 3c). *Chlorella* sp. effectively remove the 2,4-dinitrophenol (2,4-DNP) from the degradation. The possible mechanism of metabolizing phenols and its derivatives involves intracellular enzyme such as polyphenol oxidase and laccase (Klekner and Kosaric 1992). Naphthalene was found to be bioaccumulated by *Chlamydomonas angulosa* within the cell but was unable to metabolize the pollutant (Soto et al. 1975). These studies demonstrate a more diverse and elaborate molecular mechanism of bioremediation than expected owing to simple methods of bioremediation by microalgae and cyanobacteria. The molecular mechanism of bioremediation not only depends on the uniqueness of algal species under study but also the pollutant to be bioremediated.

3 Phycoremediation of Persistent Organic Pollutants

3.1 Polycyclic Aromatic Hydrocarbons

PAHs are highly recalcitrant, a ubiquitous group of environmental pollutants, which is mainly composed of carbon and hydrogen with two or more fused ring structures in linear, cluster, and angular arrangements. Basically, they are accumulating in the natural habitats, viz., air, soil, and water, mainly due to anthropogenic activity (Hussain et al. 2016; Bamforth and Singleton 2005). PAHs are divided into two classes on the basis of a number of rings present in compound: (i) low molecular weight (LMW) PAH (up to 3 fused ring) and (ii) high molecular weight (HMW) PAH (more than 4 fused ring structures). The LMW PAHs are more soluble, easy to degrade, and volatile as compared to HMW PAHs; thus HMW PAHs are a major concern among us because they are highly recalcitrant and can persist in the environment due to their low water solubility and biodegradability (Cerniglia 1992; Bisht et al. 2015). PAHs are primarily produced during the incomplete combustion of organic materials (coal, wood, and petroleum) and result in a complex mixture of LMW and HMW-PAHs such as fluoranthene (Fla), pyrene (Pyr), benzo [a] anthracene (BaA), benzo [b] fluoranthene (BaF), benzo [k] fluoranthene (BkF), benzo (g, h, i) perylene (BghiP), and others (Hussain et al. 2016) and are common in the aquatic environment. Microorganisms have exhibited a strong ability to degrade PAHs. The bacteria isolated from oil contaminated site and the non-oil contaminated site have been already known, which have the ability to degrade PAHs efficiently (Wilson and Jones 1993). Besides this, the use of rhizospheric and endophytic bacteria was also reported for PAHs remediation from the soil (Bisht et al. 2015). Earlier, the ability of algae to degrade PAHs has been also noticed by a various group of scientists around the globe, and it was fully documented that a number of PAHs compounds such as Phe, Fla., naphthalene, BaP, etc. were completely or partially degraded/transformed into the nontoxic compound by a number of algal sp., viz., *Skeletonema costatum*, *Nitzschia* sp., *Chlorella sorokiniana, Chlamydomonas* sp., *Selenastrum capricornutum*, and various others (Table 2).

Cerniglia et al. (1980) performed a detailed study on the oxidation of naphthalene (carcinogenic PAH) using a number of algal species and concluded that oxidation ability of naphthalene under photoautotrophic condition is widely distributed among algae. Hong et al. (2008) studied the removal of two PAHs compound (Fla and Phe) by using two diatoms sp. *Skeletonema costatum* and *Nitzschia* sp. and concluded that (i) mixture of Phe/Fla. showed higher removal efficiency as compared to Phe/Fla. alone. This may be due to the presence of one type of PAH stimulates the degradation of other, and (ii) Fla. was more recalcitrant as compared to Phe.

Benzo[a]pyrene (BaP) is a HMW-PAH which is considered a dangerous pollutant because of its carcinogenicity, teratogenicity, and toxicity (Table 2). Kirso and Irha (1998) explore the importance of marine and freshwater microalgae in the fate of BaP and concluded that the transformation of BaP by fresh/marine algae is species specific and depends on the activity of enzyme, viz., aero-diphenoloxidase, cytochrome 450, and peroxidase present in algal cells. The unicellular green microalgae *S. capricornutum* and *S. acutus* were able to remove BaP from aqueous environment, and it was observed that the physical sorption and degradation was the basic mechanism used by algae to remove BaP (Fig. 3a) (de Llasera et al. 2016). Muñoz et al. (2003) observed that the green microalgae *C. sorokiniana* with *Pseudomonas migulae* (Bacteria) had the ability to efficiently degrade PHE under photoautotrophic conditions and suggested the use of algae and bacterial consortium for the detoxification of priority PAHs. This could be strategies in future to combat PAHs pollution.

3.2 Polychlorinated Biphenyls

PCBs are a hydrophobic chlorine containing organic compound and widely used in a variety of industrial applications such as coolants and lubricants due to its insulating and non-flammable properties (Subashchandrabose et al. 2013; Ben Chekroun et al. 2013). However, use of PCBs stopped in the United States in 1977, due to their toxicity, carcinogenicity, and slow biodegradation (Ben Chekroun et al. 2013). PCBs have a very high affinity toward lipid of both animal and plants which results in its magnification in the food chain (Subashchandrabose et al. 2013).

Microbial degradation of PCBs has been explored to a greater extent, but phycoremediation of PCBs is still under infancy. Harding and Phillips (1978) suggested the absorption and accumulation of several chlorinated hydrocarbons by marine organism including phytoplankton. The accumulation of hydrophobic PCBs by algae depends on its physicochemical properties, cell density, nutrient status and exudates produced by the algae (Subashchandrabose et al. 2013; Lynn et al. 2007). In a few types of research work, *Desmarestia* sp. and *Caepidium antarcticum* have shown the ability to associate their exudates with PCBs (Lara et al. 1989). Sijm et al. (1998) observed in *Chlorella pyrenoidosa* that rate constants for uptake of PCBs (PCB-15, PCB-52, and PCB-153) increase with hydrophobicity of PCBs and varied from 200 to 7.1×10^5 liter kg^{-1} day^{-1} and supported the previous finding of Stange and Swackhamer (1994) that the PCBs accumulation also depends upon the algal growth stage, membrane permeability and hydrophilicity, and chemical nature of PCBs.

3.3 Pesticides, Organochlorines, and Insecticides

Pesticides find their usage in post-harvested crops and agricultural products to preserve crop from various pests and rodents. Pesticides are broadly classified by two criteria: (1) usage and 2) chemical composition (synthetic). They are further classified based on their toxicity to a specific class of organisms and may be grouped into fungicide, insecticide, nematicide, molluscicide, rodenticide, herbicide, and acaricide (Rani et al. 2017). Synthetic pesticides may also be classified based on their chemical nature, toxicity, target organism, and environmental effects in to; organochlorine, organophosphorus, pyrethroids and carbamates. It is very evident that use of pesticide in agriculture and manufacturing may exert a severe threat on animal and human health. Every pesticide has its half-life, several of them are easily degradable, and some takes a long time to degrade. Slow and long exposure of such pesticides is usually detrimental to the ecosystem and eventually accumulates in sea water, birds, and other organisms (Arienzo et al. 2013; Hong et al. 2014). Primarily POPs enter into the living system (animal and humans) through food and water (about 95%), and only trace amount is reported to enter via air. Accumulation of toxic substance in the body imbalances the metabolic processes and reduces the immunity, which leads to various secondary infections to the host. This is evident with studies supporting the fact that in highly polluted areas, chances of the immune disorders are more in comparison to unpolluted areas (Ruzzin 2012; Gascon et al. 2013).

POPs also have a significant role in neurotropism and fertility-related issues in animals. Population-based studies are in favor of the neurotoxic behaviour of POPs, they promote neuro-circulatory vegetative dystonia, organic pathologies of the nervous system, polyneuritis vegetative as the depressant syndrome and encephalopathies etc. (Ljunggren et al. 2014; Vested et al. 2014).

Bioremediation of pesticide is a complex process and depends upon the biochemical behavior of the organism used under study. They are mainly affected by the type of pesticide, chemical structure, chemical load, pH, and temperature of the environment in which they are being used. Cyanobacteria may offer an eco-friendly and sustainable way of the bioremediation process. Cyanobacteria undergo numerous physiological and genetic adaptations to bioremediate POPs from water bodies. Usually, they grow in low fluid potential, variable pH, and tolerate with arid and variable saline environment. Most importantly they acquire the ability to degrade a wide range of pollutants (González et al. 2012).

Organophosphorus are commonly used against sucking, chewing, and boring types of the pest in crop protections; they may be hydrolyzed easily, but a sign of excessive use is primarily detected in the water bodies (Majewski and Capel 1995). Megharaj and co-workers established the importance of cyanobacteria in the decomposition of organophosphorus pesticides. The problem associated with this pesticide is slow degradation in nonenzymatic environmental condition. The slow degradation of organophosphates results in the generation of more reactive and carcinogenic p-nitrophenol (Megharaj et al. 1994). However, cyanobacteria such as *Nostoc*, *Oscillatoria*, and *Phormidium* have been extensively studied in the decomposition of organophosphorus (methyl parathion). Barton et al. (2004), reported that *Anabaena* sp. under aerobic conditions and light has the potential to transform methyl parathion to, o,o- dimethyl o-p-aminophenyl thiophosphate ($C_2H_7O_2PS_2$). Several authors have stated that cyanobacteria grow very efficiently in pesticide supplemented medium and exploit the phosphorous from the pesticide for their establishment and maintenance. *Phormidium valderianum* was tested for chlorpyrifos (O, O-diethyl-O-[3,5,6-trichloro-2-pyridyl] phosphorothioate) degradation at various concentration; result suggested an increase in activation of pesticide metabolism cascades including enzymes such as polyphenol oxidase, catalase, superoxide dismutase, esterase, and glutathione S-transferase, catalase, and superoxide dismutase, which suggests a possible mechanism of ROS-mediated degradation pathway of chlorpyrifos (Palanisami et al. 2009). Thengodkar and Sivakami (2010) reported that *Spirulina platensis* has immense growth ability in chlorpyrifos (up to 80 ppm) supplemented medium. They further isolated alkaline phosphatase from a crude cell-free extract of *Spirulina platensis*, which was able to potentially degrade 100 ppm chlorpyrifos into 20 ppm within 1 hour. In another study, biosorption on to *Chlorella vulgaris*, using short and long time exposures, has shown to promote bioremoval of a mixture of various pesticides including atrazine, molinate, simazine, isoproturon, propanil, carbofuran, dimethoate, pendimethalin, metolachlor, and pyriproxin (Hussein et al. 2017).

Anabaena azotica is well known for its nitrogen-fixing ability in paddy soil; moreover, it has immense ability to bioremediate lindane (γ-hexachlorocyclohexane). In a recent study, *Anabaena* PD1, an isolate from PCB contaminated soil, demonstrated the potential to degrade PCB mixture. Genomic and proteomic profiling of Anabaena PD1 in PCB mixture revealed the elevated level of about 25 proteins, directly related to the pathway involved in the PCB degradation, transport, and another cellular process. These results were further confirmed by real-time PCR analysis which suggests that during PCB degradation, genes related to dioxygenase, electron transporter, ABC transporters, transmembrane proteins, and energetic metabolism are upregulated several folds (Zhang et al. 2015).

Glyphosate is widely applied in agriculture and may become a major source of phosphorous for various microbes and phytoplankton. There are several species reported to utilize glyphosate directly. *Prorocentrum donghaiense* has been shown to grow well in glyphosate in the coastal region. Glyphosate selectively promotes the growth of *P. donghaiense* in association with microbial community present in that environment (Wang et al. 2017a, 2017b). *Spirulina platensis* in association with *Streptomyces* sp. have the ability to use glyphosate as a sole phosphorous source. *Anabaena* sp., *L. boryana*, *M. aeruginosa*, *N. punctiforme*, and *Trichodesmium erythraeum* are well reported for increased growth in glyphosate supplemented medium (Dyhrman et al. 2006; Forlani et al. 2008; Lipok et al. 2010).

3.4 Phenolics

Phenol (hydroxybenzene) and its derivatives are an aromatic hydrocarbon categorized under alcohol. Phenol serves as an important raw material in a wide range of industries including oil refinery, pharmaceuticals, chemical industries, and leather (Senthilvelan et al. 2014). Phenol or its derivatives are released in the wastewaters of aforementioned industries which could be attributed to its heavy use. Due to the high water solubility (8.3 mg mL^{-1}), it easily leaches into downstream water sources. Phenol being toxic is placed under the priority list of pollutants that need to be treated before being released into the wastewater. There are several reports of phenol biodegradation ability by algal strains such as *Chlorella* sp., *Scenedesmus obliquus* and *Spirulina maxima* (Klekner and Kosaric 1992), *Ochromonas danica* (Semple and Cain 1996), *Ankistrodesmus braunii* and *Scenedesmus quadricauda* (Pinto et al. 2003; Pinto et al. 2002), *Chlorella vulgaris* (Scragg 2006; El-Sheekh et al. 2012), *Chlorella* VT-1 (Scragg 2006), *Volvox aureus*, *Lyngbya lagerlerimi*, *Nostoc linckia*, and *Oscillatoria rubescens* (El-Sheekh et al. 2012). Algae are competent to metabolize phenols and its derivatives via inducible metabolic enzymes like polyphenol oxidase and laccase (Semple and Cain 1996; Megharaj et al. 1994). The phenol degradation ability of microalgae is influenced by various culture conditions such as light and dark regime, mode of cultivation, presence and absence of oxygen, and microalgal species. Lima et al. (2004) demonstrated that microalgae consortium including axenic cultures of *Chlorella vulgaris* and *Coenochloris pyrenoidosa* bioremediated 50 mg pentachlorophenol (PCP) L^{-1} within 5 days in the presence of light. However, under the dark conditions, the PCP degradation rate was reduced. Contrary to this, Pinto et al. (2003) showed phenol-resistant microalgae, *Ankistrodesmus braunii* and *Scenedesmus quadricauda*, were found to remove low molecular weight phenolic compounds present in the olive oil wastewater, under dark conditions. *Ankistrodesmus braunii* and *Scenedesmus quadricauda* were capable of biodegrading almost 100% of phenolic compounds including hydroxytyrosol, catechol, ferulic acid, and sinapic acid. Other phenolics such as tyrosol, 4-hydroxybenzoate, p-coumaric acid, caffeic acid, and vanillic acid were also removed efficiently (70%) when *Ankistrodesmus braunii* and *Scenedesmus quadricauda* were grown for 5 days under dark conditions (Pinto et al. 2003). Lika and Papadakis (2009) proposed a dynamic mechanistic model for aerobic degradation of phenols and suggested that the presence of other carbon sources like glucose may have an inhibitory effect on the removal of phenols due to competition between phenol and glucose for oxygen. The mixotrophic cultivation of microalgae reduces the toxic effect of the phenolic compounds compared to either phototrophic or heterotrophic cultivation methods (Megharaj et al. 1994). In the similar context, it has also been reported that the toxicity of PNP and MNP to *C. vulgaris* was reduced by adding glucose (Megharaj et al. 1988). On a similar note, Tikoo et al. (1997) also reported improved degradation of phenol by microalgae grown under mixotrophic conditions. In another contemporary study, *Ochromonas danica* has been demonstrated to grow heterotrophically in the medium containing o- and p-cresols and xylenols (Semple 1998; Semple and Cain 1995, 1997). Further to this in the same study, *O. danica* was grown in [$U^{14}C$] rich phenol, and carbon assimilation was traced down. ^{14}C was found to be assimilated in the protein, nucleic acids, and lipid of *O. danica* which suggest that phenol was metabolized by the algae and carbon was incorporated in form of proteins, lipids, and nucleic acids (Semple and Cain 1996). In a similar study, Lima et al. (2003) found that p-nitrophenol was efficiently removed by *Chlorella pyrenoidosa* within 4 days at the removal rate of 12.5 mg L^{-1} d^{-1}. The removal rate of PNP was enhanced by 1.32-fold (16.5 mg $L^{-1}d^{-1}$) when co-cultivated with *Chlorella vulgaris* at 3:1 ratio within 2 days.

Some derivatives of phenolic compounds, viz., bisphenol A (BPA), 17 α-ethinylestradiol, and 4-octylphenol, are capable of disrupting the endocrine system, while some are carcinogenic too (Crain et al. 2007). In a recent report, it was found that 17 α-ethinylestradiol (EE2), a phenolic derivative, at very low concentration can induce feminization in male fish (fathead minnow, *Pimephales promelas*) (Sole and Matamoros 2016). Among all endocrine disruptors (EDCs), BPA has been extensively studied. BPA can interfere in cell division machinery of the cell as well as induce feminization in *Xenopus laevis* tadpoles. Other phenolic derivatives such as 4-octylphenol (OP) has also feministic effect on the *Xenopus laevis* (Levy et al. 2004; Huang et al. 1999; Gattullo et al. 2012). These toxic compounds can spread to humans via the food chain (Jobling and Sumpter 1993) and therefore need to be remediated. Industries in India typically use BPA as a monomer in the manufacture of epoxy resin and polycarbonate plastic. Moreover, it is also utilized in various food and drink packaging, baby bottles, and dental sealant which eventually poses threat to humans and animals (Staples et al. 1998). Many reports suggest that EDCs can bioaccumulate due to their high stability and lipophilicity and eventually can contaminate the aquatic environment. Gattullo et al. (2012) suggested the use of a coccal green microalga *Monoraphidium braunii* for the phytoremediation of BPA contaminated aquatic environments. However, *M. braunii* growth was not remarkably affected by the lower concentration of BPA (2–4 mg L^{-1}) but was strongly inhibited at a higher concentration (10 mg L^{-1}).

3.5 Petroleum Hydrocarbons

Petroleum is considered to be a major pollutant due to its widespread usage. Petroleum refineries convert crude petroleum to an array of hydrocarbon products including naphtha, waxes, fuels, LPG, etc. (Souza et al. 2014; Varjani and Upasani 2017). Hence, leakage of petroleum hydrocarbons is bound to happen during production, refining, transport, and storage of these petroleum products (Bachmann et al. 2014; Varjani, 2017). This further leads to environmental pollution from petroleum products (Guo et al. 2012; Ishak et al. 2012; Waigi et al. 2015). Algae are an important microbial community having the potential to biodegrade petroleum hydrocarbons; however, the reports are scanty. The past four decades have witnessed a variety of algae spp. such as *Prototheca zopfi* (Walker et al. 1975), *Oscillatoria quadricauda* (Joseph and Joseph 2001), *Microcystis aeruginosa*, *Spirogyra mirabilis*, *Ulothrix subtilissima*, *Mougeotia scalaris*, *Pediastrum sp.*, *Scenedesmus quadricauda*, *Tetraedron minimum*, *Ankistrodesmus acicularis* (Ibrahim and Gamila 2004), *Scenedesmus obliquus* (Kneifel et al. 1997), *Naviculla* sp. (Headley et al. 2008), and *Nitzschia linearis* (Gamila and Ibrahim 2004) having potential to biodegrade petroleum/oil hydrocarbons. *Prototheca zopfi* isolated from Colgate Creek in Baltimore was shown to poses ability to degrade motor oil and crude oil (Walker et al. 1975). *P. Zopfi* degraded 17.9% of aromatics, relative to saturates (12.1%), from the motor oil compared to the crude oil (Walker et al. 1975). Up to 40% of dissolved solids present in petroleum effluents rich in phenols, sulfides, and aromatic hydrocarbons were effectively adsorbed by a cyanobacterium

Oscillatoria quadricauda, within 4 days (Joseph and Joseph 2001). 4-Methyl cyclohexaneacetic acid (5.5 mg L^{-1}) was removed within 14 days by using a diatom, *Navicula* sp. (Headley et al. 2008). The axenic culture of green microalga *Scenedesmus obliquus* effectively utilized 1-naphthalene sulfonic acid when grown under limited sulfate condition. Bioremediation of 1-naphthalene sulfonic yields 1-hydroxy-2-naphthalene sulfonic acid, 1-naphthol, and 1-naphthyl β-D-glucopyranoside (Kneifel et al. 1997). Besides single compounds, researchers have also investigated the capability of algae to degrade the whole matrix of oil. Gamila and Ibrahim (2004) evaluated the potential of freshwater microalgae in bioremediation of crude oil (specific gravity 0.85) through an algal bioassay, and they found that diatom strain *Nitzschia linearis* and green microalgae *Scenedesmus obliquus* showed almost similar capability for degradation of petroleum hydrocarbon. *Nitzschia linearis* removed PAHs at a higher rate than n-alkane. *Scenedesmus obliquus* efficiently grew and degraded seven out of eight types of n-alkane and approximately 80% of different types of PAHs within 6 weeks of incubation. El-Sheekh et al. (2012) studied the biodegradation of crude oil by a mixed culture of *Chlorella vulgaris* and *Scenedesmus obliquus.* It was observed that both the algae grew efficiently under heterotrophic conditions and used crude oil as a sole carbon source. In a contemporary study, Tang et al. (2010) suggested that *Scenedesmus obliquus* GH2 can also be used to construct an artificial microalgal-bacterial consortium for crude oil degradation.

3.6 Antibiotics

Antibiotics are any organic chemicals/compounds that kill or control the growth of microorganisms. They have a peculiar mechanism of interaction with the target molecule present in the host cellular system and inhibit its growth. These are not limited to the microbial origin. Presently numerous antibiotics are in the market that is exclusively chemically synthesized. Administration of antibiotic in clinical and health system has saved myriad lives and has remarkable regulation over a number of diseases. Industrial production of antibiotics and their overuse in clinical practices in humans, animals, horticulture, and the flesh industry have emancipated a huge amount of antibiotics into the environment (Pruden et al. 2013). This may generate severe health problems for various living forms including humans and animals. There are various reports that indicate the presence of antibiotics in the wastewater and soils at a variable concentration ranging from μg Kg^{-1} to mg Kg^{-1}. It is well known that antibiotics hamper the growth of microbial flora; hence in natural conditions antibiotics are not deteriorated by normal microbes which ultimately increase the antibiotic load in the surrounding environment making these new and emerging pseudo POPs.

Antibiotics can contaminate the aquatic environment by various ways. Antibiotics are dumped into our environment because of their heavy use in livestock treatments, animal treatments, aquaculture treatments, manufacturing process, storage of manure/slurry, inappropriate disposal of used medicine containers, and wastewater treatments. Once released these pharmaceuticals are transported and distributed to air, water, soil, or sediment (Boxall 2004). For example, concentrations of trimethoprim have been reported to range from less than 3.4×10^{-5} μmol L^{-1} in surface waters of United Kingdom to 0.0061 μmol L^{-1} in the United States (Ashton et al. 2004). The presence of lincomycin in surface water has been recorded from less than 2.46×10^{-6} μmol L^{-1} to 1.8×10^{-3} μmol L^{-1} in the United States (Monteiro and Boxall 2010). Similar reports of contaminants such as tylosin were found at 5.46×10^{-5} μmol/L downstream of agricultural land in the United States (Boxall et al. 2012). Freshwater cyanobacteria *Microcystis aeruginosa, Aphanizomenon gracile, Chrysosporum bergii, and Planktothrix agardhii* have the ability to grow on different concentrations of antibiotics (amoxicillin, ceftazidime, kanamycin, ceftriaxone, gentamicin, tetracycline, trimethoprim, nalidixic acid, and norfloxacin). The increased resistance of these cyanobacteria to antibiotics suggests that they are naturally resistant to antibiotics (Dias et al. 2015). In another study, Guo et al. (2016) studied tylosin, lincomycin, and trimethoprim on the growth of two chlorophytes, *Pseudokirchneriella subcapitata and Desmodesmus subspicatus* and *Anabaena flos-aqua* and evaluated chlorophyll content and oxygen evolution rate. They reported that antibiotics have a remarkable effect on photosynthesis rate and growth (Guo et al. 2016). Fenton algal treatment is proposed as a powerful method for removal of antibiotics such as amoxicillin and cefradine from wastewater using *chlorella pyrenoidosa*, and it does not have any adverse effect on algal biomass. Additionally, combined methods can achieve a higher removal rate in a short period (Li et al. 2015). Ge and Deng (2015) reported the removal of fluoroquinolone antibiotics (enrofloxacin and ciprofloxacin hydrochloride) by two marine algae (*Platymonas subcordiformis* and *Isochrysis galbana*) in a photoinduced system. The system efficiently degraded the antibiotics in the presence of FeIII.

4 Recent Approach to Enhance Phycoremediation

Our understanding of algal cell biochemistry and molecular biology, along with knowledge of latest omics technology (genomics, transcriptomics, proteomics, and metabolomics) has increased immensely. In-depth knowledge of algal metabolism, genetic engineering, and functional proteomics has a paved platform for modern bioremediation methods. Their knowledge has huge application in phycoremediation research. Recently, engineering of plants by introducing bacterial or animal xenobiotic degrading genes has been successfully tested for bioremediation of xenobiotics such as explosives and hydrocarbons (Dhankher et al. 2012; Suresh and Ravishankar 2004). Plants harboring bacterial gene constitutes a new generation of the designer plants for efficient and environment-friendly remediation of soil and aquatic ecosystems (Aken et al. 2010). Specialized cloning vectors, gene sequences, and methods of mutagenesis provide plenty of opportunities to generate algae with designer traits for sustainable bioremediation process (Koksharova and Wolk 2002).

Transgenic microalgae have received serious attention in various biotechnological industries (Walker et al. 2005). Nuclear and chloroplast transformation in algae for high-level protein expression have been demonstrated well in previous studies (Rosenberg et al. 2008). Kuritz et al. (1997) successfully introduced a lindane degrading gene (Lin A) of *P. paucimobilis* UT26 into *Anabaena* sp. PCC7120 and found lindane degrading capabilities in the transgenic *Anabaena* sp. Similar results were also observed in *Nostoc ellipsosporum* harboring Lin A gene. Rajamani et al. (2007) showed that these algal species have more tolerance to pollutants and can be exploited for bioremediation.

5 Phycoremediation Boundaries

The removal and detoxification of man-made organic pollutant by algae is advantageous because it presents an eco-friendly process with no secondary pollution, and, above all, it is cost-effective. Phycoremediation is time-dependent technology as they need enough time (up to few days) to achieve desired cell density, followed by absorption process for facilitating the uptake and degradation of contaminants. The degree of bioaccumulation and remediation of organic pollutant by algae is unique to species and depends on the initial number of cells of algae (Lei et al. 2002). The mode of cultivation also influences the degradation potential of microalgae. Yan et al. (2002) observed that the green microalgae *Chlorella protothecoides* removed more (33.53%) anthracene under heterotrophic conditions as compared to autotrophic conditions (20%). Mixotrophic mode of cultivation is always found superior over auto- and heterotrophic conditions (Subashchandrabose et al. 2013). Besides these, the algal bioremediation is also affected by other factors such as surface area and cellular biovolume, concentration of pollutant, lipid content, algal cell density (Subashchandrabose et al. 2013), algal sp., light regime (Hirooka et al. 2003), chemical preferences (Chan et al. 2006), external carbon source (Papazi and Kotzabasis 2007), as well as contaminants and presence of other microbes (Rastogi et al. 2014). Algal biotransformation not necessarily always yields nontoxic (as compared to parent molecule) intermediate compound. Under these conditions, the consortium of algae and bacteria is effective since bacteria can utilize the algal metabolites (Subashchandrabose et al. 2011; Muñoz and Guieysse 2006). *P. subcapitata* and *Chlorococcum* sp. accumulated and transformed fenamiphos (an organophosphorus nematicide) to its oxides (fenamiphos sulfoxide) and fenamiphos sulfone and phenols, which are considerably more toxic than fenamiphos (Cáceres et al. 2008). Such problems have to be overcome by integrating other microbes, which can further metabolize the toxic intermediates and end product.

More research work is desirable to develop cost-effective and efficient algal bioremediation methods in the case of wastewater containing POPs. The challenge is to combine basic knowledge related to growth kinetics and unique POP metabolizing characteristics of algal strains.

6 Conclusions and Future Scenarios

Persistent organic pollutants (POPs) are serious global threats as they augment the risk to humans and aquatic organisms. Industrial effluents, radioactive and pesticide waste, oils, and their by-products are the main sources of POPs. For clean, safe, and sustainable environment, there is an urgent need to develop an effective, sustainable, eco-friendly, and efficient method of pollutant remediation. The number of traditional methods exists to treat such pollutants, but these methods mostly can only transform, and therefore the next-generation water treatment techniques are in high demand. This chapter discussed in detail the potential of microalgae including cyanobacteria and eukaryotic algae for biodegradation and bioaccumulation of POPs that are usually contaminate our natural and wastewaters. The eukaryotic algae appear to be possibly valuable when initial sorption/accumulation is required because of properties such as high cell density, low nutrient requirement, and simple cultivation methods. Algal cell density could be initially increased by autotrophic cultivation. Sorption process would then start to accumulate pollutants followed by biotransformation of the compound. These phototrophs are able to enhance the bioavailability of POPs to the microbiota present in the environment and thus contribute to the elimination of the pollutant from the respective ecosystem. At the end we conclude that phycoremediation is a well-established, eco-friendly, cost-effective, sustainable, and sensible method for resolving problems generated by POP contamination. Various cyanobacteria and microalgae have been reviewed in this chapter showing great potential to degrade, transform, or accumulate innumerable classes of POPs (Table 2). We also touched upon use of genetic engineering techniques to create transgenic microalgae for remediating certain class of pollutants. This could be the future of bioremediation methods where microalgae are the method of choice. However, further research is needed to focus at the molecular and biochemical level to elucidate the remediation mechanism so that improved and cost-effective modern methods of bioremediation could be developed.

Acknowledgment Authors (A.P. and M.P.S.) would like to give their sincere thanks to Ministry of Human Resource Development (MHRD), New Delhi (India), for the financial support.

References

Abou-Shanab RAI (2011) Bioremediation: new approaches and trends. In: Khan MS, Zaidi A, Goel R, Musarrat J (eds) Biomanagement of metal-contaminated soils, vol 20., Environmental pollution. Springer, Dordrecht

Aken BV, Correa PA, Schnoor JL (2010) Phytoremediation of polychlorinated biphenyls: new trends and promises. Environ Sci Technol 44:2767–2776

Andrate MR, Costa JAV (2007) Mixotrophic cultivation of microalga Spirulina platensis using molasses as organic substrate. Aquaculture 264:130–134

Aoki Y (2001) Polychlorinated biphenyls, polychloronated dibenzo-*p*-dioxins, and polychlorinated dibenzofurans as endocrine disrupters—what we have learned from Yusho disease. Environ Res 86(1):2–11

Arienzo M, Masuccio AA, Ferrara L (2013) Evaluation of sediment contamination by heavy metals, organochlorinated pesticides, and polycyclic aromatic hydrocarbons in the Berre coastal lagoon (Southeast France). Arch Environ Contam Toxicol 65(3):396–406

Ashton D, Hilton M, Thomas KV (2004) Investigating the environmental transport of human pharmaceuticals to streams in the United Kingdom. Sci Total Environ 333(1–3):167–184

Bachmann R, Johnson A, Edyvean R (2014) Biotechnology in the petroleum industry: an overview. Int Biodeterior Biodegradation 86:225–237

Bahadar H, Mostafalou S, Abdollahi M (2014) Current understandings and perspectives on non-cancer health effects of benzene: a global concern. Toxicol Appl Pharmacol 276(2):83–94

Baker JA, Entsch B, Neilan BA, McKay DB (2002) Monitoring changing toxigenicity of a cyanobacterial bloom by molecular methods. Appl Environ Microbiol 68(12):6070–6076

Bamforth SM, Singleton I (2005) Bioremediation of polycyclic aromatic hydrocarbons: current knowledge and future directions. J Chem Technol Biotechnol 80(7):723–736

Barton JW, Kuritz T, O'Connor LE, Ma CY, Maskarinec MP, Davison BH (2004) Reductive transformation of methyl parathion by the cyanobacterium Anabaena sp. strain PCC7120. Appl Microbiol Biotechnol 65(3):330–335

Ben Chekroun K, Moumen A, Rezzoum N, Sanchez E, Baghour M (2013) Role of macroalgae in biomonitoring of pollution in 'Marchica', the Nador lagoon. Phyton 82:31–34

Bergman B, Matveyev A, Rasmussen U (1996) Chemical signalling in cyanobacterial-plant symbioses. Trends Plant Sci 1(6):191–197

Bilanovic D, Andargatchew A, Kroeger T, Shelef G (2009) Freshwater and marine microalgae sequestering of CO2 at different C and N concentrations-response surface methodology analysis. Energy Convers Manag 50(2):262–267

Bisht S, Pandey P, Bhargava B, Sharma S, Kumar V, Sharma KD (2015) 2015. Bioremediation of polyaromatic hydrocarbons (PAHs) using rhizosphere technology. Braz J Microbiol 46(1):7–21

Borowitzka MA (1999) Commercial production of microalgae: ponds, tanks, tubes and fermenters. J Biotechnol 70:313–321

Boxall ABA (2004) The environmental side effects of medication. EMBO Rep 5(12):1110–1116

Boxall ABA, Rudd MA, Brooks BW, Caldwell DJ, Choi K, Hickmann S, Innes E, Ostapyk K, Staveley JP, Verslycke T, Ankley GT, Beazley KF, Belanger SE, Berninger JP, Carriquiriborde P, Coors A, DeLeo PC, Dyer SD, Ericson JF, Gagné F, Giesy JP, Gouin T, Hallstrom L, Karlsson MV, Larsson DGJ, Lazorchak JM, Mastrocco F, McLaughlin A, McMaster ME, Meyerhoff RD, Moore R, Parrott JL, Snape JR, Murray-Smith R, Servos MR, Sibley PK, Straub JO, Szabo ND, Topp E, Tetreault GR, Trudeau VL, Van Der Kraak G (2012) Pharmaceuticals and personal care products in the environment: what are the big questions? Environ Health Perspect 120(9):1221–1229

Cáceres T, Megharaj M, Naidu R (2008) Toxicity and transformation of fenamiphos and its metabolites by two micro algae Pseudokirchneriella subcapitata and Chlorococcum sp. Sci Total Environ 398:53–59

Castiglioni B, Rizzi E, Frosini A, Sivonen K, Rajaniemi P, Rantala A, Mugnai MA, Ventura S, Wilmotte A, Boutte C, Grubisic S, Balthasart P, Consolandi C, Bordoni R, Mezzelani A, Battaglia C, De Bellis G (2004) Development of a universal microarray based on the ligation detection reaction and 16S rrna gene polymorphism to target diversity of cyanobacteria. Appl Environ Microbiol 70(12):7161–7172

Cerniglia CE (1992) Biodegradation of polycyclic aromatic hydrocarbons. Biodegradation 3:351–368

Cerniglia CE, van Baalen C, Gibson DT (1980) Metabolism of naphthalene by the cyanobacterium *Oscillatoria* sp., strain JCM. J Gen Microbiol 116:485–494

Cerniglia CE, Gibson DT, van Baalen C (1982) Naphthalene metabolism by diatoms from the Kachemak Bay region of Alaska. J Gen Microbiol 128:987–990

Chan SMN, Luan T, Wong MH, Tam NFY (2006) Removal and biodegradation of polycyclic aromatic hydrocarbons by Selenastrum capricornutum. Environ Toxicol Chem 25:1772–1779

Cheng Y, Zhou W, Gao C, Lan K, Gao Y, Wu Q (2009) Biodiesel production from Jerusalem artichoke (*Helianthus tuberosus* L.) tuber by heterotrophic microalgae *Chlorella protothecoides*. J Chem Technol Biotechnol 84:777–781

Chisti Y (2007) Biodiesel from microalgae. Biotechnol Adv 25:249–306

Couto RM, Simões PC, Reis A, Da Silva TL, Martins VH, Sánchez-Vicente Y (2010) Supercritical fluid extraction of lipids from the heterotrophic microalga *Crypthecodiniumcohnii*. Eng Life Sci 10:158–164

Crain DA, Eriksen M, Iguchi T, Jobling S, Laufer H, LeBlanc GA, Guillette LJ (2007) An ecological assessment of bisphenol-A: evidence from comparative biology. Reprod Toxicol 24:225–239

Crinnion W (2011) Polychlorinated biphenyls: persistent pollutants with immunological, neurological, and endocrinological consequences. Environ Med 16(1):5–13

de Llasera MPG, de Jesús Olmos-Espejel J, Díaz-Flores G, Montaño-Montiel A (2016) Biodegradation of benzo (a) pyrene by two freshwater microalgae *Selenastrum capricornutum* and *Scenedesmus acutus*: a comparative study useful for bioremediation. Environ Sci Pollut Res 23:3365–3375

Dhankher OP, Pilon-Smits EAH, Meagher RB, Doty S (2012) Biotechnological approaches for phytoremediation. In: Atman A, Hasegawa PM (eds) Plant biotechnology and agriculture, prospects for the 21st century. Academic Press, Oxford, U.K, pp 309–328

Dias E, Oliveira M, Jones-Dias D, Vasconcelos V, Ferreira E, Manageiro V, Caniça M (2015) Assessing the antibiotic susceptibility of freshwater Cyanobacteria spp. Front Microbiol 6:799

Ding T, Lin K, Yang B, Yang M, Li J, Li W, Gan J (2017) Biodegradation of naproxen by freshwater algae *Cymbella* sp. and *Scenedesmus quadricauda* and the comparative toxicity. Bioresour Technol 238:164–173

Dosnon-Olette R, Trotel-Aziz P, Couderchet M, Eullaffroy P (2010) Fungicides and herbicide removal in Scenedesmus cell suspensions. Chemosphere 79:117–123

Droop MR (1974) The nutrient status of algal cells in continuous culture. J Mar Biol Assoc UK 54:825–855

Dyhrman ST, Chappell PD, Haley ST, Moffett JW, Orchard ED, Waterbury JB, Webb EA (2006) Phosphonate utilization by the globally important marine diazotroph Trichodesmium. Nature 439(7072):68–71

Dziallas C, Grossart H-P (2011) Temperature and biotic factors influence bacterial communities associated with the cyanobacterium Microcystis sp. Environ Microbiol 13(6):1632–1641

El-Sheekh MM, Ghareib MM, EL-Souod GWA (2012) Biodegradation of phenolic and polycyclic aromatic compounds by some algae and cyanobacteria. J Bioremed Biodegr 3:133

Forlani G, Pavan M, Gramek M, Kafarski P, Lipok J (2008) Biochemical bases for a widespread tolerance of cyanobacteria to the Phosphonate herbicide glyphosate. Plant Cell Physiol 49(3):443–456

Friesen-Pankratz B, Doebel C, Farenhorst A, Goldsborough LG (2003) Influence of algae (*Selenastrum capricornutum*) on the aqueous persistence of atrazine and lindane: implications for managing constructed wetlands for pesticide removal. J Environ Sci Heal B B38:147–155

Gamila HA, Ibrahim MBM (2004) Algal bioassay for evaluating the role of algae in bioremediation of crude oil:1-isolated strains. Bull Environ Contam Toxicol 723:883–889

Gascon M, Morales E, Sunyer J, Vrijheid M (2013) Effects of persistent organic pollutants on the developing respiratory and immune systems: a systematic review. Environ Int 52:51–65

Gattullo CE, Bahrs H, Steinberg CEW, Loffredo E (2012) Removal of bisphenol A by the freshwater green alga *Monoraphidium braunii* and the role of natural organic matter. Sci Total Environ 416:501–506

Ge L, Deng H (2015) Degradation of two fluoroquinolone antibiotics photoinduced by Fe(III)-microalgae suspension in an aqueous solution. Photochem Photobiol Sci 14(4):693–699

González R, García-Balboa C, Rouco M, Lopez-Rodas V, Costas E (2012) Adaptation of microalgae to lindane: a new approach for bioremediation. Aquat Toxic 109:25–32

Greenberg MM (1997) The central nervous system and exposure to toluene: a risk characterization. Environ Res 72(1):1–7
Guo XW, Liu KY, He S, Hu HJ, Xu TW (2012) Petroleum generation and charge history of the northern Dongying Depression, Bohai Bay Basin, China: insight from integrated fluid inclusion analysis and basin modeling. Mar Pet Geol 32(1):21–35
Guo J, Selby K, Boxall ABA (2016) Effects of antibiotics on the growth and physiology of chlorophytes, cyanobacteria, and a diatom. Arch Environ Contam Toxicol 71(4):589–602
Harding LW, Phillips JH (1978) Polychlorinated Biphenyl (PCB) uptake by marine phytoplankton. Mar Biol 49:103–111
Headley JV, Du JL, Peru KM, Gurprasa N, McMartin DW (2008) Evaluation of algal phytodegradation of petroleum naphthenic acids. J Environ Sci Health A Tox Hazard Subst Environ Eng 43:227–232
Hirooka T, Akiyama Y, Tsuji N, Nakamura T, Nagase H, Hirata K, Miyamoto K (2003) Removal of hazardous phenols by microalgae under photoautotrophic conditions. J Biosci Bioeng 95:200–203
Hoffmann L, Komárek J, Kaštovský J (2005) System of cyanoprokaryotes (cyanobacteria) state in 2004. Algol Stud/Archiv Hydrobiol Supplement Volumes 117:95–115
Hong YW, Yuan DX, Lin QM, Yang TL (2008) Accumulation and biodegradation of phenanthrene and fluoranthene by the algae enriched from a mangrove aquatic ecosystem. Mar Pollut Bull 56:1400–1405
Hong SH, Shim WJ, Han GM, Ha SY, Jang M, Rani M, Hong S, Yeo GY (2014) Levels and profiles of persistent organic pollutants in resident and migratory birds from an urbanized coastal region of South Korea. Sci Total Environ 470–471:1463–1470
Huang H, Marsh-Armstrong N, Brown DD (1999) Metamorphosis is inhibited in transgenic Xenopus laevis tadpoles that overexpress type III deiodinase. Proc Natl Acad Sci U S A 296(3):962–967
Humbert JF, Quiblier C, Gugger M (2010) Molecular approaches for monitoring potentially toxic marine and freshwater phytoplankton species. Anal Bioanal Chem 397(5):1723–1732
Hussain J, Zhao Z, Pang Y, Xia L, Hussain I, Jiang X (2016) Effects of different water seasons on the residual characteristics and ecological risk of polycyclic aromatic hydrocarbons in sediments from Changdang Lake, China. J Chem 2016:1–10
Hussein IAS, Mansour MSM (2016) A review on polycyclic aromatic hydrocarbons: source, environmental impact, the effect on human health and remediation. Egyp J Petro 25:107–123
Hussein MH, Abdullah AM, Badr El-Din NI, Mishaqa ESI (2017) Biosorption potential of the microchlorophyte chlorella vulgaris for some pesticides. J Fertil Pestic 08:01
Ibrahim MBM, Gamila HA (2004) Algal bioassay for evaluating the role of algae in bioremediation of crude oil: II. freshwater phytoplankton assemblages. Bull Environ Contam Toxicol 73:971–978
Ishak S, Malakahmad A, Isa MH (2012) Refinery wastewater biological treatment: a short review. J Sci Ind Res 71:251–256
Ji PH, Song YF, Sun TH, Liu Y, Cao XF, Xu D, Yang XX, McRae T (2011) In-situ cadmium phytoremediation using *Solanum nigrum* L.: the bio-accumulation characteristics trial. Int. J Phytorem 13(10):1014–1023
Jin ZP, Luo K, Zhang S, Zheng Q, Yang H (2012) Bioaccumulation and catabolism of prometryne in green algae. Chemosphere 87:278–284
Jobling S, Sumpter JP (1993) Detergent components in sewage effluent are weakly oestrogenic to fish: An in vitro study using rainbow trout (Oncorhynchus mykiss) hepatocytes. Aquatic Toxicology 27:361–372
Johnson MB, Wen Z (2009) Production of biodiesel fuel from the microalga by direct transesterification of algal biomass. Energy Fuel 23:5179–5183
Jonsson PR, Pavia H, Toth G (2009) Formation of harmful algal blooms cannot be explained by allelopathic interactions. Proc Natl Acad Sci 106(27):11177

Joseph V, Joseph A (2001) Microalgae in petrochemical effluent: growth and biosorption of total dissolved solids. Bull Environ Contam Toxicol 66:522–527
Juneja A, Ceballos RM, Murthy GS (2013) Effects of environmental factors and nutrient availability on the biochemical composition of algae for biofuels production: a review. Energies 6:4607–4638
Jurewicz J, Hanke W (2008) Prenatal and childhood exposure to pesticides and neurobehavioral development: a review of epidemiological studies. Int J Occup Med Environ Health 21(2):121–132
Kirso U, Irha N (1998) Role of algae in fate of carcinogenic polycyclic aromatic hydrocarbons in the aquatic environment. Ecotox Environ Safety 41(1):83–89
Klekner V, Kosaric N (1992) Degradation of phenols by algae. Environ Technol 13:493–501
Kneifel H, Kerstin E, Eberhard H, Carl JS (1997) Biotransformation of 1-naphthalene sulfonic acid by the green alga Scenedesmus obliquus. Arch Microbiol 167:32–37
Kobayashi H, Rittman BE (1982) Microbial removal of hazardous organic compounds. Environ Sci Technol 16:170–183
Komárek J (2016) A polyphasic approach for the taxonomy of cyanobacteria: principles and applications. Eur J Phycol 51(3):346–353
Kraut A, Lilis R, Marcus M, Valciukas JA, Wolff MS, Landrigan PJ (1988) Neurotoxic effects of solvent exposure on sewage treatment workers. Arch Environ Health 43(1):263–268
Kurade MB, Kim JR, Govindwar SP, Jeon BH (2016) Insights into microalgae mediated biodegradation of diazinon by *Chlorella vulgaris*: microalgal tolerance to xenobiotic pollutants and metabolism. Algal Res 20:126–134
Kuritz T, Bocanera LV, Rivera NS (1997) Dechlorination of lindane by the cyanobacterium *Anabaena* sp. strain PCC7120 depends on the function of the air operon. J Bacteriol 179:3368–3370
Koksharova OA, Wolk CP (2002) Genetic tools for cyanobacteria Appl Microbiol Biotechnol. 58(2):123–137
Lam MK, Lee KT (2012) Microalgae biofuels: a critical review of issues, problems and the way forward. Biotechnol Adv 30(3):673–690
Lara RJ, Wiencke C, Ernst W (1989) Association between exudates of Brown algae andpolychlorinated biphenyls. J Appl Phycol 1:267–270
Lauze JF, Hable WE (2017) Impaired growth and reproductive capacity in marine rockweeds following prolonged environmental contaminant exposure. Bot Mar 60:137–148
Lei A, Wong Y, Tam N (2002) Removal of pyrene by different microalgal species. Water Sci Technol 46:195–201
Leston S, Nunes M, Viegas I, Ramos F, Pardal MA (2013) The effects of chloramphenicol on *Ulva lactuca*. Chemosphere 91:552–557
Levy G, Lutz I, Kruger A, Kloas W (2004) Bisphenol a induces feminization in Xenopus laevis tadpoles. Environ Res 94:102–111
Li H, Pan Y, Wang Z, Chen S, Guo R, Chen J (2015) An algal process treatment combined with the Fenton reaction for high concentrations of amoxicillin and cefradine. RSC Adv 5(122):100775–100782
Liang Y, Sarkany N, Cui Y (2009) Biomass and lipid productivities of *Chlorella vulgaris* under autotrophic, heterotrophic and mixotrophic growth conditions. Biotechnol Lett 31:1043–1049
Lika K, Papadakis IA (2009) Modeling the biodegradation of phenolic compounds by microalgae. J Sea Res 62:135–146
Lima SAC, Castro PML, Morais R (2003) Biodegradation of p-nitrophenol by microalgae. J Appl Phycol 15:137–142
Lima SAC, Raposo MFJ, Castro PML, Morais RM (2004) Biodegradation of p-chlorophenol by a microalgae consortium. Water Res 38:97–102
Lin TS, Wu JY (2015) Effect of carbon sources on growth and lipid accumulation of newly isolated microalgae cultured under mixotrophic condition. Bioresour Technol 184:100–107

Lindquist B, Warshawsky D (1985) Identi¢cation of the 11,12-dihydro-11,12-dihydroxybenzo(a) pyrene as a major metabolite by the green alga, *Selenastrum capricornutum*. Biochem Biophys Res Commun 130:71–75

Lipok J, Studnik H, Gruyaert S (2010) The toxicity of Roundup® 360 SL formulation and its main constituents: glyphosate and isopropylamine towards non-target water photoautotrophs. Ecotoxicol Environ Saf 73(7):1681–1688

Ljunggren SA, Helmfrid I, Salihovic S, van Bavel B, Wingren G, Lindahl M, Karlsson H (2014) Persistent organic pollutants distribution in lipoprotein fractions in relation to cardiovascular disease and cancer. Environ Int 65:93–99

Lynn SG, Price DJ, Birge WJ, Kilham SS (2007) Effect of nutrient availability on the uptake of PCB congener 2,20,6,60-tetrachlorobiphenyl by a diatom (*Stephanodiscus minutulus*) and transfer to a zooplankton (*Daphnia pulicaria*). Aquat Toxicol 83:24–32

Mahdavi H, Prasad V, Liu Y, Ulrich AC (2015) In situ biodegradation of naphthenic acids in oil sands, tailings pond water using indigenous algae–bacteria consortium. Bioresour Technol 187:97–105

Majewski MS, Capel PD (1995) Pesticides in the atmosphere; distribution, trends, and governing factors. Report:94–506. Online: http://pubs.er.usgs.gov/publication/ofr94506

Megharaj M, Venkateswarlu K, Rao AS (1988) Influence of glucose amendment on the algal toxicity of nitrophenols. Ecotoxicol Environ Saf 15:320–323

Megharaj M, Madhavi DR, Sreenivasulu C, Umamaheswari A, Venkateswarlu K (1994) Biodegradation of methyl parathion by soil isolates of microalgae and cyanobacteria. Bull Environ Contam Toxicol 53:2

MistraPharma (2011) Collaborating to reduce the environmental risks of pharmaceuticals. MistraPharma researchers and stakeholders. MistraPharma, Stockholm

Monteiro SC, Boxall ABA (2010) Occurrence and fate of human pharmaceuticals in the environment. Rev Environ Contam Toxicol 202:53–154

Muñoz R, Guieysse B (2006) Algal-bacterial processes for the treatment of hazardous contaminants: a review. Water Res 40:2799–2815

Muñoz R, Guieysse B, Mattiasson B (2003) Phenanthrene biodegradation by an algal-bacterial consortium in two-phase partitioning bioreactors. Appl Microbiol Biotechnol 61:261–267

Montero-Rodríguez D, Andrade RFS, Ribeiro DLR, Rubio-Ribeaux D, Lima RA, Araújo HWC, Campos-Takaki GM (2015) Bioremediation of Petroleum Derivative Using Biosurfactant Produced by Serratia marcescens UCP/WFCC 1549 in Low-Cost Medium. Int J Curr Microbiol App Sci 4(7):550–562

Narro ML, Cerniglia CE, Baalen CV, Gibson DT (1992a) Evidence for an NIH shift in oxidation of naphthalene by the marine cyanobacterium *Oscillatoria* sp. strain JCM. Appl Envron Microbiol 58:1360–1363

Narro ML, Cerniglia CE, Baalen CV, Gibson DT (1992b) Metabolism of phenanthrene by the marine cyanobacterium *Agmenellum quadrulicatum* PR-6. Appl Environ Microbiol 58:1351–1359

Naturvårdsverket (2008) Avloppsreningsverkens förmåga att ta hand om läkemedelsrester och andra farliga ämnen. Report 5794 http://www.naturvardsverket.se/Documents/publikationer/620-5794-7.pdf

Navarro AE, Portales RF, Sun-Kou MR, Llanos BP (2008) Effect of pH on phenol biosorption by marine seaweeds. J Hazard Mater 156:405–411

Net S, Henry F, Rabodonirina S, Diop M, Merhaby D, Mahfouz C, Amara R, Ouddane B (2015) Accumulation of PAHs, Me-PAHs, PCBs and total mercury in sediments and marine species in coastal areas of Dakar, Senegal: contamination level and impact. Int J Environ Res 9:419–432

Ogbonna JC, Yoshizawa H, Tanaka H (2000) Treatment of high strength organic wastewater by a mixed culture of photosynthetic microorganisms. J Appl Phycol 12:277–284

Olette R, Couderchet M, Biagianti S, Eullaffroy P (2010) Fungicides and herbicide removal in Scenedesmus cell suspensions. Chemosphere 79:117–123

Palanisami S, Prabaharan D, Uma L (2009) Fate of few pesticide-metabolizing enzymes in the marine cyanobacterium Phormidium valderianum BDU 20041 in perspective with chlorpyrifos exposure. Pestic Biochem Physiol 94(2–3):68–72

Papazi A, Kotzabasis K (2007) Bioenergetic strategy of microalgae for the biodegradation of phenolic compounds — exogenously supplied energy and carbon sources adjust the level of biodegradation. J Biotechnol 129:706–716

Pavoni B, Caliceti M, Sperni L, Sfriso A (2003) Organic micropollutants (PAHs, PCBs, pesticides) in seaweeds of the lagoon of Venice. Ocean Acta 26:585–5960

Pinto G, Pollio A, Previtera L, Temussi F (2002) Biodegradation of phenols by microalgae. Biotechnol Lett 24(24):2047–2051

Pinto G, Pollio A, Previtera L, Stanzione M, Temussi F (2003) Removal of low molecular weight phenols from olive oil mill wastewater using microalgae. Biotechnol Lett 25:1657–1659

Priyadarshani I, Sahu D, Rath B (2011) Microalgal bioremediation: current practices and perspectives. J Biochem Technol 3(3):299–304

Pruden A, Larsson DGJ, Amézquita A, Collignon P, Brandt KK, Graham DW, Lazorchak JM, Suzuki S, Silley P, Snape JR, Topp E, Zhang T, Zhu Y-G (2013) Management options for reducing the release of antibiotics and antibiotic resistance genes to the environment. Environ Health Perspect 121:8, 878–885

Qiu YW, Zeng EY, Qiu H, Yu K, Cai S (2017) Bioconcentration of polybrominated diphenyl ethers and organochlorine pesticides in algae is an important contaminant route to higher trophic level. Sci Total Environ 579:1885–1893

Quintana N, Kooy FV, Miranda DVR, Gerben PV, Robert V (2011) Renewable energy from Cyanobacteria: energy production optimization by metabolic pathway engineering. Appl Microbiol Biotechnol 91:471–490

Rajamani S, Siripornadulsil S, Falcão V, Torres M, Colepicolo P, Sayre R (2007) Phycoremediation of heavy metals using transgenic microalgae. Adv Exp Med Biol 616:99–109

Rani M, Shanker U, Jassal V (2017) Recent strategies for removal and degradation of persistent & toxic organochlorine pesticides using nanoparticles: a review. J Environ Manag 190:208–222

Rastogi R, Sinha R, Incharoensakdi A (2014) The cyanotoxin-microcystins: a current overview. Rev Environ Sci Bio Technol 13:215–249

Rosenberg JN, Oyler GA, Wilkinson L, Betenbaugh MJ (2008) A green light for engineered algae: redirecting metabolism to fuel a biotechnology revolution. Curr Opinion Biotechnol 19:430–436

Rubin BS (2011) Bisphenol A: an endocrine disruptor with widespread exposure and multiple effects. J Steroid Biochem Mol Biol 127(1–2):27–34

Ruzzin J (2012) Public health concern behind the exposure to persistent organic pollutants and the risk of metabolic diseases. BMC Public Health 12:298

Safonova E, Kvitko K, Kuschk P, Moder M, Reisser W (2005) Biodegradation of phenanthrene by the green alga Scenedesmus obliquus ES-55. Eng Life Sci 5:234–239

Sanborn M, Kerr KJ, Sanin LH, Cole DC, Bassil KL, Vakil C (2007) Non-cancer health effects of pesticides: systematic review and implications for family doctors. Can Fam Physician 53(10):1712–1720

Scragg AH (2006) The effect of phenol on the growth of *Chlorella vulgaris* and *Chlorella* VT-1. Enzym Microb Technol 39:796–799

Semple KT (1998) Heterotrophic growth on phenolic mixtures by *Ochromonas danica*. Res Microbiol 149:65–72

Semple KT, Cain RB (1995) Metabolism of phenols by *Ochromonas danica*. FEMS Microbiol Lett 133:253–257

Semple KT, Cain RB (1996) Biodegradation of phenols by the alga *Ochromonas danica*. Appl Environ Microbiol 62(4):1265–1273

Semple KT, Cain RB (1997) Biodegradation of phenol and its methylated homologues by *Ochromonas danica*. FEMS Microbiol Lett 152:133–139

Semple KT, Cain RB, Schmidt S (1999) Biodegradation of aromatic compounds by microalgae. FEMS Microbiol Lett 170:291–300

Senthilvelan T, Kanagaraj J, Panda RC, Mandal AB (2014) Biodegradation of phenol by mixed microbial culture: an ecofriendly approach for pollution reduction. Clean Techn Environ Policy 16:113–126

Sethunathan N, Megharaj M, Chen ZL, Williams BD, Lewis G, Naidu R (2004) Algal degradation of a known endocrine-disrupting insecticide, alpha-endosulfan, and its metabolite, endosulfan sulfate, in liquid medium and soil. J Agric Food Chem 52:3030–3035

Sole A, Matamoros V (2016) Removal of endocrine disrupting compounds from wastewater by microalgae co-immobilized in alginate beads. Chemosphere 164:516–523

Soto C, Hellebust JA, Hutchinson TC (1975) Effect of naphthalene and crude oil extracts on the green flagellate *Chlamydomonas angulosa*. II. Photosynthesis and the uptake and release of naphthalene. Can J Bot 53:118–126

Souza EC, Vessoni-Penna TC, Oliveira RPDS (2014) Biosurfactant-enhanced hydrocarbon bioremediation: an overview. Int Biodeterior Biodegradation 89:88–94

Stange K, Swackhamer DL (1994) Factors affecting phytoplankton species-specific differences in accumulation of 40 polychlorinated biphenyls (PCBs). Environmental Toxicology and Chemistry 13(11):1849–1860

Stanier RY, Sistrom WR, Hansen TA, Whitton BA, Castenholz RW, Pfennig N, Gorlenko VN, Kondratieva EN, Eimhjellen KE, Whittenbury R (1978) Proposal to place the nomenclature of the cyanobacteria (blue-green algae) under the rules of the international code of nomenclature of bacteria. Int J Syst Evol Microbiol 28(2):335–336

Staples CA, Dorn PB, Klecka GM, Branson DR, O'Block ST, Harris LR (1998) A review of the environmental fate, effects and exposures of Bisphenol A. Chemosphere 36:2149–2173

Subashchandrabose SR, Ramakrishnan B, Megharaj M, Venkateswarlu K, Naidu R (2011) Consortia of cyanobacteria/microalgae and bacteria: biotechnological potential. Biotechnol Adv 29:896–907

Subashchandrabose SR, Ramakrishnan B, Megharaj M, Venkateswarlu K, Naidu R (2013) Mixotrophic cyanobacteria and microalgae as distinctive biological agents for organic pollutant degradation. Environ Int 51:59–72

Suresh B, Ravishankar GA (2004) Phytoremediation—a novel and promising approach for environmental cleanup. Crit Rev Biotechnol 24:97–124

Sijm D, Broersen KW, de Roode DF, Mayer P (1998) Bioconcentration kinetics of hydrophobic chemicals in different densities of Chlorella pyrenoidosa. Environ Toxicol Chem. 17:1695–704

Tang X, He LY, Tao XQ, Dang Z, Guo CL, Lu GN, Yi XY (2010) Construction of an artificial microalgal-bacterial consortium that efficiently degrades crude oil. J Hazard Mater 181:1158–1162

Thengodkar RRM, Sivakami S (2010) Degradation of Chlorpyrifos by an alkaline phosphatase from the cyanobacterium *Spirulina platensis*. Biodegradation 21(4):637–644

Tikoo V, Scragg AH, Shales SW (1997) Degradation of pentachlorophenol by microalgae. J Chem Technol Biotechnol 68:425–431

Turner JT (2014) Planktonic marine copepods and harmful algae. Harmful Algae 32:81–93

Valério E, Chambel L, Paulino S, Faria N, Pereira P, Tenreiro R (2009) Molecular identification, typing and traceability of cyanobacteria from freshwater reservoirs. Microbiology 155(Pt 2):642–656

Varjani SJ (2017) Microbial degradation of petroleum hydrocarbons. Bioresource Technology 223:277–286

Varjani SJ, Upasani VN (2017) A new look on factors affecting microbial degradation of petroleum hydrocarbon pollutants. Int Biodeter Biodegr 120:71–83

Vested A, Giwercman A, Bonde JP, Toft G (2014) Persistent organic pollutants and male reproductive health. Asian J Androl 16(1):71–80

Waigi MG, Kang F, Goikavi C, Ling W, Gao Y (2015) Phenanthrene biodegradation by sphingomonads and its application in the contaminated soils and sediments: a review. Int Biodeterior Biodegradation 104:333–349

Walker JD, Colwell RR, Petrakis L (1975) Degradation of petroleum by an alga, Prototheca zopfii. Appl Environ Microbiol 30:79–81

Walker TL, Becker DK, Dale JL, Collet C (2005) Towards the development of a nuclear transformation system for *Dunaliella tertiolecta*. J Appl Phycol 17:363–368

Wang B, Li Y, Wu N, Lan QC (2008) CO2 bio-mitigation using microalgae. Applied Microbiology and Biotechnology 79(5):707–718

Wang J, Yang H, Wang F (2014) Mixotrophic Cultivation of Microalgae for Biodiesel Production: Status and Prospects. Applied Biochemistry and Biotechnology 172(7):3307–3329

Wang C, Lin X, Li L, Lin L, Lin S (2017a) Glyphosate shapes a dinoflagellate-associated bacterial community while supporting algal growth as sole phosphorus source. Front Microbiol 8:2530

Wang R, Wang S, Tai Y, Tao R, Dai Y, Guo J, Yang Y, Duan S (2017b) Biogenic manganese oxides generated by green algae *Desmodesmus* sp. WR1 to improve bisphenol A removal. J Hazard Mater 339:310–319

Warshawsky D, Keenan TH, Reilman R, Cody TE, Radike MJ (1990) Conjugation of benzo(a) pyrene by freshwater green alga *Selenastrum capricornutum*. Chem Biol Interact 74:93–105

Wilson SC, Jones KC (1993) Bioremediation of soil contaminated with polynuclear aromatic hydrocarbons (PAHs): a review. Environ Pollut 81:229–249

Winneke G (2011) Developmental aspects of environmental neurotoxicology: lessons from lead and polychlorinated biphenyls. J Neurol Sci 308(1–2):9–15

Xiong W, Li X, Xiang J, Wu Q (2008) High-density fermentation of microalga *Chlorella protothecoides* in bioreactor for microbial-diesel production. Appl Microbiol Biotechnol 78:29–36

Yan X, Yang Y, Li Y, Sheng G, Yan G (2002) Accumulation and biodegradation of anthracene by Chlorella protothecoides under different trophic conditions. Chin J Appl Ecol 13:145–150

Yang S, Wu RSS, Kong YC (2002) Biodegradation and enzymatic responses in the marine diatom Skeletonema costatum upon exposure to 2,4-dichlorophenol. Aquat Toxicol 59:191–200

Zaini MAA, Amano Y, Machida M (2010) Adsorption of heavy metals onto activated carbons derived from polyacrylonitrile fiber. J Hazar Mater 180:552–560

Zhang H, Jiang X, Lu L, Xiao W (2015) Biodegradation of polychlorinated biphenyls (PCBs) by the novel identified cyanobacterium Anabaena PD-1, H.-J. Lehmler. PLoS One 10(7):e0131450

Feasibility of Microalgal Technologies in Pathogen Removal from Wastewater

Rouf Ahmad Dar, Nishu Sharma, Karamjeet Kaur, and Urmila Gupta Phutela

1 Background

Wastewater can be defined as raw, untreated, spent water which can potentially pollute the environment. Wastewater contains impurities that were present either originally or are added by anthropogenic activities. Wastewater cannot be discharged to the receiving water body, which may be a river, lake, or sea, unless they have been treated to reduce the concentration of polluting substances to safe levels. Wastewater can originate from many sources such as homes, businesses, and industries. The source of wastewater determines its characteristics and the treatment process that wastewater should undergo. The entire wastewater treatment process involves primary, secondary, and tertiary stages which constitute physical, chemical, and biological processes. Due to the insufficiency of these processes to remove pathogens from wastewater, microalgae-mediated wastewater treatment, phycoremediation, is another paradigm for wastewater treatment. Phycoremediation involves the utilization of algae for the removal of contaminants from wastewater. Coliforms, heavy metals, and xenobiotics are effectively removed by phycoremediation, and this reduces the chemical and biological oxygen demand of wastewater (Olguín et al. 2003; Rawat et al. 2011; Abdel-Raouf et al. 2012; Cai et al. 2013). Microalgae-mediated wastewater treatment is advantageous over conventional techniques in terms of better pathogen removal, decreased sludge formation, reduced greenhouse

R. A. Dar (✉) · N. Sharma
Department of Microbiology, Punjab Agricultural University, Ludhiana, Punjab, India
e-mail: roufdar-mb@pau.edu

K. Kaur
College of Medicine, The Ohio State University, Columbus, OH, USA

U. G. Phutela
Department of Renewable Energy Engineering, Punjab Agricultural University, Ludhiana, Punjab, India

S. K. Gupta, F. Bux (eds.), *Application of Microalgae in Wastewater Treatment*,
https://doi.org/10.1007/978-3-030-13913-1_12

gas emission, and parallel generation of energy-rich algal biomass (Cai et al. 2013; Batista et al. 2015). This chapter furnishes an overview of conventional processes and the applicability of microalgae-mediated pathogen removal from wastewater.

2 Wastewater

An insight into the characteristics of wastewater is crucial for determining the type of treatment it requires. Industries (industrial wastewater) and household activities (domestic wastewater) are majorly responsible for wastewater generation. Centralized sewage treatment plants (STPs) collect wastewater through sewage systems (underground sewage pipes), and STPs are the sites where sewage water is treated.

2.1 Wastewater Types: The two common types of wastewaters are briefed below.

Industrial Wastewater

It can be segregated into two classes as follows:

Inorganic Industrial Wastewater: It is generally produced by coal and steel industry and comprises huge amount of suspended matter. It also consists of harmful solutes like cyanides. Due to the extremely harmful nature of the effluent, these industries are so situated that they discharge their wastewater directly into municipal wastewater system after treating the effluent, in compliance with local regulations (Shi 2009).

Organic Industrial Wastewater: It contains organic waste flow from chemical industries using organic substances. This sort of wastewater is majorly produced by tanneries, leather factories, textile industries, paper manufacturing factories, oil refineries, breweries and industries manufacturing pharmaceuticals, cosmetics, organic dyes, soaps, detergents, pesticides, and herbicides. Due to the myriad of manufacturing processes, the type of effluent varies widely.

Domestic/Residential Wastewater

Domestic wastewater is generated in the residencies like houses, hotels, restaurants, offices, schools, theaters, shopping centers, commercial laundries, etc. This kind of wastewater is less toxic than industrial wastewater, and the effluent generated is also less varied as compared to industrial wastewater.

2.2 Wastewater Characterization

Physical Characteristics

Color: Fresh wastewater is usually slight gray, while septic sewage is dark gray or black. Industrial wastes containing coloring substances may affect the color of the wastewater.

Odor: Fresh wastewater has a distinctive disagreeable odor. Industrial wastewater may also add up to the odor of the wastewater by the dissemination of odorous compounds or compounds that produce odors during the process of wastewater treatment. Hydrogen sulfide is commonly responsible for the wastewater odor. The fear of generation of potential odors during treatment is so intense that implementation of wastewater treatment can be stalled.

Solids: Total solids are the total residues left after evaporation at 105 °C. Suspended solids constitute a major part of total solids and are removed from by membrane filtration. Suspended solids increase turbidity and silt load in the receiving water (Muttamara 1996).

Temperature: Geographic location governs the average temperature of wastewater. The temperature of wastewater affects chemical and biological reaction rates. Undesirable planktonic species and fungi grow fast at higher temperatures. At the same time, the effectiveness of treatment decreases at low temperatures (Muttamara 1996).

Chemical Characteristics

Organic materials: The main organic constituents in wastewater are proteins (40–60%), carbohydrates (25–50%), and fats and oils (10%) (Muttamara 1996). Urea is another key organic compound present in wastewater. The presence of easily biodegradable organic materials reduces the oxygen demand, and the presence of non-biodegradable organic material impedes the wastewater treatment processes.

Inorganic materials: Chloride, nitrogen, phosphorus, sulfur, and heavy metals are the regular inorganic constituents present in wastewater. Phosphorus is present in appreciably lower concentrations than nitrogen or carbon in natural waters. Wastewater organisms are adversely affected by the trace concentrations of inorganic materials, as these substances limit the growth of organisms present in water. The inorganics can be efficiently utilized by algae, and macroscopic plant forms their metabolism.

Gases: Nitrogen, oxygen, carbon dioxide, hydrogen sulfide, ammonia, and methane are the major gases which constitute wastewater. The maintenance of aerobic state is essential in order to annihilate problematic conditions in the wastewater treatment technology and in the natural waters receiving the effluent (Muttamara 1996). However, in anaerobic system, oxidation is carried out by the reduction of inorganic salts like sulfates or through the action of methane-forming bacteria.

Biological Characteristics

Bacteria: Wastewater makes an ideal medium for growth of both aerobic and anaerobic microbes. Among the numerous types of bacteria in wastewater, the most common types are fecal coliforms, which originate in human intestines and travel via human discharges. *Acinetobacter*, *Clostridium*, *Aeromonas*, *Enterococcus*, *Campylobacter*, *Enterobacter*, *Klebsiella*, *Escherichia*, *Mycobacterium*, *Shigella*, *Pantoea*, *Serratia*, *Staphylococcus*, *Salmonella*, *Pseudomonas*, and *Vibrio* are the most prevalent bacterial species in wastewater (Korzeniewska 2011). The bacteria are the key to the biological unit processes. In the presence of adequate dissolved oxygen, the soluble organic matter is converted to new cells and inorganic elements which act as substrates for higher orders of living beings, thus causing a decline in the organic loading.

Viruses: Viruses found in human excreta are a major public health hazard and enter the water stream via fecal contamination. Pathogenic viruses that majorly exist in wastewater are polio and hepatitis. Huge amount (10,000–100,000) of infectious particles of viruses are discharged per gram of feces from hepatitis-positive patients. The titer of plant and animal viruses in wastewater is comparatively small though bacterial viruses may be present (Akpor et al. 2014; Okoh et al. 2007; Gomez et al. 2000; Toze 1997). Most of the viruses are persisters and are resistant to treatment processes.

Fungi: A number of filamentous fungi are found naturally in wastewater as spores or vegetative cells. Various fungi are reported to have the ability to break down organic matter and adsorb the suspended solids in wastewater through their hyphae (Molla et al. 2004; Akpor et al. 2014). *Alternaria*, *Aspergillus*, *Cladosporium*, *Penicillium*, and *Trichoderma* are some fungi commonly found in wastewater (Eva 2011).

Protozoa: The presence of pathogenic protozoa in wastewater is comparatively higher than other environmental sources. *Giardia intestinalis*, *Entamoeba histolytica*, and *Cryptosporidium parvum* are the prevalent protozoans, frequently detected in wastewater due to fecal contamination. Some protozoa, which are obligate aerobes, are able to survive up to 12 h in anoxic conditions and are thus excellent indicators of an aerobic environment.

Helminths: Helminths are usual intestinal parasites which, like protozoans, are spread by fecal-oral route. Wastewater is highly contaminated with these nematodes and tapeworms. Intestinal nematodes have been reported by the World Health Organization (WHO) as the most health risk comprising aquacultural/agricultural utilization of wastewater and untreated excreta (WHO 1989).

3 Conventional Technologies for Wastewater Treatment

For reuse of wastewater, nutrient conservation and pathogen removal are essential steps. The pathogen profile of wastewater varies widely with the type of wastewater (Jiménez 2003). Therefore, choice of treatment process is critically dependent on the type of wastewater (Mohiyaden et al. 2016). Various wastewater treatment stages include preliminary, primary, secondary, and tertiary treatment (Shrestha 2013; Topare et al. 2011) (Fig. 1), and every stage comprises of physical, chemical, and biological treatment processes separately or in association. A brief discussion of each of these treatment stages is given below:

Preliminary Treatment
This step removes large solids, abrasive grit, rags, and high levels of organic content (Mohiyaden et al. 2016). In preliminary treatment, bars placed at 20–60 mm are used for removing large floating objects, and retained substances are raked from the bars periodically (Tebbutt 1983). Abrasive grit material is removed by reduction in the flow speed to the level of 0.2–0.4 m/s at which sediment will settle but organic material remain suspended (Gray 1989). However, this step does not affect pathogen and nutrient concentration (Jiménez et al. 2010).

Primary Treatment
After the preliminary treatment, wastewater is treated in primary settling tanks where BOD is decreased by 40% in the form of settable solids (Horan 1990). For the partial reduction of suspended solids and organic matter, physical unit

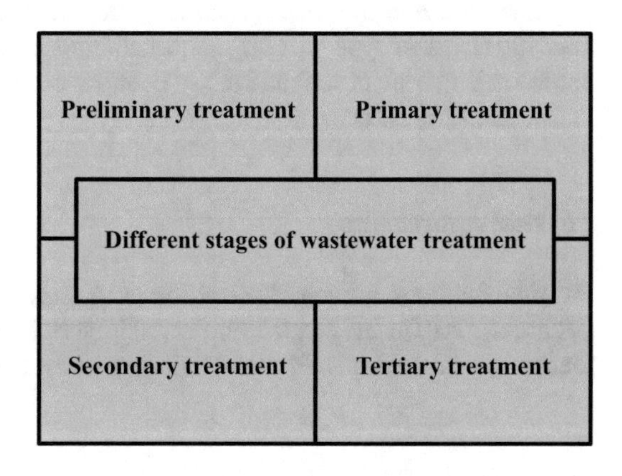

Fig. 1 Stages of wastewater treatment. (Source: Shrestha 2013; Topare et al. 2011)

operations such as sedimentation and screening or some chemicals are primarily used in primary treatment (Mohiyaden et al. 2016). In this step suspended solids (70%), BOD_5 (50%), grease and oil (65%), heavy metals, some organic nitrogen, and phosphorus are removed. The effluent leaving the primary sedimentation unit is called primary effluent (FAO 2006).

Secondary Treatment
After this, wastewater is subjected to secondary treatment for the elimination of solubilized, suspended, and colloidal matter through various biological approaches such as lagoon system, fixed-film reactors, activated sludge, etc. In this step, wastewater is treated in reactor succeeded by treatment in a secondary sedimentation tank where separation of biomass produced by the oxidation of organic matter occurs (Jiménez et al. 2010). A significant decline in BOD takes place by reduction of organic matter mediated by consortium of heterotrophic bacteria (Abdel-Raouf et al. 2012). Many workers have found that about 90% of pathogenic bacteria can be eliminated by this treatment and viruses are removed by adsorption, but rate of removal varies with the type of the reactor (Gray 1989; Kott et al. 1974; Lloyd and Morris 1983).

Tertiary/Advanced Wastewater Treatment
In this advance stage of wastewater treatment, inorganic nutrients like phosphorus and nitrogen, fine suspended particles, heavy metals, and pathogenic microorganisms are removed (Prabu et al. 2011). It can be done through rapid sand filtration (RSF), post-precipitation, reverse osmosis, chemical oxidation, carbon adsorption, ultrafiltration, microfiltration, and dissolved air flotation (DAF) (Hamoda et al. 2002; Jolis et al. 1996; Nieuwstad et al. 1988; Ødegaard 2001; Pinto Filho and Brandão 2001). Tertiary treatment is approximately four times costlier as compared to primary treatment (de la Noüe et al. 1992).

3.1 Types of Conventional Wastewater Treatment Methods

Wastewater is mainly treated physically, chemically, and biologically (Amoatey and Bani 2011). The type of unit operations and processes in wastewater treatment shown in Fig. 2 are described below (Economic and Social Commission for Western Asia (ESCWA) 2003)

Physical Approaches

Physical methods employ physical forces to remove contaminants from wastewater (Bhargava 2016). Suspended and settable solids, oil, and grease are removed by these physical methods. Physical unit operations commonly used are:

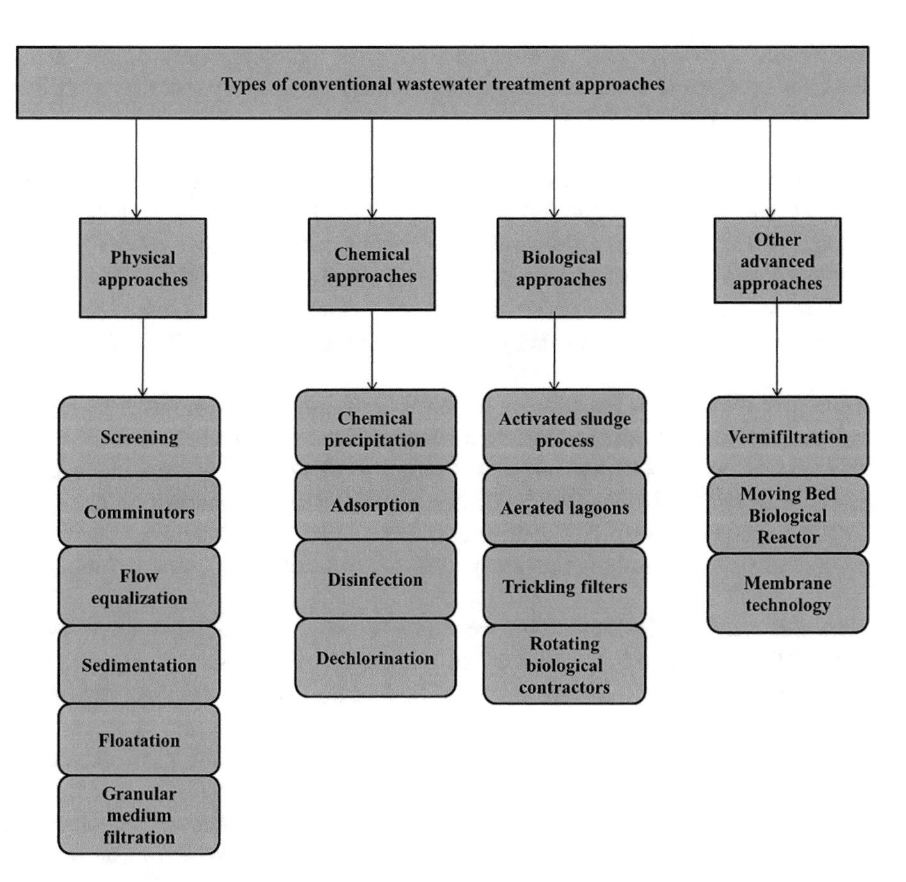

Fig. 2 Different approaches of conventional wastewater treatment. (Source: Anusha and Sham Sundar 2015; Borkar et al. 2013; Doumenq 2017; Misal and Mohite 2017; Morão 2008; Mulder 1996; Rawat et al. 2011; Shon et al. 2009)

Screening: This step employs the sieving of gross pollutants from the wastewater using devices such as parallel bars, wire mesh, rods, perforated plates, etc. After cleaning of bar screens either manually or mechanically, retained material is called screenings. This protects downstream equipment from damage (ESCWA 2003).

Comminutors: Comminutors are positioned in the middle of grit chamber and primary settling tanks and consist of rotating or oscillating cutters. These are used for reducing odors, flies, and unsightliness and for crushing the large suspended material in the wastewater flow (ESCWA 2003).

Flow equalization: Flow equalization levels out the process parameters like flow, temperature, and amount of pollutant over a period of time for ameliorating the efficacy of wastewater treatment processes like secondary and tertiary/advanced. In a wastewater treatment plant, flow equalization can be applied at many places.

Intermittent flow diversion, alternating flow diversion, completely mixed mixed flow, and completely mixed combined flow are the four basic types of flow equalization processes (ESCWA 2003).

Sedimentation: Sedimentation involves separation of suspended particles through gravity separation (WEF 2008). Particulate matter, biological flocs, and chemicals present in wastewater are eliminated in the primary settling basin, activated sludge settling basin, and chemical coagulation, respectively. Sedimentation occurring in settling tank is known as clarifier. Solid contact clarifiers and horizontal-flow and inclined-surface basins are the main designs of sludge collectors (ESCWA 2003).

Flotation: Flotation is the removal of solids or liquids from wastewater by injecting air bubbles which either attach to the liquid or get confined in suspended particles, increasing the particles' buoyant. As the particles float to the top, they can be easily removed (Koivunen 2007). Dispersed air flotation, dissolved air flotation, electroflotation (Edzwald 1995; Rubio et al. 2002), precipitate flotation, mineral flotation, and colloid flotation (Koivunen 2007) are some of the flotation techniques.

Granular medium filtration: This technique is used for the additional removal of chemically precipitated phosphorus and suspended solids from the effluent from biological and chemical treatment units. The filtration process employs two steps: filtration and cleaning/backwashing. In filtration, the waste effluent is passed to a filter bed made of granular medium with or without the addition of chemicals. Suspended materials present in wastewater are then removed by different processes like interception, adsorption, flocculation, impaction, and sedimentation. Cleaning or backwashing can be either continuous involving simultaneous filtering and cleaning operations or semicontinuous including sequential filtering and cleaning operations (ESCWA 2003).

Chemical Approaches

Chemical methods require the use of chemicals for wastewater treatment by means of chemical reactions to remove dissolved solids, nutrients, and heavy metals. Chemical unit processes are employed in synchrony with physical unit and biological unit processes (Bhargava 2016).

Chemical precipitation: In this approach, finely divided solids are flocculated into settable flocs. Coagulation-flocculation is used for the treatment of wastewater in chemical precipitation. Common coagulants used for wastewater treatment include lime ($Ca(OH)_2$), ferrous sulfate ($FeSO_4.7H_2O$), ferric chloride ($FeCl_3.6H_2O$), and alum ($Al_2(SO_4)_3.14H_2O$) (Jiménez et al. 2010). Colloidal substances responsible for the color and turbidity of the wastewater are treated through coagulation/flocculation (Arvanitoyannis and Ladas 2008). This method eliminates heavy metals and phosphorus effectively, but large amount of sludge is generated that can be dewatered and used for land filling (WEF 2008).

Adsorption with activated carbon: It involves accumulation of soluble particles present within a liquid on an appropriate interface. Activated alumina, hydroxides, activated charcoal, and resins are some of the common examples of adsorbents which are used for removal of substances like detergents and toxic compounds (Samer 2015). Activated carbon is a commonly used absorbent, and powdered activated carbon (PAC) and granular activated carbon (GAC) are its two common types (ESCWA 2003). Unlike GAC, powder activated carbon is added to wastewater using feed equipment instead of being carrying in column or bed (Corbitt 1998; Weber 1972).

Disinfection: Disinfection is the last step of wastewater treatment process for the conservation of ecosystem and human health (Sun et al. 2009). A good disinfectant should be easy to handle, inexpensive, and reliable and have potential bactericidal action (Samer 2015). Several factors affect the process of disinfection which include pH, type of disinfectant, temperature, exposure time, and type of effluent and pathogen (WEF 1996). Most commonly used disinfectants are physical agents such as light and heat, radiations (ionizing as well as nonionizing radiations), UV light, and chemical substances like chlorine and its compounds, bromine, peracetic acid (PAA), iodine, ozone, soaps and detergents, heavy metals, phenols, alcohols, etc. (Koivunen 2007; Russell 2006).

Dechlorination: For wastewater disinfection, chlorine and its derivative compounds are most commonly used, but it undergoes certain chemical reactions with the organic compounds in wastewater and produces disinfection by-products (DBPs) which have carcinogenic and mutagenic properties (Sun et al. 2009) which necessitate dechlorination (Amin et al. 2013). In dechlorination process, chlorine residues (in free and combined form) are removed from wastewater effluent (ESCWA 2003). It is done by using reducing agents such as sodium sulfite (Na_2SO_3), sulfur dioxide (SO_2), or sodium metabisulfite ($Na_2S_2O_5$) or by activated carbon (Bagchi and Kelley 1991).

Biological Approaches

Biodegradable organic matter in dissolved or colloidal form can be removed by using biological approach (Rosen et al. 1998). Contaminants are removed by the biological activity of microorganisms which degrade the organic matter in wastewater into gases (Topare et al. 2011).

Activated sludge process: Municipal wastewater is commonly treated with this process. It is an aerobic process for the elimination of BOD and suspended solids by using suspended bacterial flocs. A variety of factors which include temperature, pH, concentration of available oxygen and organic matter, waste rates, and aeration period influence the activated sludge system (Amoatey and Bani 2011). The main principle behind this process is that vigorous aeration of waste effluent generates activated sludge (flocs of bacteria) which degrades organic compounds. Activated sludge is recycled for the maintenance of concentration of active bacteria. Settling tanks are equipped with accessories like waste pumps, blowers providing aeration, and a device for measurement of flow rate. In this process, degradation occurs mainly through three main processes including microbial processes, volatilization, and sorption onto sludge flocs (Grandclement et al. 2017).

Aerated lagoons: It is a basin of about 1–4 meter depth wherein treatment of wastewater occurs either by solids recycling or flow-through basis. The aerators provide aeration, dissolved oxygen, and suspended microbial biomass for achieving maximum aerobic activity. Based on the strength and temperature of waste effluent and level of treatment, the hydraulic retention time (HRT) varies from 5 to 8 days (Samer 2015). One study reported that in household water for HRT of 5 days, 85% reduction in BOD was achieved, but BOD value decreased to 65% at 10 °C temperature (Gray 2005).

Trickling filters: A trickling filter is a basin packed with an inert carriers like volcano rock, gravels, or other synthetic material in which wastewater is supplied from the top tickles through the filter medium where organic compounds in wastewater are absorbed by microorganisms that are attached to medium as a slime layer having thickness of approximately 0.1–0.2 mm. In the outer part of slime, breakdown of organic material occurs by the aerobic microorganisms. Further, growth of anaerobic microorganisms occurs due to oxygen deprivation which makes thick layer of microbial growth. Until the microorganisms present near the surface cannot adhere to media, continuous development of biological film occurs. A section of the biological slime layer repeatedly falls off by a process called sloughing. Removal of the sloughed off portions occurs by the drain system by transferring to a clarifier (EPA 2000).

Rotating biological contractors: These consist of plastic media with diameter ranging from 2 to 4 m mounted vertically on a horizontal rotating shaft (Peavy et al. 1985). As the shaft rotates slowly with about 40% submerged media, the media coated with biomass are exposed alternately to wastewater and oxygen. Biomass oxidize the organic matter present, and excess biomass is shredded off in a downstream clarifier automatically (Amoatey and Bani 2011). These are best suited for treatment of municipal wastewater (Peavy et al. 1985). Due to their ability of quick recovery from unfavorable conditions, these have been installed in many petroleum facilities (Schultz 2005).

Other Advanced Approaches

Vermifiltration: It is a new technology that is a combination of traditional process of filtration with vermicomposting, i.e., using earthworms for wastewater bioremediation (Anusha and Sham Sundar 2015). It is a simple filtration apparatus consisting of lower layer of gravels covered with aggregates and sand layer covered with cow dung clay and a population of earthworms. As the wastewater passes through the filter bed, earthworms use fats and oils for their metabolism from it, and the leftover water percolating from bottom is collected in another vessel (Misal and Mohite 2017). The body of earthworms acts as biofiltering agent, and body wall absorbs compounds from wastewater, and reduction in wastewater COD by 80–90%, BOD_5 by over 90%, total dissolved solids (TDS) by 90–92%, and the total suspended solids (TSS) by 90–95% have been observed (Sinha et al. 2008).

Moving bed biological reactor (MBBR): A moving bed biological reactor (MBBR) is integration of activated sludge and trickling filters where biomass exists as suspended congregation of microorganisms and biofilms attached to carriers made of materials like high-density polyethylene or polypropylene (Borkar et al. 2013). The advantages of moving bed biological reactor is that it is not sensitive to load variations and other types of disturbances (Delenfort and Thulin 1997; Odegaard et al. 1994), slight head loss, and no recycling of biomass is required (Xiao et al. 2007).

Membrane technology: Membrane technology is a broad term used for different processes for transportation of substances from one phase to another phase with the aid of permeable membranes allowing passage of some specific substances while retaining others (Mulder 1996). A gradient of concentration, electric potential, temperature, and pressure acts as major driving force for solute transportation (Mulder 1996). The technology depends on physical forces, and no addition of chemicals is required (Morão 2008). Based on the driving force, membrane processes can be divided into four main types: ultrafiltration (UF), microfiltration (MF), nanofiltration (NF), and reverse osmosis (RO) (Shon et al. 2009).

3.2 Limitations of Conventional Techniques for Pathogen Removal

Though commonly used, conventional techniques are not able to remove variety of chemicals and pathogenic microorganisms from wastewater. Limitations of various conventional wastewater techniques are mentioned below:

Physical approach limitations

- Manual cleaning of different types of screen is laborious task, and overflowing may occur due to clogging. Mechanically cleaned screens operate well but jam due to obstructions (WEF 2008).
- Moreover, a substantial amount of dissolved and colloidal material is still present in waste effluent after physical treatment of wastewater (Samer 2015).

Chemical approach limitations

- Various studies have shown that physicochemical processes like coagulation and flocculation are ineffective for removing various pollutants like pharmaceuticals and endocrine disrupting compounds (EDCs) (Petrovic et al. 2003; Vieno et al. 2006; Westerhoff et al. 2005).

- Also, coagulation-flocculation generate complex sludge and are costlier (Ghoreishi and Haghighi 2003; Sirianuntapiboon et al. 2006).
- The processes of chemical unit are additive which result in net increase in the constituents of wastewater (ESCWA 2003).
- Although residual protection is provided by chlorination, against regrowth of pathogens (Szewzyk et al. 2000; Zhang and DiGiano 2002), it produces undesirable tastes and odors (Suffet et al. 1995) and forms different disinfection by-products (Becher 1999; Hozalski et al. 2001; Gopal et al. 2007; Sadiq and Rodriguez 2004). Furthermore, enteric viruses, spores of bacteria, and protozoan cysts in sewage are also not removed efficiently (Sobsey 1989).
- Chlorine and ozone are inefficacious against helminth eggs and protozoan cysts, and certain viruses like adenoviruses show high resistance against UV light (Jiménez et al. 2010).

Biological approach limitations

- The complex polluted waters consisting of pharmaceuticals, surfactants, and various industrial products cannot be treated by traditional technologies like activated sludge (Amin et al. 2014).
- Most of the contaminants remain soluble in waste effluent which cannot be removed by activated sludge and tickling filters (Servos et al. 2005; Urase and Kikuta 2005).
- The main limitations of trickling filter are having limited flexibility and problem of operation at low temperature (Metcalf and Eddy 1991; Reynolds 1982).
- Rotating biological contractors may give problem in conditions of high organic load and temperature below 13 °C (WEF 2008).

4 Suitability of Wastewater for Algal Growth and Water Quality Indicators

Microalgae are unicellular or multicellular simple structured and primordially photosynthetic organisms having a large surface-to-volume body ratio. These can thrive and grow expeditiously in severe conditions. This bestows them to take considerable proportion of nutrients from the environment where they grow. These absorb sunlight, assimilate atmospheric CO_2, and obtain nutrients from the aquatic habitat under their natural conditions. Apart from phototrophic mode of nutrition, these can be cultivated heterotrophically (i.e., utilization of organic carbon as the source of energy and carbon), mixotrophically (cultivated under both phototrophic and heterotrophic conditions), and photoheterotrophically (using light, organic carbon as carbon and energy source). Algae can be cultivated according to the availability of the resources and for the purpose to be used for (Christenson and Sims 2011). Various kinds of wastewaters can be exploited for growing microalgae, thus improving the water quality apart from reducing the demand of water and fertilizer appreciably (Prajapati et al. 2013). A number of factors are responsible for the substantial microalgal growth in wastewater. These crucial factors are temperature and pH of cultivation medium, concentration of N, P, and carbon (organic), light, CO_2, and O_2. The concentration of N and P in wastewater is higher compared to other cultivation media. Mostly the N present in it is found in the state of ammonia, and this may impede growth of algae (Konig et al. 1987; Wrigley and Toerien 1990; Pittman et al. 2011). However, it differs with the wastewater type and its treatment sites. In addition to this, the capability to sustain in different wastewater conditions varies from species to species. For example, the chlorophytic unicellular microalgal species efficiently uptake nutrients from wastewater and thus thrive in many wastewater conditions (Aslan and Kapdan 2006; Ruiz-Marin et al. 2010). Still, the efficiency of nutrient accumulation among various chlorophyte species varies. For example, Travieso et al. (1992) described that *Chlorella vulgaris* was more efficient in nutrient accumulation (N, P) from wastewater compared to *Chlorella kessleri*, and Ruiz-Marin et al. (2010) also noticed that compared to *Chlorella vulgaris*, *Scenedesmus obliquus* showed appreciable growth in municipal wastewater. In high-rate algal and oxidation ponds, the dominant phytoplanktonic communities are generally *Chlorella* and *Scenedesmus* (Masseret et al. 2000).

Microalgal species in suspension or immobilized form were found to be effective accumulators of nitrogen and phosphorus from sewage-based wastewater. Many *Scenedesmus* and *Chlorella* species can extensively eliminate (>80%) nitrate, ammonia, and total phosphorus from secondary treated wastewater (Ruiz-Marin et al. 2010; Zhang et al. 2008), thus depicting the capability of these microalgal species for sewage treatment. In case of agricultural wastewater, the N and P content is very high despite which efficient microalgal growth has been achieved in it (An et al. 2003; Wilkie and Mulbry 2002). Industrial wastewater has also been tried out for microalgal cultivation, but the algal production has been found to be less as it mostly contains high toxin concentrations (zinc, cadmium, hydrocarbons, chromium, etc.) and low phosphorus and nitrogen concentration (Ahluwalia and Goyal 2007; de-Bashan and Bashan 2010). Therefore, utilization of industrial wastewater for algal cultivation is less feasible. However, one recent study suggests potential of industrial effluent from carpet mill in furnishing nutrients for the significant algae biomass production (Chinnasamy et al. 2010). Moreover, wider availability and uniformity in composition make the agricultural wastewater and municipal more feasible for algae cultivation than the variable composition of various industrial wastewaters. Researchers have utilized various kinds of wastewater for the microalgae cultivation (Table 1).

Microalgae as Water Quality Indicators

Bioindicators consist of microorganisms or biological processes. Bioindicators assess the cumulative effect of various pollutants on water quality and how it alters with time and to what time period it may prevail. However, there is a range of indicator organisms, but algae are potential indicators for evaluating quality of water due to the following reasons:

- Easy availability of the nutrients required for growth.
- Faster growth rate.
- Shorter life cycle.
- Wider geographical distribution.

Table 1 Algal biomass production from wastewater (Rawat et al. 2016; Kumar et al. 2015; Show and Lee 2014; Rawat et al. 2011; Pittman et al. 2011)

Type of wastewater	Algae species	Biomass productivity	References
Municipal wastewater	*Chlorella* sp.	0.948 d^{-1} (growth rate)	Wang et al. (2010).
Drain wastewater	*Chroococcus* sp. 1	1.05 g L^{-1}	Prajapati et al. (2013)
Livestock wastewater	*Chroococcus* sp. 1	4.44 g L^{-1}	Prajapati et al. (2014)
Wastewater from metro plant	*Chlamydomonas reinhardtii*	2.00 g $L^{-1}d^{-1}$	Kong et al. (2013)
Urban wastewater	*Desmodesmus communis*	0.138–0.227 g $L^{-1}d^{-1}$	Samorì et al. (2013).
Piggery wastewater	*Arthrospira* sp.	11.8 g $L^{-1}d^{-1}$	Olguín et al. (2003)
Domestic wastewater with urea supplementation	*Chlorella sorokiniana*	0.2 g $L^{-1}d^{-1}$	Ramanna et al. (2014)
Piggery wastewater	*Botryococcus braunii*	–	An et al. (2003)
Sewage wastewater	*Chlorella minutissima*	0.073–0.379 g L^{-1}	Bhatnagar et al. (2010)
Carpet mill	*Scenedesmus* sp.	0.126 g $L^{-1}d^{-1}$	Chinnasamy et al. (2010)
Industrial wastewater	*Desmodesmus* sp. TAI-1 and *Chlamydomonas*	1.5–1.8 g L^{-1}	Wu et al. (2012)
Campus sewage	*Scenedesmus quadricauda*	0.052–0.082 g $L^{-1}d^{-1}$	Han et al. (2015)
Artificial wastewater	*Scenedesmus* sp.	0.996–0.119 g $L^{-1}d^{-1}$	Voltolina et al. (1999)
Anaerobically digested dairy manure	Mix of *Ulothrix zonata*, *Ulothrix aequalis*, *Microspora willeana*, *Oedogonium* sp., *Rhizoclonium hieroglyphicum*	5.5 g $L^{-1}d^{-1}$	Wilkie and Mulbry (2002)

- Bulk availability of diverse groups.
- Quick response to qualitative and quantitative changes in the environment due to pollution.
- Easier detection and sampling (Gökçe 2016).

Algae have demonstrated to be appropriate indicators of water quality. Microalgae are essential and probable bioindicators of eutrophication because of their immediate response to variations of environmental conditions resulting from the different anthropogenic activities (Kelly-Gerreyn et al. 2004; Álvarez-Góngora and Herrera-Silveira 2006; Livingston 2001). Microalgae thrive in almost all aquatic habitats besides dwelling on rocks, macroalgae, or submerged surfaces, where both planktonic and microphytobenthic assemblages are utilized for characterization of aquatic ecosystems with the use of biological, physicochemical, or hydromorphological indicators (Hermosilla Gomez 2009). Various microalgal species like *Oscillatoria*, *Chlamydomonas*, *Scenedesmus*, and *Chlorella* are used as indicators of water pollution (Padisák et al. 2006).

5 Role of Algae in Pathogen Removal

Wastewater poses many threats to the public health as it contains pathogenic microorganisms. So to attenuate this problem and to make this water usable, removal of such pathogenic microorganisms is necessary and must be a primary concern in treatment process (Jiménez et al. 2010). As there are various waterborne human pathogens (Wu et al. 2016), their assessment would be very cost-intensive. Hence, the assessment is done by monitoring of bacterial indicator organisms (like *Escherichia coli*, total coliforms, or fecal coliforms) in treated wastewater. The utilization of algae for wastewater treatment has been in trend for approximately >50 years. Oswald and Gotaas (1957) were the first to demonstrate the application of algae in treatment process. The basic principle underlying the biological treatment is to boost the removal of pathogens, nutrients, and heavy metals and to provide oxygen for the mineralization of organic pollutants by heterotrophic aerobic bacteria which ultimately leads to the production of CO_2 valuable for the agents carrying biological treatment like algae (Munoz and Guieysse 2008). The dissolved oxygen (DO) and pH of wastewater increase due to the algal activity. It has been investigated that growth of algae can facilitate the removal and inactivation of both *Escherichia coli* and total coliforms. The mechanisms and factors responsible for this have been discussed ahead in the chapter.

Removal or biotransformation of pollutants from wastewater like xenobiotics and nutrients and CO_2 from polluted air by the utilization of macroalgae or microalgae is known as phycoremediation (Mulbry et al. 2008; Moreno-Garrido 2008; Olguın 2003; Olguın et al. 2004). Microalgae either aerobically or anaerobically can treat wastewater, industrial effluents, and solid wastes through various processes. Microalgae being effective converters of solar energy can generate massive blooms and also can produce different kinds of valuable secondary metabolites (Moreno-Garrido 2008; Lebeau and Robert 2006) and are thus potential treating candidates for wastewater treatment.

6 Mechanisms Involved in Pathogen Removal by Microalgae

The various mechanisms of pathogen removal from wastewater by algae are as under:

- Competition of nutrients.
- Elevation of pH and dissolved oxygen.

- Algal toxins.
- Adhesion and sedimentation of pathogens.

Competition of Nutrients

Algae consume nutrients and carbon sources needed by the bacterial cells for their survival. This increases their retention time in water. This diminution of the sources of carbon in water may lead to the starvation of fecal bacteria due to unavailability of its energy sources, thus ultimately resulting in their death (Van der Steen et al. 2000).

Elevation of pH and Dissolved Oxygen

The photosynthetic activity of microalgae has been found to increase the pH and dissolved oxygen (DO) content in the wastewater. The elevated levels of these two factors result in the deactivation of the pathogens present in water (Muñoz and Guieysse 2006). Actually, the combined action of sunlight, pH, and oxygen through photosensitizers, in a process called photooxidation, results in the removal of pathogens from wastewater. These photosensitizers both present inside (porphyrins) and outside of the bacterial cells (dissolved organic matter) help in the absorption of light of wavelengths (400–700 nm), thereby splitting the oxygen and resulting in the formation of singlet oxygen and hydrogen peroxides, the potential agents responsible for the damage of DNA of cell membrane (Ansa et al. 2015; Curtis et al. 1992). In aquatic environments, hydrogen ion is pivotal for many metabolic reactions in microbial cells like ion transport and energy generation (Mitchell 1992). This is fundamental in major phases of water and wastewater treatment. The substantial usage of dissolved carbon dioxide by microalgae for its growth is generally responsible for the elevation of pH and DO. Algae utilize dissolved inorganic carbon through photosynthesis and liberate oxygen as a photosynthetic by-product, as given in Eq. (1).

$$6CO_2 + 12H_2O \xrightarrow{\text{light, pigment receptor}} C_6H_{12}O_6 + 6H_2O + 6O_2 \tag{1}$$

Under sufficient availability of light and nutrients, rate of removal CO_2 by algae is higher as compared to the generation of respiratory CO_2 by heterotrophic microorganisms. The resulting change in CO_2 equilibrium is illustrated in Eqs. (2, 3, and 4) (Mayes et al. 2009).

$$H_2CO_3 \leftrightarrow CO_2 + H_2O \tag{2}$$

$$HCO^-{}_3 + H_2O \leftrightarrow H_2CO_3 + OH^- \tag{3}$$

$$CO^{2-}{}_3 + H_2O \leftrightarrow HCO_{3-} + OH^- \tag{4}$$

Uptake of CO_2 from the system will shift Eqs. (2, 3, and 4) to the right to generate more CO_2 to maintain equilibrium. Due to this, pH will get increased by generation of hydroxide ions. Hence, elevated DO and pH levels are generally seen in algae-grown wastewater ponds. Warmer climate particularly daylight hours favors this type of effect (Gschlößl et al. 1998).

Algal Toxins

The microalgae like *Chlorella vulgaris* under stress and high pH have been found to produce toxins of long-chain fatty acids. These toxins have been found to be pathogen destructive in nature (Awuah 2006). A toxin called microcystin-LR produced by *Synechocystis* sp. was found to be harmful for fecal bacteria. These toxins could harm algal communities as well, but microalgae like *Scenedesmus quadricauda* and *Chlorella vulgaris* protect themselves from these toxins by producing huge amount of polysaccharides (Mohamed 2008). Also, with the elevation in the levels of chlorophyll-a, the inactivation of fecal coliform increases. The green algae remove fecal coliforms by secreting substances harmful to fecal coliforms (Ansa et al. 2012). The pathogen removal by algal toxins is still under debate. This needs the development and modification of rapid detection methods for the detection and assessment of algal toxin role in the removal of pathogens in wastewater (Litaker et al. 2008).

Adhesion and Sedimentation of Pathogens

The pathogens may attach to the solid matter that sinks as sediment and on the surface of algae (Awuah 2006). The attachment of fecal bacteria to algae in algal ponds is essential as it exposes the fecal bacteria in close proximity to the production site of severe environmental conditions like high pH and dissolved oxygen for more effect to be felt.

The rate of sedimentation is higher in aggregated bacteria compared to the planktonic form (Characklis et al. 2005). The aggregation of suspended matter is determined by the availability of polysaccharides (acid soluble) in the solution having the potential of protonation, i.e., formation of positively charged amino groups. The microalgae *Chlorella* bears a negative zeta potential or surface charge (Liu et al. 2009). Thus, these positively charged polymers neutralize the negative algal surface charge resulting into the bridging between particles. This leads to the formation of high cell density bacterial flocs which are bigger in size with quicker sedimentation rate (Henderson et al. 2008).

7 Factors Affecting Pathogen Removal by Algae

Temperature

Most microalgae species grow in the temperature range from 15 to 35 °C, and the temperatures above and below this are not favorable for microalgal growth. Because at low temperatures, rate of growth is slower, while at higher temperatures growth rate decreases due to oxidative stress. The removal efficiency was observed to have doubled on elevating the temperature from 25 to 30 °C by utilizing a symbiotic microcosm of *Chlorella sorokiniana* and a *Ralstonia basilensis* strain (Munoz et al. 2004).

pH

The photosynthetic activity apart from the algal respiration, wastewater composition, and the kind of metabolites determine the pH of the algal cultivation medium. The rise in pH during photosynthesis is due to the uptake of CO_2, and this could increase pH up to 10–11. This rise in pH could impede the activity of both bacteria and microalgae (Posadas et al. 2014). The decrease in pH by the activity of nitrifying bacteria due to the release of H^+ also decreases the removal of pathogens from wastewater (Posadas et al. 2017).

Light

Intensity of sunlight changes significantly throughout the day and the year. Light intensity of 200–400 $mEm^{-2} s^{-1}$ increases the algal activity (Ogbonna and Tanaka 2000). The microalgal growth and photoperiod have been found to be directly related to each other, but with high irradiance and longer photoperiod, photoinhibition and damage will occur (Molinuevo-Salces et al. 2016). Photoinhibition is prominent after noon as the flux of radiant energy per unit area can go up to 4000 $mEm^{-2} s^{-1}$. It is mostly observed when algal concentration is low, like during start-up (Göksan et al. 2003), because there is not enough shading from irradiance due to other microalgal cells (Contreras-Flores et al. 2003; Richmond 2000). Homogenous distribution of light in microalgal cultures is a must to obtain high biomass productivity. Microalgae grown under field conditions, for wastewater treatment, are exposed to seasonal and daily variations of irradiation which ultimately affects the microalgal waste removal potential.

Dissolved Oxygen Concentration (DOC)

Dissolved oxygen and solar irradiance are correlated to each other. As the solar irradiance increases, O_2 production also increases and vice versa. It has been illustrated that under maximal rates of photosynthesis, DOC can reach to 40 mg L^{-1}; in fact sometimes supersaturation of oxygen occurs in closed photobioreactors or on the top of open bioreactors (Posadas et al. 2015). Even oxygen concentrations above 20 mg L^{-1} have been found to be detrimental to many microalgal species, and it reduces the photosynthetic production by 98% (Matsumoto et al. 1996). The high oxygen concentration damages the microalgal cells by a process known as photo-oxidation. This damage of microalgal cells ultimately reduces the microalgal waste removal efficiency (Suh and Lee 2003).

Predators

Due to invasion by *Chytridium* sp. or any parasitic fungi, various food chain formations in the cultivation system led to unforeseen failure of process (Abeliovich and Dikbuck 1977). Microalgae in wastewater treatment process are subjected to various inhibitory products produced by other algae, phages, protozoa, bacteria, and nematodes. These can also hamper the process of removal of pathogens by microalgae (Mawdsley et al. 1995). These can be tackled by running the process for a short period of time (1 h) at low concentrations of O_2 on daily basis in order to quell the growing ability of higher aerobic organisms (Abeliovich 1986).

Operational Conditions

Apart from the abovementioned parameters, other parameters like mixing and penetration of light are of utmost importance. Mixing is the main factor as it provides proper turbulence and homogeneity in the growth medium, thus avoids the sinking of microalgal cells. It prevents the formation of gas, nutrient, and heat gradients. Mixing also leads to the shifting of microalgae from dark and light zones so the cells can perform photosynthesis actively without any problem of light saturation and light inhibition and also increases the mass transfer between the algal cells and environment, thus increasing the removal efficiency (Grobbelaar 2000; Eriksen 2008). However, mixing beyond certain frequency limit causes shear stress which has negative impact on microalgal cells.

Microalgae being photosynthetic in nature use light energy to carry out various metabolic activities like CO_2 and nutrient uptake, synthesis of biomass which actually define the wastewater treatment efficiency. Wastewater also contains various suspended particles and compounds which limit the penetration of light to the microlagal cultures. This, in turn, lowers the biomass productivity and subsequently hampers the treatment of wastewater (Markou 2015).

8 Case Studies of Removal of Several Pathogens from Wastewater by Algae

Ansa et al. (2012) evaluated varying-strength wastewater (low, medium) and a mixture of 10-day treated wastewater and raw wastewater for the effect of varying density of *Chlorella* sp. on the fecal coliform (FC) decay rate under light and dark conditions. Under dark conditions, it was found that the decay rate of FC fluctuated with chlorophyll-a concentration and for the maximum FC destruction optimum chlorophyll-a concentration was 10 ± 2 mg L^{-1}. It was further reported that under both light and dark conditions, at algal densities of ≥ 13.9 mg L^{-1}, decay rate was faster in medium-strength wastewater compared to low-strength wastewater. While under light conditions, addition of second feed of wastewater to already operating wastewater treatment process decreased the FC decay rate for varying algal densities in the range of 0.6–19.6 mg L^{-1}.

Mezzari et al. (2017) investigated the elimination of *Salmonella enterica* serovar *Typhimurium* by *Scenedesmus* sp. in swine wastewater. Photobioreactors filled with 3 L of diluted swine wastewater with and without microalgae *Scenedesmus* sp. (30% v/v, 70 mg L^{-1} dry weight) inoculated with *S. enterica* (10^5 CFU mL^{-1}) were subjected to mixotrophic cultivation using red light emission diode at 630 nm and 121.5 μmol m^{-2} s^{-1} at room temperature under continuous mixing conditions. Cell count was taken by plate count method, and qPCR amplifications of the *Salmonella* invasion gene activator, *hilA*, were executed. It was found that *S. enterica* was removed completely in the presence of microalgae within 48 h of treatment, while in the absence of microalgae, concentration of *S. enterica* increased 1.5 log CFU

mL^{-1} in 96 h. However, in photobioreactor with controlled pH *S. enterica* concentration remained constant (2.8 ± 0.2 log CFU mL^{-1}) throughout 96 h.

Ansa et al. (2011) evaluated the effect of algae *Chlorella* on pathogenic *Escherichia coli* in eutrophic lake and the significance of attachment of *E. coli* to suspended matter as well as algae. *E. coli* die-off rate in dialysis tube at different depths and locations in Weija Lake was evaluated. A significant decay of *E.coli* was reported which was attributed to increase in concentration of dissolved oxygen (DO) and pH. It was found that at chlorophyll-a concentration ≤0.08 mgL^{-1}, there exist a direct relation between chlorophyll concentration and decay rate of *E. coli*. They further reported that as concentration of chlorophyll increases with light, concentration of chlorophyll-a reaches at optimal value (0.24 mg/L) and *E. coli* decay rate decreases.

Rhizoclonium implexum (an algal species) has been reported to be efficient in the removal of coliform bacteria as well as total suspended solids, total dissolved solids, COD, BOD, total Kjeldahl nitrogen, and total phosphorus. Algal wastewater treatment is amiable in terms of its economic and environment considerations (Ahmad et al. (2014).

9 Utilization of Algal Biomass Obtained from Wastewater

Various useful products can be derived from the microalgae biomass like biofuels, bioactive compounds, etc. It can be converted to biofuels through different routes like biogas can be produced through anaerobic digestion, ethanol, acetone, and butanol by fermentation, biohydrogen by biophotolysis and dark fermentation, biodiesel through transesterification of lipids derived from it, and hydrocarbon and biocrude oils through gasification/pyrolysis (Heubeck et al. 2007).

Biogas

Microalgae can serve as an efficient fuel for biogas generation. Mixed microalgal cultures show comparable biogas quality and productivity as that of sewage sludge. Higher temperatures (55 °C) have been demonstrated to enhance biogas production (1020 L kg^{-1} VS) as compared to mesophilic range (986 L kg^{-1} VS at 35 °C) with CH_4 content ranging from 61% to 63%. At the same time, various algal species directly affect biogas production due to varied cell wall structure and composition (Mussgnug et al. 2010; Zamalloa et al. 2012). *Chlamydomonas reinhardtii* has been found to produce up to 390 L CH_4 kg^{-1} VS which is higher compared to methane obtained (100 L CH_4 kg^{-1} VS) from *Scenedesmus* lipid extraction leftovers. Cell wall structure governs the susceptibility of algal species to anaerobic digestion. Algal species such as *Arthrospira platensis*, *Chlamydomonas reinhardtii*, and *Epicrates gracilis* constitute proteinaceous cell walls lacking cellulose and hemicellulose (Mussgnug et al. 2010). The cellulose-free cell walls make these species undergo easier hydrolysis than that of carbohydrate-based cell wall (arduous to hydrolyze) species like *Scenedesmus obliquus* and *Chlorella kessleri*.

Biodiesel

Microalgae, the huge lipid reservoirs, are important renewable substrates for biodiesel production. Recently, lipids have lured the attention of scientists to alleviate the conventional fuel adversity. The lipid content is dependent on algal species, cultural conditions like nitrogen limitations, etc. (Brennan and Owende 2010). However, the condition of biomass also governs the lipid content like dried biomass of *Nannochloropsis oculata*, lyophilized biomass of *Chlorella pyrenoidosa*, algal cake of *Chlorella vulgaris* ESP-31, wet biomass of *Chlorella vulgaris* ESP-31, and dried biomass of *Chlorella pyrenoidosa* and has been observed to be 26.8, 47, 26.3, 14–63, and 56.3%, respectively (Li et al. 2011; Cao et al. 2013; Tran et al. 2013).

Bioethanol

Bioethanol production from microalgae is of substantial interest (Harun and Danquah 2011). Bioethanol production from algal biomass is less due to the limited availability of carbohydrate content (~13% dry matter) compared to rest bioethanol crops (~65% carbohydrate content of dry matter for maize) (Sheehan et al. 1998). Bioethanol can be generated from either the whole biomass or the biomass left after lipid extraction. Due to the lack of lignin, polysaccharide-rich microalgal biomass is easier to convert to fermentable sugars and then to bioethanol. The hydrolysis of starch storing microalgal species like *Chlorella vulgaris* and *Chlamydomonas reinhardtii* UTEX 90 to glucose via chemical or enzymatic processes is easy and attainable (Choi et al. 2010; Brányiková et al. 2011). Guo et al. (2013) have reported production of 0.103 g of ethanol/g of dry weight of *Scenedesmus abundans* PKUAC 12 biomass after treating with dilute acid and cellulose.

Acetone-butanol-ethanol (ABE)

There are various substrates for the production of ABE like microalgae and macroalgae (Ellis et al. 2012; Potts et al. 2012). Carbohydrate fermentation of algal biomass by saccharolytic *Clostridium* sp. leads to the production of ethanol, acetone, and butanol (Efremenko et al. 2012). Dilute acid and heat pretreated cyanobacteria resulted in the production of ethanol and butanol at concentrations of 0.29 g/L and 0.43 g/L (Efremenko et al. 2012).

Bio-oil

Bio-oil is produced from various algal species by thermo-conversion. Gasification, direct combustion, and pyrolysis are the major processes that cause thermo-conversion of algal biomass. As pyrolysis is executed out at comparatively lower temperatures than gasification and direct combustion, it is more favorable and results in the formation of products in all three states (solid, liquid, and gas) (Zhang et al. 2007). Bio-oil, the liquid product of pyrolysis, can be utilized in the transportation sector, thereby reducing the emission of greenhouse gases. The composition of bio-oil generated through pyrolysis from different microalgae species like *Chaetoceros muelleri* (Grierson et al. 2009), *Spirulina platensis* (Vardon et al. 2012), *Synechococcus* (Grierson et al. 2009), *Nannochloropsis* sp. (Borges et al. 2014), *Chlorella vulgaris* (Belotti et al. 2014; Wang et al. 2015), *Scenedesmus* sp. (Kim et al. 2014), *Dunaliella tertiolecta* (Grierson et al. 2009), *Tetraselmis chui*

(Grierson et al. 2011), and *Chlorella protothecoides* (Demirbaş 2006) has been widely studied.

Hydrogen Production

Another renewable energy source is hydrogen which has zero CO_2 emission during combustion (Nasr et al. 2013a) and produces extra energy per unit weight (Nasr et al. 2013b). It can be produced from microalgae through two biological methods, namely, biophotolysis and dark fermentation. Biophotolysis involves the utilization of light energy to generate hydrogen from water, whereas dark fermentation uses various bacteria that can ferment microalgal carbohydrates, proteins, and lipids to yield hydrogen (Das and Veziroglu 2008). *Chlamydomonas reinhardtii* has been found to be the most promising H_2 producing microalga. Table 2 presents the biohydrogen production from various microalgal species.

Feeds

High-protein feed supplements for livestock and aquaculture (Becker 1988) can be obtained substantially from algal biomass as it contains more than 50% crude protein which is manifold higher than the conventional protein sources (de la Noue and de Pauw 1988).

High-Value Products

A wide variety of high-value products like carotenoids (e.g., β-carotene), astaxanthin, long-chain polyunsaturated fatty acids (eicosapentaenoic acid (EPA) and docosahexaenoic acid (DHA)), etc. can be commercially produced by various microalgae. These are utilized as human nutritional supplements (Borowitzka 2013).

Table 2 Biohydrogen production from microalgal biomass through dark fermentation (Buitrón et al. 2017; Khetkorn et al. 2017; Roy and Das 2015; Pandey et al. 2013)

Microalgae substrate	Pretreatment applied	H_2 production	References
Scenedesmus obliquus	15 min ultrasonication at 45 °C	56 mL/g biomass	Jeon et al. (2013)
Chlorella vulgaris	1.6% HCl, 35 min	36.5 mL/g total xolids	Yun et al. (2013)
Chlorella pyrenoidosa	15 min exposure of 1% H_2SO4 at 135 °C	56.1 mL/g volatile solids	Xia et al. (2014)
Nannochloropsis oceanica	15 min exposure of 1% H_2SO4 at 140 °C	39 mL/g volatile solids	Xia et al. (2013)
Arthrospira platensis	1% H_2SO4, 140 °C microwave for 15 min, glucoamylase degradation	96.6 mL/g total solids	Cheng et al. (2012)
Cyanobacterial blooms	pH 13 for 30 min	94 mL/g volatile solids	Cai et al. (2015)
Scenedesmus obliquus	Autoclaved (121 °C, 15 min) and dried (80 °C, 16 h)	90.3 mL/g total solids	Batista et al. (2014)
Synechocystis sp. PCC 6803	Mutagenesis	190 nmol H_2 mg $chla^{-1}$ min^{-1}	Cournac et al. (2004)

10 Conclusion and Key Challenges

Conventional technologies of wastewater treatment have not proven to be enough successful in significant pathogen removal from wastewater, whereas microalgae-based wastewater treatment has shown quite a success at laboratory scale. The key challenge is to bring the technology to the field successfully. To accomplish that, robust techniques for bulk production of microalgae are required to be developed and cold weather issues need to be urgently addressed. The bigger challenge, after making the wastewater pathogen-free, is to develop cohesive wastewater treatment system, biomass generation and harvesting, and effective biomass processing to algae-based biofuels thereby utilizing all valuable components of microalgae.

References

Abdel-Raouf N, Al-Homaidan AA, Ibraheem IBM (2012) Microalgae and wastewater treatment. Saudi J Biol Sci 19:257–275

Abeliovich A (1986) Algae in wastewater oxidation ponds. In: Richmond A (ed) Handbook of microalgal mass culture. CRC Press, Boca Raton, FL, pp 331–338

Abeliovich A, Dikbuck S (1977) Factors affecting infection of *Scenedesmus obliquus* by a *Chytridium sp.* in sewage oxidation ponds. Appl Environ Microbiol 34:832–836

Ahluwalia SS, Goyal D (2007) Microbial and plant derived biomass for removal of heavy metals from wastewater. Bioresour Technol 98:2243–2257

Ahmad F, Iftikhar A, Ali AS, Shabbir SA, Wahid A, Mohy-u-Din N, Rauf A (2014) Removal of coliform bacteria from municipal wastewater by Algae. Proc Pakistan Acad Sci 51(2):129–138

Akpor OB, Ogundeji MD, Olaolu TD, Aderiye B (2014) Microbial roles and dynamics in wastewater treatment systems: an overview. Int J Pure Appl Biosci 2(1):156–168

Álvarez-Góngora CC, Herrera-Silveira J (2006) Variations of phytoplankton community structure related to water quality trends in a tropical karstic coastal zone. Mar Pollut Bull 52:48–60

Amin MM, Hashemi H, Bovini AM, Hung YT (2013) A review on wastewater disinfection. International. Int J Environ Health Eng 2:1–9

Amin MT, Alazba AA, Manzoor U (2014) A review of removal of pollutants from water/wastewater using different types of nanomaterials. Adv Mater Sci Eng 2014:1–24

Amoatey P, Bani R (2011) Wastewater management. In: García Einschlag FS (ed) Waste water - evaluation and management. InTech, London, SE19SG-United Kingdom, pp 379–398

An JY, Sim SJ, Lee JS et al (2003) Hydrocarbon production from secondarily treated piggery wastewater by the green alga *Botryococcus braunii*. J Appl Phycol 15:185

Ansa EDO, Lubberding HJ, Ampofo JA, Gijzen HJ (2011) The role of algae in the removal of *Escherichia coli* in a tropical eutrophic lake. Ecol Eng 37:317–324

Ansa EDO, Lubberding HJ, Gijzen HJ (2012) The effect of algal biomass on the removal of faecal coliform from domestic wastewater. Appl Water Sci 2:87–94

Ansa EDO, Awuah E, Andoh A, Banu R, Dorgbetor WHK, Lubberding HJ, Gijzen HJ (2015) A review of the mechanisms of faecal coliform removal from algal and duckweed waste stabilization pond systems. Am J Environ Sci 11(1):28–34

Anusha V, Sham Sundar KM (2015) Application of Vermifiltration in domestic wastewater treatment. Int J Innov Res Sci Eng Technol 4(8):7301–7304

Arvanitoyannis IA, Ladas D (2008) Meat waste management: treatment methods and potential uses of treated waste. In: Taylor SL (ed) Waste management for food industries. Academic Press, 30 Corporate Drive, Suite 400, Burlington, MA 01803, USA, pp 765–799

Aslan S, Kapdan IK (2006) Batch kinetics of nitrogen and phosphorus removal from synthetic wastewater by algae. Ecol Eng 28:64–70

Awuah E (2006) Pathogen removal mechanisms in Macrophyte and algal waste stabilization ponds: PhD: UNESCO-IHE Institute. CRC Press, Delft, p 160

Bagchi D, Kelley RT (1991) In: Hatcher KJ (ed) Selecting a dechlorinating chemical for a wastewater treatment plant in Georgia, Proceedings of the 1991 Georgia Water Resources Conference, Athens

Batista AP, Moura P, Marques PASS, Ortigueira J, Alves L, Gouveia L (2014) *Scenedesmus obliquus* as feedstock for biohydrogen production by *Enterobacter aerogenes* and *Clostridium butyricum*. Fuel 117:537–543

Batista AP, Ambrosano L, Graca S, Sousa C, Marques PA, Ribeiro B, Botrel EP, Castro Neto P, Gouveia L (2015) Combining urban wastewater treatment with biohydrogen production—An integrated microalgae-based approach. Bioresour Technol 184:230–235

Becher G (1999) Drinking water chlorination and health. Acta Hydrochim Hydrobiol 27:100–102

Becker EW (1988) Micro-algae for human and animal consumption. In: Borowitzka MA, Borowitzka LJ (eds) Micro-algal biotechnology. Cambridge University Press, London, pp 222–256

Belotti G, De Caprariis B, De Filippis P, Scarsella M, Verdone N (2014) Effect of *Chlorella vulgaris* growing conditions on bio-oil production via fast pyrolysis. Biomass Bioenergy 61:187–195

Bhargava A (2016) Physico-chemical waste water treatment technologies: an overview. Int J Sci Res Edu 4:5308–5319

Bhatnagar A, Bhatnagar M, Chinnasamy S, Das K (2010) *Chlorella minutissima* – a promising fuel alga for cultivation in municipal wastewaters. Appl Biochem Biotechnol 161:523–536

Borges FC, Xie Q, Min M, Muniz LAR, Farenzena M, Trierweiler JO, Chen P, Ruan R (2014) Fast microwave-assisted pyrolysis of microalgae using microwave absorbent and HZSM-5 catalyst. Bioresour Technol 166:518–526

Borkar RP, Gulhane ML, Kotangale AJ (2013) Moving bed biofilm reactor – a new perspective in wastewater treatment. IOSR J Environ Sci Toxicol Food Technol 6(6):15–21

Borowitzka MA (2013) High-value products from microalgae-their development and commercialisation. J Appl Phycol 25:743–756

Brányiková I, Maršálková B, Doucha J, Brányik T, Bišová K, Zachleder V, Vítová M (2011) Microalgae-novel highly efficient starch producers. Biotechnol Bioeng 108:766–776

Brennan L, Owende P (2010) Biofuels from microalgae—a review of technologies for production, processing, and extractions of biofuels and co-products. Renew Sust Energ Rev 14(2):557–577

Buitrón G, Carrillo-Reyes J, Morales M, Faraloni C, Torzillo G (2017) Biohydrogen production from microalgae. In: Fernandez CG, Mũnoz R (eds) Microalgae-based biofuels and bioproducts from feedstock cultivation to end-products. Woodhead Publishing House, 50 Hampshire Street, 5th Floor, Cambridge, MA 02139, United States.pp. 209–234

Cai T, Park SY, Li Y (2013) Nutrient recovery from wastewater streams by microalgae: status and prospects. Renew Sust Energ Rev 19:360–369

Cai J, Chen M, Wang G, Pan G, Yu P (2015) Fermentative hydrogen and polyhydroxybutyrate production from pretreated cyanobacterial blooms. Algal Res 12:295–299

Cao H, Zhang Z, Wu X, Miao X (2013) Direct biodiesel production from wet microalgae biomass of *Chlorella pyrenoidosa* through in situ transesterification. Biomed Res Int 2013:930686

Characklis GW, Dilts MJ, Simmons OD, Likirdopulos CA, Krometis LH et al (2005) Microbial partitioning to settleable particles in stormwater. Water Res 39:1773–1782

Cheng J, Xia A, Liu Y, Lin R, Zhou J, Cen K (2012) Combination of dark- and photo-fermentation to improve hydrogen production from *Arthrospira platensis* wet biomass with ammonium removal by zeolite. Int J Hydrog Energy 37:13330–13337

Chinnasamy S, Bhatnagar A, Hunt RW, Das KC (2010) Microalgae cultivation in a wastewater dominated by carpet mill effluents for biofuel applications. Bioresour Technol 101:3097–3105

Choi SP, Nguyen MT, Sim SJ (2010) Enzymatic pretreatment of *Chlamydomonas reinhardtii* biomass for ethanol production. Bioresour Technol 101:5330–5336

Christenson L, Sims R (2011) Production and harvesting of microalgae for wastewater treatment, biofuels, and bioproducts. Biotechnol Adv 29:686–702
Contreras-Flores C, Pena-Castro JM, Flores-Cotera LB, Cañizares-Villanueva RO (2003) Advances in conceptual design of photobioreactors for microalgal culture. Interciencia 28:450–456
Corbitt RA (1998) Standard handbook of environmental engineering, vol 6, 2nd edn. McGraw-Hill, New York, pp 202–203
Cournac L, Guedeney G, Peltier G, Vignais PM (2004) Sustained photoevolution of molecular hydrogen in a mutant of Synechocystis sp. strain PCC 6803 deficient in the type I NADPH-dehydrogenase complex. J Bacteriol 186:1737–1746
Curtis TP, Mara DD, Silva SA (1992) The effect of sunlight on faecal coliforms in ponds: implications for research and design. Water Sci Technol 26(7/8):1729–1738
Das D, Veziroglu TN (2008) Advances in biological hydrogen production processes. Int J Hydrog Energy 33:6046–6057
de la Noue J, de Pauw N (1988) The potential of microalgal biotechnology: a review of production and uses of microalgae. Biotechnol Adv 6:725–770
de la Noüe J, Laliberete G, Proulx D (1992) Algae and wastewater. J Appl Phycol 4:247–254
de-Bashan LE, Bashan Y (2010) Immobilized microalgae for removing pollutants: review of practical aspects. Bioresour Technol 101:1611–1627
Delenfort E, Thulin P (1997) The use of Kaldnes suspended carrier process in treatment of wastewaters from the forest industry. Water Sci Technol 35(2–3):123–130
Demirbaş A (2006) Hydrogen from mosses and algae via pyrolysis and steam gasification. Energ Source Part A 28:933–940
Doumenq P (2017) From the conventional biological wastewater treatment to hybrid processes, the evaluation of organic micropollutant removal: a review. Water Res 111:297–317
Economic and Social Commission for Western Asia (ESCWA) (2003) Waste-water treatment technologies: a general review. United Nation Publication, New York
Edzwald JK (1995) Principles and applications of dissolved air flotation. Water Sci Technol 31:1–23
Efremenko EN, Nikolskaya AB, Lyagin IV, Senko OV, Makhlis TA, Stepanov NA et al (2012) Production of biofuels from pretreated microalgae biomass by anaerobic fermentation with immobilized *Clostridium acetobutylicum* cells. Bioresour Technol 114:342–348
Ellis JT, Hengge NN, Sims RC, Miller CD (2012) Acetone, butanol, and ethanol production from wastewater algae. Bioresour Technol 111:491–495
EPA US (2000) Wastewater technology fact sheet: Trickling filter. EPA 832-F-00-014, Office of Water, Environmental Protection Agency U S Washington, DC. Available online at https://www3.epa.gov/npdes/pubs/trickling_filter.pdf
Eriksen NT (2008) The technology of microalgal culturing. Biotechnol Lett 30:1525–1536
Food and Agricultural Organisation (2006) Wastewater Treatment. http://www.fao.org/docrep/t0551e/t0551e06.htm#TopOfPage
Ghoreishi SM, Haghighi R (2003) Chemical catalytic reaction and biological oxidation for treatment of non-biodegradable textile effluent. J Chem Eng 95:163–169
Gökçe D (2016) Algae as an Indicator of water quality. In: Dhanasekaran D (ed) Algae-organisms for imminent biotechnology. InTech. https://doi.org/10.5772/62916
Göksan T, Dumaz Y, Gokpinar S (2003) Effect of light paths lengths and initial culture density on the cultivation of *Chaetoceros muelleri* (Lemmermann, 1898). Aquaculture 217:431–436
Gomez MA, Gonzalez-Lopez J, Hontoria-Garcia E (2000) Influence of carbon source on nitrate removal of contaminated groundwater in a denitrifying submerged filter. J Hazard Mater 80(1):69–80
Gopal K, Tripathy SS, Bersillon JL, Dubey SP (2007) Chlorination byproducts, their toxicodynamics and removal from drinking water. J Hazard Mater 140:1–6
Grandclement C, Seyssiecq I, Piram A, Wong-Wah-Chung P, Vanot G, Tiliacos N, Roche N, Doumenq P (2017) From the conventional biological wastewater treatment to hybrid processes, the evaluation of organic micropollutant removal: a review. Water Res 111:297–317

Gray NF (1989) Biology of wastewater treatment. Oxford Univ. Press, Oxford
Gray NF (2005) Water technology: an introduction for environmental scientists and engineers, 2nd edn. Elsevier Science & Technology Books, ISBN 0750666331, Amsterdam
Grierson S, Strezov V, Ellem G, Mcgregor R, Herbertson J (2009) Thermal characterisation of microalgae under slow pyrolysis conditions. J Anal Appl Pyrolysis 85:118–123
Grierson S, Strezov V, Shah P (2011) Properties of oil and char derived from slow pyrolysis of *Tetraselmis chui*. Bioresour Technol 102:8232–8240
Grobbelaar JU (2000) Physiological and technological considerations for optimising mass algal cultures. J Appl Phycol 12:201–206
Gschlößl T, Steinmann C, Schleypen P, Melzer A (1998) Constructed wetlands for effluent polishing of lagoons. Water Res 32:2639–2645
Guo S, Zhao X, Tang Y et al (2013) Establishment of an efficient genetic transformation system in *Scenedesmus obliquus*. J Biotechnol 163:61–68
Hamoda MF, Al-Ghusain I, Al-Mutairi NZ (2002) Tertiary filtration of wastewater for effluent reuse in irrigation. IWA Regional Symposium on Water Recycling in Mediterranean Region, Iraklio, Greece, September 26–29, 2002. Symposium Preprint Book 2, National Foundation for Agricultural Research. Eds. Angelakis A N, Tsagarakis K P, Paranychianakis N V, Asano T, pp. 225–33
Han L, Pei H, Hu W, Jiang L, Ma G, Zhang S, Han F (2015) Integrated campus sewage treatment and biomass production by *Scenedesmus quadricauda* SDEC-13. Bioresour Technol 175:262–268
Harun R, Danquah MK (2011) Influence of acid pre-treatment on microalgal biomass for bioethanol production. Process Biochem 46:304–309
Henderson R, Parsons SA, Jefferson B (2008) The impact of algal properties and pre-oxidation on solid-liquid separation of algae. Water Res 42:1827–1845
Hermosilla Gomez Z (2009) Methodological development for the correct evaluation of the ecological status of the coastal waters of the Valencian Community, within the framework of the Water Framework Directive, using chlorophyll a as an indicator parameter of quality Universitat Politècnica de València. https://doi.org/10.4995/Thesis/10251/6064
Heubeck S, Craggs RJ, Shilton A (2007) Influence of CO_2 scrubbing from biogas on the treatment performance of a high rate algal pond. Water Sci Technol 55(11):193–200
Horan NJ (1990) Biological Wastewater Treatment Systems. Theory and operation. John Wiley and Sons Ltd. Baffins Lane, Chickester. West Sussex, PO, UK
Hozalski RM, Zhang L, Arnold WA (2001) Reduction of haloacetic acids by Fe0: implications for treatment and fate. Environ Sci Technol 35:2258–2263
Jeon BH, Choi JA, Kim HC, Hwang JH, Abou-Shanab RAI, Dempsey BA, Regan JM, Kim JR (2013) Ultrasonic disintegration of microalgal biomass and consequent improvement of bioaccessibility/bioavailability in microbial fermentation. Biotechnol Biofuels 6:37
Jiménez B (2003) Health risks in aquifer recharge with recycled water. In: Aertgeerts R, Angelakis A (eds) Aquifer recharge using reclaimed water. WHO Regional Office for Europe, Copenhagen, pp 54–172
Jiménez B, Mara D, Carr R, Brissaud F (2010) Wastewater treatment for pathogen removal and nutrient conservation: suitable systems for use in developing countries. In: Drechsel P, Scott CA, Raschid-Sally L, Redwood M, Bahri A (eds) Wastewater irrigation and health. Assessing and mitigating risk in low-income countries. International Water Management Institute and International Development Research Centre (IDRC), London, pp 149–169
Jolis D, Hirano RA, Pitt PA, Müller A, Mamais D (1996) Assessment of tertiary treatment technology for water reclamation in San Francisco, California. Water Sci Technol 33:181–192
Kelly-Gerreyn BA, Anderson TR, Holt JT, Gowen RJ, Proctor R (2004) Phytoplankton community structure at contrasting sites in the Irish Sea: a modelling investigation. Estuar Coast Shelf Sci 59:363–383
Khetkorn W, Rastogi RP, Incharoensakdi A, Lindblad P, Madamwar D, Pandey A, Larroche C (2017) Microalgal hydrogen production – a review. Bioresour Technol 243:1194–1206

Kim SW, Koo BS, Lee DH (2014) A comparative study of bio-oils from pyrolysis of microalgae and oil seed waste in a fluidized bed. Bioresour Technol 162:96–102

Koivunen J (2007) Effects of conventional treatment, tertiary treatment and disinfection processes on hygienic and physico-chemical quality of municipal wastewaters. Dissertation, University of Kuopio

Kong B, Shanks JV, Vigil RD (2013) Enhanced algal growth rate in a Taylor vortex reactor. Biotechnol Bioeng 110:2140–2149

Konig A, Pearson HW, Silva SA (1987) Ammonia toxicity to algal growth in waste stabilization ponds. Water Sci Technol 19:115–122

Korzeniewska E (2011) Emission of bacteria and fungi in the air from wastewater treatment plants - a review. Front Biosci. 1(3):393–407

Kott Y, Rose N, Sperber S, Betzer N (1974) Bacteriophages as viral pollution indicators. Water Res 8:165–171

Kumar K, Mishra SK, Choi G (2015) CO2 sequestration through algal biomass production. In: Das D (ed) Algal biorefinery: an integrated approach. Springer International Publishing, Cham

Lebeau T, Robert JM (2006) Biotechnology of immobilized micro-algae: a culture technique for the future? In: Rao S (ed) Algal cultures, analogues of blooms and applications. Science Publishers, Enfield, NH, pp 801–837

Li P, Miao X, Li R, Zhong J (2011) In situ biodiesel production from fast-growing and high oil content *Chlorella pyrenoidosa* in rice straw hydrolysate. J Biomed Biotechnol 2011:141207

Litaker RW, Stewart TN, Eberhart BL, Wekell JC, Trainer VL et al (2008) Rapid enzyme-linked immunosorbent assay for detection of the algal toxin domoic acid. J Shellfish Res 27:1301–1310

Liu D, Li D, Zhang B (2009) Removal of algal bloom in freshwater using magnetic polymer. Water Sci Technol 59:1085–1091

Livingston RJ (2001) Eutrophication processes in coastal systems: origin and succession of plankton blooms and effects on secondary production in Gulf Coast estuaries. CRC Press, New York

Lloyd BJ, Morris R (1983) Effluent and water treatment before disinfection. In: Bulter M, Medlen AR, Morris R (eds) Viruses and disinfection of water and wastewater. Univ. of Surrey Print Unit, Guild Ford, pp 154–189

Markou G (2015) Fed-batch cultivation of *Arthrospira* and *Chlorella* in ammonia-rich wastewater: optimization of nutrient removal and biomass production. Bioresour Technol 193:35–41

Masseret E, Amblard C, Bourdier G, Sargos D (2000) Effects of a waste stabilization lagoon discharge on bacterial and phytoplanktonic communities of a stream. Water Environ Res 72:285–294

Matsumoto H, Hamasaki A, Shioji N, Ikuta Y (1996) Influence of dissolved oxygen on photosynthetic rate of microalgae. J Chem Eng Jpn 29:711–714

Mawdsley JL, Bardgett RD, Merry RJ, Pain BF, Theodorou MK (1995) Pathogens in livestock waste, their potential for movement through soil and environmental pollution. Appl Soil Ecol 2:1–15

Mayes WM, Batty LC, Younger PL, Jarvis AP, Kõiv M, Vohla C, Mander U (2009) Wetland treatment at extremes of pH: a review. Sci Total Environ 407:3944–3957

Metcalf and Eddy (1991) Wastewater engineering treatment, disposal, and reuse. McGraw-Hill, New York

Mezzari MP, Prandini JM, Kich JD, Silva MLB (**2017**) Elimination of antibiotic multi-resistant salmonella typhimurium from swine wastewater by microalgae-induced antibacterial mechanisms. J Bioremed Biodegr 8:1–4

Misal N, Mohite NA (2017) Community wastewater treatment by using vermifiltration technique. Int J Eng Res Technol 10(1):363–365

Mitchell R (1992) Environmental microbiology. Wiley-liss Inc, New York, p 411

Mohamed ZA (2008) Polysaccharides as a protective response against microcystin-induced oxidative stress in *Chlorella vulgaris* and *Scenedesmus quadricauda* and their possible significance in the aquatic ecosystem. Ecotoxicology 17:504–516

Mohiyaden HA, Sidek LM, Salih GHA, Birima AH, Basri H, Sabri AFM, Noh MD (2016) Conventional methods and emerging technologies for urban river water purification plant: a short review. ARPN J Eng Appl Sci 11(4):2547–2556

Molinuevo-Salces B, Mahdy A, Ballesteros M, González-Fernández C (2016) From piggery wastewater nutrients to biogas: microalgae biomass revalorization through anaerobic digestion. Renew Energy 96:1103–1110

Molla AH, Fakhru'l-Razi A, Alam MZ (2004) Evaluation of solid-state bioconversion of domestic wastewater sludge as a promising environmental friendly disposal technique. Water Res 38(19):4143–4152

Morão AEC (2008) Transport mechanisms governing the nanofiltration of multicomponent solutions – application to the isolation of clavulanic acid. Universidade Téchica de Lisboa, 1649-004 Lisboa

Moreno-Garrido I (2008) Microalgae immobilization: current techniques and uses. Bioresour Technol 99:3949–3964

Mulbry W, Kondrad S, Pizarro C, Kebede-Westhead E (2008) Treatment of dairy manure effluent using freshwater algae: algal productivity and recovery of manure nutrients using pilot-scale algal turf scrubbers. Bioresour Technol 99:8137–8142

Mulder M (1996) Basic principles of membrane technology. J Memb Sci 72(3):564

Muñoz R, Guieysse B (2006) Algal–bacterial processes for the treatment of hazardous contaminants: a review. Water Res 40(15):2799–2815

Munoz R, Guieysse B (2008) Algal–bacterial processes for the treatment of hazardous contaminants: a review. Water Res 40:2799–2815

Munoz R, Kollner C, Guieysse B, Mattiasson B (2004) Photosynthetically oxygenated salicylate biodegradation in a continuous stirred tank photobioreactor. Biotechnol Bioeng 87(6):797–803

Mussgnug JH, Klassen V, Schlüter A, Kruse O (2010) Microalgae as substrates for fermentative biogas production in a combined biorefinery concept. J Biotechnol 150:51–60

Muttamara S (1996) Wastewater characteristics. Resour Conserv Recycl 16:145–159

Nasr M, Tawfik A, Ookawara S, Suzuki M (2013a) Biological hydrogen production from starch wastewater using a novel up-flow anaerobic staged reactor. Bio Resources 8:4951–4968

Nasr M, Tawfik A, Ookawara S, Suzuki M (2013b) Environmental and economic aspects of hydrogen and methane production from starch wastewater industry. J Water Environ Technol 11:463–475

Nieuwstad TJ, Mulder EP, Havelaar AH, van Olphen M (1988) Elimination of microorganisms from wastewater by tertiary precipitation and simultaneous precipitation followed by filtration. Water Res 22:1389–1397

Ødegaard H (2001) The use of dissolved air flotation in municipal wastewater treatment. Water Sci Technol 43:75–81

Odegaard H, Rusten B, Swestrum T (1994) A new moving bed biofilm reactor – applications and results. Water Sci Technol 29(10–11):157–165

Ogbonna JC, Tanaka H (2000) Light requirement and photosynthetic cell cultivation—development of processes for efficient light utilization in photobioreactors. J Appl Phycol 12:207–218

Okoh AT, Odjadjare EE, Igbinosa EO, Osode AN (2007) Wastewater treatment plants as a source of microbial pathogens in receiving water sheds. Afr J Biotechnol 6(25):2932–2944

Olguın EJ (2003) Phycoremediation: key issues for cost-effective nutrient removal processes. Biotechnol Adv 22:81–91

Olguín EJ, Galicia S, Mercado G, Pérez T (2003) Annual productivity of *Spirulina* (*Arthrospira*) and nutrient removal in a pig wastewater recycling process under tropical conditions. J Appl Phycol 15:249–257

Olguın EJ, Sanchez G, Mercado G (2004) Cleaner production and environmentally sound biotechnology for the prevention of upstream nutrient pollution in the Mexican coast of the Gulf of Mexico. Ocean Coast Manag 47:641–670

Oswald WJ, Gotaas HB (1957) Photosynthesis in sewage treatment. Trans Am Soc Civil Eng 122:73–105

Padisák J, Borics G, Grigorszky I, Soróczki-Pintér É (2006) Use of phytoplankton assemblages for monitoring ecological status of lakes within the water framework directive: the assemblage index. Hydrobiologia 553:1–14

Pandey A, Chang JS, Patrick H, Christian L (2013) Biohydrogen. Elsevier Science & Technology

Peavy SH, Rowe DR, Tchobanoglous G (1985) Environmental Engineering. International Edition. MacGraw-Hill pp. 207–322

Petrovic M, Diaz A, Ventura F, Barceló D (2003) Occurrence and removal of estrogenic short-chain ethoxy nonylphenolic compounds and their halogenated derivatives during drinking water production. Env Sci Technol 37:4442–4448

Pinto Filho ACT, Brandão CCS (2001) Evaluation of flocculation and dissolved air flotation as an advanced wastewater treatment. Water Sci Technol 43:83–90

Pittman JK, Dean AP, Osundeko O (2011) The potential of sustainable algal biofuel production using wastewater resources. Bioresour Technol 102:17–25

Posadas E, Bochon S, Coca M, García-González MC, Garcıá-Encina PA, Muñoz R (2014) Microalgae-based agro-industrial wastewater treatment: a preliminary screening of biodegradability. J Appl Phycol 26:2335–2345

Posadas E, Morales MM, Gómez C, Acén FG, Muñoz R (2015) Influence of pH and CO_2 source on the performance of microalgae-based secondary domestic wastewater treatment in outdoors pilot raceways. Chem Eng J 265:239–248

Posadas E, Alcántara C, Garcá-Encina PA, Gouveia L, Guieysse B, Norvill Z, Acién FG, Markou G, Congestri R, Koreiviene J, Muñoz R (2017) Microalgae cultivation in wastewater. In: Fernandez GC, Muñoz R (eds) Microalgae-based biofuels and bioproducts from feedstock cultivation to end-products. Woodhead Publishing House, 50 Hampshire Street, 5th Floor, Cambridge, MA 02139, United States. pp 67–91

Potts T, Du J, Paul M, May P, Beitle R, Hestekin J (2012) The production of butanol from Jamaica bay macro algae. Environ Prog Sustain 31:29–36

Prabu LS, Suriyaprakash TNK, Ashok Kumar J (2011) Wastewater treatment technologies: a review. Pharma Times 43:55–62

Prajapati SK, Kaushik P, Malik A, Vijay VK (2013) Phycoremediation and biogas potential of native algal isolates from soil and wastewater. Bioresour Technol 135:232–238

Prajapati SK, Choudhary P, Malik A, Vijay VK (2014) Algae mediated treatment and bioenergy generation process for handling liquid and solid waste from dairy cattle farm. Bioresour Technol 167:260–268

Ramanna L, Guldhe A, Rawat I, Bux F (2014) The optimization of biomass and lipid yields of *Chlorella sorokiniana* when using wastewater supplemented with different nitrogen sources. Bioresour Technol 168:127–135

Rawat I, Kumar RR, Mutanda T, Bux F (2011) Dual role of microalgae: phycoremediation of domestic wastewater and biomass production for sustainable biofuels production. Appl Energy 88:3411–3424

Rawat I, Gupta SK, Shriwastav A, Singh P, Kumari S, Bux F (2016) Microalgae applications in wastewater treatment. In: Bux F, Chisti Y (eds) Algae biotechnology products and processes. Springer International Publishing, Switzerland, p.249–268

Reynolds TD (1982) Unit operations and processes in environmental engineering. Thomson-Engineering, Toronto

Richmond A (2000) Microalgal biotechnology at the turn of the millennium: a personal view. J Appl Phycol 12:441–451

Rosen M, Welander T, Lofqvist A (1998) Development of a new process for treatment of a pharmaceutical wastewater. Water Sci Technol 37:251–258

Roy S, Das D (2015) Gaseous fuels production from algal biomass. In: Das D (ed) Algal biorefinery: an integrated approach. Springer International Publishing, Cham

Rubio J, Souza ML, Smith RW (2002) Overview of flotation as a wastewater treatment technique. Minerals Eng 15:139–155

Ruiz-Marin A, Mendoza-Espinosa LG, Stephenson T (2010) Growth and nutrient removal in free and immobilized green algae in batch and semi-continuous cultures treating real wastewater. Bioresour Technol 101:58–64

Russell DL (2006) Practical wastewater treatment. John Wiley and Sons, Inc, ISBN-13:978–0–471-78044-1, Hoboken, NJ

Sadiq R, Rodriguez MJ (2004) Disinfection by-products (DBPs) in drinking water and predictive models for their occurrence: a review. Sci Total Environ 321:21–46

Samer M (2015) Biological and chemical wastewater treatment processes. In Samer M (ed.) InTech. https://doi.org/10.5772/61250. Available from https://www.intechopen.com/books/wastewater-treatment-engineering/biological-and-chemical-wastewater-treatment-processes

Samorì G, Samorì C, Guerrini F, Pistocchi R (2013) Growth and nitrogen removal capacity of *Desmodesmus communis* and of a natural microalgae consortium in a batch culture system in view of urban wastewater treatment: part I. Water Res 47:791–801

Schultz T E (2005) Biotreating process wastewater: airing the options. Chemical Engineering Magazine

Servos MR, Bennie DT, Burnison BK (2005) Distribution of estrogens, 17β-estradiol and estrone, in Canadian municipal wastewater treatment plants. Sci Total Environ 336:155–170

Sheehan J, Dunahay T, Benemann J, Roessler P (1998) A look back at the US Department of Energy's aquatic species program: biodiesel from algae. National Renewable Energy Laboratory, Golden, CO

Shi H (2009) *Industrial wastewater* - Types, amounts and effects. Point Sources of Pollution: Local Effects and Its Control. Vol. I

Shon HK, Vigneswaran S, Kandasamy J, Cho J (2009) Membrane technology for organic removal in wastewater. In: Vigneswaran S (ed) Water and wastewater treatment technologies, in encyclopedia of life support systems (EOLSS) developed under the auspices of the UNESCO. Eolss Publishers, Oxford. Available at http://www.eolss.net. Retrieved at 24 Aug 2011

Show KY, Lee DJ (2014) Production of biohydogen from microalgae. In: Ashok P, Duu-Jong L, Yusuf C, Carlos SR (eds) Biofuels from algae. Elsevier, Burlington, MA

Shrestha A (2013) Specific moving bed biofilm reactor in nutrient removal from municipal wastewater. Thesis. University of Technology, Sydney

Sincero AP, Sincero GA (2003) Physical–chemical treatment of water and wastewater. CRC Press, Florida

Sinha RK, Bharambe G, Chaudhari U (2008) Sewage treatment by vermifiltration with synchronous treatment of sludge by earthworms: a low cost sustainable technology over conventional systems with potential for decentralization. Springer Science 28:409–420

Sirianuntapiboon S, Chairattanawan K, Jungphungsukpanich S (2006) Some properties of a sequencing batch reactor system for removal of vat dyes. Bioresour Technol 97:1243–1252

Sobsey MD (1989) Inactivation of health-related microorganisms in water by disinfection processes. Water Sci Technol 21(3):179–195

Suffet IH, Ho J, Chou D, Khiari D, Mallevialle J (1995) Taste and odor problems observed during drinking water treatment. In: Suffet IH, Mallevialle J, Kawczynski E (eds) Advances in taste-and-odor treatment and control. American Water Works Association. 1199 North Fairfax Street, Suite 900, Alexandria, Virginia

Suh IS, Lee CG (2003) Photobioreactor engineering: design and performance. Biotechnol Bioprocess Eng 8:313–321

Sun YX, Wu QY, Hu HY, Tian J (2009) Effects of operating conditions on THMs and HAAs formation during wastewater chlorination. J Hazard Mater 168:1290–1295

Szewzyk U, Szewzyk R, Manz W, Schleifer KH (2000) Microbiogical safety of drinking water. Ann Rev Microbiol 54:81–127

Tebbutt THY (1983) Principles of water quality control. Pergammon Press, Oxford

Topare NS, Attar SJ, Manfe MM (2011) Sewage/wastewater treatment technologies: a review. Sci Revs Chem Commun 1:18–24

Toze S (1997) Microbial pathogens in wastewater. CSIROL and Water Technical Report

Tran DT, Chen CL, Chang JS (2013) Effect of solvents and oil content on direct transesterification of wet oil-bearing microalgal biomass of *Chlorella vulgaris* ESP-31 for biodiesel synthesis using immobilized lipase as the biocatalyst. Bioresour Technol 135:213–221

Travieso L, Benitez F, Dupeiron R (1992) Sewage treatment using immobilized microalgae. Bioresour Technol 40:183–187

Urase T, Kikuta T (2005) Separate estimation of adsorption and degradation of pharmaceutical substances and estrogens in the activated sludge process. Water Res 39:1289–1300

Van der Steen P, Brenner A, Shabtai Y, Oron G (2000) The effect of environmental conditions on faecal coliform decay in post-treatment of UASB reactor effluent. Water Sci Technol 42:111–118

Vardon DR, Sharma BK, Blazina GV, Rajagopalan K, Strathmann TJ (2012) Thermochemical conversion of raw and defatted algal biomass via hydrothermal liquefaction and slow pyrolysis. Bioresour Technol 109:178–187

Vieno N, Tuhkanen T, Kronberg L (2006) Removal of pharmaceuticals in drinking water treatment: effect of chemical coagulation. Environ Technol 27:183–192

Voltolina D, Cordero B, Nieves M, Soto LP (1999) Growth of *Scenedesmus sp.* in artificial wastewater. Bioresour Technol 68:265–268

Wang L, Min M, Li Y, Chen P, Chen Y, Liu Y et al (2010) Cultivation of green algae Chlorella sp. in different wastewaters from municipal wastewater treatment plant. Appl Biochem Biotechnol 162:1174–1186

Wang N, Tahmasebi A, Yu J, Xu J, Huang F, Mamaeva A (2015) A comparative study of microwave-induced pyrolysis of lignocellulosic and algal biomass. Bioresour Technol 190:89–96

Weber WJ (1972) Physicochemical processes for water quality control. Wiley & Sons, New York

WEF (1996) Wastewater Disinfection: Manual of Practice No. FD-10, Water Environment Federation, Alexandria, Virginia

WEF (2008) Industrial wastewater management, treatment and disposal. 3rd ed., Manual of Practice No. FD-3 Water Environment Federation: Alexandria, Virginia

Westerhoff P, Yoon Y, Snyder S, Wert E (2005) Fate of endocrine-disruptor, pharmaceutical, and personal care product chemicals during simulated drinking water treatment processes. Environ Sci Technol 39:6649–6663

WHO (1989) Health guidelines for the use of waste water in agriculture and aquaculture. Technical Report. Series No. 74, World Health Organization, Geneva

Wilkie AC, Mulbry WW (2002) Recovery of dairy manure nutrients by benthic freshwater algae. Bioresour Technol 84:81–91

Wrigley TJ, Toerien DF (1990) Limnological aspects of small sewage ponds. Water Res 24:83–90

Wu LF, Chen PC, Huang AP, Lee CM (2012) The feasibility of biodiesel production by microalgae using industrial wastewater. Bioresour Technol 113:14–18

Wu X, Lu Y, Zhou S, Chen L, Xu B (2016) Impact of climate change on human infectious diseases: Empirical evidence and human adaptation. *Environ Int.* 1(86):14–23

Xia A, Cheng J, Lin R, Lu H, Zhou J, Cen K (2013) Comparison in dark hydrogen fermentation followed by photo hydrogen fermentation and methanogenesis between protein and carbohydrate compositions in *Nannochloropsis oceanica* biomass. Bioresour Technol 138:204–213

Xia A, Cheng J, Ding L, Lin R, Song W, Zhou J, Cen K (2014) Effects of changes in microbial community on the fermentative production of hydrogen and soluble metabolites from *Chlorella pyrenoidosa* biomass in semi-continuous operation. Energy 68:982–988

Xiao LW, Rodgers M, Mulqueen J (2007) Organic carbon and nitrogen removal from a strong wastewater using a denitrifying suspended growth reactor and a horizontal-flow biofilm reactor. Bioresour Technol 98:739–744

Yun YM, Jung KW, Kim DH, Oh YK, Cho SK, Shin HS (2013) Optimization of dark fermentative H_2 production from microalgal biomass by combined (acid + ultrasonic) pretreatment. Bioresour Technol 141:220–226

Zamalloa C, Boon N, Verstraete W (2012) Anaerobic digestibility of *Scenedesmus obliquus* and *Phaeodactylum tricornutum* under mesophilic and thermophilic conditions. Appl Energ 92:733–738

Zhang W, DiGiano FA (2002) Comparison of bacterial regrowth in distribution systems using free chlorine and chloramine: a statistical study of causative factors. Water Res 36:1469–1482

Zhang Q, Chang J, Wang T, Xu Y (2007) Review of biomass pyrolysis oil properties and upgrading research. Energy Convers Manag 48:87–92

Zhang ED, Wang B, Wang QH, Zhang SB, Zhao BD (2008) Ammonia-nitrogen and orthophosphate removal by immobilized *Scenedesmus* sp isolated from municipal wastewater for potential use in tertiary treatment. Bioresour Technol 99:3787–3793

Remediation of Domestic Wastewater Using Algal-Bacterial Biotechnology

Shashi Bhushan, Halis Simsek, Aswin Krishna, Swati Sharma, and Sanjeev Kumar Prajapati

1 Introduction

Management and treatment of wastewater is one of the key challenges in the current scenario of environmental protection. Domestic wastewater consists of two types of water input: gray water and blackwater. The water excreted out from all sources except toilet outputs are defined as gray water, while blackwater originates from toilets. Domestic wastewater contain nutrients, organics, and minerals including ammonia, nitrate, nitrite, organic nitrogen, phosphorous (P), carbon (C), volatile fatty acids (VFAs), and other trace elements (iron, manganese, etc.). In some cases, presence of undesirable components such as heavy metals and emerging pollutants (personal care products, surfactants, pharmaceuticals, etc.) can be observed. The quality of domestic wastewater depends on many factors such as location of the residence, number of resident, economical and climatic conditions, etc. Generally the wastewater excreted out of the household contains all necessary nutrients to support the growth and optimal activity of inherent microbes in the sewage which eventually help in bioremediation of the available nutrients. Considering the composition of wastewater, many research groups inspired to develop treatment technologies which are based on microbes. Over the past decades, many technologies come up which is driven by microbial consortia. Bacteria and algae are the major player in this regard. These technologies have been known promising because of their self-sustaining ability, low operational and capital cost, and generating less toxic compounds at the end of the process.

S. Bhushan · A. Krishna · S. K. Prajapati (✉)
BioResource Engineering Laboratory (BREL), Chemical and Biochemical Engineering, Indian Institute of Technology (IIT) Patna, Patna, Bihar, India
e-mail: sanjukec@iitp.ac.in

H. Simsek · S. Sharma
Agricultural and Biosystems Engineering, North Dakota State University, Fargo, ND, USA

S. K. Gupta, F. Bux (eds.), *Application of Microalgae in Wastewater Treatment*,
https://doi.org/10.1007/978-3-030-13913-1_13

Bacteria being the large domain of prokaryotic microorganisms follow heterotrophic mode of nutrition. Aerobic, anaerobic respiration as well as fermentation plays an important role in biological/secondary treatment of sewage. Bacterial systems are robust as it can maintain high cell density in the form of biofilm and have higher growth rate. Throughout the world, biological treatment of sewage is facilitated by activated sludge. Analyzing the microbial population, it is found that bacterial community contributes the most in term of performance. Many research groups also attempted to work on tailor made consortia (Saha et al. 2018) or even tried with single isolates (Castro-Barros et al. 2017; Leung et al. 2000).

Algae are microscopic eukaryotes which can perform photosynthesis for their sustenance. Algae can switch from phototropic to heterotrophic mode based on availability of the nutrients (Hammed et al. 2016). They can be found in fresh and marine water system and can live on surface and in sediments. Over the past decades, algae-driven technology has been proposed and commonly applied for domestic wastewater treatments (Muñoz and Guieysse 2006; Olguín 2012; Oswald 1988; Park et al. 2011). Several successful attempts were documented where wastewater treatment was coupled with methane production through anaerobic digestion (Prajapati et al. 2013a; Ward et al. 2014). However, these technologies are still in its budding stage and demand innovative scientific approach and technological breakthroughs to make it feasible at industrial-scale operations. In nature, algae and bacteria grow together and share number of nutrients and compounds that required for their growth. The symbiotic relationship of algae and bacteria also dominates in wastewater treatment. Several studies highlighted that use of algal-bacteria results in significant improvement in the rate of pollutant removal (De-Bashan et al. 2004; Muñoz and Guieysse 2006; Shen et al. 2017). Because of several inherent advantages, algal-bacterial consortium is now being preferred for treatment range of wastewater and biomass production over other systems involving either only-algae or only-bacteria.

This chapter reviews the quality of domestic wastewater and algal technologies available for the wastewater treatment and gives accounts on algal-bacterial interactions. It also explains about how these interactions would help in developing the wastewater technologies which overcome the present-day challenges and can be used at optimal performance in the near future.

2 Domestic Wastewater and Major Pollutants

Domestic wastewater consists of two type of water input: gray water and blackwater. The water excreted out from all sources, i.e., bath, kitchen basin, etc., except toilet outputs is defined as gray water. It has very less organic content but contributes to 50–80% of the total domestic wastewater. Although it has very less organic content and pathogenic microbes, it has array of contaminants. These contaminants originate from the products being used by the human beings on regular basis. The major categories of contaminants come under pharmaceuticals, personal care products, steroid hormones, surfactants, industrial chemicals, and pesticides (Luo et al. 2014).

Blackwater only consists of toilet output. It contributes half the load of organic material in domestic wastewater; nitrogen and phosphorous share the major fraction of nutrients. Pathogens, hormones, and pharmaceutical residues can also be seen in this water. The quantity of blackwater produce depends on the type of toilet and the amount of water needed for flushing (de Graaff et al. 2010). Besides, domestic wastewater contains large amount of inorganicpollutants including nitrogen, phosphate, and other trace compounds. Some of the major characteristics of domestic wastewater are listed in Table 1.

As reflected from the above discussion, the wastewaters generated from the domestic sources are full of pollutants and cause severe environmental problems including contamination of fresh water bodies and eutrophication and toxicity to aquatic life. Hence, proper treatment of domestic wastewater is crucial before its discharge to the environment. However, wastewater treatment facilities are costly to construct, require substantial land, and demand high energy input. It can be also noted that the plant operational cost is big and requires qualified manpower for functioning.

3 Methods Available for Domestic Wastewater Treatment

Quality water is a limited resource on planet earth. Wastewater treatment is common practice, which helps in neutralizing the harmful contaminants and enables the reuse of discarded water which results in sustainable use of water resources. Basically three modules were deployed for the treatment and categorized as physical, chemical, and biological methods. The selection of methods depends on the quality of the wastewater, which is assessed on some initial parameters like pH,

Table 1 Range of the major constituents of domestic wastewater

S. No.	Contaminants	Concentration range (in mg/L)
1.	Total solids (TS)	350–1200
2.	Total dissolved solid (TDS)	250–850
3.	Total suspended solids (TSS)	100–350
4.	Total volatile solids (TVS)	105–325
5.	Total organic carbon (TOC)	80–300
6.	Chemical oxygen demand (COD)	250–1000
7.	Total nitrogen (as N)	20–85
8.	Total phosphorous (as P)	4–15
9.	Chlorides	30–100
10.	Sulfate	20–50

Adopted and modified from Rawat et al. (2011)

turbidity, chemical oxygen demand (COD), biochemical oxygen demand (BOD), total dissolved solid (TDS), etc. The treatment modules were briefly discussed below.

3.1 Physical Treatment Methods

Most of the physical treatment methods are employed form separation of insoluble solids and debris from the wastewater. Traditional physical treatment methods include sedimentation, screening, floatation and skimming, aeration, and filtration which are quite effective in removing most of the macro-sized particles. Some chemical substances can also be removed physically by adsorption to suitable adsorbents. For example, methylene blue dyes have been purified from wastewater by adsorbing to synthetic zeolite MCM22. Activated carbon (granular being the more effective form compared to powdered) is another common adsorbent for a wide variety of substances including metal ions, phenols, dyes, pesticides, detergents, and humic acids (De Gisi et al. 2016). Attempts are being made by researchers to shift from high- to low-cost adsorbents (LCAs) – biosorption using microorganisms by Aksu (2005) and use of sawdust to remove dyes and metals by Shukla et al. (2002).

3.2 Chemical Treatment Methods

A wide array of chemical treatment methods can be employed including oxidation, floatation, coagulation, electrolysis, ion exchange, stabilization, and other physicochemical processes. Suitable oxidants can be added to oxidize specific impurities and convert them into less toxic substances through a series of steps. Advanced oxidation processes (some of which are mentioned below) use hydroxyl radicals as the oxidants which are much more powerful compared to normal chemical oxidizing agents like $KMnO_4$ or hydrogen peroxide. Cavitation process uses microbubbles that occur only for a fraction of a second. These cavities are present throughout the reactor and release a large amount of energy. Here, the oxidation is explained to be as a result of combined free radical attack and pyrolysis (Gogate and Pandit 2004). Photocatalysis is another chemical treatment method in which the energetic photons from ultraviolet radiations can release free radicals from chemical species (catalysts may be required, e.g., oxides of titanium and zirconium and sulfides of zinc and cadmium) and oxidize the pollutants (Agustina et al. 2005). Treatment with ozone is also an option since it can break multiple bonds simultaneously, but the quality of oxidation is compromised. High-quality oxidation by ozone followed by sand filtration has been proved to be effective in eliminating a wide range of chemicals depending on its chemical properties (Hollender et al. 2009). The use of ozone in municipal wastewater treatments however has been relatively new.

3.3 Biological Treatment Methods

Biological treatment methods of wastewaters involve growth of range of microbes including bacteria, fungi, and algae. These microbes play a pivotal role in the biodegradation of nutrients and organic matter present in wastewater and reduction in COD. The activated sludge process (ASP) is one of the most commonly used systems for biological treatment of wastewater. In a basic ASP, wastewater is aerated in the presence of microbes, mainly heterotrophic bacteria (but may also include protozoa and rotifers which improve floc formation and separation), which metabolize all the matter, and then settling of flocs (clumps rich in microorganisms) is done after which the effluent is discharged and a part of the activated sludge from the underflow is recycled (RAS) back to the aeration tank. Nowadays, several modifications have been done to enhance the performance of ASP system. Apart from activated sludge, another common type is the trickling filter process. It uses microorganisms attached to some surface (e.g., on biofilms) for wastewater purification. It has also been shown that when algal biomass is produced along with wastewater treatment, then nitrogen and phosphorus treatment is enhanced along with production of useful by-products (Choudhary et al. 2017). Apart from biofilm systems, algal-bacterial consortia are also used as suspended culture from treatment of wastewater.

4 Algae in Wastewater Treatment: Utilizing Waste Nutrient for Growth

Mainly, algae require C, N, and P for its growth. C is the primary source for its growth followed by N. Wastewater which is rich in C, N, and P serves as potential growth media for mass-scale algae cultivation. In fact, in past few decades, algae have found its significant place in the removal of wastewater-derived organic and inorganic pollutants coupled with biomass production. Consequently, algae are being used in tertiary treatment process in a number of treatment plants around the world.

Major N sources for algal metabolism are ammonium, nitrate, and urea. Ammonium is an important nitrogen source for algal growth. However, excess amount of ammonium (more than 100 mg/L) in wastewaters could be toxic to algae (Muñoz and Guieysse 2006). Concentration of ammonium and pH greatly impacts algal growth when ammonium is the primary source of nitrogen (Becker 1994; Shi et al. 2000). Determining the suitable algal species for wastewater treatment is crucial for sustainable and practical application of algal biotechnology on wastewater treatment. The criteria for selecting the best algal species for a domestic wastewater treatment are (i) adaptability of algae to various wastewater loading conditions, (ii) fast-growing capacity and biomass productivity of algae, (iii) settling and flock forming capacity for easy harvesting, (iv) symbiotic relationship with the common bacteria found in the biological treatment system, (v) occurrence of extracellular polymeric substances (EPS) and soluble microbial products (SMP) during the biological activity, (vi) nutrient utilization capacity of algae, and finally (vii) lipid content in algal cell.

Several studies are conducted to understand the removal mechanism and fate of nutrients using algae. Some types of algae species, including cyanobacteria*Phormidium* sp., *Chlamydomonas* and *Chlorella vulgaris*, and *Selenastrum capricornutum*, have high nutrient removal rate and biofuel-generating potential (Kong et al. 2010; Olguín 2012; Simsek et al. 2013). Some of the most popular species of algae used in the wastewater treatment are found to be *Scenedesmus*, *Chlorella*, and *Chlamydomonas* sp. A study conducted by Tam and Wang demonstrated the efficiency of algae in the removal of N and P and the feasibility of algae cultivation in domestic sewage wastewater. Algae *Chlorella* and *Scenedesmus* species were compared in the study to understand the efficiency of their nutrient removal and growth while they were utilizing organic and inorganic N and P. The amount of nutrient removed by algae depends on the rate of algal growth. The study suggested that *Chlorella* sp. is most suitable algae in the removal of wastewater nutrients (Tam and Pollution 2000). Autotrophic algae achieve limited nutrient removal efficiency. This is mainly due to low algal biomass yield resulting from insufficient light caused by shading effect and limited availability of inorganic carbon. This issue was first addressed by González-Fernández et al., (González-Fernández et al. 2011) introducing heterotrophic algal cultivation for the treatment of wastewater. Heterotrophic algae has high biomass yield and can grow under lightless conditions (Liang et al. 2009; Xu et al. 2006).

In another study, four different algae cultures, including *Phormidium* sp. (cyanobacteria strain), *Chlamydomonas reinhardtii*, *Scenedesmus rubescens*, and *Chlorella vulgaris* (green algae strains), have been utilized to determine the settling capacity, biomass productivity, and organic carbon and nutrient removal performance on the effluent of the second clarifier in a WWTP. Results showed that nearly 100% of ammonia and 98% of P were removed within 6 days of incubation by all four species. During ammonia removal, nitrate level stayed the same as the initial value and started to decrease after ammonia removed completely. Removal of nitrite also followed this trend by decreasing after ammonia was completely utilized. These outcomes proved that these four types of algae prefer to utilize ammonium first, then following nitrate and nitrite when all those nutrients were available in the same aquatic sample. Overall removal performance of N and P for the algae species were as follows: *C. reinhardtii* > *C. vulgaris* > *S. rubescens* > *Phormidium* sp. The amount of algal biomass production reduced when the algal species fed by ammonium only (Su et al. 2012a). *C. vulgaris*, *C. reinhardtii*, and *S. rubescens* settled very well and very fast (in 15 min), while *Phormidium* sp. showed poor settling ability. The growth of filamentous cyanobacteria affected the settling ability in the samples. It was proved that the special cultivation strategies (mixing and non-mixing operation), cell surface properties, and the capacity of EPS and SMP that released by the algal cells affected the settleability (Shen et al. 2015; Su et al. 2012b).

Similarly, Wang et al., (Wang et al. 2010), studied the cultivation of green algae *Chlorella* sp. in different locations of municipal WWTP and its ability to remove nutrients. The study observed that algal growth was enhanced in the locations with high concentration of N, P, and COD. Locations with limiting P concentration yielded lower growth of algae. Between 73% and 82% of ammonium was removed from the primary and settling tank locations, and total nitrogen (TN) removal ranged between 58% and 80% depending on the location from where the wastewater samples were collected. *C. vulgaris, O. multisporus, S. obliquus, S. intermedius, and Nannochloris* sp. have been reported to have substantially removed TN and P from municipal wastewater (Aslan and Kapdan 2006; Ji et al. 2013a, 2013b; Jiménez-Pérez et al. 2004).

4.1 Dissolved Organic Nitrogen Removal

Recent incubation studies have proved that some portion of wastewater-derived dissolved organic nitrogen (DON) from different locations of WWTP including the final effluent is essentially biodegradable to bacterial species (Sattayatewa et al. 2009; Simsek et al. 2013), while some portion of it is bioavailable to algal and bacterial communities (Pehlivanoglu-Mantas and Sedlak 2006; Pehlivanoglu and Sedlak 2004; Sattayatewa et al. 2009; Simsek et al. 2013; Urgun-Demirtas et al. 2008). Once DON is biomineralized by bacteria to ammonia and some other low molecular weight compounds, it becomes bioavailable to algae, bacteria, and phytoplankton. Biodegradable DON (BDON) is a fraction of DON mineralized by bacteria only, while bioavailable DON (ABDON) is a fraction of DON that is directly or indirectly available as a nitrogen source for aquatic plant species [26]. Mostly, a certain portion of BDON and ABDON overlaps. There are many sources of BDON and ABDON in the receiving water systems with wastewater-derived BDON as the main one. Since excessive amounts of algal growth have negative impact on aquatic ecosystems, BDON and ABDON in the aquatic system need to be minimized. Therefore, BDON and ABDON removal should be performed in the treatment plant before discharging effluent to the aquatic ecosystem (Simsek et al. 2013; Urgun-Demirtas et al. 2008).

ABDON indicates the potential effect of wastewater effluent DON on the quality of receiving waters, because available DON promotes algal growth in the aquatic system. The Printz Algal Assay Bottle Test, a US Environmental Protection Agency (EPA) method, was adapted by Urgun-Demirtas et al. (2008) to determine ABDON by using a commercially available algae inoculum. They were successful in determining ABDON exertion in low total dissolved nitrogen (TDN) effluent samples and concluded that ABDON was bioavailable to algae in the presence of bacteria. However, they suggested that their method needs to be applied to different treatment effluents since the nature and characteristics of DON could be different.

Some studies determined ABDON in the final effluent only using algae, bacteria, and algae + bacteria seeds to address the behavior of algae in receiving waters. Samples were collected from various types of full-scale WWTPs or pilot-scale wastewater treatment processes including activated sludge (AS) system followed by separate nitrification and denitrification units, fully nitrifying membrane bioreactor pilot plant, and a 4-stage Bardenpho nitrogen removal plants (Pehlivanoglu and Sedlak 2004; Sattayatewa et al. 2009; Urgun-Demirtas et al. 2008). Other studies collected samples from various parts of the WWTPs to address the variation of ABDON along with the treatment train and in the final effluent. For instance, Simsek et al. (2013) conducted a comprehensive data collection study to investigate fate of ABDON in various stages of two different WWTPs, which were AS and trickling filter (TF) plants. Algal (*S. capricornutum*) and bacterial (mixed culture) inoculum were used to determine ABDON. It was found that from influent to effluent, 63 and 56% of ABDON were removed in TF and AS WWTPs, respectively. Similarly, 71 and 47% of effluent DON was ABDON in TF and AS plants, respectively. These results showed that high ABDON values were obtained in TF process compare to AS system. It was highlighted that ABDON values were high when algae and bacteria were used as co-inoculum in the samples.

4.2 *Carbon for Algal Growth During Wastewater Treatment*

From the above discussion, it is clear that algae have the potential for treating wastewater by utilizing the nutrients from the wastewater during their growth. This aspect has been widely studied by the researchers for the past 50 years. Besides, the carbon requirement for the cultivation of algae is high and usually costs up to 60% of the total nutrients cost. Carbon compounds in domestic wastewaters are determined as COD, BOD_5 (5-day BOD), or TOC (total organic carbon). However, the common sources of carbon for algae include (i) atmospheric CO_2; (ii) CO_2 from industrial exhaust gases (e.g., flue gas and flaring gas); and (iii) chemically fixed CO_2 in form of soluble carbonates. Majority of work reported earlier involves the supply of CO_2 gas in order to meet the carbon requirement of the algal culture in media or during algae-mediated wastewater treatment. Interestingly, some algal strains show heterotrophic/mixotrophic growth on organic carbon present in wastewater (Bhatnagar et al. 2011). However, majority of algal strain required inorganic carbon and usually have low biomass productivity if it is not supplied. Therefore, ensuring sufficient availability of inorganic carbon is a major challenge for algal cultivation. Systems that rely on conventional carbon supply strategies are carbon limited due to (i) low transfer efficiency and (ii) relatively low CO_2 content of the atmospheric air (Bai 2015). However, this problem may be solved by employing algal-bacterial interaction biotechnology.

5 Algal-Bacterial Interaction: General Concept

Microbial interactions in ecosystem play a crucial role in their vital life process, i.e., growth, survival, etc. Algae and bacteria naturally bound to each other so well that they influence each other in evolution. Algal-bacterial consortium shows its presence in different ecosystem and gives substantial contribution in global carbon cycle and climate. It is hard to separate native bacteria from an algal culture in order to get an axenic culture. Algal and bacterial interactions have been studied exclusively in near past. They represent nearly all modes of interaction, ranging from mutualism to parasitism. They switch the interaction based on availability of nutrients and environmental/culturing conditions (Amin et al. 2015; Ramanan et al. 2016). Schematic of the algae bacteria interaction in the wastewater is shown in Fig. 1.

Alga shows four modes of nutrition: phototrophic (sunlight as energy and CO_2 as carbon source), heterotrophic (organic carbon as the energy and carbon source), photo-heterotrophic (light as energy and organic carbon as carbon source), and mixotrophic (growth under heterotrophic and/or phototrophic mode). The mode of nutrition depends on availability of nutrients in the wastewater. Bacteria show the heterotrophic mode of nutrition in which it takes up the organic nutrients and dissolved oxygen (DO) for their growth. In positive interactions bacteria as well as algae produce certain substrates which is mutual beneficial for them. Algae produce dissolved organic matter (DOM), and oxygen which is taken up by bacteria. DOM includes dissolved organic carbon (DOC), DON, and dissolved organic phosphorous (DOP). Inorganic carbon, growth promoters, hormones, vitamins, and EPS secreted by bacteria are utilized by algae. In negative interactions, each partners produces certain metabolites which suppress/inhibit the growth of the other partner.

6 Symbiotic Interaction of Algae and Bacteria in Wastewater Treatment

A symbiotic relationship between photoautotrophic algae and heterotrophic bacteria concerning the exchange of O_2 and CO_2 is well known for decades (Fig. 2) (Gutzeit et al. 2005; Oswald 1988). It is not reasonable to use pure culture algae on wastewater treatment. Hence, the feasibility and applicability of algal-bacterial symbiosis has been investigated for various types of wastewaters. The interaction between wastewater associated microbes and algae may either enhance or resist the occurrence of pathogens. These microbes directly or indirectly interact with algae via different mechanisms such as commensalism, mutualism, parasitism, or antagonism.

Algae facilitate bacterial growth by providing either organic compound during cell growth or nutrients during cell decay. Decomposition of algae contributes to generate dissolved organic substances that act as a major component for bacterial

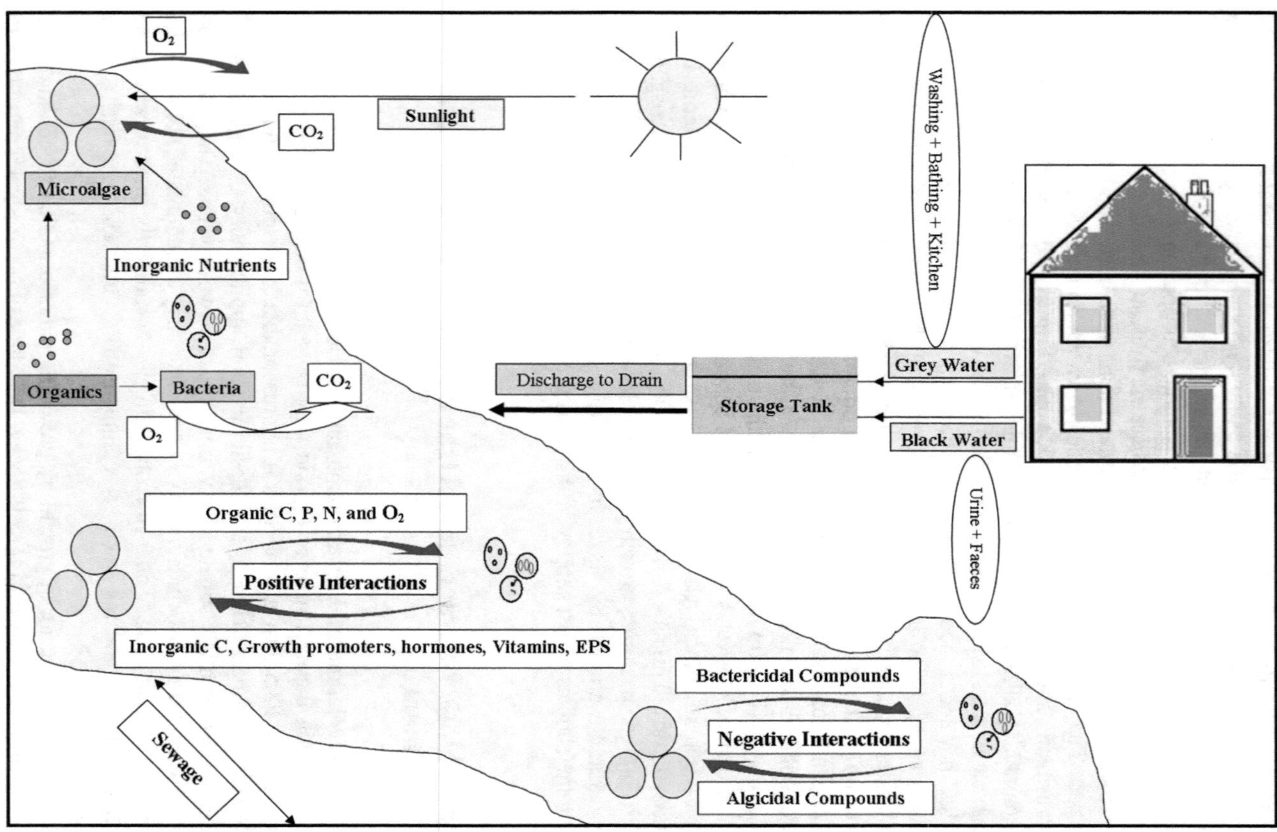

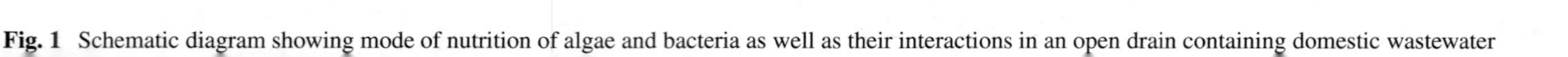
Fig. 1 Schematic diagram showing mode of nutrition of algae and bacteria as well as their interactions in an open drain containing domestic wastewater

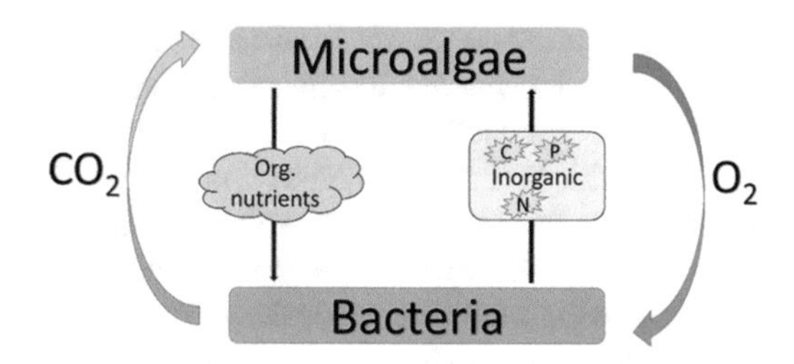

Fig. 2 Schematic diagram of algal-bacterial interaction in biological wastewater treatment process

growth. Algae produce oxygen during photosynthesis. These oxygen molecules are taken up by heterotrophic bacteria to break down organic substances in wastewater. CO_2 released by bacteria is taken up by algae for photosynthesis (Fig. 2) (Choudhary et al. 2015; Rawat et al. 2011). Algae also provides a protective habitat for bacteria against harsh environmental conditions (Byappanahalli et al. 2009).

Bacteria cells are responsible for oxidation of organic matters to readily bioavailable inorganic species including nitrite, nitrate, ammonium, C, and P. DO concentration in the system increases during the nutrients and C assimilation by algae through photosynthesis while using CO_2 (Acién et al. 2016; Muñoz and Guieysse 2006; Su et al. 2012a). DO in the algal process helps to reduce energy requirement that used to supply DO from other sources. Produced DO is used by heterotrophic bacteria during the nitrification process. Results showed that oxygen production rate by algae (during the day time) were higher than its consumption by bacteria (Acién et al. 2016).

As a result of symbiotic relationship, both partners in the consortia get benefit. This results in high growth and also enhances the nutrient removal from the wastewater. Recently, Ashok et al. (2014) studied removal of N, P, and COD from synthetic wastewater using photobioreactors seeded with mixed culture of algae and bacteria. The initial N and P concentrations were 110 and 25 mg/L, respectively, and COD concentration was 45 mg/L before the treatment. It was observed that more than 90% N and P and 80% of COD were removed at a hydraulic retention time (HRT) of 2 days. Conversely, some of the studies showed that around 8–12 days HRT is required for proper growth and 90–100% N and P removal from wastewater when pure culture of algae only are used. This clearly shows the potential beneficial role of bacteria in algal-mediated wastewater treatment. In line with this, Bai et al. (2015) have attempted to qualify the contribution of bacteria to the algal growth through supply of inorganic carbon during cultivation in lab-scale PBR.

6.1 Using Algal-Based Biotechnology in Waste Stabilization Pond System

Algal-based wastewater treatment using pond system is proved to be simple, efficient, and cost-effective especially in rural areas with year-round solar irradiation over the past few decades (Gutzeit et al. 2005). Conventional stabilization pond systems required large amount of area to support algal-bacterial growth. Evaporation of water during the hot weather could be a negative impact of the pond systems since algae and bacteria need certain amount of water along with nutrient, oxygen, and sunlight. When flocculation and coagulation processes of algal and bacterial biomass are not successfully performed, separating the algal biomass in the pond becomes difficult. Some algal species may settle down to the bottom of the pond due to gravity, while some species may not settle. Hence, technical algal-bacterial separation methods should be investigated to increase the flocculation and the following coagulation to overcome the settlement problem.

In some aspects, algal-bacterial stabilization pond treatment system is superior to bacterial-only stabilization pond system. Algal-bacterial system has the following advantages: (i) about 25–30% of reduction on the pond area can be obtained, (ii) mechanical aeration is not required, (iii) wastewater treatment will be held in aerobic condition (no need to create anaerobic conditions), (iv) nutrient removal will be higher, (v) nutrients are assimilated by algae to support their growth during the treatment, and (vi) it produces valuable biomass with high nutrient content (Gutzeit et al. 2005).

6.2 Using Algal-Based Biotechnology in Conventional or Advanced WWTP

Algal-based biotechnology can be integrated with several different parts of a WWTP from influent (raw wastewater) to effluent (final discharge). Raw wastewater in WWTP mainly contains high ammonium, P, and organic pollutants (COD and BOD). Primary clarifier wastewaters contain about 30 and 60% less BOD and TSS, compare to raw wastewater, respectively. Secondary clarifier wastewaters contain very less organic contaminant. Anaerobic digestion effluent contains high amount (concentrated) of pollutant, and final effluent mainly contains nitrate and less amount of organic substances. Depending on the location in a WWTP, algal treatment technology can be adapted. It was found that primary clarifier effluent wastewater sources are very suitable to grow algae (Acién et al. 2016). The pros and cons of conventional, advanced, and algal-based wastewater treatment processes are presented in Table 2.

7 Process Conditions for Algal-Bacterial-Based Wastewater

In laboratory conditions, the algal-bacterial consortium has proved to be suitable for removal of low concentration of organic matters and high concentration of nutrient compounds in wastewater. However, the environmental conditions such as temperature, diurnal cycles, sunlight intensity, and types of nutrients present in the

Table 2 Pros and cons of conventional, advanced, and algal-based wastewater treatment technologies

Pros	Cons
Conventional wastewater treatment technologies: completely mixed or extended aeration activated sludge	
Simple operational and control systems	Excess energy consumption for aeration
Less sludge production	High capital cost
Less chemical usage	Cannot achieve denitrification process
Can achieve low N, P, and C level	Tend to grow filament bacteria
Resistant to varies types of wastewaters	Sludge bulking problems
Resistant against abrupt and toxic loads	Tend to cause low F/M
High-quality effluent water	Big tank volume for aeration
Easy construction and operation	
Advanced wastewater treatment technologies: A^2O or Bardenpho processes	
Simple operational and control systems	High energy cost for oxygen supply
Can achieve low N, P, and C level	The return sludge contains NO_3 and it affects P removal
Effluent P is under 2 mg/l	Required qualified operators
High sludge settling capability	Low-efficiency elimination of P
Easy operation	Big tank volume for aeration
Energy saving (if there is digester)	Required quality operator
	Excess energy consumption for aeration
Algal-based wastewater treatment technologies	
Simple operational and control systems	Energy cost when mixing is needed
Natural systems (inexpensive to facilitate)	Excess sludge accumulation
High sludge settling capability	Cost of dewatering the sludge
Can achieve low N, P, and C level	High capital cost
Produce valuable biomass	Harvesting the algal blooms

wastewater significantly impact the performance of algal-bacterial systems in real world. Studies have shown that various factors need to be considered to assess the efficiency of algal-bacterial interaction in treatment of wastewaters (Olguín 2012; Rawat et al. 2011). The factors affecting the symbiotic efficiency are pH, DO, dissolved CO_2, illumination, and dark/light cycle. The pH of algae and bacteria cultures is important in the removal of nutrients from wastewater. Acidic pH was reported to decrease the N and P removal efficiency. However, higher pH conditions resulted in reduced bacterial population in the algal-bacterial biofilm system. This produced negative impact on N removal (Schumacher et al. 2003). Neutral pH condition in algal-bacterial combined systems has achieved higher removal of nutrients such as ammonia and P from wastewaters (Liang et al. 2015). Adequate levels of DO and dissolved CO_2 are crucial for proper growth of bacteria and algae, respectively, in wastewater treatment systems. However, in case of algae-bacteria symbiosis, external supply of CO_2 and O_2 is not required as these are produced and consumed within the system itself.

Most of the algae (particularly microalgae) and cyanobacteria use light as only energy source, and, therefore, illumination is one of the critical requirements for algal cultivation. The part of the radiation in photosynthetically active radiation (PAR), i.e., λ = 400–700 nm, is used by algae as energy source (Prajapati et al. 2013b). The processes involving algal-bacterial consortium also need light conditions for proper growth of algae. However, high light intensity may reduce the growth of algae by inhibiting photosynthesis. Conversely, if the light intensities are low, it will limit the algal growth by reducing the rate photosynthesis. The optimal light intensity for high growth rate is species dependent and usually lies in the range of 62.5–2000 μ mol m^{-2} s^{-1} (Wang et al. 2014). Furthermore, the dark/light cycle also affect algal growth. Several studies have tested different dark/light cycle in the range of 8/16–16/8. However, dark/light period of 8/16 is considered optimal for growth of the algae (Wang et al. 2014). Hence, to keep the algal partner in the consortium live and metabolically active, the process conditions are required to be maintained during cultivation in media or wastewater.

8 Growth System Options for Algal-Bacterial Co-Culture

Algal productivity depends on two properties: intrinsic and extrinsic. Intrinsic includes algae strain and its nutrient requirements and extrinsic deals with the design of the cultivation system. The growth system, which is used for algal culture, can also be useful for algal-bacterial consortia. Algal cultivation system can be open pond, close pond and in a controlled growth systems known as photobioreactor (PBR). Pictorial view or schematic of some of growth systems, the recently tested for range of algae cultivation, are shown in Fig. 3. Among the conventional systems, open pond is easy to construct and has low operating and production cost. It is preferred for the production of algal consortia. For wastewater treatment, open pond system such as high-rate algal pond (HRAP) is preferred as closed PBR may not be suitable for handing large volume of wastewater. Furthermore during wastewater treatment, algal-bacterial system works more efficiently as compared with axenic algal culture. There often, pond provides proper conditions to support both algal and bacterial growth during wastewater treatment. However, open ponds do not provide any control over temperature, light intensity, and strain purity and thus have low biomass productivity. To overcome the major hurdles researchers develop close pond system, in which pond is enclosed in a green house. The temperature and level of carbon dioxide can be managed in this module, which lead to higher production of biomass, but the construction and operational cost become higher than open pond. Conversely, a PBR is an enclosed illuminating vessel used for optimal productivity. The parameters, i.e., temperature, pH, carbon dioxide concentration, light intensity and its duration, etc., can easily be controlled by the PBR in real time. It provides aseptic growth condition, favors higher cell concentration, allows axenic cultivation, and prevents water loss through evaporation (Singh and Sharma 2012). Consequently, PBR is preferred when pure algal biomass is to be produced for

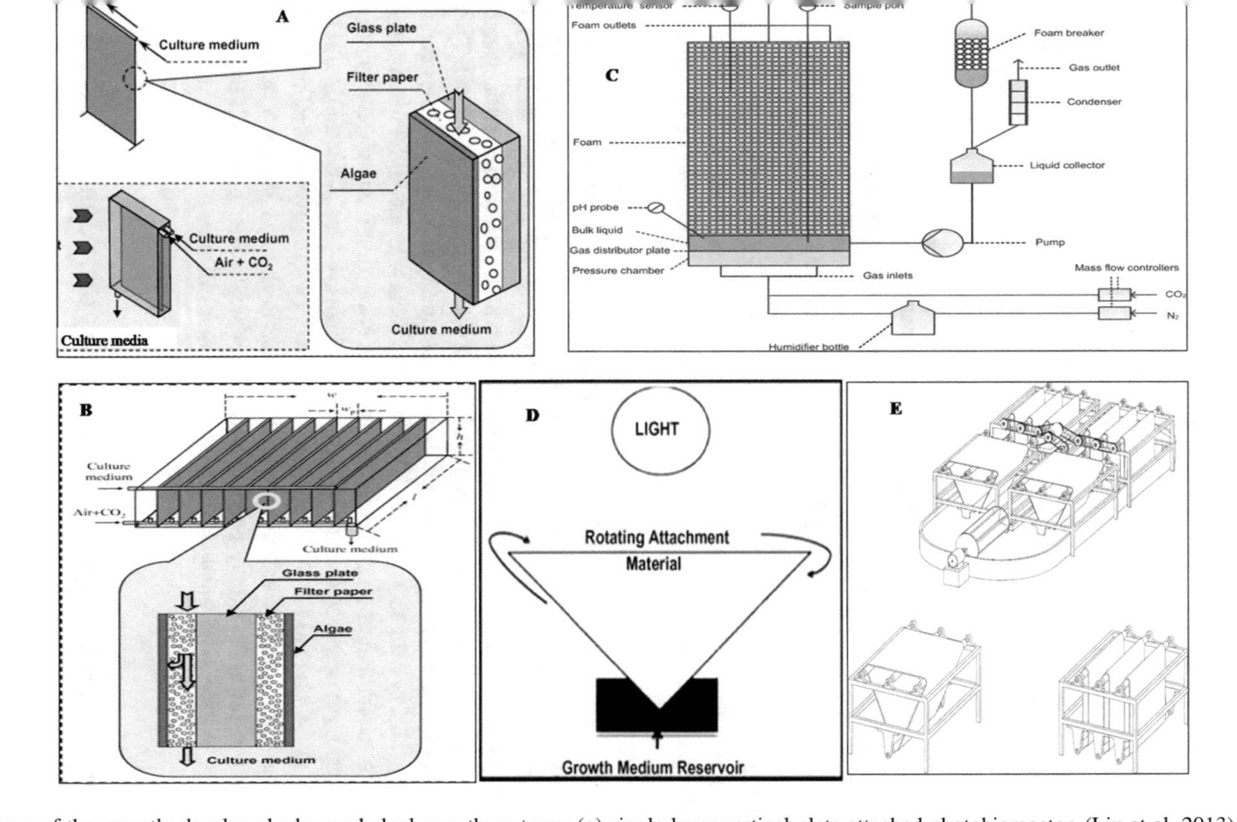

Fig. 3 Some of the recently developed advanced algal growth systems: (**a**) single layer vertical plate attached photobioreactor. (Liu et al. 2013), (**b**) multi-plates attached photobioreactor. (Liu et al. 2013), (**c**) liquid foam-bed photobioreactor. (Janoska et al. 2017), (**d**) laboratory-scale rotating algal biofilm growth, and (**e**) pilot-scale revolving algal biofilm (RAB) cultivation system. (Gross and Wen 2014)

various applications. A number of PBR variants were tested to date like α-type tubular PBR (Lee et al. 1995), cone-shaped helical tubular PBR (Watanabe and Hall 1996), and hollow-fiber photobioreactor (Markov et al. 1995).

Conventional PBR has limitation with mass transfer rates. To improve the rate of mass transfer, recently a liquid foam-bed PBR for algae production was tested. Continuous foam breaking method was established, and a growth rate of 0.1 h^{-1} for *Chlorella sorokiniana* was achieved (Janoska et al. 2017). There are few studies on wastewater treatment in PBR, but most of them are limited to lab scale only. All the available PBR designs can be tested and optimized for algal-bacterial consortium-based wastewater treatment. However, serious systematic efforts are to be made to reduce the capital and operational cost of these systems to be used in wastewater treatment.

Biofilm-based attached growth system is also being tested in many studies. The main advantage of this type of system is easy biomass recovery. In these modules algae are used to grow on solid support. Interestingly, bacteria are good in biofilm formation, and hence the attached system may be more effective in the growth of algal-bacterial consortia. There have been systematic studies in this direction and continuous research and development in process. In these recent studies, cotton-based duct canvas and ropes (Gross et al. 2013), nonwoven fabric (Choudhary et al. 2017), and artificial supporting material (Liu et al. 2013) have been used as solid support for algal growth. In a pilot-scale study conducted on revolving algal biofilm (RAB), cultivation system fabricated with cotton-based duct canvas and ropes showed 300% increase in the biomass productivity compared to raceway ponds (Gross and Wen 2014). In a similar study, nonwoven fabric was tested for the algal biofilm reactor. It was successfully tested for domestic gray water treatment. Maximum productivity of 4 g m^{-2} d^{-1} was achieved in neat livestock wastewater (Choudhary et al. 2017). Liu et al. (2013) develop a noble attached growth system in which algae grew as a thin film on a vertically artificial supporting material. This design dilutes the solar irradiation and helps in acquiring the high photosynthetic efficiency. This system showed 400–700% higher biomass productivity when compared to conventional open ponds. Many attempts have been made throughout the world to make the algal cultivation system work at industrial scale, but this is still in evolving stage. Overall for WW, open pond is always useful over controlled PBR due to various techno-economic issues. However, in the specific case of algal-bacterial consortia, ABR may become suitable for wastewater treatment as these systems have several advantages over open ponds and PBRs.

9 Advantages of Algal-Bacterial Process over Conventional Methods

Conventional methods have requirement of high energy and capital cost for proper functioning. Activated sludge system needs continuous air or oxygen purging for the optimal performance. Sludge is a concentrate of harmful substances and pathogenic microbes which can eventually give rise to potential health risks. To get rid of the produced sludge, it is transported and dumped to fill low area. Conversely, in phycoremediation, algae are unable to degrade organic compounds present in wastewater and hence require supply of inorganic carbon as CO_2 for proper growth. This makes algal wastewater treatment unfeasible as large amount of energy is consumed during CO_2 purging in the wastewater treatment systems. On the other hand, algal-bacterial system has no energy requirement, produces no sludge, and has ability of self-sustainability. The algal-bacterial synergic relation has potential to tackle all these hurdles together. As both partners share O_2/CO_2 and other nutrients between them (Ramanan et al. 2016), the requirement of energy for O_2/CO_2 purging can be removed from the systems. Furthermore, algae-mediated processes produce biochemical-rich biomass (Kumar et al. 2017; Nicol et al. 2014). The biomass can be utilized for the synthesis of number of products including high-value biofuel precursors (Sharma et al. 2011; Williams and Laurens 2010; Kumar et al. 2019). In fact, integration of algae-mediated wastewater treatment with biofuel production is considered most feasible option at large scale (Choudhary et al. 2015; Prajapati et al. 2013b; Zhang et al. 2013).

10 Conclusion and Recommendations

Water is a limited resource on planet earth. It is one of the major commodities used exhaustively in anthropogenic activities. As the human population is rising at higher pace, the quality of water is being compromised. Ever increasing demand of water leads to water stress conditions in number of countries. This event compels to develop efficient wastewater treatment technologies throughout the world.

The domestic wastewater is being treated by physical, chemical, and biological methods. Among these, biological treatment is most promising because of many reasons, i.e., it has no additional nutrient requirements, shows low energy consumption and capital and operational cost, and leads to sustenance. In this regard studies on microalga-driven technologies show a viable option for wastewater treatment as well as energy generation in terms of biogas, biodiesel, etc.

In natural systems, alga is always associated with bacterial strains and shows number of interacting mode. It is well reported that they can exist in as a cooperative or competitive mode depending on the surrounding conditions. Mainly nutrient availability and metabolites govern the surrounding conditions. Advance PBRs are being deployed to work on wastewater treatment with the help of algal-bacterial strains. Many researchers claim deigns which support industrial scale-up feasibility. However, the cost associated is relatively high. In order to make this technology economically suitable, algal-bacterial interactions are strongly recommended.

Acknowledgment Funding for this research was provided by (i) Science and Engineering Research Board (SERB), Government of India, through the research grant received under Early Career Research Award Scheme (Ref. ECR/2015/000065) and (ii) North Dakota Agricultural Experiment Station (NDAES). Any opinions, findings, and conclusions or recommendations expressed in this material are those of the author(s) and do not necessarily reflect the views of SERB and NDAES.

References

Acién FG, Gómez-Serrano C, Morales-Amaral MM, Fernández-Sevilla JM, Molina-Grima E (2016) Wastewater treatment using microalgae: how realistic a contribution might it be to significant urban wastewater treatment? Appl Microbiol Biotechnol 100:9013–9022. https://doi.org/10.1007/s00253-016-7835-7

Agustina TE, Ang HM, Vareek VK (2005) A review of synergistic effect of photocatalysis and ozonation on wastewater treatment. J Photochem Photobiol C: Photochem Rev 6:264–273. https://doi.org/10.1016/J.JPHOTOCHEMREV.2005.12.003

Aksu Z (2005) Application of biosorption for the removal of organic pollutants: a review. Process Biochem 40:997–1026. https://doi.org/10.1016/J.PROCBIO.2004.04.008

Amin SA, Hmelo LR, Van Tol HM, Durham BP, Carlson LT, Heal KR, Morales RL, Berthiaume CT, Parker MS, Djunaedi B, Ingalls AE, Parsek MR, Moran MA, Armbrust EV (2015) Interaction and signalling between a cosmopolitan phytoplankton and associated bacteria. Nature 522:98–101. https://doi.org/10.1038/nature14488

Ashok V, Shriwastav A, Bose P (2014) Nutrient Removal Using Algal-Bacterial Mixed Culture. Appl Biochem Biotechnol 174:2827–2838. doi:10.1007/s12010-014-1229-z.

Aslan S, Kapdan IK (2006) Batch kinetics of nitrogen and phosphorus removal from synthetic wastewater by algae. Ecol Eng 28:64–70. https://doi.org/10.1016/J.ECOLENG.2006.04.003

Bai X (2015) Enhancing algal biomass and biofuels recovery from open culture systems. PhD Thesis. School of Chemical Engineering, The University of Queensland, Australia

Bai X, Lant P, Pratt S (2015) The contribution of bacteria to algal growth by carbon cycling. Biotechnol Bioeng 112:688–695. https://doi.org/10.1002/bit.25475

Becker E (1994) Microalgae: biotechnology and microbiology. Cambridge University Press, Cambridge

Bhatnagar A, Chinnasamy S, Singh M, Das KCC (2011) Renewable biomass production by mixotrophic algae in the presence of various carbon sources and wastewaters. Appl Energy 88:3425–3431. https://doi.org/10.1016/j.apenergy.2010.12.064

Byappanahalli MN, Sawdey R, Ishii S, Shively DA, Ferguson JA, Whitman RL, Sadowsky MJ (2009) Seasonal stability of Cladophora-associated Salmonella in Lake Michigan watersheds. Water Res 43:806–814. https://doi.org/10.1016/J.WATRES.2008.11.012

Castro-Barros CM, Jia M, van Loosdrecht MCM, Volcke EIP, Winkler MKH (2017) Evaluating the potential for dissimilatory nitrate reduction by anammox bacteria for municipal wastewater treatment. Bioresour Technol 233:363–372. https://doi.org/10.1016/J.BIORTECH.2017.02.063

Choudhary P, Bhattacharya A, Prajapati SK, Kaushik P, Malik A (2015) Phycoremediation-coupled biomethanation of microalgal biomass. In: KIM PS-K (ed) Handbook of marine microalgae: biotechnology advances. Elsevier, pp 483–499. https://doi.org/10.1016/B978-0-12-800776-1.00032-7

Choudhary P, Prajapati SK, Kumar P, Malik A, Pant KK (2017) Development and performance evaluation of an algal biofilm reactor for treatment of multiple wastewaters and characterization of biomass for diverse applications. Bioresour Technol 224:276–284. https://doi.org/10.1016/j.biortech.2016.10.078

De Gisi S, Lofrano G, Grassi M, Notarnicola M (2016) Characteristics and adsorption capacities of low-cost sorbents for wastewater treatment: a review. Sustain Mater Technol 9:10–40. https://doi.org/10.1016/J.SUSMAT.2016.06.002

de Graaff MS, Temmink H, Zeeman G, Buisman CJN (2010) Anaerobic treatment of concentrated black water in a UASB reactor at a short HRT. Water 2:101–119. https://doi.org/10.3390/w2010101

De-Bashan LE, Hernandez JP, Morey T, Bashan Y (2004) Microalgae growth-promoting bacteria as "helpers" for microalgae: a novel approach for removing ammonium and phosphorus from municipal wastewater. Water Res 38:466–474. https://doi.org/10.1016/j.watres.2003.09.022

Gogate PR, Pandit AB (2004) A review of imperative technologies for wastewater treatment I: oxidation technologies at ambient conditions. Adv Environ Res 8:501–551. https://doi.org/10.1016/S1093-0191(03)00032-7

González-Fernández C, Molinuevo-Salces B, García-González MC (2011) Nitrogen transformations under different conditions in open ponds by means of microalgae–bacteria consortium treating pig slurry. Bioresour Technol 102:960–966. https://doi.org/10.1016/J.BIORTECH.2010.09.052

Gross M, Wen Z (2014) Yearlong evaluation of performance and durability of a pilot-scale Revolving Algal Biofilm (RAB) cultivation system. Bioresour Technol 171:50–58. https://doi.org/10.1016/J.BIORTECH.2014.08.052

Gross M, Henry W, Michael C, Wen Z (2013) Development of a rotating algal biofilm growth system for attached microalgae growth with in situ biomass harvest. Bioresour Technol 150:195–201. https://doi.org/10.1016/j.biortech.2013.10.016

Gutzeit G, Lorch D, Weber A, Engels M, Neis U (2005) Bioflocculent algal-bacterial biomass improves low-cost wastewater treatment. Water Sci Technol 52:9–18

Hammed AM, Prajapati SK, Simsek S, Simsek H (2016) Growth regime and environmental remediation of microalgae. Algae 31:189–204. https://doi.org/10.4490/algae.2016.31.8.28

Hollender J, Zimmermann SG, Koepke S, Krauss M, McArdell CS, Ort C, Singer H, von Gunten U, Siegrist H (2009) Elimination of organic micropollutants in a municipal wastewater treatment plant upgraded with a full-scale post-ozonation followed by sand filtration. Environ Sci Technol 43:7862–7869. https://doi.org/10.1021/es9014629

Janoska A, Lamers PP, Hamhuis A, van Eimeren Y, Wijffels RH, Janssen M (2017) A liquid foam-bed photobioreactor for microalgae production. Chem Eng J 313:1206–1214. https://doi.org/10.1016/J.CEJ.2016.11.022

Ji M-KK, Abou-Shanab RAII, Kim S-HH, Salama E-SS, Lee S-HH, Kabra AN, Lee Y-SS, Hong S, Jeon B-HH (2013a) Cultivation of microalgae species in tertiary municipal wastewater supplemented with CO2 for nutrient removal and biomass production. Ecol Eng 58:142–148. https://doi.org/10.1016/j.ecoleng.2013.06.020

Ji M-K, Kim H-C, Sapireddy VR, Yun H-S, Abou-Shanab RAI, Choi J, Lee W, Timmes TC, Inamuddin, Jeon B-H (2013b) Simultaneous nutrient removal and lipid production from pretreated piggery wastewater by *Chlorella vulgaris* YSW-04. Appl Microbiol Biotechnol 97:2701–2710. https://doi.org/10.1007/s00253-012-4097-x

Jiménez-Pérez M, Sánchez-Castillo P, Romera O, Fernández-Moreno D, Pérez-Martınez C (2004) Growth and nutrient removal in free and immobilized planktonic green algae isolated from pig manure. Enzym Microb Technol 34:392–398. https://doi.org/10.1016/J.ENZMICTEC.2003.07.010

Kong QX, Li L, Martinez B, Chen P, Ruan R (2010) Culture of microalgae *chlamydomonas reinhardtii* in wastewater for biomass feedstock production. Appl Biochem Biotechnol 160:9–18. https://doi.org/10.1007/s12010-009-8670-4

Kumar P, Prajapati SK, Malik A, Vijay VK (2017) Cultivation of native algal consortium in semi-continuous pilot scale raceway pond for greywater treatment coupled with potential methane production. J Environ Chem Eng 5:5581. https://doi.org/10.1016/j.jece.2017.10.044

Kumar P, Prajapati SK, Malik A, Vijay VK (2019) Evaluation of biomethane potential of waste algal biomass collected from eutrophied lake: effect of source of inocula, co-substrate, and VS loading. J Appl Phycol. 31: 533. https://doi.org/10.1007/s10811-018-1585-0

Lee Y-K, Ding S-Y, Low C-S, Chang Y-C, Forday WL, Chew P-C (1995) Design and performance of an α-type tubular photobioreactor for mass cultivation of microalgae. J Appl Phycol 7:47–51. https://doi.org/10.1007/BF00003549

Leung WC, Wong M-F, Chua H, Lo W, Yu PHF, Leung CK (2000) Removal and recovery of heavy metals by bacteria isolated from activated sludge treating industrial effluents and municipal wastewater. Water Sci Technol 41:233–240

Liang Y, Sarkany N, Cui Y (2009) Biomass and lipid productivities of *Chlorella vulgaris* under autotrophic, heterotrophic and mixotrophic growth conditions. Biotechnol Lett 31:1043–1049. https://doi.org/10.1007/s10529-009-9975-7

Liang Z, Liu Y, Ge F, Liu N, Wong M (2015) A pH-dependent enhancement effect of co-cultured *Bacillus licheniformis* on nutrient removal by *Chlorella vulgaris*. Ecol Eng 75:258–263

Liu T, Wang J, Hu Q, Cheng P, Ji B, Liu J, Chen Y, Zhang W, Chen X, Chen L, Gao L, Ji C, Wang H (2013) Attached cultivation technology of microalgae for efficient biomass feedstock production. Bioresour Technol 127:216–222. https://doi.org/10.1016/J.BIORTECH.2012.09.100

Luo Y, Guo W, Ngo HH, Nghiem LD, Hai FI, Zhang J, Liang S, Wang XC (2014) A review on the occurrence of micropollutants in the aquatic environment and their fate and removal during wastewater treatment. Sci Total Environ 473–474:619–641. https://doi.org/10.1016/J.SCITOTENV.2013.12.065

Markov SA, Bazin MJ, Hall DO (1995) Hydrogen photoproduction and carbon dioxide uptake by immobilized *Anabaena variabilis* in a hollow-fiber photobioreactor. Enzym Microb Technol 17:306–310. https://doi.org/10.1016/0141-0229(94)00010-7

Muñoz R, Guieysse B (2006) Algal-bacterial processes for the treatment of hazardous contaminants: a review. Water Res 40:2799–2815. https://doi.org/10.1016/j.watres.2006.06.011

Nicol RW, Lamers A, Lubitz WD, McGinn PJ (2014) Evaluation of algal biomass and biodiesel Co-products for bioenergy applications. J Biobaased Mater Bioenergy 8:429–436. https://doi.org/10.1166/jbmb.2014.1454

Olguín EJ (2012) Dual purpose microalgae-bacteria-based systems that treat wastewater and produce biodiesel and chemical products within a Biorefinery. Biotechnol Adv 30:1031–1046. https://doi.org/10.1016/j.biotechadv.2012.05.001

Oswald WJ (1988) Large-scale algal culture systems (engineering aspects), Micro-alga biotechnology. Cambridge University Press, Cambridge

Park JBK, Craggs RJ, Shilton AN (2011) Wastewater treatment high rate algal ponds for biofuel production. Bioresour Technol 102:35–42. https://doi.org/10.1016/j.biortech.2010.06.158

Pehlivanoglu E, Sedlak DL (2004) Bioavailability of wastewater-derived organic nitrogen to the alga *Selenastrum Capricornutum*. Water Res 38:3189–3196. https://doi.org/10.1016/J.WATRES.2004.04.027

Pehlivanoglu-Mantas E, Sedlak DL (2006) Wastewater-derived dissolved organic nitrogen: analytical methods, characterization, and effects—a review. Crit Rev Environ Sci Technol 36:261–285. https://doi.org/10.1080/10643380500542780

Prajapati SK, Kaushik P, Malik A, Vijay VK (2013a) Phycoremediation and biogas potential of native algal isolates from soil and wastewater. Bioresour Technol 135:232–238. https://doi.org/10.1016/j.biortech.2012.08.069

Prajapati SK, Kaushik P, Malik A, Vijay VK (2013b) Phycoremediation coupled production of algal biomass, harvesting and anaerobic digestion: possibilities and challenges. Biotechnol Adv 31:1408–1425. https://doi.org/10.1016/j.biotechadv.2013.06.005

Ramanan R, Kim BH, Cho DH, Oh HM, Kim HS (2016) Algae-bacteria interactions: evolution, ecology and emerging applications. Biotechnol Adv 34:14–29. https://doi.org/10.1016/j.biotechadv.2015.12.003

Rawat I, Ranjith Kumar R, Mutanda T, Bux F (2011) Dual role of microalgae: Phycoremediation of domestic wastewater and biomass production for sustainable biofuels production. Appl Energy 88:3411–3424. https://doi.org/10.1016/j.apenergy.2010.11.025

Saha A, Bhushan S, Mukherjee P, Chanda C, Bhaumik M, Ghosh M, Sharmin J, Datta P, Banerjee S, Barat P, Thakur AR, Gantayet LM, Mukherjee I, Ray Chaudhuri S (2018) Simultaneous sequestration of nitrate and phosphate from wastewater using a tailor-made bacterial consortium in biofilm bioreactor. J Chem Technol Biotechnol 93:1279–1289. https://doi.org/10.1002/jctb.5487

Sattayatewa C, Pagilla K, Pitt P, Selock K, Bruton T (2009) Organic nitrogen transformations in a 4-stage Bardenpho nitrogen removal plant and bioavailability/biodegradability of effluent DON. Water Res 43:4507–4516. https://doi.org/10.1016/J.WATRES.2009.07.030

Schumacher G, Blume T, Sekoulov I (2003) Bacteria reduction and nutrient removal in small wastewater treatment plants by an algal biofilm. Water Sci Technol 47:195

Sharma YC, Singh B, Korstad J (2011) A critical review on recent methods used for economically viable and eco-friendly development of microalgae as a potential feedstock for synthesis of biodiesel. Green Chem 13:2993–3006. https://doi.org/10.1039/c1gc15535k

Shen Z, Zhou Y, Liu J, Xiao Y, Cao R, Wu F (2015) Enhanced removal of nitrate using starch/PCL blends as solid carbon source in a constructed wetland. Bioresour Technol 175:239–244. https://doi.org/10.1016/J.BIORTECH.2014.10.006
Shen Y, Gao J, Li L (2017) Municipal wastewater treatment via co-immobilized microalgal-bacterial symbiosis: microorganism growth and nutrients removal. Bioresour Technol 243:905–913. https://doi.org/10.1016/J.BIORTECH.2017.07.041
Shi X-M, Zhang X-W, Chen F (2000) Heterotrophic production of biomass and lutein by *Chlorella protothecoides* on various nitrogen sources. Enzym Microb Technol 27:312–318. https://doi.org/10.1016/S0141-0229(00)00208-8
Shukla A, Zhang Y-H, Dubey P, Margrave J, Shukla SS (2002) The role of sawdust in the removal of unwanted materials from water. J Hazard Mater 95:137–152. https://doi.org/10.1016/S0304-3894(02)00089-4
Simsek H, Kasi M, Ohm J-B, Blonigen M, Khan E (2013) Bioavailable and biodegradable dissolved organic nitrogen in activated sludge and trickling filter wastewater treatment plants. Water Res 47:3201–3210. https://doi.org/10.1016/J.WATRES.2013.03.036
Singh RN, Sharma S (2012) Development of suitable photobioreactor for algae production – a review. Renew Sust Energ Rev 16:2347–2353. https://doi.org/10.1016/J.RSER.2012.01.026
Su Y, Mennerich A, Urban B (2012a) Coupled nutrient removal and biomass production with mixed algal culture: impact of biotic and abiotic factors. Bioresour Technol 118:469–476. https://doi.org/10.1016/j.biortech.2012.05.093
Su Y, Mennerich A, Urban B (2012b) Comparison of nutrient removal capacity and biomass settleability of four high-potential microalgal species. Bioresour Technol 124:157–162. https://doi.org/10.1016/J.BIORTECH.2012.08.037
Tam N, Pollution YW-E (2000) Effect of immobilized microalgal bead concentrations on wastewater nutrient removal. Environ Pollut 107:145
Urgun-Demirtas M, Sattayatewa C, Pagilla KR (2008) Bioavailability of dissolved organic nitrogen in treated effluents. Water Environ Res 80:397–406
Wang L, Min M, Li Y, Chen P, Chen Y, Liu Y, Wang Y, Ruan R (2010) Cultivation of green algae Chlorella sp. in different wastewaters from municipal wastewater treatment plant. Appl Biochem Biotechnol 162:1174–1186. https://doi.org/10.1007/s12010-009-8866-7
Wang S-K, Stiles AR, Guo C, Liu C-Z (2014) Microalgae cultivation in photobioreactors: an overview of light characteristics. Eng Life Sci 14:550–559. https://doi.org/10.1002/elsc.201300170
Ward AJ, Lewis DM, Green FB (2014) Anaerobic digestion of algae biomass: a review. Algal Res 5:204–214. https://doi.org/10.1016/j.algal.2014.02.001
Watanabe Y, Hall DO (1996) Photosynthetic production of the filamentous cyanobacterium Spirulina platensis in a cone-shaped helical tubular photobioreactor. Appl Microbiol Biotechnol 44:693–698. https://doi.org/10.1007/BF00178604
Williams PJLB, Laurens LML (2010) Microalgae as biodiesel & biomass feedstocks: review & analysis of the biochemistry, energetics & economics. Energy Environ Sci 3:554. https://doi.org/10.1039/b924978h
Xu H, Miao X, Wu Q (2006) High quality biodiesel production from a microalga *Chlorella protothecoides* by heterotrophic growth in fermenters. J Biotechnol 126:499–507. https://doi.org/10.1016/j.jbiotec.2006.05.002
Zhang Y, White MA, Colosi LM (2013) Environmental and economic assessment of integrated systems for dairy manure treatment coupled with algae bioenergy production. Bioresour Technol 130:486–494. https://doi.org/10.1016/j.biortech.2012.11.123

Phycoremediation of Textile Wastewater: Possibilities and Constraints

Steffi Jose and S. Archanaa

1 Introduction

The textile industry is one of the oldest industries of the world. The need for clothing and the fascination for fabric and colours gave root to this traditional industry several hundred years ago. Since then, increasing population and improved standard of living have spurred on the growth of the industry. The textile industry, however, uses large volumes of water. In a manual issued by the United States Environmental Protection Agency (USEPA) in 1996, it was estimated that approximately 72 kg water is utilized to generate less than 0.5 kg of a textile product (USEPA 1996). As a result, very large amount of wastewater is generated by the industry. The industry is also chemical intensive and releases high concentrations of dyes, solvents, detergents, acids, alkalis, heavy metals, insecticides, pesticides and organic and inorganic compounds as waste effluent (Archana 2013; Khandare and Govindwar 2015). Most of these effluent components have poor biodegradability and are recalcitrant. They are also extremely toxic to all life forms (Khandare and Govindwar 2015). When released to waterbodies, the wastewater can cause severe changes in water colour, clarity, temperature and pH. The impaired physical properties of the waterbodies alone can severely hamper aquatic life. The toxic pollutants can further cause several diseases in humans and animals.

Treatment of these effluents prior to release is thus imperative. The textile industry generates different kinds of wastes including toxic air emissions, wastewater streams and complex solid waste. This leads to air, water and land pollution (Chavan 2001). This chapter will, however, concentrate on the treatment of wastewater. Several conventional methods are employed to treat textile effluents. These are, however, associated with disadvantages and can often give rise to secondary pollution. Incorporation of biological treatments has improved the characteristics of

S. Jose (✉) · S. Archanaa
Department of Biotechnology, Bhupat and Jyoti Mehta School of Biosciences building, Indian Institute of Technology Madras, Chennai, India

S. K. Gupta, F. Bux (eds.), *Application of Microalgae in Wastewater Treatment*,
https://doi.org/10.1007/978-3-030-13913-1_14

textile wastewater. Phycoremediation, the removal of pollutants using microalgae, macroalgae and cyanobacteria, has recently emerged as an effective mode of biological treatment with several advantages over the conventional methods (de La Noüe et al. 1992; Lim et al. 2010; Pathak et al. 2015). This chapter discusses the potential of phycoremediation in textile wastewater treatments.

2 Textile Effluents: Composition and Hazards

The composition of the textile effluent varies depending on the nature of the product manufactured by the industry. It carries a large number of pollutants such as dyes, detergents, surfactants, heavy metals, inorganic salts, pesticides, insecticides, solvents, oils, etc. (Archana 2013). The three parameters that can best describe the quality of textile wastewater are colour, total suspended solids and chemical oxygen demand (COD) (Nawaz and Ahsan 2014). The effluent constituents that primarily affect these three parameters are briefed below:

Dyes: Dyes, the major source of colour in the effluent, are some of the key contributors to the pollutants in textile effluents. As much as 50% of the dye initially used for the dyeing process may be expelled into the waste stream (Punzi et al. 2012). Release of wastewater containing high concentration of dyes into waterbodies not only leads to aesthetic pollution but also significantly hampers aquatic life (Sarayu and Sandhya 2012). Presence of dark dyes decrease sunlight penetration and retard aquatic photosynthesis, thereby leading to decreased dissolved oxygen (DO) levels. This in turn would negatively affect aquatic life. It also increases the chemical and biological oxygen demand (COD and BOD). Industries often employ synthetic water soluble dyes (Sarayu and Sandhya 2012; Sinha et al. 2016). Azo dyes are the most commonly used and account for 60–70% of all dyes used. Further, these dyes are extremely hazardous to human health as their breakdown products have been found to be carcinogenic and mutagenic (Lim et al. 2010; Sarayu and Sandhya 2012; Sinha et al. 2016).

Other Chemicals: Several chemicals, such as detergents, surfactants, salts, acids, bases, etc., are used for the multiple operations in the industry (Khandare and Govindwar 2015). Majority of these chemicals are persistent and have poor biodegradability (Sarayu and Sandhya 2012). They significantly increase the COD and BOD. Chlorine compounds and hydrogen peroxide used in the bleaching process are found in high concentrations and can cause skin irritation and other diseases. Chromium, zinc and copper are the major metal pollutants that arise from complex dyes and salts (Chavan 2001). Nutrients such as nitrogen from ammonia and urea, phosphorus from certain detergents and buffers and sulphur from sulphur dyes are also found in high concentration (Kumar et al. 2017). Finishing products such as biocides, insecticides, etc. are also found in the effluent (Chavan 2001).

Dissolved and Suspended Solids: Salts that are used in the different operations contribute to the dissolve solid concentration. Fibrous substrates, remnants of printing gum, pulp, starch and cellulose give rise to suspended solids (Kumar et al. 2017). Although compounds like starch and cellulose are biodegradable, they significantly increase the turbidity, density as well as the COD and BOD of the effluent.

It is the combined presence of high concentration of recalcitrant, salts and heavy organic load that leads to the complex and hazardous nature of textile waste (Punzi et al. 2015). The presence of these pollutants grossly disturbs the characteristics of the waterbodies that they are discharged into. High temperature of the effluent elevates the average temperature of the waterbody causing harm to both flora and fauna. The use of alkalis in operations such as scouring and mercerization results in high pH of the effluent (between 10 and 11) (Kumar et al. 2017). The presence of the nutrients – nitrogen, phosphorous and sulphur – in the effluent can lead to eutrophication of the waterbodies. Waterbodies contaminated by these pollutants are consequently rendered unfit for further agricultural, industrial or domestic use.

3 Environmental Regulations

Several legislations impose standards on industrial wastewater. The ISO 14000 series are international standards that issue guidelines for the effective management of the environment by organizations. In addition, wastewater effluents are regulated by country-specific regulations. Some of these are listed in Table 1. While some countries have generic regulations that impose standards for wastewater effluents across all industries, others such as India, China and Bangladesh that have huge textile sectors have regulations specific to the textile industry.

In addition to these, several international organizations have recognized the need for a uniform set of guidelines across nations and have thus defined standards for the textile industry. Some examples are the Global Organic Textile Standard (GOTS), the Environmental, Health and Safety guidelines issued by the International Finance Corporation (IFC-EHS) and bluesign® (ZDHC 2015). The Zero Discharge of Hazardous Chemicals (ZDHC) foundation – a collaboration of leading textile brands and associates – has also established a set of regulations (ZDHC 2015). The programme recommends and aims to achieve 'zero' discharge of 11 chemical groups: azo dyes, heavy metals, APEOs/NPEs, organotin compounds, brominated and chlorinated flame retardants, perfluorinated chemicals, chlorobenzenes, chlorophenols, chlorinated solvents, phthalates and short-chained chlorinated paraffins.

Table 1 Environmental regulations applicable to textile effluents in various countries

Country/region	Regulation
Bangladesh	The Environment Conservation Rules 1997; schedule 10 for industrial units/ projects and schedule 12B for textile units (ECR 1997).
Canada	Wastewater systems effluent regulations (SOR/2012-139 2015).
China	GB 8978–1996 Integrated Wastewater Discharge Standard (China Ministry of Environmental Protection, 1996) (Environmental Protection Law 1998) GB 4287–2012 Discharge Standards of Water Pollutants for Dyeing and Finishing of Textile Industry (China Ministry of Environmental Protection, 2015) (SGS 2015).
European Union (EU)	Urban Wastewater Treatment Directive 91/271/EEC (EEC Council 1991).
India	The Environment (Protection) Rules, 1986; (Schedule-VI 1986)
Malaysia	Environmental Quality (Industrial Effluent) Regulations (Malaysia Department of Environment Ministry of Natural Resources and Environment, 2009; schedule 5 for industrial effluent and schedule 7 for textile industry (Environmental Quality Act 1974 2009).
Sri Lanka	National Environmental (Protection and Quality) Regulations, No. 1 of 2008 (Sri Lanka Central Environmental Authority, 2008); Schedule 1, List V for textile wastewater (National Environmental Act 2008).
United States of America	US EPA Code of Federal Regulations (CFR) Title 40 Part 410 - Textile Mills Point Source Category (USEPA 2012a). US EPA CFR Title 40 Part 425 -Leather Tanning and Finishing Point Source Category (USEPA 2012b).

4 Conventional Treatment Methodologies and Their Limitations

Textile wastewater, as described above, is characterized by strong colour, high COD, inorganic and organic nutrients and toxic heavy metals (Bisschops and Spanjers 2003; Mantzavinos and Psillakis 2004; Sivakumar et al. 2011; Pathak et al. 2015) and therefore needs adequate treatment prior to disposal. Several methods have been used for the treatment of textile wastes. However, it has been established in the literature that no single method is capable of satisfactorily removing all pollutants (Nawaz and Ahsan 2014). The treatment process, thus, often involves combinations of physical, physico-chemical and biological methods. The different conventional methods used and their associated limitations are discussed below:

4.1 Physical Methods

The physical methods of treatment include screening, sedimentation, floatation and membrane treatment. These methods may be employed for solid and colour removal in the primary treatment stage and for ionic impurities and heavy metal removals in the tertiary treatment stage (Kadirvelu and Goal 2007; Wang and Chen 2009; Liang et al. 2014; Elumalai and Saravanan 2016).

Screening, sedimentation and floatation often form part of the primary treatment in most conventional wastewater treatment processes. Screening makes use of meshes, sieves and bars to remove large debris and other undissolved particulate matter from the waste stream (Kumar et al. 2017). Sedimentation is allowed to separate suspended solids and coagulated mass from the remaining effluent. In floatation, compressed air is passed through the waste stream. It is used to remove small, light suspended particles, oil etc., which adhere onto the gas bubbles and float to the top. The top layer is then skimmed off (Kumar et al. 2017).

Membrane filtration methods include microfiltration, ultrafiltration, nanofiltration and reverse osmosis (RO) (Al-Bastaki 2004; Fersi et al. 2005; Archana 2013). These processes are major unit operations in tertiary treatments. They have good potential for colour and ion removal and decrease the COD/BOD of textile wastewater (Chollom et al. 2015). Ultrafiltration has been used to remove hydrophobic pollutants arising from the 'textile fibre rinsing' operation. These filtration units, however, clog rapidly (Joshi et al. 2004). Fouling of the membrane eventually decreases process efficiency (Koyuncu and Güney 2013). Nanofiltration units have been reported to facilitate complete dye removal. Microfiltration is effective in the treatment of dye as well as rinsing baths. These two methods are also often used as pretreatment to the RO. RO membranes effectively remove most ionic compounds such as mineral salts and reactive dyes and produce high-quality permeate. The method is however energy intensive and thus expensive.

4.2 *Physico-Chemical Methods*

The major physico-chemical treatment methods employed for the treatment of textile waste are coagulation, flocculation, adsorption, advanced oxidation processes (AOPs) and electrochemical techniques (Robinson et al. 2001; Jonstrup et al. 2011; Nawaz and Ahsan 2014; Asghar et al. 2015; Holkar et al. 2016). These methods are used for colour removal (Lin and Chen 1997) in the primary treatment stage and for reducing BOD/COD in the secondary treatment stage. In general, chemical methods result in the degradation of dyes, dissolved or colloidal organic contaminants via oxidation (Jadhav et al. 2015).

Coagulating agents, such as lime, alum, ferrous sulphate, etc., are added to the effluent to facilitate floc formation or flocculation of pollutants (Kumar et al. 2017). The flocs are then allowed to settle and the settled mass separated from the rest of the effluent. Coagulation/flocculation is used for removing colour. Though this process offers easy operation (Khouni et al. 2011), it is slow and has low decolorization efficiency. Further, coagulation can be expensive depending on the nature and amount of the coagulating chemical used. Water soluble dyes, for example, require higher concentration of coagulants (Singh and Arora 2011).

Adsorption involves removal of organic waste from the effluent stream using suitable adsorbents such as Fuller's earth and activated charcoal (Singh and Arora 2011; Kumar et al. 2017). This is an efficient process that removes dissolved organics from the effluent. It is also employed for colour removal from wastewater and is the most acceptable technique with considerable efficiency (Robinson et al. 2001; Vasanth Kumar et al. 2006; Kamaruddin et al. 2013). Activated charcoal is, however, expensive and requires periodic regeneration once the pores on its surface are clogged with pollutants (Fernández et al. 2010; Galán et al. 2013). Further, the regeneration process leads to 10–15% loss of the adsorbent (Joshi et al. 2004). To overcome the cost barrier, several low-cost adsorbents such as bentonite clay, fly ash and peat have been proposed as alternatives. However, these result in sludge generation (Gupta et al. 2016).

Ion exchange resins have been used to remove cationic/anionic pollutants such as reactive dyes (Singh and Arora 2011). Chitosan, quaternized sugarcane bagasse and cellulose are examples of resins used. The hydrodynamic properties of these resins are, however, very poor in comparison to activated charcoal thereby leading to low efficiency. The resins also require recharging which again increases the cost associated.

All the above methods lead to the formation of complex sludge that require further disposal strategies like incineration, land filling, compaction and anaerobic digestion (Chavan 2001).

Advance oxidative processes (AOPs) are cleaner and more efficient methods that lead to the complete degradation of several pollutants (Jonstrup et al. 2011). AOPs involve the generation of highly reactive hydroxyl radicals to disintegrate complex organic compounds to simple inorganic molecules (Punzi et al. 2015). The different AOPs available are ozonation, Fenton processes, photo-Fenton processes, photocatalysis and hydrogen peroxide + UV irradiation treatment. Iron sludge generation (Babuponnusami and Muthukumar 2014) was characteristic of older techniques. Ozonation resulted in the formation of toxic by-products (Gosavi and Sharma 2014; Miralles-Cuevas et al. 2017). These by-products could again contribute to increase in COD/BOD causing secondary by-products (Henze et al. 2001; Renuka et al. 2015). However, improved AOPs do not give rise to toxic sludge. For example, H2O2+ UV irradiation has been shown to degrade dyes effectively with no residual sludge (Soares et al. 2014; Yen 2016). Further, the technique is still highly expensive and chemical intensive (Punzi et al. 2012; Bagal and Gogate 2014). The use of chemicals for treatment is also said to increase conductivity of water. This would require additional processing (Lin and Chen 1997).

Electrochemical techniques, used for the removal of reactive dyes and levelling agents, are also clean methods that lead to the formation of minimal secondary sludge (Dogan Dogan and Turkdemir 2012). Electrocoagulation and electrochemical oxidation are two commonly evaluated techniques (Singh and Arora 2011). Electrochemical oxidation facilitates degradation of pollutants such as dyes and metals at the cathode of an electrochemical cell. Electrocoagulation involves the self-generation of coagulants during the electrochemical process such that additional chemicals do not have to be added for the treatment (Singh and Arora 2011). Electrochemical destruction offers rapid decolorization with concurrent reduction in COD. However, the process results in the formation of non-settling iron flocs that interfere with colour removal and COD measurements (Lin and Chen 1997).These techniques are energy intensive and therefore costly.

4.3 Biological Methods

Biological methods are claimed to be more economical than the physical and physico-chemical methods (Archana 2013). These methods use microorganisms that can take up pollutants in the effluent or degrade them into smaller molecules that can be easily managed. Both aerobic and anaerobic modes of treatment are used, and significant reduction in BOD and COD of the effluent is achieved (Punzi et al. 2015). The treatments include enzymatic degradation of dyes (Picot et al. 1992; Abadulla et al. 2000), colour and heavy metal removal through biosorption (Charumathi and Das 2012), biological oxidation and degradation of dyes and organic/inorganic chemicals (Wang et al. 2007b; Morillo et al. 2009; Holkar et al. 2016). Several bacteria of the *Clostridium* sp., *Eubacterium* sp. and fungi of the *Aspergillus* sp. have been identified that can effectively remove colour/dyes from the effluents (Archana 2013). The chief problem encountered in these methods is the availability of organisms that can withstand the high toxicity and harsh conditions (extreme pH, high temperature, etc.) that are often characteristic of textile effluents (Punzi et al. 2015). Enzymatic reactions require optimum conditions of pH, temperature, etc. Additional substrates for the growth of the organism and enzymes for detoxification may have to be supplied. The aerobic processes often require external oxygen supply for degrading the contaminants. This directly increases the cost (Pacheco et al. 2015). It has also been reported that the methods do not lead to effective removal of nitrogen, phosphorous, xenobiotics and heavy metals (Yuan et al. 2011; Olguín and Sánchez-Galván 2012; Boelee et al. 2014). The effluent is also faintly coloured and sometimes of high conductivity (Lin and Chen 1997).The treatment also results in large amount of sludge that must be adequately disposed (Feng et al. 2003). Further, these methods lead to the release of high concentration of CO_2 into the atmosphere (Brar et al. 2017).

5 Phycoremediation of Textile Wastewater

As discussed in the previous sections, the conventional methods of treatments suffer from several limitations. Further, it is extensively agreed upon in the literature that combinations of the different methods have to be used for effective waste treatment. Biological treatment is considered an eco-sustainable process (Tchobanoglous and Burton 1991) with the potential to eliminate a broad spectrum of pollutants (Paraskeva and Diamadopoulos 2006). However, the existing biological treatments have several limitations as discussed earlier (Feng et al. 2003). To overcome these limitations, use of algae in bioremediation of wastewater has been suggested. Algae can be utilized for low-cost and eco-sustainable treatment processes, known as phycoremediation (de la Noüe et al. 1992; Lim et al. 2010; Pathak et al. 2015). Algal systems have already been employed in tertiary treatments (Martin et al. 1985; Oswald 1988). It also has immense potential in secondary treatment (Tam and Wong 1989) due to its ability to resist and grow in polluted environment (Jais et al. 2017). In addition, algae can also be employed as biosorbent in primary treatment stage of textile water for colour removal owing to their excellent sorption capacity.

5.1 Phycoremediation: Definition and Advantages

Phycoremediation (from Greek 'phykos' meaning algae and latin 'remedium' meaning restoring balance) is a biological method of treatment, which can be generally defined as the removal or degradation of pollutants (nutrients, heavy metals and xenobiotics) from wastewater using algae, either micro or macro and cyanobacteria (Olguín 2003; Olguín and Sánchez-Galván 2012). Though the term 'phycoremediation' was coined recently (John 2000), the idea of using algae in wastewater remediation was illustrated six decades ago (Oswald et al. 1957). Since then, phycoremediation has been intensively studied (Shelef et al. 1980; Oswald 1988; Zhu et al. 2008; Abdel-Raouf et al. 2012) and is considered promising (Olguín and Sánchez-Galván 2012). Algae with their range of diversity and ability to adapt to extreme environment led scientists to screen and identify suitable and promising strains for wastewater treatment (Fouilland 2012). Some of the benefits of using algae in wastewater remediation are:

- The primary advantage of algae over the other organisms is its ability to perform photosynthesis (Brar et al. 2017). Thus, they can mitigate CO_2 levels (Renuka et al. 2015). In addition, they do not release high concentrations of CO_2 into the atmosphere.
- Wastewater containing organic and inorganic carbon, nitrogen, phosphorous and certain other compounds like heavy metals such as that in textile wastewater (Sivakumar et al. 2011) offers suitable conditions for algal growth (Liang 2013).
- Algae have high potential to acclimatize and utilize inorganic nutrients such as nitrogen and phosphorous (Talbot and de la Noüe 1993; Blier et al. 1995; Farhadian et al. 2008).
- Algae are also known to uptake and accumulate high amount of metal ions (Brierley et al. 1986) that are known to be rich in textile effluents (Sponza 2002). This is due to their large surface area and high binding affinity (Gupta and Suhas 2009) facilitated by assemblage of polymers similar to pectin, cellulose, hemicelluloses and lignin (Domozych et al. 2012). This structure provides them with multiple functional groups such as carboxyl, hydroxyl and amino (Abdel-Monem et al. 2010; Al-Gheethi et al. 2014) that help them function like typical ion exchangers (Kuyucak and Volesky 1990).
- Algal remediation 'often' results in no residual sludge (Brar et al. 2017).

5.2 Phycoremediation: A Multifaceted Technology

Phycoremediation is a multifaceted technology involving wastewater treatment, CO_2 mitigation and simultaneous biomass generation (Renuka et al. 2015). The biomass could be used as animal feed additive, feedstock for biofuel and biogas, bio-ore for heavy metals (Spolaore et al. 2006; Gupta et al. 2013) and, for extracting vitamins, β carotenes and fine chemicals like antioxidants (Mata et al. 2010). However wastewater-grown algae are not suitable as animal feed (Pathak et al. 2015) and fine chemicals extraction. But its application as energy source such as biofuel or biogas still holds good. The various roles of phycoremediation are summarized in Fig. 1.

The role of phycoremediation in improving textile wastewater effluent characteristics by removal of nutrients and simultaneous biomass production is discussed in the following sections.

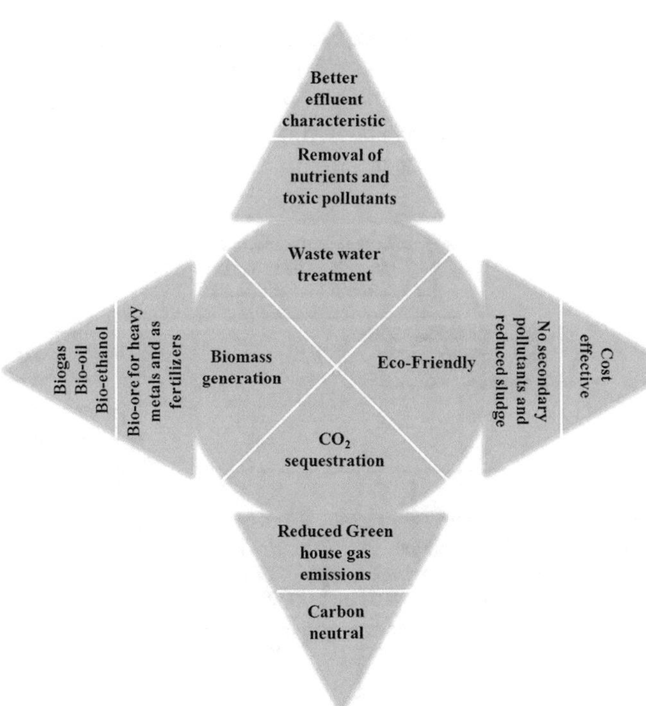

Fig. 1 The multifaceted phycoremediation technology

Improved Effluent Characteristics

Algae and cyanobacteria known as prokaryotic algae (Mata et al. 2010) have been proven to be efficient in reducingCOD/BOD and removing nitrogen, phosphorous and heavy metals from different wastewater including textiles (Colak and Kaya 1988; Sturm and Lamer 2011; Zhou et al. 2012) thereby improving the effluent quality. Depending on the needs and the strain potential, both untreated and partially treated wastewater can be processed. The efficiency of treatment depends on the nature of species and their adaptability to wastewater environment (Holkar et al. 2016; Jais et al. 2017). As already mentioned, algae are versatile organisms with wide range of diversity and easy adaptability. The challenge lies in selecting the suitable species. Both viable and non-viable biomass of algae can be employed for removal of dye and toxic metals (Volesky 2001; Aravindhan et al. 2007; Khalaf 2008), while for nutrient removal it is always the viable biomass. Depending on the physiological state of biomass used (live/dead), the mechanism of dye and heavy metal removal differs (Schematic 1).

Removal of COD and BOD

COD and BOD are important gauge factors for textile wastewater like any other one (Bisschops and Spanjers 2003; Mostafa 2015). High COD/BOD can deplete the dissolved oxygen content of the receiving aquatic system (Colak and Kaya 1988). Almost all the process in textile manufacturing such as desizing, scouring, bleaching, dyeing and printing contribute to high COD, while BOD arises from mostly

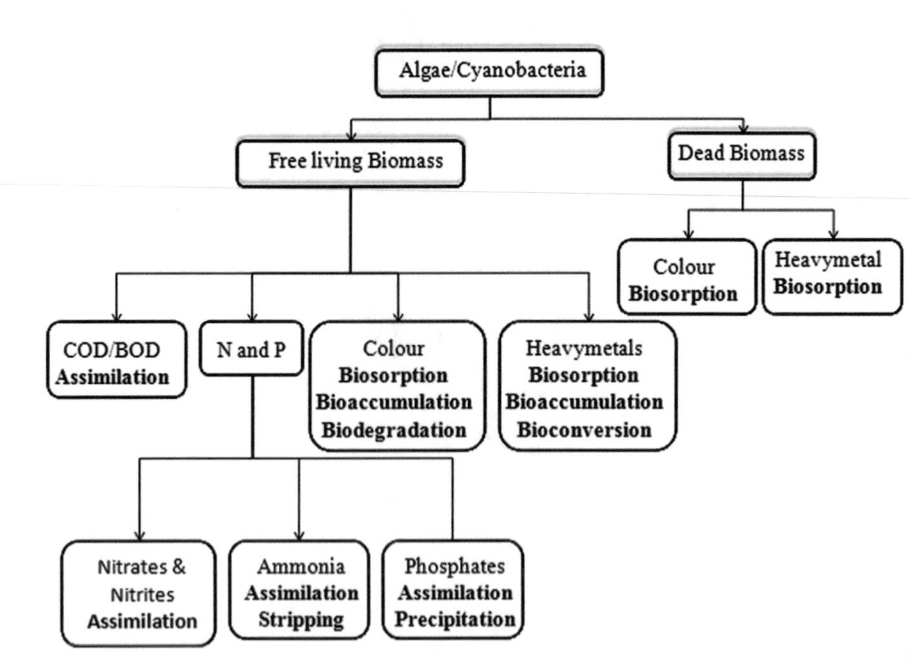

Schematic 1 Phycoremediation mechanism (text in bold) of various wastewater contaminants

scouring process followed by desizing and dyeing (Bisschops and Spanjers 2003). Algae, by assimilation of the organic and inorganic contaminants (Ota et al. 2011), contributes to decrease in COD/BOD. The azo dyes that are responsible for high COD in textile wastewater (Joshi et al. 2004) can also serve as carbon and nitrogen source for algae (Urushigawa and Yonezawa 1977; Kulla et al. 1983).

The heterotrophic mode of organic and inorganic contaminant uptake can happen either in the presence or absence of light. For certain species such as *C. vulgaris* (Subramanian et al. 2016) and *Platimonas convolutae* (Pacheco et al. 2015), light becomes necessary for organic carbon uptake and is known as photoheterotrophic mode. This mode of cultivation is advantageous in a way that the light dependency of algae can be avoided resulting in high cell density (Ceron Garcia et al. 2000; Venkata Mohan et al. 2015). Further, CO_2 released by heterotrophic respiration of algae in turn can be fixed through photosynthesis in photoheterotrophic mode, thereby avoiding the greenhouse gas release. The photoheterotrophic cultivation of *C. pyrenoidosa* in 75% textile water reduced the BOD by 81% (Pathak et al. 2014). Algae are also capable of growing in raw textile water. *C. vulgaris*, when grown in raw textile water, was capable of reducing COD by 62% (Lim et al. 2010). The efficiency of COD/BOD removal varies with species. In a textile wastewater remediation study conducted with four different cyanobacterial species which included *A. variabilis*, *O. salina*, *N. muscorum* and *L. majuscula*, *A. variabilis* had the highest COD removal of 75%, and *O. salina* had the highest BOD removal of 86%. COD removal by other species such as *N. muscorum*, *L. majuscule* and *O. salina* was 55.5%, 62.4% and 66.4%, respectively. Similarly, BOD removal of *N. muscorum*, *A. variabilis* and *L. majuscula* was 73%, 83.2% and 79.3%, respectively (David Noel and Rajan 2014). In a different study conducted with *N. muscorum*, the strain was capable of reducing COD of textile wastewater by 90.04% (Ghazal et al. 2016). The COD removal efficiency also depends on pH and the presence of other nutrients in the wastewater. As per the study conducted with *Chlorella* sp. G23, the algae had good COD removal efficiency which was higher than 60% for the pH range of 7–9, while at higher alkaline pH of 11, the COD removal decreased significantly (Wu et al. 2017). Since pH of most textile effluents lie in the range of 7–8 (Lau et al. 1995; Chinnasamy et al. 2010b; David Noel and Rajan 2014; Pathak et al. 2014), it provides a suitable environment to algae for COD removal. While COD removal is not affected by the type of nitrogen source present in the water as studied with *Chlorella* sp. G23, it does depend on the available concentration of phosphates. Phosphate level in abundance can interfere with COD removal efficiency (Wu et al. 2017). Though macroalgae are also known for their role in phycoremediation, in general they are employed only as biosorbent for dye and heavy metal removal from textile wastewater (Aravindhan et al. 2007; Daneshvar et al. 2012) and not in the secondary treatment for COD/BOD removal.

Removal of Nitrogen and Phosphorous

Textile wastewater is known to contain high levels of nitrogen (N) and phosphorous (P) (Sivakumar et al. 2011), which when discarded into waterbodies can cause eutrophication (Mahapatra et al. 2013). The main source of nitrogen in textile effluent is from dye bath additives (Delée et al. 1998). Some of the operations that release ammonia are printing, coating preparation and dyeing. Large amounts of urea can also be added into the effluent through printing process. Similarly phosphate buffers used in textile wet processing contributes to the phosphate content in effluents (Bisschops and Spanjers 2003). As mentioned already, conventional biological methods are not efficient in removing N and P. Since algae have high capacity for inorganic nutrient uptake (Talbot and de la Noüe 1993), phycoremediation has high ability to reduce N and P in wastewater (Abdel-Raouf et al. 2012; Arbib et al. 2014). Apart from ammonia and urea, the principle form of nitrogen in wastewater is nitrite (NO_2^-) and nitrate (NO_3^-), and phosphate is present as orthophosphates (PO_4^{3-}). Algae by the process of assimilation covert inorganic nitrogen to organic form by the action of enzymes such as nitrate and nitrite reductase (Cai et al. 2013). The organic nitrogen is further used for the synthesis of proteins, enzymes, chlorophylls, ATP and other biomolecules (Barsanti and Gualtieri 2006). Similarly orthophosphates is incorporated as organic phosphates in nucleic acids and energy molecules through phosphorylation (Martínez et al. 1999). Apart from assimilation, ammonia and phosphorous can also be removed by 'stripping' (Ceron Garcia et al. 2000) and precipitation, respectively (Cai et al. 2013). This occurs as a result of the increase in pH and dissolved oxygen from the photosynthetic activity of algae (Cai et al. 2013; Pires et al. 2013) (Schematic 1). While lots of studies are available on N and P removal by algae in several other wastewaters, studies with textile wastewater in particular are limited. The algae *C. vulgaris* was found to reduce ammonia and phosphates in textile wastewater by 45% and 33%, respectively, when cultivated in an open pond. The variations in parameters such as pH, temperature, DO and solar irradiance observed in the course of study (Lim et al. 2010) might have resulted in moderate nutrient removal efficiency. In controlled environment of light and temperature, *Chlorella* sp. G23 displayed high ammonia removal efficiency of 78% in textile wastewater (Wu et al. 2017). Further, N and P removal efficiency of algae depends on their initial inoculum density. As studied with *C. vulgaris* growth in textile wastewater, when starting with an inoculum size of 10^6 cells ml^{-1}, the removal efficiency of ammonia and total Kjeldahl nitrogen (TKN) was in the range of 90% and 80%, respectively. The removal efficiency of total phosphorous and PO_4^{3-} was also in similar ranges. When the initial algal density was increased by a factor of 10 (10^7 cells ml^{-1}), it increased the rate of removal of N and P. Also the removal efficiency of ammonia reached up to 99%. This is because the nutrient removal capacity of algae is dependent on its physiology and growth which in turn changes with initial inoculum size (Lau et al. 1995). The optimal removal efficiency of N and P requires that they are removed simultaneously, for which the N/P ratio in wastewater should be in optimal range (Xin et al. 2010). An optimal N/P ratio ensures that neither of the nutrients is limiting the algal growth. As per the empirical formula for microalgae ($C_{106}H_{263}O_{116}N_{16}P$), the N/P ratio is around 7.2, which means that the removal rate of nitrogen should be faster than phosphorous (Cai et al. 2013) as algae require more nitrogen than phosphorous for their growth (Lau et al. 1995). Practically, the optimal N/P ratio required for algal growth is around 10 (Martin et al. 1985). For *C. vulgaris*, when N/P ratio was maintained within a narrow range of 8–10 in textile wastewater, simultaneous removal of N and P occurred, resulting in their better uptake (Lau et al. 1995). However, the optimal N/P ratio varies by species and may not be necessarily be around 10. For example, *Scenedesmus* sp. requires an optimal N/P ratio of 30, for simultaneous removal of N and P (Martin et al. 1985). Cyanobacteria, when compared to algae, may prefer a low N/P ratio (Laliberté et al. 1997). The removal efficiency of N and P can be improved by change in cultivation techniques such as immobilization. Studies have shown that immobilizing algae enhanced the removal rate of N and P when compared to suspension cultures (Hoffmann 1998; Mallick 2002; Johnson and Wen 2010). However, immobilization is associated with decreased biomass productivity (Hoffmann 1998; Zeng et al. 2015), high cost (Cai et al. 2013) and scalability issues (Zeng et al. 2015). Due to above mentioned reasons, immobilizing algae may not be a suitable technique for phycoremediation, as one of the advantages of using algae is its process economy.

Removal of Heavy Metals

The effluents from textile industries have high levels of toxic heavy metals (Chen et al. 2001; Sponza 2002). The main source of heavy metals in textile wastewater is the dyeing process, since most dyes contain chromium, cadmium and zinc, among other metals (Bisschops and Spanjers 2003). Though many of the newly developed dyes are metal free (Delée et al. 1998), there are other sources of metals as well. These includes metal parts such as pumps, pipes, etc., oxidizing and reducing agents, electrolytes and maintenance chemicals (Smith 1988). Algae are considered suitable for heavy metal sequestration owing to their fibrous structure and the amorphous matrix of their cell wall that contains various functional groups (Bayramoğlu et al. 2006). Using algae is considered economical for metal removal (Akhtar et al. 2008; Pandi et al. 2009). Both live and dead cells can be employed for metal removal (Abdel-Raouf et al. 2012) based on which the removal mechanism varies. In case of dead cells, the metal removal from wastewater occurs via biosorption (Schematic 1), during which the metal ions are attached to functional groups of the cell membrane via complexation, ion exchange, chelation and micro precipitation (Çetinkaya Dönmez et al. 1999; Jais et al. 2017). Dead cells are also capable of storing metals, in which the metals enter through the pores of damaged cells via passive transport (Jais et al. 2017). In case of viable or live cells, the removal is through biosorption followed by bioaccumulation (Schematic 1) that involves active transport of heavy metals across cell membrane (Yee et al. 2004; Han et al. 2007; Al-Gheethi et al. 2014). Dead biomass used for biosorption is often spent biomass from some process. While using dead biomass, can be economical, live biomass is advantageous in that it can reduce the toxicity of heavy metals (Silver 1996; Perales-Vela et al. 2006; Jaishankar et al. 2014). Detoxification can happen via precipitation of heavy metal ions as a carbonate, phosphate or sulphide (Silver 1996).The toxicity can also be neutralized by formation of organometallic complexes of heavy metals with polypeptides produced by algae, which are then partitioned within vacuoles (Perales-Vela et al. 2006). Thus, algae are capable of bioconversion of heavy metals into its less toxic form. Free-living suspension culture of naturally colonized algae showed a chromium removal rate of 98% from textile wastewater. It was also observed that changes in parameters such as pH, dissolved oxygen and light regimes did not cause variation in metal removal efficiency (Sekomo et al. 2012). However, metal removal efficiency is strongly dependent on the initial algal density (Çetinkaya Dönmez et al. 1999). The magnitude of binding for a metal will differ with species. Similarly, the same species would exhibit different binding capacity for different metals. For example, dead biomass of *C. vulgaris* showed highest binding for nickel (42.3%), followed by copper (38.2%) and chromium (23%). Also *C. vulgaris* exhibited better binding capacity than cyanobacteria *Synechocystis* sp. (Çetinkaya Dönmez et al. 1999). However, live cells of cyanobacterial species such as *N. muscorum*, *A. variabilis*, *L. majuscula* and *O. salina* when grown in textile wastewater showed high nickel removal efficiencies of 74%, 63%, 76.1% and 76.1%, respectively. The corresponding removal efficiencies for zinc were 57%, 67%, 73% and 73.5%, respectively (David Noel and Rajan 2014). Non-viable biomass of macroalgae has also been used as potent biosorbent (Sandau et al. 1996; Park and Lee 2002). Non-viable biomass of green seaweed when used as biosorbent in textile wastewater showed a removal efficiency of 86.8% of chromium, 87.5% of iron and 100% of silver in 1 hour (Latinwo et al. 2015). Non-viable biomass of seaweed is also cheaper and an efficient biosorbent when compared to non-viable microalgae and cyanobacteria. The biosorption capacity of the seaweed *Fucus vesiculosus* for cadmium was 97.7%, while with *C. vulgaris* it was 84.1% (Sandau et al. 1996). As mentioned already, non-viable biomass, while cheaper, is not capable of heavy metal detoxification, while viable cells are. Viable cells of algae *C. vulgaris* were capable of reducing Cr (VI) to its less toxic form of Cr (III) (Shen et al. 2013). As with N and P removal, immobilizing the algae also improves heavy metal removal (Rangsayatorn et al. 2004; Dixit and Singh 2014). However, there is this high cost factor involved with respect to immobilization that needs strong consideration.

Removal of Dyes

In case of textile wastewater, dyes and their colour are of major concern (Holkar et al. 2016). Synthetic dyes are one of the most widely used chemicals in textile manufacturing (Elumalai and Saravanan 2016). Even small amounts of dye in effluent are visible as colour, which affects the aesthetics and water transparency (Bisschops and Spanjers 2003). This consequently affects aquatic organisms. Since algae are capable of effective dye decolorization and degradation (Acuner and Dilek 2004; Ertuğrul et al. 2008), phycoremediation is a viable technique for textile water decolorization. Algae are capable of utilizing dye as sole carbon and nitrogen source (Kulla et al. 1983; Jinqi and Houtian 1992). As with heavy metal removal, colour removal of dyes can also be done with both live and dead biomass, and the mechanism includes biosorption, bioaccumulation and bioconversion (Lim et al. 2010) (Schematic 1). In case of bioconversion or biodegradation, dyes are transformed into non-colour intermediates or even reduced to CO_2 and H_2O (Kulla et al. 1983). The degradation happens by the action of enzyme azo reductase that breaks down the dyes into less toxic aromatic amines (Kulla et al. 1983; Meng et al. 2014). Few alga species are also capable of further degrading aromatic amines into CO_2 (Jinqi and Houtian 1992). Enzyme laccase was also found to have possible roles in dye degradation (KIlIç et al. 2011). In studies conducted with algae, the dye removal from wastewater is measured by monitoring the disappearance of colour from the medium and expressed as decolorization efficiency. The dye degradation is elucidated by visualizing the colour of biomass after incubating them with the dye or by measuring the increase in activity of enzyme azo reductase or by analysing the degradation products by HPLC.

Colour removal: The non-viable biomass of *Spirogyra* sp. displayed a high biosorption capacity of 80% for the reactive dye Synozol found in textile wastewater. It was also found that biosorption was better with non-viable biomass than live biomass (Khalaf 2008). The de-oiled algal biomass (DAB), which is the spent biomass after oil extraction, is also a potential biosorbent. The DAB of *Micorspora* sp. was capable of completely removing methylene blue in 24 h (Maurya et al. 2014). The non-viable biomass of *Pithophora* sp. was found to be effective as a sorbent for malachite green (Vasanth Kumar et al. 2006). Free live biomass of certain algal species also had shown promising colour removal efficiency. Free live biomass of *Spirogyra* sp. and *Oscillatoria* sp. showed a colour removal efficiency of 78% and 76%, respectively, for a blue dye present in textile effluent (Brahmbhatt and Jasrai 2016). However, free live biomass of cyanobacterial species such as *Synechocystis* sp. and *Phormidium* sp. showed only moderate removal efficiency of 37.5% and 25.5%, respectively, for Remazol blue (Karacakaya et al. 2009). *Phormidium* sp. when immobilized achieved a better colour removal of 88% for Remazol blue (Ertuğrul et al. 2008). Similar to DAB of algae, spent biomass of hydrogen-producing cyanobacteria *N. linckia* has been studied for its sorption capacity. When immobilized, spent biomass of *N. linckia* displayed a good colour removal efficiency of 72% for crystal violet (Mona et al. 2011). Though the sorption capacity of non-viable biomass is better than that of live biomass (Khalaf 2008; Pathak et al. 2015), it is to be noted that in case of the former, the toxic dye still persists in the environment, whereas, live/viable biomass is capable of degradation and detoxification of the dye. In such cases, measuring the disappearance of the dye colour in the medium will not be an accurate measure of the bioremediation potential of the organism.

Dye degradation: The free live culture of algae *Gonium* sp. displayed the highest reactive blue 220 removal efficiency of 96.8%. It was also observed that *Gonium* sp. was capable of degrading the dye, as the pelleted biomass after its incubation with dye exhibited no colour. The laccase activity was considered as a possible reason for dye degradation (Boduroğlu et al. 2014). The dye degradation ability of algae has also been studied by measuring the increased activity of azo reductase. The algae *C. vulgaris* was capable of degrading G-Red, as it had increased azo reductase activity after its incubation with the dye. Similarly *N. linckia* showed good degradation activity for methyl red (El-Sheekh et al. 2009). The degradation of dye by a particular species depends on the molecular structure of the dye (Acuner and Dilek 2004). It was proved that colour removal and degradation ability of a species towards a particular dye can be improved by acclimatizing the species to that dye. The algae *C. vulgaris* showed improved TY2G removal efficiency of 88% upon its acclimation with the dye, while for unacclimatized *C. vulgaris*, TY2G removal efficiency was only 69%. HPLC analysis showed was that unacclimatized *C. vulgaris* was capable of degrading TY2G to aniline, while acclimatized *C. vulgaris* was found to have degraded the aniline even further. This suggests complete degradation of the dye (Acuner and Dilek 2004). Algae are also capable of degrading aniline – the dye's end product by means of light-assisted degradation. Photodegradation of aniline has been demonstrated with *C. vulgaris*, *C. sajao*, *A. cylindrical* and *N. hantzschiana*. Generation of reactive oxygen species such as hydroxyl radical and singlet oxygen was observed and suspected to have caused the degradation (Wang et al. 2007a). Thus, using free live culture of algae helps in reducing the toxicity of dyes present in textile wastewater.

Biomass Production

The algal biomass produced through phycoremediation of textile wastewater could be used in several ways. The biomass becomes an energy feedstock for obtaining biofuel or biogas, or the dried biomass could be used as fertilizers. For energy generation from algae, accumulated triacylglycerols (TAG) can be converted into biodiesel, or they can be anaerobically digested to produce biogas (Brune et al. 2009). By using wastewater as a medium for algal biomass production, cost associated with raw material could be reduced significantly. The alga *C. vulgaris* when grown in raw textile wastewater produced 106.67 mg l^{-1} of biomass. However, their lipid yield was not reported (Lim et al. 2010). Another alga *Chlorella* sp. G23, when grown in textile wastewater, produced a biomass of 137 mg l^{-1} in a day. The corresponding lipid productivity was 8.6 mg l^{-1} day^{-1}(Wu et al. 2017). *B. braunii*, *C. saccharophila*, *D. tertiolecta* and *P. carterae*, when grown in carpet textile wastewater produced biomass of 34, 28, 28 and 33 mg $l^{-1}d^{-1}$, respectively. The corresponding lipid productivities were 4.5, 4.2, 4.3 and 4 mg l^{-1} day^{-1}. When a consortium of 15 different algal species was used to remediate textile wastewater, the maximum biomass productivity achieved was 17.8 tons ha^{-1} $year^{-1}$. However, the lipid yield was very less – 6.82% of cell dry weight (CDW). In such cases of low lipid content, the energy present in the biomass could be converted to methane via anaerobic digestion (Chinnasamy et al. 2010b). Similarly, *C. variabilis* when grown in textile wastewater produced 74.96 g m^{-2} of biomass in a day. The lipid yield was found to be 20.1% CDW. The total microalgal biomass obtained was 495 g. The use of this biomass for production of γ-linolenic acid and ε-polylysine was also demonstrated. Based on material analysis and economic assessment, phycoremediation and resulting biomass production were found to have good scope for scalability (Bhattacharya et al. 2017).

5.3 Consortia for Textile Wastewater Remediation

In an ecosystem, no organism is capable of sustaining on its own (Renuka et al. 2015). Often the microalgal and cyanobacterial species are found associated with other aerobic or anaerobic microorganisms in natural environment. The nature of association could be either competitive or mutualistic. The bacterial colonies are known to influence the algal bloom development positively (Fukami et al. 1997). Laboratory studies have shown that algal cultures maintain a symbiotic relationship with bacteria (Park et al. 2008). Such relationships are known to influence the remediation of wastewater. In case of cyanobacteria/algae – bacteria consortia – O_2 released by the photosynthetic activity of cyanobacteria/algae will be used as an electron acceptor by bacteria to degrade the organic contaminants. The CO_2 released by bacterial respiration is used for photosynthetic carbon fixation by cyanobacteria/algae. Further algae secrete extracellular biosurfactants into the medium, which enhances the availability of organic contaminants for bacteria. The growth factors released by bacteria would also accelerate the development of algae (Muñoz et al. 2009; Tang et al. 2010; González-Fernández et al. 2011). Monocultures of algae/cyanobacteria may not be efficient in removing all contaminants present in wastewater. While one strain is effective in removing one type of pollutant, other strains would perform well for a different contaminant (Chinnasamy et al. 2010a). Developing a consortium using different strains with promising potential can render a synergistic and self-sustaining biodegradation system (Subashchandrabose et al. 2011). The alga *C. pyrenoidosa* in consortium with indigenous microbes present in textile wastewater was capable of removing nitrate by 81%. Its monoculture, on the other hand, achieved 62% nitrate removal (Pathak et al. 2014). The consortium formed with 15 different isolates of alga and cyanobacteria from carpet textile wastewater was capable of removing nitrate levels by 99% in a duration of 24 h and phosphate levels by 96% in a duration of 72 h (Chinnasamy et al. 2010b). As already mentioned, favourable range of N/P ratio should exist in the wastewater medium, to facilitate algal growth (Xin et al. 2010). Thus, in case of consortium, it is ideal to select species with similar N/P range. Otherwise, one species can outgrow another resulting in disturbed consortia harmony. The presence of bacteria in consortia can help towards maintaining the N/P ratio and favours algal development. The algae/bacteria consortia are also capable of enhancing the species performance in open pond algal treatment systems (Lau et al. 1995). This cohabitation concept had been adopted in the high rate algal pond (HRAP) system proposed by Oswald, which can be used for combined secondary/tertiary wastewater treatment (Elumalai and Saravanan 2016). By using algal consortium, improvement in heavy metal removal has also been achieved. While monocultures of *N. muscorum* and *A. subcylindrica* removed 64.4% and 33.3% of Cu and 84.6% and 86.2% of Pb, respectively, as a coculture they were capable of removing Cu by 75% and Pb by 100% (El-Sheekh et al. 2009). The degree of azo dye degradation depends on the molecular structure of the dye and also the species. While dyes containing hydroxyl or amine groups are readily degraded, dyes containing groups such as methyl, methoxy, nitro or sulfo derivatives might resist degradation by a species (Jinqi and Houtian 1992). For example, *C. vulgaris* that was capable of degrading amino or hydroxyl group containing dyes such as basic fuchsin failed to degrade methyl red. However, *N. linckia* was capable of degrading methyl red (El-Sheekh et al. 2009). Thus, their cocultures are capable of degrading dye mixtures containing different functional groups. Instead of engineering a single organism to do multiple things, developing a consortium with versatile species having individual potential could be cheaper and relatively easier for developing a robust treatment system (Subashchandrabose et al. 2011).

5.4 Phycoremediation: Challenges and Constraints

Based on the numerous reports available, it is worthwhile to state that phycoremediation is an efficient technique for treating textile wastewater. However, most of the available studies are based on laboratory scale results and need proof for reliability in the actual local environment, where multiple factors such as pH, temperature, light intensity and nature of wastewater and its composition would influence the process efficiency. Field studies are required to ascertain the efficiency of the process. Further, optimizing the process of phycoremediation for a given textile sample would be necessary. Although several common pollutants exist across different textile effluents, the exact composition of the effluent would vary depending upon the product being manufactured. As a result, independent studies would be required for the different textile effluents. Efficiency obtained with one effluent cannot be directly extrapolated to another. The primary challenge often lies in choosing a suitable species (Şentürk et al. 2017). Further, scale-up (Zeng et al. 2015; Brar et al. 2017) and design of cost-effective algal cultivation systems are additional hurdles that need to be overcome (Cai et al. 2013). Management of spent biomass is additionally required. Although generation of large volumes of the biomass can be coupled with other applications such as biofuel production, biomass used for uptake and bioaccumulation of heavy metals may not be suitable for this purpose. Adequate disposal of spent biomass is once again necessary here.

5.5 Phycoremediation: Proof of Concept

In spite of the constraints mentioned above, scientists have been successful in setting up commercial phycoremediation plants. Some examples are discussed below:

SNAP Alginates and Natural Products

SNAP alginates and natural products (Ranipet, Tamil Nadu, India) are pioneer in manufacturing alginates from seaweeds for its applications in various industries such as textiles (PERC n.d.-a, n.d.-b). The liquid discharge generated by alginate extraction process is acidic in nature (pH 1.8) with total dissolved solids (TDS) of about 40, 000 mg l^{-1}. The wastewater used to be treated by conventional physio-chemical methods (Schematic 2). This resulted in generation of large amount of solid discharge. This method was later replaced by an improved biological method developed by a research team from RKM Vivekananda College (Chennai, Tamil Nadu, India) headed by Dr. V. Sivasubramaniam (Director, PERC). In the improved biological method, the treatment tank is charged with blue-green alga – *Chroococcus turgidus*. The alga would utilize the TDS present in wastewater as nutrients and produce alkalinity that neutralizes the water. Based on the promising results from a pilot scale study, the world's first phycoremediation plant was set up by SNAP in September 2006 and has been running successfully since then. The advantages of their phycoremediation technique over physio-chemical methods are shown in Schematic 2.

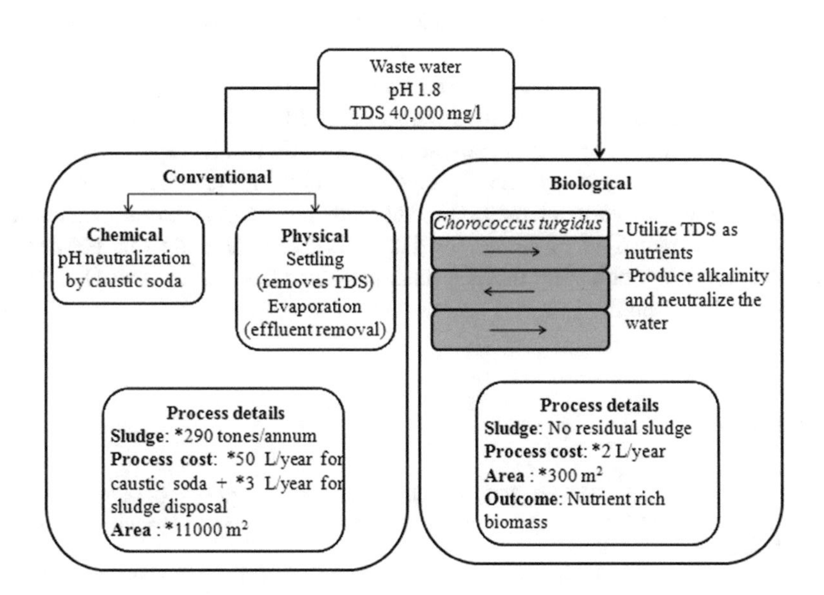

Schematic 2 SNAP's phycoremediation vs physio-chemical methods. *The figures are based on experience from SNAP's research team

SUNTEX Processing Mills

After the success in algal remediation technology by SNAP, field trials were conducted at SUNTEX processing mills, a textile dyeing industry located in Gummidipoondi (Chennai, Tamil Nadu, India). The characteristics of wastewater from SUNTEX included dark colour, alkalinity (pH 11.4) with TDS of 6.8 g l^{-1}. The conventional chemical treatment reduced the pH to 9.6. However, due to the usage of chemicals, the TDS of treated water increased to 10.28 g l^{-1} and also resulted in concentrated sludge production. By using algae for remediation, excellent colour removal was achieved, pH was corrected to 9.5, and TDS was reduced to 1.7 g l^{-1}. The sludge production was also reduced by 80%. After treatment, the algal biomass was harvested by the settling process, and its possible use as fuel for boilers was also explored (PERC n.d.-a, n.d.-b).

6 Conclusion

The generation of large volumes of wastewater by the textile industry and the high concentration of toxic contaminants in the effluents necessitate advanced treatment procedures. While conventional methods have been available for treatment, complete removal of pollutants is still a major challenge due to the complex composition of the wastewater. Phycoremediation has proved itself a viable biological treatment in numerous studies due to effective reduction in BOD, COD, heavy metals and nutrients and complete removal of colours from the effluent. It is also carbon neutral and therefore an environment friendly alternative. Phycoremediation can be used as both post- and pretreatment methods. Although the method has its limitations, effective implementation of phycoremediation has been shown to be possible by various companies indicating that the method indeed has great potential in textile wastewater treatment.

References

Abadulla E, Tzanov T, Costa S et al (2000) Decolorization and detoxification of textile dyes with a laccase from Trametes hirsuta. Appl Environ Microbiol 66:3357–3362. https://doi.org/10.1128/AEM.66.8.3357-3362.2000

Abdel-Monem MO, Al-Zubeiry AHS, Al-Gheethi AAS (2010) Biosorption of nickel by Pseudomonas cepacia 120S and Bacillus subtilis 117S. Water Sci Technol 61:2994–3007. https://doi.org/10.2166/wst.2010.198

Abdel-Raouf N, Al-Homaidan AA, Ibraheem IBM (2012) Microalgae and wastewater treatment. Saudi J Biol Sci 19:257–275. https://doi.org/10.1016/j.sjbs.2012.04.005

Acuner E, Dilek FB (2004) Treatment of tectilon yellow 2G by *Chlorella vulgaris*. Process Biochem 39:623–631. https://doi.org/10.1016/S0032-9592(03)00138-9

Akhtar N, Iqbal M, Zafar SI, Iqbal J (2008) Biosorption characteristics of unicellular green alga *Chlorella sorokiniana* immobilized in loofa sponge for removal of Cr(III). J Environ Sci 20:231–239. https://doi.org/10.1016/S1001-0742(08)60036-4

Al-Bastaki NM (2004) Performance of advanced methods for treatment of wastewater: UV/TiO2, RO and UF. Chem Eng Process Process Intensif 43:935–940. https://doi.org/10.1016/j.cep.2003.08.003

Al-Gheethi AAS, Norli I, Lalung J et al (2014) Biosorption of heavy metals and cephalexin from secondary effluents by tolerant bacteria. Clean Techn Environ Policy 16:137–148. https://doi.org/10.1007/s10098-013-0611-9

Aravindhan R, Rao JR, Nair BU (2007) Removal of basic yellow dye from aqueous solution by sorption on green alga Caulerpa scalpelliformis. J Hazard Mater 142:68–76. https://doi.org/10.1016/j.jhazmat.2006.07.058

Arbib Z, Ruiz J, Álvarez-Díaz P et al (2014) Capability of different microalgae species for phytoremediation processes: wastewater tertiary treatment, CO2 bio-fixation and low cost biofuels production. Water Res 49:465–474. https://doi.org/10.1016/j.watres.2013.10.036

Archana LKN (2013) Biological methods of dye removal from textile effluents - a review. J Biochem Technol 3:177–180

Asghar A, Raman AAA, Daud WMAW (2015) Advanced oxidation processes for in-situ production of hydrogen peroxide/hydroxyl radical for textile wastewater treatment: a review. J Clean Prod 87:826–838. https://doi.org/10.1016/j.jclepro.2014.09.010

Babuponnusami A, Muthukumar K (2014) A review on Fenton and improvements to the Fenton process for wastewater treatment. J Environ Chem Eng 2:557–572. https://doi.org/10.1016/j.jece.2013.10.011

Bagal MV, Gogate PR (2014) Wastewater treatment using hybrid treatment schemes based on cavitation and Fenton chemistry: a review. Ultrason Sonochem 21:1–14. https://doi.org/10.1016/j.ultsonch.2013.07.009

Barsanti L, Gualtieri P (2006) Algae: anatomy, biochemistry, and biotechnology. CRC Press\Taylor & Francis Group, Boca Raton, FL

Bayramoğlu G, Tuzun I, Celik G et al (2006) Biosorption of mercury(II), cadmium(II) and lead(II) ions from aqueous system by microalgae *Chlamydomonas reinhardtii* immobilized in alginate beads. Int J Miner Process 81:35–43. https://doi.org/10.1016/j.minpro.2006.06.002

Bhattacharya S, Pramanik SK, Gehlot PS et al (2017) Process for preparing value-added products from microalgae using textile effluent through a biorefinery approach. ACS Sustain Chem Eng 5:10019–10028. https://doi.org/10.1021/acssuschemeng.7b01961

Bisschops I, Spanjers H (2003) Literature review on textile wastewater characterisation. Environ Technol (UK) 24:1399–1411. https://doi.org/10.1080/09593330309385684

Blier R, Laliberté G, de la Noüe J (1995) Tertiary treatment of cheese factory anaerobic effluent with Phormidium bohneri and *Micractinium pusillum*. Bioresour Technol 52:151–155. https://doi.org/10.1016/0960-8524(95)00014-6

Boduroğlu G, Kiliç NK, Dönmez G (2014) Bioremoval of reactive blue 220 by gonium sp. biomass. Environ Technol (UK) 35:2410–2415. https://doi.org/10.1080/09593330.2014.908240

Boelee NC, Temmink H, Janssen M et al (2014) Balancing the organic load and light supply in symbiotic microalgal-bacterial biofilm reactors treating synthetic municipal wastewater. Ecol Eng 64:213–221. https://doi.org/10.1016/j.ecoleng.2013.12.035

Brahmbhatt NH, Jasrai RT (2016) The role of algae in bioremediation of textile effluent. Int J Eng Res Gen Sci 4:443–453

Brar A, Kumar M, Vivekanand V, Pareek N (2017) Photoautotrophic microorganisms and bioremediation of industrial effluents: current status and future prospects. 3. Biotech 7:1–8. https://doi.org/10.1007/s13205-017-0600-5

Brierley JA, Brierley CL, Goyak GM (1986) AMT-BIOCLAIMTM: a new waste water treatment and metal recovery technology. In: Lawrence RW, Branion RMR, Ebner HG (eds) Fundamental and applied biohydrometallurgy. Elsevier, Amsterdam, pp 291–304

Brune DE, Lundquist TJ, Benemann JR (2009) Microalgal biomass for greenhouse gas reductions: potential for replacement of fossil fuels and animal feeds. J Environ Eng 135:1136–1145. https://doi.org/10.1061/(ASCE)EE.1943-7870.0000100CE

Cai T, Park SY, Li Y (2013) Nutrient recovery from wastewater streams by microalgae: status and prospects. Renew Sust Energ Rev 19:360–369. https://doi.org/10.1016/j.rser.2012.11.030

Ceron Garcia MC, Fernandez Sevilla JM, Acien Fernandez FG et al (2000) Mixotrophic growth of *Phaeodactylum tricornutum* on glycerol : growth rate and fatty acid profile. J Appl Phycol 12(12):239–248. https://doi.org/10.1002/bit.22257

Charumathi D, Das N (2012) Packed bed column studies for the removal of synthetic dyes from textile wastewater using immobilised dead *C. tropicalis*. Desalination 285:22–30. https://doi.org/10.1016/j.desal.2011.09.023

Chavan R (2001) Indian textile industry-environmental issues. Indian J Fibre Text Res 26:11–21

Chen CM, Shih ML, Lee SZ, Wang JS (2001) Increased toxicity of textile effluents by a chlorination process using sodium hypochlorite. Water Sci Technol 43:1–8

Chinnasamy S, Bhatnagar A, Claxton R, Das KC (2010a) Biomass and bioenergy production potential of microalgae consortium in open and closed bioreactors using untreated carpet industry effluent as growth medium. Bioresour Technol 101:6751–6760. https://doi.org/10.1016/j.biortech.2010.03.094

Chinnasamy S, Bhatnagar A, Hunt RW, Das KC (2010b) Microalgae cultivation in a wastewater dominated by carpet mill effluents for biofuel applications. Bioresour Technol 101:3097–3105. https://doi.org/10.1016/j.biortech.2009.12.026

Chollom MN, Rathilal S, Pillay VL, Alfa D (2015) The applicability of nanofiltration for the treatment and reuse of textile reactive dye effluent. Water SA 41:398–405. https://doi.org/10.4314/wsa.v41i3.12

Colak O, Kaya Z (1988) A study on the possibilities of biological wastewater treatment using algae. Doga Biyolji Serisi 12:18–29

Çetinkaya Dönmez G, Aksu Z, Öztürk A, Kutsal T (1999) A comparative study on heavy metal biosorption characteristics of some algae. Process Biochem 34:885–892. https://doi.org/10.1016/S0032-9592(99)00005-9

de la Noüe J, Laliberté G, Proulx D (1992) Algae and waste water. J Appl Phycol 4:247–254. https://doi.org/10.1007/BF02161210

Daneshvar E, Kousha M, Sohrabi MS et al (2012) Biosorption of three acid dyes by the brown macroalga Stoechospermum marginatum: isotherm, kinetic and thermodynamic studies. Chem Eng J 195–196:297–306. https://doi.org/10.1016/j.cej.2012.04.074

David Noel S, Rajan M (2014) Cyanobacteria as a potential source of Phycoremediation from textile industry effluent. J Bioremed Biodegr 5:10–13. https://doi.org/10.4172/2155-6199.1000260

Delée W, O'Neill C, Hawkes FR, Pinheiro HM (1998) Anaerobic treatment of textile effluents: a review. J Chem Technol Biotechnol 73:323–335. https://doi.org/10.1002/(SICI)1097-4660(199812)73:4<323::AID-JCTB976>3.0.CO;2-S

Dixit S, Singh DP (2014) Role of free living, immobilized and non-viable biomass of *Nostoc muscorum* in removal of heavy metals: An impact of physiological state of biosorbent. Cell Mol Biol 60:110–118. https://doi.org/10.14715/cmb/2014.60.5.18

Dogan D, Turkdemir H (2012) Electrochemical treatment of actual textile indigo dye effluent. Polish J Environ Stud 21:1185–1190

Domozych DS, Ciancia M, Fangel JU et al (2012) The cell walls of green algae: a journey through evolution and diversity. Front Plant Sci 3:1–8. https://doi.org/10.3389/fpls.2012.00082

ECR (1997) The Environment Conservation Rules, 1997. https://www.elaw.org/system/files/Bangladesh+%2D%2D+Environmental+Conservation+Rules,+1997.pdf. Accessed 21 Jan 2018

EEC Council (1991) Council directive of 21 May 1991 concerning urban waste-water treatment. Off J Eur Communities L 135:0040–0052. doi: http://eur-lex.europa.eu/legal-content/en/ALL/?uri=CELEX:31991L0271

El-Sheekh MM, Gharieb MM, Abou-El-Souod GW (2009) Biodegradation of dyes by some green algae and cyanobacteria. Int Biodeterior Biodegrad 63:699–704. https://doi.org/10.1016/j.ibiod.2009.04.010

Elumalai S, Saravanan GK (2016) The role of microalgae in textile dye industrial waste water recycle (phycoremediation). Int J Pharma Bio Sci 7:B662–B673
Environmental Protection Law (1998) Integrated wastewater discharge standard GB 8978–1996. http://english.mep.gov.cn/standards_reports/standards/water_environment/Discharge_standard/200710/t20071024_111803.htm. Accessed 21 Jan 2018
Environmental Quality Act 1974 (2009) Environmental Quality (Industrial Effluent) Regulations 2009. https://www.doe.gov.my/portalv1/wp-content/uploads/2015/01/Environmental_Quality_Industrial_Effluent_Regulations_2009_-_P.U.A_434-2009.pdf. Accessed 21 Jan 2018
Ertuğrul S, Bakir M, Dönmez G (2008) Treatment of dye-rich wastewater by an immobilized thermophilic cyanobacterial strain: Phormidium sp. Ecol Eng 32:244–248. https://doi.org/10.1016/j.ecoleng.2007.11.011
Farhadian M, Vachelard C, Duchez D, Larroche C (2008) In situ bioremediation of monoaromatic pollutants in groundwater: a review. Bioresour Technol 99:5296–5308. https://doi.org/10.1016/j.biortech.2007.10.025
Feng C, Sugiura N, Shimada S, Maekawa T (2003) Development of a high performance electrochemical wastewater treatment system. J Hazard Mater 103:65–78. https://doi.org/10.1016/S0304-3894(03)00222-X
Fernández C, Larrechi MS, Callao MP (2010) An analytical overview of processes for removing organic dyes from wastewater effluents. TrAC - Trends Anal Chem 29:1202–1211. https://doi.org/10.1016/j.trac.2010.07.011
Fersi C, Gzara L, Dhahbi M (2005) Treatment of textile effluents by membrane technologies. Desalination 185:399–409. https://doi.org/10.1016/j.desal.2005.03.087
Fouilland E (2012) Biodiversity as a tool for waste phycoremediation and biomass production. Rev Environ Sci Biotechnol 11:1–4. https://doi.org/10.1007/s11157-012-9270-2
Fukami K, Nishijima T, Ishida Y (1997) Stimulative and inhibitory effects of bacteria on the growth of microalgae. Hydrobiologia 358:185–191. https://doi.org/10.1023/A:1003139402315
Galán J, Rodríguez A, Gómez JM et al (2013) Reactive dye adsorption onto a novel mesoporous carbon. Chem Eng J 219:62–68. https://doi.org/10.1016/j.cej.2012.12.073
Ghazal FM, Battah MG, El-Aal AAA et al (2016) Studies on the efficiency of cyanobacteria on textile wastewater treatment. Res J Pharm Biol Chem Sci 7:2925–2931
González-Fernández C, Molinuevo-Salces B, García-González MC (2011) Nitrogen transformations under different conditions in open ponds by means of microalgae-bacteria consortium treating pig slurry. Bioresour Technol 102:960–966. https://doi.org/10.1016/j.biortech.2010.09.052
Gosavi V, Sharma S (2014) A General Review on Various Treatment Methods for Textile Wastewater. Journal of Environmental Science, Computer Science and Engineering & Technology 3:29–39
Gupta VK, Suhas (2009) Application of low-cost adsorbents for dye removal - a review. J Environ Manag 90:2313–2342. https://doi.org/10.1016/j.jenvman.2008.11.017
Gupta V, Ratha SK, Sood A et al (2013) New insights into the biodiversity and applications of cyanobacteria (blue-green algae)-prospects and challenges. Algal Res 2:79–97. https://doi.org/10.1016/j.algal.2013.01.006
Gupta N, Kushwaha AK, Chattopadhyaya MC (2016) Application of potato (*Solanum tuberosum*) plant wastes for the removal of methylene blue and malachite green dye from aqueous solution. Arab J Chem 9:S707–S716. https://doi.org/10.1016/j.arabjc.2011.07.021
Han X, Wong YS, Wong MH, Tam NFY (2007) Effects of anion species and concentration on the removal of Cr(VI) by a microalgal isolate, *Chlorella miniata*. J Hazard Mater 146:65–72. https://doi.org/10.1016/j.jhazmat.2008.02.024
Henze M, Harremoes P, La Cour JJ, Arvin E (2001) Waste water treatment: biological and chemical processes. Springer, Berlin Heidelberg
Hoffmann JP (1998) Wastewater treatment with suspended and nonsuspended algae. J Phycol 34:757–763. https://doi.org/10.1046/j.1529-8817.1998.340757.x
Holkar CR, Jadhav AJ, Pinjari DV et al (2016) A critical review on textile wastewater treatments: possible approaches. J Environ Manag 182:351–366. https://doi.org/10.1016/j.jenvman.2016.07.090

Jadhav AJ, Holkar CR, Karekar SE et al (2015) Ultrasound assisted manufacturing of paraffin wax nanoemulsions: process optimization. Ultrason Sonochem 23:201–207. https://doi.org/10.1016/j.ultsonch.2014.10.024
Jais NM, Mohamed RMSR, Al-Gheethi AA, Hashim MKA (2017) The dual roles of phycoremediation of wet market wastewater for nutrients and heavy metals removal and microalgae biomass production. Clean Techn Environ Policy 19:37–52. https://doi.org/10.1007/s10098-016-1235-7
Jaishankar M, Tseten T, Anbalagan N et al (2014) Toxicity, mechanism and health effects of some heavy metals. Interdiscip Toxicol 7:60–72. https://doi.org/10.2478/intox-2014-0009
Jinqi L, Houtian L (1992) Degradation of azo dyes by algae. Environ Pollut 75:273–278. https://doi.org/10.1016/0269-7491(92)90127-V
John J (2000) A self-sustainable remediation system for acidic mine voids. In: Proceedings of the 4th international conference of diffuse pollution, Bangkok, Thailand, pp 506–511
Johnson MB, Wen Z (2010) Development of an attached microalgal growth system for biofuel production. Appl Microbiol Biotechnol 85:525–534. https://doi.org/10.1007/s00253-009-2133-2
Jonstrup M, Punzi M, Mattiasson B (2011) Comparison of anaerobic pre-treatment and aerobic post-treatment coupled to photo-Fenton oxidation for degradation of azo dyes. J Photochem Photobiol A Chem 224:55–61. https://doi.org/10.1016/j.jphotochem.2011.09.006
Joshi M, Bansal R, Purwar R (2004) Colour removal from textile effluents. Indian J Fibre Text Res 29:239–259. https://doi.org/10.1016/0043-1354(91)90006-C
Kadirvelu K, Goal J (2007) Eco-friendly technologies for removal of hazardous heavy metals from water and industrial wastewater. In: Lewinsky AA (ed) Hazardous Materials and wastewater. Nova Science Publishers, Nature, pp 127–148
Kamaruddin MA, Yusoff MS, Aziz HA, Akinbile CO (2013) Review paper recent developments of textile waste water treatment by adsorption process : a review. International Journal of Scientific Research in Knowledge 1:60–73. https://doi.org/10.12983/ijsrk-2013-p060-073
Karacakaya P, Kiliç NK, Duygu E, Dönmez G (2009) Stimulation of reactive dye removal by cyanobacteria in media containing triacontanol hormone. J Hazard Mater 172:1635–1639. https://doi.org/10.1016/j.jhazmat.2009.08.037
Khalaf MA (2008) Biosorption of reactive dye from textile wastewater by non-viable biomass of Aspergillus Niger and Spirogyra sp. Bioresour Technol 99:6631–6634. https://doi.org/10.1016/j.biortech.2007.12.010
Khandare RV, Govindwar SP (2015) Phytoremediation of textile dyes and effluents: current scenario and future prospects. Biotechnol Adv 33:1697–1714. https://doi.org/10.1016/j.biotechadv.2015.09.003
Khouni I, Marrot B, Moulin P, Ben Amar R (2011) Decolourization of the reconstituted textile effluent by different process treatments: enzymatic catalysis, coagulation/flocculation and nanofiltration processes. Desalination 268:27–37. https://doi.org/10.1016/j.desal.2010.09.046
KIlIç NK, Karatay SE, Duygu E, Dönmez G (2011) Potential of Gonium spp. in synthetic reactive dye removal, possible role of laccases and stimulation by triacontanol hormone. Water Air Soil Pollut 222:297–303. https://doi.org/10.1007/s11270-011-0824-7
Koyuncu I, Güney K (2013) Membrane-based treatment of textile industry wastewaters. Encyclopedia of Membrane Science and Technology. https://doi.org/10.1002/9781118522318.emst127
Kulla HG, Klausener F, Meyer U et al (1983) Interference of aromatic sulfo groups in the microbial degradation of the azo dyes Orange I and Orange II. Arch Microbiol 135:1–7. https://doi.org/10.1007/BF00419473
Kumar PS, Narayan AS, Dutta A (2017) Nanochemicals and Effluent Treatment in Textile Industries. In: Muthu S. (eds) Textiles and Clothing Sustainability. Textile Science and Clothing Technology. Springer, Singapore
Kuyucak N, Volesky B (1990) Biosorption of algal Biomass. In: Volesky B (ed) Biosorption of heavy metals. CRC Press, Boca Raton\Ann Arbor\Boston, pp 7–44
Laliberté G, Lessard P, De la Noüe J, Sylvestre S (1997) Effect of phosphorus addition on nutrient removal from wastewater with the cyanobacterium *Phormidium bohneri*. Bioresour Technol 59:227–233. https://doi.org/10.1016/S0960-8524(96)00144-7

Latinwo GK, Jimoda LA, Agarry SE, Adeniran JA (2015) Biosorption of some heavy metals from textile wastewater by green seaweed Biomass. Univers J Environ Res Technol 5:210–219

Lau PS, Tam NFY, Wong YS (1995) Effect of algal density on nutrient removal from primary settled wastewater. Environ Pollut 89:59–66. https://doi.org/10.1016/0269-7491(94)00044-E

Liang Y (2013) Producing liquid transportation fuels from heterotrophic microalgae. Appl Energy 104:860–868. https://doi.org/10.1016/j.apenergy.2012.10.067

Liang CZ, Sun SP, Li FY et al (2014) Treatment of highly concentrated wastewater containing multiple synthetic dyes by a combined process of coagulation/flocculation and nanofiltration. J Memb Sci 469:306–315. https://doi.org/10.1016/j.memsci.2014.06.057

Lim SL, Chu WL, Phang SM (2010) Use of *Chlorella vulgaris* for bioremediation of textile wastewater. Bioresour Technol 101:7314–7322. https://doi.org/10.1016/j.biortech.2010.04.092

Lin SH, Chen ML (1997) Treatment of textile wastewater by chemical methods for reuse. Water Res 31:868–876

Mahapatra DM, Chanakya HN, Ramachandra TV (2013) Treatment efficacy of algae-based sewage treatment plants. Environ Monit Assess 185:7145–7164. https://doi.org/10.1007/s10661-013-3090-x

Mallick N (2002) Biotechnological potential of immobilised algae for wastewater N, P and metal removal: a review. Biometals 15:377–390

Mantzavinos D, Psillakis E (2004) Enhancement of biodegradability of industrial wastewaters by chemical oxidation pre-treatment. J Chem Technol Biotechnol 79:431–454. https://doi.org/10.1002/jctb.1020

Martin C, de la Noüe J, Picard G (1985) Intensive cultivation of freshwater microalgae on aerated pig manure. Biomass 7:245–259. https://doi.org/10.1016/0144-4565(85)90064-2

Martínez ME, Jiménez JM, El Yousfi F (1999) Influence of phosphorus concentration and temperature on growth and phosphorus uptake by the microalga *Scenedesmus obliquus*. Bioresour Technol 67:233–240. https://doi.org/10.1016/S0960-8524(98)00120-5

Mata TM, Martins AA, Caetano NS (2010) Microalgae for biodiesel production and other applications: A review. Renew Sustain Energy Rev 14:217–232. https://doi.org/10.1016/j.rser.2009.07.020

Maurya R, Ghosh T, Paliwal C et al (2014) Biosorption of methylene blue by de-oiled algal biomass: equilibrium, kinetics and artificial neural network modelling. PLoS One 9:1–13. https://doi.org/10.1371/journal.pone.0109545

Meng X, Liu G, Zhou J, Fu QS (2014) Effects of redox mediators on azo dye decolorization by Shewanella algae under saline conditions. Bioresour Technol 151:63–68. https://doi.org/10.1016/j.biortech.2013.09.131

Miralles-Cuevas S, Oller I, Agüera A et al (2017) Combination of nanofiltration and ozonation for the remediation of real municipal wastewater effluents: acute and chronic toxicity assessment. J Hazard Mater 323:442–451. https://doi.org/10.1016/j.jhazmat.2016.03.013

Mona S, Kaushik A, Kaushik CP (2011) Waste biomass of *Nostoc linckia* as adsorbent of crystal violet dye: optimization based on statistical model. Int Biodeterior Biodegrad 65:513–521. https://doi.org/10.1016/j.ibiod.2011.02.002

Morillo JA, Antizar-Ladislao B, Monteoliva-Sánchez M et al (2009) Bioremediation and biovalorisation of olive-mill wastes. Appl Microbiol Biotechnol 82:25–39. https://doi.org/10.1007/s00253-008-1801-y

Mostafa M (2015) Waste water treatment in chemical industries: the concept and current technologies. J Biodivers Environ Sci 7:2222–3045

Muñoz R, Köllner C, Guieysse B (2009) Biofilm photobioreactors for the treatment of industrial wastewaters. J Hazard Mater 161:29–34. https://doi.org/10.1016/j.jhazmat.2008.03.018

National Environmental Act (2008) National Environmental (protection and quality) regulations, No. 1 of 2008. http://www.cea.lk/web/images/pdf/envprotection/G_1534_18.pdf. Accessed 21 Jan 2018

Nawaz MS, Ahsan M (2014) Comparison of physico-chemical, advanced oxidation and biological techniques for the textile wastewater treatment. Alexandria Eng J 53:717–722. https://doi.org/10.1016/j.aej.2014.06.007

Olguín EJ (2003) Phycoremediation: key issues for cost-effective nutrient removal processes. Biotechnol Adv 22:81–91. https://doi.org/10.1016/S0734-9750(03)00130-7
Olguín EJ, Sánchez-Galván G (2012) Heavy metal removal in phytofiltration and phycoremediation: the need to differentiate between bioadsorption and bioaccumulation. New Biotechnol 30:3–8. https://doi.org/10.1016/j.nbt.2012.05.020
Oswald WJ (1988) Micro-algae and waste water treatment. In: Borowitzka MA, Borowitzka LJ (eds) Micro-algal biotechnology. Cambridge University Press, Cambridge, pp 305–328
Oswald WJ, Gotaas HB, Golueke CG, Kellen WR (1957) Algae in wastewater treatment. Sew Ind Waste 29:437–455
Ota M, Kato Y, Watanabe M et al (2011) Effects of nitrate and oxygen on photoautotrophic lipid production from Chlorococcum littorale. Bioresour Technol 102:3286–3292. https://doi.org/10.1016/j.biortech.2010.10.024
Pacheco MM, Hoeltz M, Moraes MSA, Schneider RCS (2015) Microalgae: cultivation techniques and wastewater phycoremediation. J Environ Sci Heal - Part A Toxic/Hazardous Subst Environ Eng 50:585–601. https://doi.org/10.1080/10934529.2015.994951
Pandi M, Shashirekha V, Swamy M (2009) Bioabsorption of chromium from retan chrome liquor by cyanobacteria. Microbiol Res 164:420–428. https://doi.org/10.1016/j.micres.2007.02.009
Paraskeva P, Diamadopoulos E (2006) Technologies for olive mill wastewater (OMW) treatment: a review. J Chem Technol Biotechnol 81:1475–1485. https://doi.org/10.1002/jctb.1553
Park EK, Lee SE (2002) Cadmium uptake by non-viable biomass from a marine brown alga Ecklonia radiata turn. Biotechnol Bioprocess Eng 7:221–224. https://doi.org/10.1007/BF02932974
Park Y, Je KW, Lee K et al (2008) Growth promotion of *Chlorella ellipsoidea* by co-inoculation with Brevundimonas sp. isolated from the microalga. Hydrobiologia 598:219–228. https://doi.org/10.1007/s10750-007-9152-8
Pathak VV, Singh DP, Kothari R, Chopra AK (2014) Phycoremediation of textile wastewater by unicellular microalga *Chlorella pyrenoidosa*. Cell Mol Biol 60:35–40. https://doi.org/10.14715/cmb/2014.60.5.7
Pathak VV, Kothari R, Chopra AK, Singh DP (2015) Experimental and kinetic studies for phycoremediation and dye removal by *Chlorella pyrenoidosa* from textile wastewater. J Environ Manag 163:270–277. https://doi.org/10.1016/j.jenvman.2015.08.041
Perales-Vela HV, Peña-Castro JM, Cañizares-Villanueva RO (2006) Heavy metal detoxification in eukaryotic microalgae. Chemosphere 64:1–10. https://doi.org/10.1016/j.chemosphere.2005.11.024
PERC Phycospectrum energy research centre (PERC) (n.d.-a). http://www.phycoremediation.in/moresnap.html. Accessed 15 Jan 2018
PERC Phycospectrum energy research centre (PERC) (n.d.-b). http://drvsiva.com/consultancy.html
Picot B, Bahlaoui A, Moersidik S et al (1992) Comparison of the purifying efficiency of high rate algal pond with stabilization pond. Water Sci Technol 25:197–206
Pires JCM, Alvim-Ferraz MCM, Martins FG, Simões M (2013) Wastewater treatment to enhance the economic viability of microalgae culture. Environ Sci Pollut Res 20:5096–5105. https://doi.org/10.1007/s11356-013-1791-x
Punzi M, Mattiasson B, Jonstrup M (2012) Treatment of synthetic textile wastewater by homogeneous and heterogeneous photo-Fenton oxidation. J Photochem Photobiol A Chem 248:30–35. https://doi.org/10.1016/j.jphotochem.2012.07.017
Punzi M, Anbalagan A, Aragão Börner R et al (2015) Degradation of a textile azo dye using biological treatment followed by photo-Fenton oxidation: evaluation of toxicity and microbial community structure. Chem Eng J 270:290–299. https://doi.org/10.1016/j.cej.2015.02.042
Rangsayatorn N, Pokethitiyook P, Upatham ES, Lanza GR (2004) Cadmium biosorption by cells of Spirulina platensis TISTR 8217 immobilized in alginate and silica gel. Environ Int 30: 57–63. https://doi.org/10.1016/S0160-4120(03)00146-6

Renuka N, Sood A, Prasanna R, Ahluwalia AS (2015) Phycoremediation of wastewaters: a synergistic approach using microalgae for bioremediation and biomass generation. Int J Environ Sci Technol 12:1443–1460. https://doi.org/10.1007/s13762-014-0700-2

Robinson T, McMullan G, Marchant R, Nigam P (2001) Remediation of dyes in textile effluent: a critical review on current treatment technologies with a proposed alternative. Bioresour Technol 77:247–255. https://doi.org/10.1016/S0960-8524(00)00080-8

Sandau E, Sandau P, Pulz O, Zimmermann M (1996) Heavy metal sorption by marine algae and algal by-products. Acta Biotechnol 16:103–119. https://doi.org/10.1002/abio.370160203

Sarayu K, Sandhya S (2012) Current technologies for biological treatment of textile wastewater-a review. Appl Biochem Biotechnol 167:645–661. https://doi.org/10.1007/s12010-012-9716-6

Schedule-VI (1986) General Standards for Discharge of Environmental Pollutants. http://cpcb.nic.in/GeneralStandards.pdf. Accessed 21 Jan 2018

Sekomo CB, Rousseau DPL, Saleh SA, Lens PNL (2012) Heavy metal removal in duckweed and algae ponds as a polishing step for textile wastewater treatment. Ecol Eng 44:102–110. https://doi.org/10.1016/j.ecoleng.2012.03.003

SGS (2015) Discharge Standards Of Water Pollutants For Dyeing And Finishing Of Textile Industry – GB 4287–2012. http://www.sgs.com/en/news/2015/07/safeguards-12115-china-discharge-standards-of-water-pollutants-for-dyeing-and-finishing-of-textile. Accessed 21 Jan 2018

Shelef G, Azov Y, Moraine R, Oron G (1980) Algal mass production as an integral part of a wastewater treatment and reclamation system. In: Shelef G, Soeder CJ (eds) Algae Biomass. Elsevier, North-Holland Biomedical Press, Amsterdam, pp 163–189

Shen QH, Zhi TT, Cheng LH et al (2013) Hexavalent chromium detoxification by nonliving *Chlorella vulgaris* cultivated under tuned conditions. Chem Eng J 228:993–1002. https://doi.org/10.1016/j.cej.2013.05.074

Silver S (1996) Bacterial resistances to toxic metal ions - a review. Gene 179:9–19. https://doi.org/10.1016/S0378-1119(96)00323-X

Singh K, Arora S (2011) Removal of synthetic textile dyes from wastewaters: a critical review on present treatment technologies. Crit Rev Environ Sci Technol 41:807–878. https://doi.org/10.1080/10643380903218376

Sinha S, Singh R, Chaurasia AK, Nigam S (2016) Self-sustainable *Chlorella pyrenoidosa* strain NCIM 2738 based photobioreactor for removal of direct Red-31 dye along with other industrial pollutants to improve the water-quality. J Hazard Mater 306:386–394. https://doi.org/10.1016/j.jhazmat.2015.12.011

Sivakumar KK, Balamurugan C, Ramakrishnan D, Bhai LH (2011) Assessment studies on wastewater pollution by textile dyeing and bleaching industries at Karur, Tamil Nadu. Rasayan J Chem 4:264–269

Smith B (1988) A workbook for pollution prevention by source reduction in textile wet processing. Pollution Prevention Program, North Carolina Department of Environment Health and Natural Resources, Raleigh, North Carolina, USA

Soares PA, Silva TFCV, Manenti DR et al (2014) Insights into real cotton-textile dyeing wastewater treatment using solar advanced oxidation processes. Environ Sci Pollut Res 21:932–945. https://doi.org/10.1007/s11356-013-1934-0

SOR/2012-139 (2015) Wastewater Systems Effluent Regulations. http://laws-lois.justice.gc.ca/PDF/SOR-2012-139.pdf. Accessed 21 Jan 2018

Spolaore P, Joannis-Cassan C, Duran E, Isambert A (2006) Commercial applications of microalgae. J Biosci Bioeng 101:87–96. https://doi.org/10.1263/jbb.101.87

Sponza DT (2002) Necessity of toxicity assessment in Turkisk industrial discharges (examples from metal and textile industry effluents). Environ Monit Assess 73:41–66. https://doi.org/10.1023/A:1012663213153

Sturm BSM, Lamer SL (2011) An energy evaluation of coupling nutrient removal from wastewater with algal biomass production. Appl Energy 88:3499–3506. https://doi.org/10.1016/j.apenergy.2010.12.056

Subashchandrabose SR, Ramakrishnan B, Megharaj M et al (2011) Consortia of cyanobacteria/microalgae and bacteria: biotechnological potential. Biotechnol Adv 29:896–907. https://doi.org/10.1016/j.biotechadv.2011.07.009

Subramanian G, Yadav G, Sen R (2016) Rationally leveraging mixotrophic growth of microalgae in different photobioreactor configurations for reducing the carbon footprint of an algal biorefinery: a techno-economic perspective. RSC Adv 6:72897–72904. https://doi.org/10.1039/C6RA14611B

Şentürk T, Çamlı Ç, Yıldız Ş (2017) Comparing the phytoremediation efficiency of three different algae for the nutrient removal of Gediz River in Manisa/Turkey. Celal Bayar Üniversitesi Fen Bilim Derg 13:737–743. https://doi.org/10.18466/cbayarfbe.339348

Talbot P, de la Noüe J (1993) Tertiary treatment of wastewater with *Phormidium bohneri* (Schmidle) under various light and temperature conditions. Water Res 27:153–159. https://doi.org/10.1016/0043-1354(93)90206-W

Tam NFY, Wong YS (1989) Wastewater nutrient removal by Chlorella pyrenoidosa and Scenedesmus sp. Environ Pollut 58:19–34. https://doi.org/10.1016/0269-7491(89)90234-0

Tang X, He LY, Tao XQ et al (2010) Construction of an artificial microalgal-bacterial consortium that efficiently degrades crude oil. J Hazard Mater 181:1158–1162. https://doi.org/10.1016/j.jhazmat.2010.05.033

Tchobanoglous G, Burton FL (1991) Wastewater engineering: Treatment, Disposal and Reuse, 3rd edn. Medcalf & Eddy, McGraw-Hill, New York

Urushigawa Y, Yonezawa Y (1977) Chemo-biological interactions in biological purification system II – biodegradation of azo compound by activated sludge. Bull Environ Contam Toxicol 17:214–218

USEPA (1996) Manual: best management practices for pollution prevention in the textile industry. https://nepis.epa.gov/EPA/html/DLwait.htm?url=/Exe/ZyPDF.cgi/30004Q2U.PDF?Dockey=30004Q2U.PDF. Accessed 21 Jan 2018

USEPA (2012a) Part 410—Textile Mills Point Source Category. https://www.gpo.gov/fdsys/pkg/CFR-2012-title40-vol30/pdf/CFR-2012-title40-vol30-part410.pdf. Accessed 21 Jan 2018

USEPA (2012b) Part 425—Leather Tanning And Finishing Point Source Category. https://www.gpo.gov/fdsys/pkg/CFR-2012-title40-vol31/pdf/CFR-2012-title40-vol31-part425.pdf. Accessed 21 Jan 2018

Vasanth Kumar K, Ramamurthi V, Sivanesan S (2006) Biosorption of malachite green, a cationic dye onto Pithophora sp., a fresh water algae. Dyes Pigments 69:102–107. https://doi.org/10.1016/j.dyepig.2005.02.005

Venkata Mohan S, Rohit MV, Chiranjeevi P et al (2015) Heterotrophic microalgae cultivation to synergize biodiesel production with waste remediation: Progress and perspectives. Bioresour Technol 184:169–178. https://doi.org/10.1016/j.biortech.2014.10.056

Volesky B (2001) No TitleDetoxification of metal-bearing effluents: biosorption for the next century. Hydrometallurgy 59:203–216

Wang J, Chen C (2009) Biosorbents for heavy metals removal and their future. Biotechnol Adv 27:195–226. https://doi.org/10.1016/j.biotechadv.2008.11.002

Wang L, Zhang C, Wu F, Deng N (2007a) Photodegradation of aniline in aqueous suspensions of microalgae. J Photochem Photobiol B Biol 87:49–57. https://doi.org/10.1016/j.jphotobiol.2006.12.006

Wang X, Gu X, Lin D et al (2007b) Treatment of acid rose dye containing wastewater by ozonizing - biological aerated filter. Dyes Pigments 74:736–740. https://doi.org/10.1016/j.dyepig.2006.05.009

Wu JY, Lay CH, Chen CC, Wu SY (2017) Lipid accumulating microalgae cultivation in textile wastewater: environmental parameters optimization. J Taiwan Inst Chem Eng 79:1–6. https://doi.org/10.1016/j.jtice.2017.02.017

Xin L, Hong-ying H, Ke G, Ying-xue S (2010) Effects of different nitrogen and phosphorus concentrations on the growth, nutrient uptake, and lipid accumulation of a freshwater microalga Scenedesmus sp. Bioresour Technol 101:5494–5500. https://doi.org/10.1016/j.biortech.2010.02.016

Yee N, Benning LG, Phoenix VR, Ferris FG (2004) Characterization of metal-cyanobacteria sorption reactions: a combined macroscopic and infrared spectroscopic investigation. Environ Sci Technol 38:775–782. https://doi.org/10.1021/es0346680

Yen HY (2016) Energy consumption of treating textile wastewater for in-factory reuse by H2O2/UV process. Desalin Water Treat 57:10537–10545. https://doi.org/10.1080/19443994.2015.1039599

Yuan X, Kumar A, Sahu AK, Ergas SJ (2011) Impact of ammonia concentration on Spirulina platensis growth in an airlift photobioreactor. Bioresour Technol 102:3234–3239. https://doi.org/10.1016/j.biortech.2010.11.019

ZDHC (2015) Textile industry wastewater discharge quality standards: literature review. https://www.roadmaptozero.com/fileadmin/pdf/WastewaterQualityGuidelineLitReview.pdf. Accessed 15 Jan 2018.

Zeng X, Guo X, Su G et al (2015) Bioprocess considerations for microalgal-based wastewater treatment and biomass production. Renew Sust Energ Rev 42:1385–1392. https://doi.org/10.1016/j.rser.2014.11.033

Zhou W, Min M, Li Y et al (2012) A hetero-photoautotrophic two-stage cultivation process to improve wastewater nutrient removal and enhance algal lipid accumulation. Bioresour Technol 110:448–455. https://doi.org/10.1016/j.biortech.2012.01.063

Zhu X-G, Long SP, Ort DR (2008) What is the maximum efficiency with which photosynthesis can convert solar energy into biomass? Curr Opin Biotechnol 19:153–159. https://doi.org/10.1016/j.copbio.2008.02.004

Potential and Application of Diatoms for Industry-Specific Wastewater Treatment

Archana Tiwari and Thomas Kiran Marella

1 Introduction

The composition of the water body is greatly influenced by the anthropogenic activities in the vicinity. The nature of effluents entering the water body can be from diverse sources but can be broadly categorized as anthropogenic waste, agricultural waste, and industrial waste. The physical and chemical changes occur after the introduction of the pollutants into the water body, thereby contributing toward remarkable alterations in the structure of water body, severely affecting the aquatic flora and fauna. The intervention of myriad pollutants into the water system leads to the enhancement in the concentration of inorganic nutrients like phosphate, nitrate, ammonium, etc. triggering a sequence of consequences that adversely effects the entire inhabiting aquatic population (Thomas et al. 2016).

The nutrient accumulation enhances the algal growth due to which there is depletion of oxygen in the water and secretion of toxins and secondary metabolites, which might be fatal for the fish and other aquatic organisms. The nature of toxins varies from hepatotoxins, neurotoxins, dermatotoxins, etc. depending upon the nature of cyanobacteria (Tiwari and Pandey 2014). Often an obnoxious smell is observed in the surrounding area, and the water becomes unsuitable for consumption even for animals.

Diatoms are microscopic photosynthetic algae commonly classified under *Bacillariophyceae*, and they inhabit a wide range of aquatic niches. They are the integral part of the aquatic food web and constitute 40% of the primary producers

A. Tiwari (✉)
Amity Institute of Biotechnology, Amity University, Noida, U.P., India

T. K. Marella
International Crops Research Institute for Semi -arid Tropics (ICRISAT), Hyderabad, India

S. K. Gupta, F. Bux (eds.), *Application of Microalgae in Wastewater Treatment*,
https://doi.org/10.1007/978-3-030-13913-1_15

(Thomas et al. 2015). Diatoms absorb the atmospheric carbon dioxide through photosynthesis and transform the carbon into carbohydrates, which I further utilized for the formation of different biomolecules (proteins, lipids, nucleic acids). Stimulated growth of diatoms in water body can aid in eradication of multiple problems related to the pollution of water by diverse sources. The occurrence of harmful algal blooms (HAB) is a common problematic condition evident in eutrophic water body, but copious diatom growth can result in nullifying or curbing the growth of cyanobacteria (blue-green algae) leading to the prevention in the formation of HAB.

Diatoms are the noteworthy algae in the phycoremediation of diverse wastewaters by virtue of their extraordinary cellular machinery. They are experts in utilizing of nitrate, phosphate, iron, copper, molybdenum, and silica; in addition, they are capable of remediation of heavy metals like lead, cadmium, chromium, copper, etc. Diatoms show high degree of flexibility in varied culture conditions that could be useful for their use in challenging conditions. Diatom algae can dominate under nutrient limiting and excess conditions. Diatom produced oxygen during photosynthesis which acts as stimulant for heterotrophic bacterial growth which in turn can enhance bacterial degradation and oxidation of organic pollutants and heavy metals. The remediation of wastewater is concomitant with its usage as source of macronutrients and macronutrients for growth of diatoms. The diatom biomass grown on the wastewater can be utilized for the generation of a range of value-added products like biofuels, nutraceuticals, antimicrobial substances, omega-3 fatty acids, and aqua feed to name a few applications of diatoms (Fig. 1). This approach is a sustainable solution of wastewater management as it is coupled with generation of useful products.

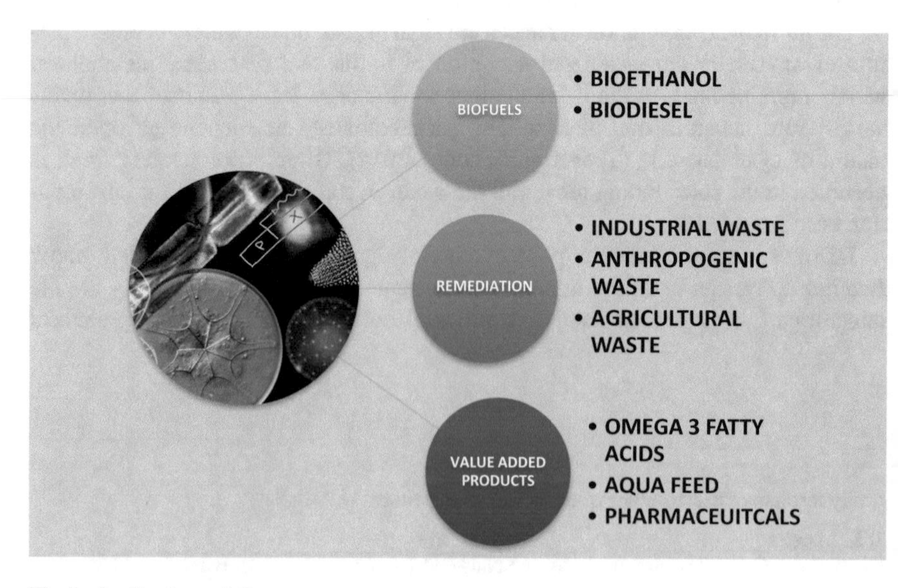

Fig. 1 Applications of diatoms

2 Physiological Advantages of Diatoms for Industrial Wastewater Treatment

Diatom algae play a significant role in controlling and biomonitoring of organic pollutants, heavy metals, hydrocarbons, PCBs, pesticides, etc. in aquatic ecosystems. Although diatoms are extensively studied for their role as indicators of different kinds of water pollution, their application in phycoremediation of polluted water bodies is in the incipient phase (Thomas et al. 2016).

Diatoms evolved dates back 180 million years, and at present, more than 100,000 species have been reported (Kroth 2007). They play a significant role in many of the earth's biogeochemical cycles like carbon, phosphate, and silicon (Falciatore and Bowler 2002). Diatoms are primary organisms in aquatic food webs. They form the basis of the most common and economically significant food web consisting of fish via copepods or to shell fish without any intermediate trophy level (Ryther 1969). They form these food webs in most productive regions and support many important economically important fish species. Diatom algae are ideal in size to be consumed by zooplankton (Ambler and Frost 1974); they also contain protein, carbohydrate, lipid, and vitamin; and they are known to be better diet than other algal species.

In coastal waters, diatom biomass contributes prominently to annual influx of organic material to benthos (Smetacek et al. 1984). Diatoms dominate under conditions optimum for phytoplankton growth like N, P, Si, and Fe concentrations (Hulburt 1990). Diatom dominance over other algae is often contributed to its silica frustules which gives protection from grazers (Hamm et al. 2003) and higher division rate (Smetacek 1999). This combination of ecological success and efficient transport of C to higher organisms can be the reason for diatoms being the base of the most productive ecosystems on the planet.

Diatom dominance in world's oceans was governed by silica availability, and distribution is due to their potential to utilize silicate for the construction of their cell walls called frustules. This makes them primary contributors for global silicon cycle. Producing silica cell walls needs less energy when compared with building with organic substances like cellulose; this gives an ecological advantage to diatoms over other algae (Raven 1983). So as long as silicate is present, diatoms will dominate other algae (Egge and Aksnes 1992). Diatom algae are main contributors to silicon pump which acts as a means for transport of silica and carbon to deep oceans (Dugdale and Wilkerson 1988). Diatom dominance in oceans altered the marine silica cycle (Racki and Cordey 2000). In modern oceans diatoms are dominant phytoplankton which utilizes N and SI and play a pivotal role in the biogeochemistry of aquatic ecosystems.

Diatom algae with their faster growth rate, nitrate uptake, and larger cell size results in faster sinking rate, so they contribute to export production (Buesseler 1998). These attributes enable diatoms to play a significant part in climate control (Traguer and Pondaven 2000). Optimum silica concentration in oceans has led to significant decrease in atmospheric pCO_2 by favoring diatom growth over coccolithophores (Archer 2006). The effect of silica-rich water in subtropics and beyond has led to diatom growth, thereby increasing the depth of organic matter remineralization; this has led to an estimated lowering of atmospheric PCO_2 by 60 ppm (Brzezinski et al. 2002). Carbon trapped inside silica frustules of diatoms acts as major components of carbon cycle on earth (Street-Perrott and Barker 2008). A silica body can sequester up to 50% of its weight of C (Elbaum et al. 2009). Diatom algae are used as indicators for climate change in lacustrine sediments due to their high temporal sensitivity, so they also act as indicators of temperature increase which is an early indication for climate change (Kilham et al. 1996).

Diatom nutrient uptake rate is significantly higher than any other group of algae (Litchman et al. 2006) and have been documented for their role in the initial uptake of nitrate at the equatorial upwelling zone of Pacific Ocean. The efficiency of nutrient uptake along with their higher growth rate makes them good candidates for nutrient accusation and transport. Diatoms are highly efficient in utilizing nutrients and are known to be responsible. This will have a significant impact on productivity and nutrient utilization. Due to their larger nutrient storage capacity, they can outcompete other algae in terms of productivity even in nutrient replete conditions. Amano et al. (2011) reported that in a eutrophic lake under low nitrate conditions, diatoms outcompete non-N-fixing cyanobacteria. Furnas (1990) reported that diatom doubling rates for both pennate and centric diatoms lie between two and four divisions per day, which is much more compared to any algae. Diatoms outcompete other phytoplankton under mixing and high turbulence (Tozzi et al. 2004). Diatoms possess higher carbon-fixing ability than other microalgae; this phenomenon is observed in both laboratory and field conditions (Thomas et al. 1978). Diatoms when compared with other algae grew better under low light conditions; this can be attributed to fucoxanthin, the major light harvesting pigment in diatoms which needs less light for saturation (Smetacek 1999). In a comparative analysis among *P. tricornutum* and *Chlorella vulgaris*, under light fluctuations, the light conversion proficiency into biomass was twice in diatoms. Diatoms store carbohydrate in the form of a chrysolaminarin (soluble form) instead of starch (insoluble form) like other algae. This gives them an advantage over other algae if we consider relative energy required to utilize soluble carbohydrate instead of insoluble carbohydrate (Libessart et al. 1995).

The presence of silica cell wall or frustule in diatoms is a unique characteristic, which not only acts as protective layer but also crowns advantageous dominance over other aquatic beings. Diatoms perform silica polymerization through an energy-efficient process even at low silica concentrations (Raven 1983). Silica cell wall plays a significant role in carbon-concentrating mechanism by acting as a pH buffer enabling enhanced carbonic anhydrase activity near diatom cell wall which results in bicarbonate to CO_2 conversion (Milligan and Morel 2002).

Diatoms with their efficient carbon fixing, nutrient utilization, and growth under varying nutrient, light, and turbulence are ideal candidates for co-processes like CO_2 sequestration and wastewater treatment. In spite of all these attributes, they are the least explored species in terms of research related to wastewater treatment and biomolecule production compared to green algae and cyanobacteria.

3 Factors Influencing Diatom Cultivation for Industrial Wastewater Treatment

The growth of algae is greatly influenced by multiple environmental factors like light, temperature, nutrients, carbon dioxide, and biological factors (Grobbelaar 2009). The utilization of wastewater for algal growth coupled with bioactive compounds requires the elucidation of factors that affect the growth and metabolism. Li et al. (2017a, b) have reported orthogonal test design for the diatom growth optimization, and it was reported that the lipid content was strongly influenced by the concentration of silica along with other factors. Elucidation of factors and their optimization can aid in better efficiency of the diatoms for wastewater management concomitant with several useful products for mankind (Fig. 2).

3.1 Light

Diatoms are photosynthetic in nature; hence, light has a profound influence on productivity. Diatoms are known to inhabit many diverse ecosystems with varying environmental conditions, making them one of the most adaptable to variations in the intensity of light, duration depending on latitude, season, and depth in order to keep growing and attain maximum productivity. In industrial wastewaters depending on the design of the treatment facility, drastic differences exist between light intensities available for algae to grow from high light in open oxidation ponds to low light in indoor effluent treatment ponds. The impact of light intensity (Falkowski and Raven 2007), relationship of light intensity and nutrient limitation (Sakshaug et al. 1989; Halsey and Jones 2015), and light fluctuations (Orefice et al. 2016) have been elucidated via competition models on diatoms (Litchman and Klausmeier 2001). In addition to the intensity of light, the growth is also effected by the dark

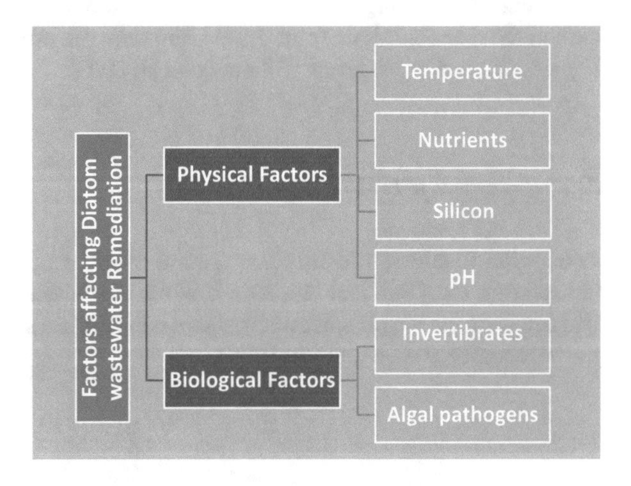

Fig. 2 Factors affecting diatom wastewater treatment

cycle. Other factors like night length, maximum irradiance, and spectral composition also influence diatom physiological response. The photosynthetic process in diatom photosynthesis comprises of intricate association with the chloroplasts and mitochondria (Bailleul et al. 2015). The cell size also plays an important role in growth of diatom at varying light intensity and photoperiods with larger cells favoring short photoperiods (Li et al. 2017a, b).

3.2 Silicon

The obtainability of silicon in wastewater is a major factor which defines diatom use for industrial wastewater treatment. Diatom metabolism and the role of silicon are quite conspicuous and perhaps account for their profound accomplishment due to their silica wall. The silicified diatom cell wall endows them additional potential and thus less energy requirement compared to other cellulose cell wall organisms. In the freshwater systems, the concentration of silicon is quite high and can sustain higher diatom productivity. This makes freshwater diatoms ideal candidates to treat fresh and brackish wastewater without the addition of silicon. In addition, many industrial wastewaters contain high amount of silica as it is used in majority of industries in a variety of production systems and in appliances, thereby making diatom-mediated remediation a good, effective, economic option coupled with other benefits associated with the further usage of residual diatom biomass (Thomas et al. 2018).

3.3 Carbon Dioxide

Diatoms can fix carbon dioxide from different sources like atmosphere and gases from industries (Wang et al. 2008). In nature diatoms are actively involved in the carbon dioxide assimilate from the air, and they can efficiently utilize substantially higher carbon dioxide levels (Bilanovic et al. 2009).

3.4 Other Nutrients

In addition to silicon diatoms also require other inorganic nutrients like nitrogen, phosphorus, etc. (Suh and Lee 2003). While cyanobacteria are capable of nitrogen fixation from the air, all other microalgae require it in a soluble form with urea being the best source (Hsieh and Wu 2009).

3.5 *Temperature*

The growth of diatoms in wastewater is greatly affected by temperature as it has a significant role in the enzymatic activities and thus metabolism. For optimum growth and remediation mediated by diatoms in wastewater, it is essential to sustain suitable temperature within constricted limits. In nature temperature variation exists on daily basis and on seasonal basis. During summer there exists huge variation in the temperature as early mornings are cooler followed by warmer climate in the daytime and then reduction in temperature at night.

In the open pond system of algal cultivation, variations in temperature are observed with change in seasons. The wastewater from various industries like cement industry not only enriches the inorganic carbons but also emits waste thermal energy, which can be utilized in open ponds in maintaining temperature particularly in cold climatic conditions. In fact the flue gas from the industrial wastewater should be cooled before entering into the wastewater cultivation system as too much of heat is not suitable for growth of diatoms (McGinn et al. 2011). The temperature has to be maintained at optimum levels to foster diatom growth, and hence effective remediation potential along with good biomass yields for valuable products (Fig. 3).

4 Cultivation Systems for Diatom-Based Industrial Wastewater Treatment

Excellent and efficient remediation of wastewater from different sources requires the systematic cultivation system for optimum removal of waste. In general the cultivation system meant for algae is open system consisting of variety of ponds or closed systems and modern hybrid systems (Fig. 4). The different cultivation systems have their own advantages and limitations. The open ponds are the simplest ways of algal cultivation under the influence of climatic conditions and controlled within limits. The closed systems require appropriate designing of photobioreactors, which can be expressive yet efficient. The hybrid systems culminate some features of open and closed system, and they are capable of attaining efficient nutrient removal from wastewater and production of biomass (Tiwari and Thomas 2018). The concept of exclusive algal cultivation began in 1950 for the application of algae as a source of protein (Brennan and Owende 2010). Later on the different algal products and most significantly wastewater treatment began to be explored. In algal cultivation systems have developed in due course of time through extensive research executed by phycologists around the world (Tan et al. 2018).

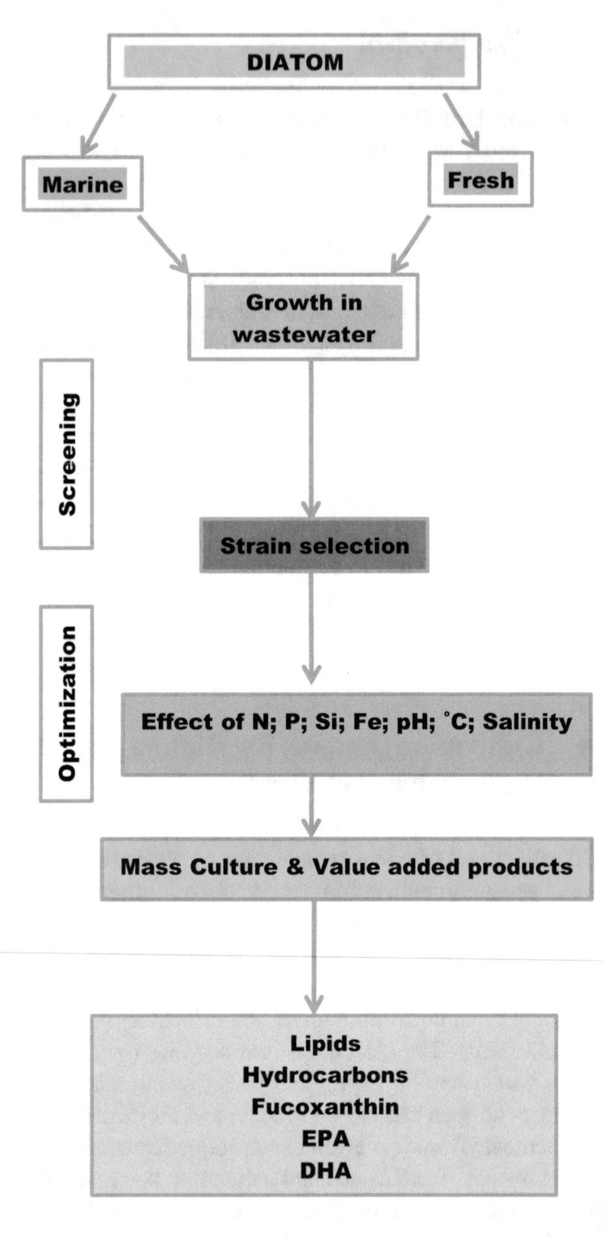

Fig. 3 Synergistic approach toward wastewater treatment

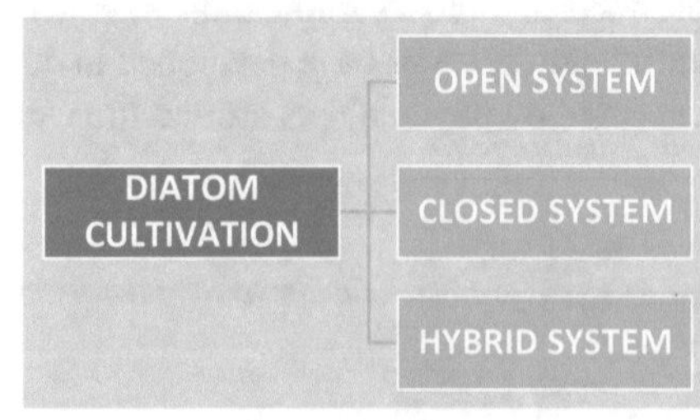

Fig. 4 Cultivation systems for remediation

4.1 Open Ponds

The open pond system has been extensively used for the cultivation of diverse algal species for wastewater remediation. Though it is an economical system of algal cultivation, there are certain constrains associated with open pond cultivation like the demand for land, climatic influence, and contaminants to name a few. The frequently used open systems include the raceway ponds, the inclined systems, circular tanks, shallow big pond, etc. In the inclined systems, the flow is in the inclined pattern to ensure the proper mixing of such systems has been reported to be successful in diatom cultures of *Phaeodactylum* and *Scenedesmus* (Fazal et al. 2018). Circular ponds are characterized by the centrally located agitator, which enables the adequate mixing of the cell suspension, and its efficiency is quite low in huge ponds.

The raceway ponds or the high-rate algal ponds (HRAP) are marked by the use of paddle wheel for uniform mixings and sedimentation prevention. These ponds are cost-effective and efficient in algal growth performance. The concept of high-rate algal ponds was conceived by Oswald and Golueke (1960), and later on, it was utilized globally in the treatment of municipal wastewater.

4.2 Closed Pond

The limitations of the open ponds are eradicated in the closed system to provide controlled culture conditions (temperature, pH, light, mixing) for optimum algal growth and suitable nutrient removal from wastewater. The photobioreactors used for algal cultivation includes:

- Tubular photobioreactors
- Vertical tank photobioreactors
- Horizontal tube photobioreactors
- Flat-plate photobioreactors
- Helical tube photobioreactors
- Airlift photobioreactors
- Vertical column photobioreactors

4.3 Hybrid Systems and Advanced Integrated Wastewater Ponds

The hybrid algal systems culminate the properties of both open and closed cultivation systems. Initially the algal culture is grown in the open pond, and later on, it is cultivated in the closed photobioreactor. The first open pond system cultivation induces nutrient stress on the algal culture which further leads to enhanced biomass and lipid productivity in the photobioreactor within the optimum set of culture conditions (Tiwari and Thomas 2018). The advanced integrated wastewater pond systems are well articulated for wastewater remediation, and it comprises of four integrated advanced ponds for rapid and effective remediation. The first pond is the facultative pond wherein the digester pit is located which flows the wastewater to the second pond called the HRAP, which eliminates the dissolved nutrients. The third pond is called the settling pond in which the sedimentations occur, and the last pond is the maturation pond which provides sunlight and sufficient oxygen (Sen et al. 2013) (Fig. 5).

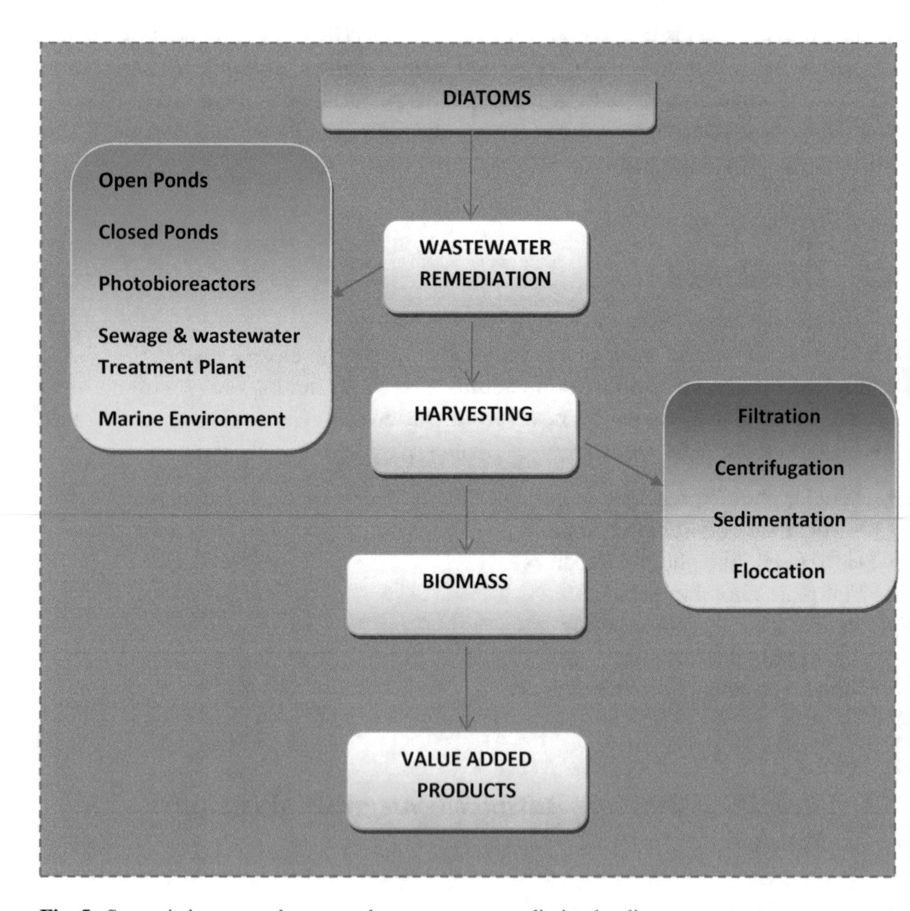

Fig. 5 Synergistic approaches toward wastewater remediation by diatoms

5 Integration of Diatom Cultivation with Industrial Wastewater Treatment

5.1 Municipal Wastewater

Municipal wastewater untreated, partially treated, and treated contains inorganic nutrients which when discharged into water bodies can lead to eutrophication. Due to rapid urbanization, many megacities are not able to treat even 50% of their domestic wastewater with their existing conventional sewage treatment infrastructure. For algae growth the main requirement is nutrients, and these are differentiated into major (C, N, P, Si) and micro nutrients (Ca, Mg, K, Fe, Mn, Cu, and Co). Any deficiency in nutrient availability can negatively affect algae growth in any type of wastewater. Depending on the strength, domestic wastewater contains 20–85 mg L^{-1} N and 4–15 mg L^{-1} P. This amount of N and P can produce 0.3–1.4 mg L^{-1} algal biomass, and micro nutrient concentration in any wastewater is always enough to sustain their growth as they are required in only trace amount (Christenson and Sims 2011). So domestic wastewater is ideal for grow microalgae. Taking this into consideration, many researchers used algae to treat wastewater from decades. In the 1960s, Oswald and Golueke (1960) proposed one of the first algae-based biological wastewater treatment system called advanced integrated wastewater pond systems (AIWPS). These systems work like present-day high-rate algae pond (HRAP) systems enabling fast growth of naturally occurring algae in paddle-wheeled ponds filled with wastewater. Subsequent research employed varied technologies like closed photobioreactors, HRAP, biofilm reactors, and raceway ponds. Of all the different microalgae species studied for their use in municipal wastewater treatment, the most studied species is green algae especially *Chlorella* sp. and *Scenedesmus* sp. followed by cyanobacteria. Compared to green and blue-green algae diatom, algae-related studies are very few. Thomas et al. (2018) studied effect of diatom growth on nutrient removal from municipal wastewater. In this study they triggered native diatom consortium in wastewater, and this resulted in 95% N, 88.9% P, 91% COD, and 51% BOD reduction in lab-scale experiments. Although many researchers prefer working with single species, working with consortium gives a distinctive advantage especially with diatoms; in diatom consortium, there are different species of diatoms which grow at different nutrient levels, so when we are treating wastewater in field-scale experiments, the inlet water nutrient strength always varies depending on factors like season, temperature, water usage, etc. To sustain this dynamic water chemistry and grow, we need robust multispecies cultures which can be possible only with diatoms. Nutrient dynamics always governs diatom species diversity in natural systems with certain species always favoring nutrient replete condition, while others prefer nutrient-depleted conditions, and this diversity and trophic flexibility of diatoms give them an edge over other algae when grown as consortium to treat wastewater. Untreated and partially municipal waste, wastewater is one of the main causes for eutrophication in urban lakes in developing countries. Traditional sewage treatment systems can remove nitrate but not phosphate due to their reliance on bacterial-based nutrient removal strategies. The main drawback with this system lies in the fact that very few phosphate-solubilizing bacteria present in these systems compared to denitrifying bacteria. This resulted in treated water with high phosphate content which is ideal for cyanobacterial growth leading to eutrophication. To counter this trend, proper system should be incorporated into existing sewage treatment infrastructure to simultaneously reduce N and P, using diatom consortium which can reduce nutrients rapidly and can create a nutrient equilibrium in treated water.

5.2 Dairy Wastewater

Dairy wastewater contains complex mix of inorganic nutrients in high concentration. One of the main challenges in treating livestock-derived wastewater is reduction of oxygen demand which can be quite high (2000–2500 mg/l for COD and 800 mg/l for BOD). It also contains high turbidity, total suspended solids (TSS), and dissolved nutrient like ammonia and phosphate. Majority of dairy waste which includes both solid and liquid is mostly treated using anaerobic digestion process, but the resultant effluent or liquid waste needs a special treatment for it to be fit to release into downstream. Improper treatment of these effluents can lead to a potential threat to watersheds, leading to eutrophication. Algae with their high nutrient assimilation rate combined with fast growth rates are extensively studied to treat dairy waste. Field-scale experiments and installations of algae-based technology for dairy effluent treatment were carried out using advanced oxidation ponds (Craggs et al. 2003) and algae scrubber technology (Mulbry and Wilkie 2001). Using a series of oxidation and settling ponds for wastewater treatment is a traditional technology used for many years which are cost-effective, but the treated water is often unsuitable for discharge due to partial treatment. Advanced pond system (APS) which is a modified version of traditional system with a series of oxidation ponds, settling ponds, high-rate ponds (HRP), and maturation ponds was successfully tested by Mulbry and coworkers to treat dairy effluents. They reported an improved effluent quality with APS with high BOD, TSS, ammonia, total phosphorus, nitrogen, and *Escherichia coli* removal. Using attached algae as a means of treating wastewaters was pioneered by Walter Adey (1989). Algae biofilm developed using attached substrate contains mainly benthic diatoms along with cyanobacteria. The algae biomass productivity in these scrubbers can reach up to 60 m^2 d^{-1} which is very high compared to other systems used for wastewater treatment. In many open pond systems, algae bacterial symbiosis remains the main mechanism for treatment. Benthic diatom and bacterial symbiosis are very strong in wastewater systems leading to maximum nutrient removal and oxygen production. Algae bacterial biofilms are highly productive leading to high assimilation and valorization of nutrient and heavy metals. Algal biomass generated using dairy effluents was tested as slow release fertilizer to vegetables and found that the growth of vegetables is same as commercial fertilizer when compared with algae fertilizer (Mulbry et al. 2005). Treating dairy effluents using microalgae is cost-effective using simple technologies like open ponds and algae biofilm scrubbers; the residual biomass can generate different value-added products ranging from biofuels to biofertilizers (Tiwari and Thomas 2016; Tiwari 2016).

5.3 Brewery Wastewater

Brewery industry generates huge amount of wastewater, although it is not very toxic in terms of heavy metals but it contains high inorganic nutrients and oxygen demand which need to be properly treated before discharge. Present treatment methods include employing anaerobic digestion using different bioreactors. Anaerobic process depends mainly on methanogenic bacteria, so it is slow time-consuming process, and it cannot remove inorganic nutrients. Biological treatment for brewery effluents was mainly confined to use of phytoremediation using macrophytes (Trivedy and Nakate 2000) and combination of macrophytes and green algae (Valderrama et al. 2002). Brewery wastewater typical N:P ratio which is critical for algae growth is 9 with total nitrogen concentration of 25–80 mg L^{-1} and phosphate at 10–50 mg L^{-1} (Basu 1975). Based on limiting nutrients, this N:P ratio can sustain theoretical algae biomass production of 42.8 g L^{-1} (Christenson and Sims 2011). In spite of the presence of nutrients required for algae growth in brewery wastewater, studies on their use are still limited. Mata et al. (2012) explored the use of green algae *Scenedesmus obliquus* for effluent treatment and reported 57.5% COD removal and 20.8% TN removal after 14 days of growth. Research articles on use of diatom algae for brewery wastewater treatment are nonexistent till now. But diatom algae have certain advantages which can make them good candidates for their use as bioremediation agents. Molasses wastewater a by-product of alcohol fermentation contains a brown pigment which hinders light penetration through the water column. This hindered light will negatively affect microalgae growth. Diatoms use fucoxanthin as the major light harvesting pigment which needs less light to reach saturation limit, so diatoms can grow even in low light. This gives them an advantage over other algae like green algae and cyanobacteria which need high irradiation as they use chlorophyll a and b as main light harvesting pigments. Diatom silica biogenesis performs more efficiently and pH 4–5 so diatoms can grow faster than other algae even at low pH wastewater from breweries. Diatoms due to their faster carbonic anhydrase activity can sequester more carbon from the atmosphere which results in higher oxygen production rate; this is ideal in reducing the huge oxygen demand of brewery wastewater. Benthic diatoms are the most productive algal community in wastewater ecosystems due to their symbiotic relationship with aerobic bacteria. This will help diatoms to survive harsh physicochemical conditions encountered in effluent treatment ponds. All these attributes makes diatoms one of the potential candidates for phycoremediation of brewery effluent treatment research.

5.4 Fish Farm Effluents

Aquaculture is one of the major industries supplying much needed nutrition to mankind. But due to excessive use of artificial food, chemicals, and antibiotics, the water after growing of cultured organisms becomes highly polluted. Indiscriminate discharge of these effluents can lead to severe pollution in downstream water bodies. Majority of the effluents from aquaculture industry which includes both production and postproduction operations contain high amount of inorganic nutrients which can be utilized by algae for their growth (Dosdat et al. 1996). Mass production of microalgae using these excess nutrients can be beneficial not only in terms of water treatment, but also the algal biomass generated can be used as high-quality nutritious supplement to aqua feed (Huntley 1995). In a study on treating fishing fish farm effluents using diatoms, Lefebvre et al. (1996) observed that nutrient addition especially silica to effluents increased diatom growth and thereby treatment efficiency with 90% nutrient removal rate in 3–5 days of outdoor culturing. Diatoms due to their ideal size and nutritional content are ideal for this purpose as they can be consumed easily by small fish, shrimp, and zooplankton. Diatoms due to their high growth rate coupled with their diversity in varied habitats can be grown using fish farm effluents. Due to their absolute requirement of silica for growth which is not the case with other algae groups, diatoms can be grown in open ponds by stimulating growth using silica dosing.

5.5 Heavy Metal and Other Pollutant Removal

Diatom algae produce oxygen during photosynthesis which acts as stimulant for heterotrophic bacterial growth which in turn can enhance bacterial degradation of organic pollutants (de Godos et al. 2010). Growth of benthic diatom *Nitzschia* sp. has resulted in enhanced aerobic bacterial activity in sediment layer which can lead to accelerated decomposition of organic matter (Yamamoto et al. 2008). Phthalate acid esters (PAEs) are commonly occurring priority pollutants and endocrine disruptors. Marine benthic diatom *Cylindrotheca closterium* has shown increased PAE removal rate in surface sediments. In bottom sediment it helped in the increase of aerobic bacterial growth by photosynthetic oxygen, thereby resulting in a combination of bacteria-diatom-dependent PAE removal. Diatom *Stephanodiscus minutulus* under optimum nutrient availability has shown increased uptake of PCB integer 2, 2′, 6, 6′- tetrachlorobiphenyl (Lynn et al. 2007). Poly-aromatic hydrocarbons (PAHs) phytoremediation has limited success rate due to their high toxicity, but diatoms *Skeletonema costatum* and *Nitzschia* sp. have shown accumulation and degradation of phenanthroline (PHE) and fluoranthene (PLA), two typical PAHs (Hong et al. 2008). Diatom algae produced O_2 that can help in bacterial degradation of PAHs and phenolic and organic solvents in benthic environments. Diatom *Amphora coffeaeformis* is known to accumulate herbicide mesotrione (Valiente Moro et al. 2012). The potential of diatom algae in biodegradation and accumulation of pollutants is enormous, but till date, little research is done in this field.

6 Diatoms Grown on Industrial Wastewater as Source of Biofuels and Nutraceuticals

Diatoms are capable of uptaking the organic and inorganic forms of carbon, nitrogen, and phosphorous accompanied by accumulation of trace elements from wastewaters and using them as a source of their growth (Li et al. 2017a, b; Xin et al. 2010; Hena et al. 2015). The growth of diatoms on the wastewater also produces good amount of biomass, which can be further processed into diverse useful components (Fig. 6) as they are great reservoirs of bioactive compounds like lipids, sterols, flavonoids, proteins, and pigments. The myriad of metabolites provide diatom robustness to act as antimicrobial, antioxidative, and therapeutic molecules in the treatment of diseases like HIV, Alzheimer, and cancer (Kuppusamy et al. 2017).

The algal biomass growing on the wastewater can find applications in the area of health food, good nutritive supplements, and also a live source of food for fishes, oysters, mollusks, mussels, and clams (Rico-Villa et al. 2008). They can also produce pigments like carotenoids, fucoxanthin, quinones, terpenes, and tocopherols (Hu et al. 2008). Diatom consortium grown on wastewater has been reported to produce biodiesel, EPA, and DHA and concluded that the silicon enriched consortium of economical for sustainable production of biodiesel and fatty acid growth on the wastewater (Thomas et al. 2018).

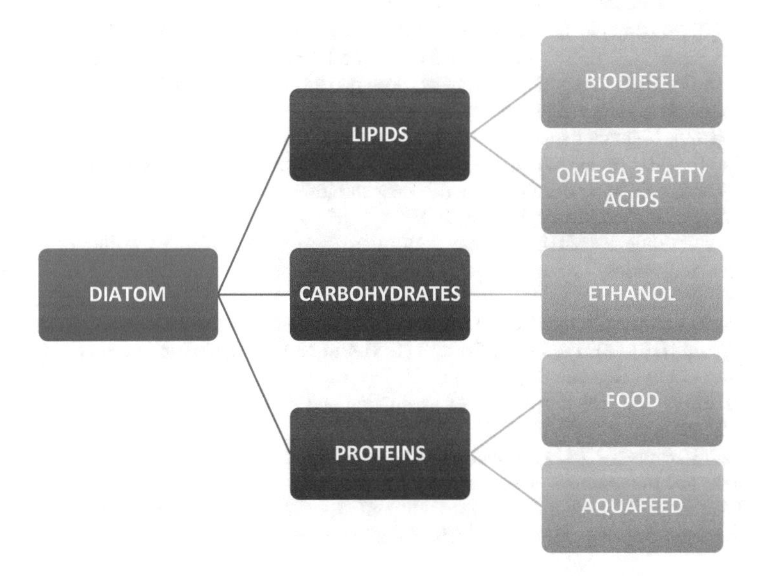

Fig. 6 Applications of diatom biomass

7 Conclusion

The efficiency of diatoms in remediation of wastewater is unparalleled and holds enormous scope in circumventing water pollution as an eco-friendly and sustainable option. More innovative approaches are indeed required to envisage novel diatoms and consortium for efficient and rapid waste remediation and biomass generation for other applications.

Diatom biorefinery approach can evoke a drastic beginning in the management of the wastewaters from different industries and simultaneous utilization of biomass for plethora of valuable products which find huge applications in the field of renewable biofuels, nutraceuticals, therapeutics, food industry, and cosmetics in the future.

Acknowledgment The research was funded by Department of Biotechnology, Ministry of Science and Technology, BT/PR15650/AAQ/3/815/2016.

References

Adey WH, Hackney L (1989) The composition and production of tropical marine algal turf in laboratory and field experiments. In: Adey W (ed) The biology, ecology and mariculture of *Mithrax spinosissimus* utilizing cultured algal turfs. Mariculture Institute, Washington

Amano Y, Takahashi K, Machida M (2011) Competition between the cyanobacterium Microcystis aeruginosa and the diatom Cyclotella sp. under nitrogen-limited condition caused by dilution in eutrophic lake. J Appl Phycol 24(4):965–971

Ambler JW, Frost BW (1974) The feeding behavior of a predatory planktonic copepod, Torlanus discaudatus. Limnol Oceanogr 19(3):446–451

Archer D (2006) Biological fluxes in the ocean. Oceans Marine Geochem 6:275

Bailleul B, Berne N, Murik O, Petroutsos D, Prihoda J, Tanaka A, Villanova V, Bligny R, Flori S, Falconet D (2015) Energetic coupling between plastids and mitochondria drives CO_2 assimilation in diatoms. Nature 524(7565):366

Basu AK (1975) Characteristics of distillery wastewater. J Water Pollut Control Fed 47:2184–2190

Bilanovic D, Andargatchew A, Kroeger T, Shelef G (2009) Freshwater and marine microalgae sequestering of CO_2 at different C and N concentrations response surface methodology analysis. Energy Convers Manag 50(2):262–267

Brennan L, Owende P (2010) Biofuels from microalgae — A review of technologies for production, processing, and extractions of biofuels and co-products. Renew Sust Energy Rev 14(2):557–577

Brzezinski MA, Pride CJ, Franck VM, Sigman DM, Sarmiento JL, Matsumoto K, Gruber N, Rau GH, Coale KH (2002) A switch from Si (OH) 4 to NO3â^' depletion in the glacial Southern Ocean. Geophys Res Lett 29(12)

Buesseler KO (1998) The decoupling of production and particulate export in the surface ocean. Glob Biogeochem Cycles 12(2):297–310

Christenson L, Sims R (2011) Production and harvesting of microalgae for wastewater treatment, biofuels, and bioproducts. Biotechnol Adv 29:686–702

Craggs RJ, Tanner CC, Sukias JP, Davies-Colley RJ (2003) Dairy farm wastewater treatment by an advanced pond system. Water Sci Technol 48:291–297

de Godos I, Vargas VA, Blanco S, González MC, Soto R, García-Encina PA, Becares E, Muoz R (2010) A comparative evaluation of microalgae for the degradation of piggery wastewater under photosynthetic oxygenation. Bioresour Technol 101(14):5150–5158
Dosdat A, Servais F, Metailler R, Huelvan C, Desbruyhres E (1996) Comparison of nitrogenous losses in five teleost fish species. Aquaculture 141:107–127
Dugdale RC, Wilkerson FP (1988) Nutrient sources and primary production in the Eastern Mediterranean. Oceanol Acta 9:179–184. Special Issue
Egge J, Aksnes D (1992) Silicate as regulating nutrient in phytoplankton competition. Mar Ecol Prog Ser 83(2):281–289
Elbaum R, Melamed-Bessudo C, Tuross N, Levy AA, Weiner S (2009) New methods to isolate organic materials from silicified phytoliths reveal fragmented glycoproteins but no DNA. Quat Int 193(1–2):11–19
Falciatore A, Bowler C (2002) Revealing the molecular secrets of marine diatoms. Annu Rev Plant Biol 53(1):109–130
Falkowski P, Raven J (2007) Photosynthesis and primary production in nature. In: Aquatic photosynthesis, pp 319–363 Princeton University Press ISBN:9780691115511
Fazal T, Mushtaq A, Rehman F, Ullah Khan A, Rashid N, Farooq W, Rehman MSU, Xu J (2018) Bioremediation of textile wastewater and successive biodiesel production using microalgae. Renew Sust Energ Rev 82:3107–3126
Furnas MJ (1990) In situ growth rates of marine phytoplankton: approaches to measurement, community and species growth rates. J Plankton Res 12(6):1117–1151
Grobbelaar JU (2009) Factors governing algal growth in photobioreactors: the open versus closed debate. J Appl Phycol 21(5):489
Halsey KH, Jones BM (2015) Phytoplankton strategies for photosynthetic energy allocation. Annu Rev Mar Sci 7:265–297
Hamm CE, Merkel R, Springer O, Jurkojc P, Maier C, Prechtel K, Smetacek V (2003) Architecture and material properties of diatom shells provide effective mechanical protection. Nature 421(6925):841
Hena S, Fatimah S, Tabassum S (2015) Cultivation of algae consortium in a dairy farm wastewater for biodiesel production. Water Resour Industry 10:1–14
Hong Y-W, Yuan D-X, Lin Q-M, Yang T-L (2008) Accumulation and biodegradation of phenanthrene and fluoranthene by the algae enriched from a mangrove aquatic ecosystem. Mar Pollut Bull 56(8):1400–1405
Hsieh C-H, Wu W-T (2009) Cultivation of microalgae for oil production with a cultivation strategy of urea limitation. Bioresour Technol 100(17):3921–3926
Hu Q, Sommerfeld M, Jarvis E, Ghirardi M, Posewitz M, Seibert M, Darzins A (2008) Microalgal triacylglycerols as feedstocks for biofuel production: perspectives and advances. Plant J 54:621–639
Hulburt EM (1990) Description of phytoplankton and nutrient in spring in the western North Atlantic Ocean. J Plankton Res 12(1):1–28
Huntley M (1995) Microalgae as a source of feeds in commercial aquaculture. Sustainable Aquaculture '95. Pacon International, Hawaii: 193–204
Kilham SS, Theriot EC, Fritz SC (1996) Linking planktonic diatoms and climate change in the large lakes of the Yellowstone ecosystem using resource theory. Limnol Oceanogr 41(5):1052–1062
Kroth P (2007) Molecular biology and the biotechnological potential of diatoms. In: Transgenic microalgae as green cell factories. Springer, pp 23–33 New York, NY
Kuppusamy P et al (2017) Potential pharmaceutical and biomedical applications of Diatoms microalgae - An overview. Indian J Geo Mar Sci 46(04):663–667
Lefebvre S, Hussenot J, Brossard N (1996) Water treatment of land-based fish farm effluents by outdoor culture of marine diatoms. J Appl Phycol 8:193–200
Li X-l, Marella TK, Tao L, Peng L, Song C-f, Dai L-l, Tiwari A, Li G (2017a) A novel growth method for diatom algae in aquaculture wastewater for natural food development and nutrient removal. Water Sci Technol 75:2777. https://doi.org/10.2166/wst.2017.156

Li X-l, Thomas KM, Tao L, Li R, Tiwari A, Li G (2017b) An Orthogonal test design for optimization of growth conditions in three fresh water diatom species. Phycol Res 65:177. https://doi.org/10.1111/pre.12174
Libessart N, Maddelein M-L, Koornhuyse N, Decq A, Delrue B, Mouille G, D'Hulst C, Ball S (1995) Storage, photosynthesis, and growth: the conditional nature of mutations affecting starch synthesis and structure in Chlamydomonas. Plant Cell 7 (8):1117–1127
Litchman E, Klausmeier CA (2001) Competition of phytoplankton under fluctuating light. Am Nat 157(2):170–187
Litchman E, Klausmeier C, Miller J, Schofield O, Falkowski P (2006) Multi-nutrient, multi-group model of present and future oceanic phytoplankton communities. Biogeosci Discuss 3(3):607–663
Lynn SG, Price DJ, Birge WJ, Kilham SS (2007) Effect of nutrient availability on the uptake of PCB congener 2, 2, 6, 6-tetrachlorobiphenyl by a diatom (*Stephanodiscus minutulus*) and transfer to a zooplankton (*Daphnia pulicaria*). Aquat Toxicol 83(1):24–32
Mata TM, Melo AC, Simões M, Caetano NS (2012) Parametric study of a brewery effluent treatment by microalgae *Scenedesmus obliquus*. Bioresour Technol 107:151–158
McGinn PJ et al (2011) Integration of microalgae cultivation with industrial waste remediation for biofuel and bioenergy production: opportunities and limitations. Phtosynth Res 109(1–3):231–247
Milligan AJ, Morel FM (2002) A proton buffering role for silica in diatoms. Science 297(5588):1848–1850
Mulbry W, Wilkie AC (2001) Growth of benthic freshwater algae on dairy manures. J Appl Phycol 13:301–306
Mulbry W, Westhead EK, Pizarro C, Sikora L (2005) Recycling of manure nutrients: use of algal biomass from dairy manure treatment as a slow release fertilizer. Bioresour Technol 96:451–458
Orefice I, Chandrasekaran R, Smerilli A, Corato F, Caruso T, Casillo A, Corsaro MM, Dal Piaz F, Ruban AV, Brunet C (2016) Light-induced changes in the photosynthetic physiology and biochemistry in the diatom *Skeletonema marinoi*. Algal Res 17:1–13
Oswald WJ, Golueke CG (1960) Biological transformation of solar energy. In: Adv Appl Microbiol, vol 2. Elsevier, pp 223–262
Racki G, Cordey F (2000) Radiolarian palaeoecology and radiolarites: is the present the key to the past? Earth Sci Rev 52(1–3):83–120
Raven JA (1983) The transport and function of silicon in plants. Biol Rev 58(2):179–207
Rico-Villa B, Woerther P, Minganta C, Lepiver D, Pouvreau S, Hamon M, Robert R (2008) A flow-through rearing system for ecophysiological studies of Pacific oyster *Crassostrea gigas* larvae. Aquaculture 282:54–60
Ryther JH (1969) Photosynthesis and fish production in the sea. Science 166(3901):72–76
Sakshaug E, Andresen K, Kiefer DA (1989) A steady state description of growth and light absorption in the marine planktonic diatom *Skeletonema costatum*. Limnol Oceanogr 34(1):198–205
Sen B, Alp MT, Sonmez F, Kocer MAT, Canpolat O (2013) Chapter 14: Relationship of algae to water pollution and waste water treatment. In: Elshorbagy W, Chowdhury RK (eds) Water treatment. Intech Open, pp 335–353
Smetacek V (1999) Diatoms and the ocean carbon cycle. Protist 150(1):25–32
Smetacek V, von Bodungen B, Knoppers B, Peinert R, Pollehne F, Stegmann P, Zeitzschel B (1984) Seasonal stages characterizing the annual cycle of an inshore pelagic system. Rapports et Proces-Verbaux des Reunions Conseil International pour l'Exploration de la Mer 183:126–135
Street-Perrott FA, Barker PA (2008) Biogenic silica: a neglected component of the coupled global continental biogeochemical cycles of carbon and silicon. Earth Surf Process Landf 33(9):1436–1457
Suh IS, Lee C-G (2003) Photobioreactor engineering: design and performance. Biotechnol Bioprocess Eng 8(6):313
Tan XB, Lam MK, Uemura Y, Lim JW, Wong CY, Lee KT (2018) Cultivation of microalgae for biodiesel production: A review on upstream and downstream processing. Chin J Chem Eng 26(1):17–30

Thomas WH, Dodson AN, Reid FM (1978) Diatom productivity compared to other algae in natural marine phytoplankton assemblages. J Phycol 14(3):250–253

Thomas KM, Tiwari A, Bhaskar MV (2015) A novel solution to grow diatom algae in large natural water bodies and its impact on CO capture and nutrient removal. J Algal Biomass Utln 6(2):22–27

Thomas KM, Bhaskar MV, Tiwari A (2016) Phycoremediation of eutrophic lakes using diatom algae. In: Lake sciences and climate change. InTechOpen

Thomas KM, Reddy Parine N, Tiwari A (2018) Potential of diatom consortium developed by Nutrient enrichment for Biodiesel production and simultaneous nutrient removal from waste water. Saudi J Biol Sci. https://doi.org/10.1016/j.sjbs.2017.05.011

Tiwari A (2016) Algal application in horticulture: novel approaches to wards sustainable agriculture. Ann Hortic 9(2):117–120. https://doi.org/10.5958/0976-4623.2016.00048.7

Tiwari A, Pandey A (2014) Toxic cyanobacterial blooms and molecular detection of hepatotoxin-microcystin. J Algal Biomass Util 5(2):33–42

Tiwari A, Thomas K (2016) Value added products from microalgae. In: Mendez-Vilas A (ed) Microbes in the spotlight: recent progress in the understanding of beneficial and harmful microorganisms. Brown Walker Press. ISBN 9781627346122

Tiwari A, Thomas K (2018) In: Nageswara-Rao M (ed) Chapter 12: Biofuels from microalgae, Advances in biofuels and bioenergy, IntechOpen, pp 239–249

Tozzi S, Schofield O, Falkowski P (2004) Historical climate change and ocean turbulence as selective agents for two key phytoplankton functional groups. Mar Ecol Prog Ser 274:123–132

Traguer P, Pondaven P (2000) Global change: silica control of carbon dioxide. Nature 406(6794):358

Trivedy RK, Nakate SS (2000) Treatment of diluted distillery waste by constructed wetlands. Ind. J Environ Prot 20:749–753

Valderrama LT, Del Campo CM, Rodriguez CM, Bashan LE, Bashan Y (2002) Treatment of recalcitrant wastewater from ethanol and citric acid using the microalga Chlorella vulgaris and the macrophyte *Lemna minuscula*. Water Res 36:4185–4192

Valiente Moro C, Bricheux G, Portelli C, Bohatier J (2012) Comparative effects of the herbicides chlortoluron and mesotrione on freshwater microalgae. Environ Toxicol Chem 31(4):778–786

Wang B, Li Y, Wu N, Lan CQ (2008) CO_2 bio-mitigation using microalgae. Appl Microbiol Biotechnol 79(5):707–718

Xin L, Hong-ying H, Ke G, Ying-xue S (2010) Effects of different nitrogen and phosphorus concentrations on the growth, nutrient uptake, and lipid accumulation of a freshwater microalga Scenedesmus sp. Bioresour Technol 101(14):5494–5500

Yamamoto T, Goto I, Kawaguchi O, Minagawa K, Ariyoshi E, Matsuda O (2008) Phytoremediation of shallow organically enriched marine sediments using benthic microalgae. Mar Pollut Bull 57(1–5):108–115

Feasibility of Using Bacterial-Microalgal Consortium for the Bioremediation of Organic Pesticides: Application Constraints and Future Prospects

James McLellan, Sanjay Kumar Gupta, and Manish Kumar

1 Introduction

A pesticide is a chemical compound designed to kill pests including weeds, fungus, insects, and rodents, as defined by the World Health Organization, but has been expanded by the Food and Agriculture Organization of the United Nations to include any chemical designed to control disease vectors for human and animals, as well as any pest threatening agricultural or industrial commodities (Li and Jennings 2017). Pesticides can be defined by both the active ingredient designed to control pests and any additional ingredients designed to improve the efficacy of the active ingredient such as emulsifiers or fumigants. Pesticides are often categorized by the targeted organism and range from avicides, rodenticides, insecticides, miticides (acaricides), molluscicides, nematicides, herbicides, fungicides, algicides, bactericides, and viricides with the prefix of each category describing the target (Uqab et al. 2016). However, this chapter focuses on four main delineations of organic pesticides, based on the chemical structure and associated mechanism of action of the active ingredient: organochlorines, organophosphates, carbamates, and pyrethrin or pyrethroids (Li and Jennings 2017). Each of these pesticide classes has been expanded to include many different isomers and related compounds, but each class has a specific mechanism affecting pests that also make them toxic to other species.

J. McLellan · M. Kumar (✉)
Department of Biology, Texas State University, San Marcos, TX, USA
e-mail: jlm413@txstate.edu; m_k135@txstate.edu

S. K. Gupta
Environmental Engineering, Department of Civil Engineering,
Indian Institute of Technology – Delhi, New Delhi, Delhi, India

S. K. Gupta, F. Bux (eds.), *Application of Microalgae in Wastewater Treatment*,
https://doi.org/10.1007/978-3-030-13913-1_16

2 Pesticide Production and Use

Pesticide use is justified through overall increases in crop yield and reductions in postharvest losses, thus improving food security and lowering overall costs (Damalas and Eleftherohorinos 2011). However, the persistence of these compounds in the ecosystem is associated with a litany of negative effects on environmental security, water security, and human health (Aktar et al. 2009; Fukuto 1990; Uqab et al. 2016). Furthermore, the unintentional by-products of the production of some of these compounds also pose ecological and human health risks, as they have historically found their way into the environment (FAO 2018).

Over time, the production and use of some pesticide classes and their associated compounds became heavily regulated in the developed world, yet many developing countries don't have the means to control or regulate their use (Alavanja 2009). Large stockpiles of expired compounds referred to as "obsolete pesticides" are causing widespread contamination of soils and surface waters in the developing world (FAO 2018). However, when one group of pesticides is outlawed, a new group soon replaces them. For instance, since the introduction of "round-up ready" genetically modified crops in the 1990s, the use of glyphosate (organophosphate herbicide) has risen dramatically and won't likely decrease any time soon; it is believed that the accumulation of this moderately persistent chemical and the associated metabolites will continue to accrue in aquatic systems and soils (Benbrook 2016; Kniss 2017).

2.1 Pesticide Types and Mechanisms

Organochlorine Pesticides

Organochlorines are chlorinated carbon compounds that were once used worldwide and are considered the first generation of pesticide chemicals. This class of compounds includes hexachlorocyclohexane (HCH), dichlorodiphenyltrichloroethane (DDT), and the endocrine disruptor endosulfan. Organochlorines are extremely resistant to environmental degradation and have been labeled as persistent organic pollutants (POPs) by the United Nations Stockholm Convention. Even though these chemicals are heavily regulated throughout the developed world, they are still used throughout parts of Asia and various members of the developing world as a public health measure against malaria spread (Jayaraj et al. 2016). Organochlorines such as DDT prevent the closing of sodium ion gates at the axon terminal of neurons resulting in a negative membrane potential causing repeated neural discharges (Coats 1990). Chlorinated cyclodienes like aldrin, chlordane, and endosulfan are neurotoxic based on the chemical binding affinity of the picrotoxin site of the γ-aminobutyric acid chloride ionophore complex (GABA), thereby disrupting neural intake of chloride anions (Coats 1990). These chemicals can wreak havoc to the endocrine system in mammals and are associated with a wide variety of health defects (Jayaraj et al. 2016). Organochlorine pesticide usage has been heavily regulated and even banned due to this off-target toxicity, prominent levels of bioaccumulation, and notorious persistence in the environment (Coats 1990; Jayaraj et al. 2016; Katagi 2010).

Organophosphates and Carbamates

The next generation of pesticides were derived from esters of phosphoric acid and are called organophosphates. The toxicity of organophosphates is identical to that of carbamate esters, derivatives of carbamic acid. Organophosphates and carbamate esters inhibit the hydrolase activity of acetylcholine hydrolase (AChE) on the neurotransmitter acetylcholine (Fukuto 1990). AChE activity is necessary in both vertebrate and invertebrate organ systems and is nearly ubiquitous in parasympathetic nervous systems and is responsible for the rapid hydrolysis of acetylcholine into acetic acid and choline (Fukuto 1990). Organophosphates include glyphosate, the most commonly used pesticide in the United States since 2001, with 1.22–1.32 × 10^5 tons applied in 2012 alone (Atwood and Paisley-Jones 2017).

Pyrethroids and Other Pesticides

Pyrethroids are synthetic derivatives of naturally occurring insecticidal compounds produced by *Chrysanthemum* spp. and have twofold mechanisms of action: (i) inhibiting sodium ion channels in neuron membranes and (ii) inhibiting the GABA complex (Coats 1990). The lipophilicity of pyrethroids increases bioaccumulation along food chains and is associated with long-term exposure problems even though they are less likely to persist in the environment abiotically (Tang et al. 2018).

Neonicotinoids were developed as a replacement for organophosphates and represent a systemic approach to controlling insects; the pesticide is taken up by the plant through root diffusion where it then spreads to all parts of the plant (Cimino et al. 2017).

2.2 Spatial and Temporal Quantification of Production and Use

The world population is predicted to exceed 9 billion by 2050; the use of pesticides is necessary and justified to ensure food security for the impending population increase (Bonner and Alavanja 2017). It is estimated that nearly 40% of agricultural production is protected through the use of pesticides (Senthil Kumar et al. 2018). Furthermore, insecticides have become an important tool in controlling insect vectors of disease including mosquitos species associated with Zika, West Nile virus, dengue, yellow fever, and malaria (Lawler 2017).

Quantifying pesticide production has proven challenging due to self-reporting issues and unauthorized use, but all figures indicate drastic increases since initial use in the middle of last century. The first pesticide use survey in the United States was conducted in 1964, and within 20 years, usage grew from 48 million pounds (21.8×10^3 tons/yr) to 430 million pounds (19.5×10^4 tons/yr) of active ingredient alone (Osteen and Fernandez-Cornejo 2016). Pesticide use is estimated to have steadily risen 11% annually, worldwide since the 1950s helping to support the exponential population growth through both disease vector control and overall crop production (Carvalho 2017). By the year 2000, an estimated 5 million tons of pesticides were being produced every year (Carvalho 2017). However, in the years that followed, the US Environmental Protection Agency's consumer-based studies show more conserved trends. An estimated 6 billion pounds (2.7×10^6 tons/yr) of pesticides were used annually, worldwide in 2011 and 2012 with the United States being responsible for 1.1 billion pounds (5.0×10^5 tons/yr) (Atwood and Paisley-Jones 2017). The most recent estimates provide that pesticide production has risen again to 3.3×10^6 tons/yr with Europe being responsible for 4.2×10^5 tons/yr (Hvězdová et al. 2018).

In the United States, herbicides such as glyphosate, atrazine, and S-metolachlor make up the bulk of current agricultural industrial use (57%), while fumigants (37%), fungicides (9%), and insecticides (5%) account for nearly the rest (Atwood and Paisley-Jones 2017). This ratio of herbicides and fungicides making up the bulk of pesticide use is mirrored throughout much of the developed world. However, 76% of India's pesticide use is attributed to insecticides, while nearly two thirds of all pesticides used were DDTs and HCHs (Yadav et al. 2015). It stands to reason that areas with more tropical climate consume higher levels of insecticides, likely for crop protection and public health measures against disease associated insect vectors.

Current paradigm of use and high rates of persistence are associated with ecological contamination often leading to human health crises. Conventional attempts at remediation are costly, environmentally hazardous, and often ineffective. Biological remediation has been investigated for decades as an efficient methodology for remediating contamination of water and soil. These methods have traditionally focused on the bacterial remediation of organic pollutants; however, microalgal-bacterial consortiums have shown great potential for the biological remediation of pesticides (Uqab et al. 2016). This chapter discusses the fate of pesticides in the environment and the associated health risks, former applications of microalgae and bacteria, inter-kingdom synergies, and factors affecting and limiting the efficacy of bacterial-algal bioremediation of pesticides.

3 Environmental Fate and Ecological Risk of Pesticides

Some studies have provided that only 0.1% of all pesticides reach their target organism, meaning the resulting 99.9% are left to enter the environment (Pimentel 1995). Even following correct application, many of these compounds enter the ecosystem through water runoff events leading to the contamination of surface water, groundwater, estuaries, marine environments, and soil deposits often persisting for prolonged periods of time. A recent study of arable soils in Europe, months after application, detected pesticides above the risk levels for the regions containing suspected carcinogens and endocrine disruptors (Hvězdová et al. 2018). These chemicals are then either broken down by photolytically, biologically, or chemically; if they aren't degraded, they persist in the water cycle or are adsorbed by other organisms, thus entering the food chain (Senthil Kumar et al. 2018). Some chemicals are highly persistent due to the chemical compound's structural resistance to abiotic or environmental degradation. For example, organochloride pesticides such as DDT and associated derivatives have half-lives ranging from 2 to 15 years (Jayaraj et al. 2016). Highly persistent chemicals undergo evaporation and condensation in the water cycle traveling immense distances (Subashchandrabose et al. 2013). This long-range atmospheric transport of persistent organic pesticides shows that pesticide pollution is not relegated to certain countries or regions; the pollutants and their associated harm are shared through geochemical processes (Yadav et al. 2015).

While some pesticides are not present in water or soils in large enough concentrations to do harm, the lipophilicity of some pesticides can lead to their accumulation in living organisms and subsequent vertical transfer through the food chain and are known as bioaccumulation and biomagnification (Katagi 2010). Even more readily degraded pesticides such as glyphosate have been shown to be accumulating in large amounts over many years, resulting risking contamination of water supply and food stuffs beyond human use (Carvalho 2017).

4 Human Health Risks

Human exposure to pesticides should be avoided because they are, by nature, hazardous as they are designed and manufactured to be toxic. Humans generally come into contact with pesticides in three ways: (i) during the production or use of pesticides, (ii) through ingestion of food or water contaminated with pesticides, and (iii) through inhalation of pesticide-contaminated air or through skin contact with contaminated water or soil. Acute exposure toxicity is well understood and more defined than low-dose long-term exposure because of the complexity of studying long-term toxicological mechanisms (Bonner and Alavanja 2017).

4.1 Acute Pesticide Toxicity

Persons at highest risk for exposure to pesticides are those who directly handle pesticides and include pesticide applicators, production workers, and farm workers (Alavanja 2009). Studies have shown that members of the developing world are disproportionately at higher risk of acute pesticide poisonings, especially in rural areas (Eddleston 2016). Contributing factors include improper handling and storage regulation, reduced access to personal protective equipment (PPE), and reduced access to adequate health care (Alavanja 2009; Yadav et al. 2015). For instance, one study compared the levels of a chlorpyrifos metabolite in urine samples between pesticide applicators without PPE to adolescents who were non-applicators finding a nearly tenfold increase (Bonner and Alavanja 2017). Organophosphate insecticides are estimated to cause nearly two million hospitalizations and resulting in nearly 100,000 deaths yearly in the developing world (Eddleston 2016). One review detailed that acute poisonings were 13-fold higher in the developing world compared to industrialized countries (Aktar et al. 2009). Another review detailed numerous studies showing the relationship between neurological disorders and exposure to HCH, as well as cardiotoxic symptoms related to methomyl application (Aktar et al. 2009), with dermal absorption likely being the method of exposure (Kim et al. 2017). While pesticide poisonings are a concern, they are predicted to decrease as safer pesticides or non-pesticidal control chemicals such as methoprene enter the market. However, long-term exposure to pesticide residues through indirect exposure is of great concern worldwide. Various regulatory bodies have prevented the developing and marketing of genotoxic pesticides through in vitro model systems, and widespread epidemiological and cohort studies have proven increased risk of cancer, diabetes, birth and development disorders, asthma, and neurodegenerative disorders (Jayaraj et al. 2016; Kamel and Hoppin 2004; Kim et al. 2017). More studies are required to assess the exposure of currently used pesticides over extended periods of time comparatively in order to work out the mechanisms of non-genotoxic carcinogenic pesticide residues, i.e., those affecting chromatin remodeling and other epigenetic effects (Alavanja 2009).

4.2 Long-Term Exposure Toxicity

Exposure to widely used pesticides has been associated with cancer, endocrine disruption, and neurological disorders (Carvalho 2017; Kamel and Hoppin 2004). Pesticides such as γ-HCH and DDT are associated with immunosuppressive effects, causing oxidative stress in blood cells, and even stimulate cancer cell propagation through in vitro studies (Bonner and Alavanja 2017). Endosulfan is associated with immunosuppression, disruption of spermatogenesis, and sperm morphology and also causes damage and mutation to DNA (Jayaraj et al. 2016). Long-term pesticide exposure meta-analysis revealed a relationship between exposure and occurrence of hematological malignancies such as Hodgkin's lymphoma (Hu et al. 2015).

The prevalence of pesticide use, their environmental fate, and associated human health effects indicate a need to develop novel approaches to remediating the environment to protect humans and ecosystems from pesticide degradation.

5 Common Biological Approaches to Remediation

Pesticide usage and the persistence of the compounds have negative implications in the realms of ecology and human health. However, conventional cleanup methods attempting to solubilize and recover organic compounds are costly and ineffective. Many of these cleanup attempts require significant investment to infrastructure and are not self-maintained, leading to a high cost with a low cost-benefit ratio (Velázquez-Fernández et al. 2012; Xia et al. 2012). Furthermore, these conventional remediation techniques are not ecologically friendly and even increase the risk of further environmental contamination. There has been a recent push toward biologically based remediation practices for the efficiency and cost-effectiveness.

The concept of bioremediation involves the exploitation of already existing mechanisms employed by bacteria, fungi, algae, and higher plants to detoxify, degrade, or accumulate pollutants to be later removed. The bioremediation of pesticides can occur in situ where the pesticides are degraded or accumulated at the location of pollution or ex situ where the contaminated soil or water is extracted and relocated to a different site for treatment (Senthil Kumar et al. 2018). Both processes *conventionally* use microbes isolated from the location of the pollution to achieve the desired biodegradation or transformation. The concept of bioaugmentation is an in situ approach where already existing flora are augmented to improve or facilitate the desired remediation. Whether the remediation occurs in situ, on site, or ex situ, the combination of bacteria and microalgae has great potential for the remediation of organic pesticides. The following section discusses the potentials and application examples.

6 Application of Bacterial-Microalgal Consortium

The efficacy of bacterial-microalgal consortia for the bioremediation of pesticides is based in the ecological associations between bacteria and algae but likely has roots in an evolutionary context as well (Ramanan et al. 2016). Microalgae and bacteria are the largest communities of primary producers across every type of aquatic ecosystem and play a major role in the aquatic carbon cycle but also play a role in terrestrial carbon cycling (Ramanan et al. 2016). Over 200 million years of coevolution have provided a number of inter-kingdom synergistic relationships including inter-kingdom quorum signaling, interspecies biofilm formation, and especially co-metabolism (Amin et al. 2012).

Microalgae have proven effective at accumulating pesticides once they've entered aquatic and terrestrial ecosystems. While some species are capable of complete mineralization or transformation of pesticides into less toxic metabolites, some species are inhibited by the toxic effect of said metabolites. Therefore, it stands to reason that by pairing these accumulators with bacterial degraders, the overall efficacy of bioaugmentation is radically improved. Here, we discuss the potentials and applications of a bacterial-microalgal consortium for the bioremediation of pesticide pollution.

6.1 Microalgae Accumulation and Transformation

One study was able to show that ten distinct species of microalgae and cyanobacteria were capable of oxidizing the organophosphate nematicide fenamiphos, while *Chlorella* sp. and *Anabaena* sp. were able to detoxify 99% of the chemical (Cáceres et al. 2008b). The same group used freshwater *P. subcapitata*, a freshwater algae and soil algae Chlorococcum sp. to accumulate and transform fenamiphos by 100% and 62%, suggesting that liquid suspension might be more effective (Cáceres et al. 2008a). *Anabaena azotica* isolated from rice paddies showed tolerance and bioremediation degradation potential of γ-HCH (Lindane) by removing nearly 50% in 5 days (Zhang et al. 2012). One study observed a random mutation that allowed a species of *S. intermedius* to develop resistance to lindane and even showed potential for its removal from aquatic systems (González et al. 2012). Both *Chlorococcum* sp. and *Scenedesmus* sp. were able to degrade endosulfan in both liquid media and soil (Sethunathan et al. 2004).

6.2 Bacterial Degradation

The basis for all degradation and transformation of pesticide pollutants begins with the necessary enzymatic activity and is separated into three classes. Enzymes modifying functional groups, enzymes associated with transfer reactions of whole groups to pollutants, and enzymes capable of translocation making pollutants unavailable to organisms are classified as phase I, II, and III enzymes, respectively (Velázquez-Fernández et al. 2012). These are typically transferase, oxidoreductase, and hydrolase enzymes. The use of consortiums to degrade complex organic molecules has been proven effective in many bioaugmentation studies (Mrozik and Piotrowska-Seget 2010). The combination of catabolic pathways of different organisms greatly enhances the overall efficacy of bioremediation. In addition, by using a consortium of bacteria from an already contaminated site, the overall efficiency of degradation is improved because there is less overall accumulation of toxic compounds and metabolic waste (Pino and Peñuela 2011). For instance, chlorpyrifos and methyl parathion were both effectively degraded by a bacterial consortium obtained from contaminated soils in Columbia (Pino and Peñuela 2011). Another study used autochthonous microbial consortiums capable of degrading organophosphates in soil, which were then inoculated with *Serratia marcescens*, thereby reducing remediation times by 8–20 days dependent upon soil types (Cycoń et al. 2013).

The combination of the transformation potentials of various microalgae with the degradation potentials of heterotrophic bacteria and the overall efficiency of bioremediation can be greatly improved. As long as there are no unintended interactions involving predation, resource competition, or metabolite toxicity, the algae-bacteria consortiums can exhibit effective remediation of organic pesticides. This has been proven effective in a few cases. The following section highlights some effective consortiums

6.3 Examples of Effective Consortiums

Microalgae-bacterial consortiums have been used for half a century for the removal of nutrients from wastewater, agro-industrial effluent, and heavy metal contamination (Ramanan et al. 2016). Surface water, groundwater, effluent, and even soils contaminated with pesticides can all be remediated with microalgal-bacterial consortiums. Many pilot scale studies have quantified the rate at which pesticides are accumulated or transformed by algal-bacterial consortiums. These organisms are often isolated from an area with high levels of pesticide contamination, are then cultured, and are used in combination with other microbes for the biodegradation or bioaugmentation of organic pollutants (Cycoń et al. 2017; Velázquez-Fernández et al. 2012; Yañez-Ocampo et al. 2009; Zhao et al. 2015). One review compiled over a decade of studies where a consortium of *Chlorella* sp., *Selenastrum* sp., *Phormidium* sp., and *Scenedesmus* sp. of microalgae were used in tandem with varieties of proteobacteria (*Ralstonia*, *Pseudomonas*, *Burkholderia*, *Sphingomonas*, *Acinetobacter*) and actinobacteria (*Rhodococcus*, *Mycobacterium*) to effectively degrade a wide variety of organic pollutants (Subashchandrabose et al. 2011). Microalgae produce O_2 which is used by the aerobic bacterial strains for the mineralization of the organic compounds (Muñoz et al. 2006). This type of co-metabolism can feasibly be exploited during the biodegradation of organic pesticides as well, with the main limitation being toxicity to microalgae strains. There are many factors that affect the rate at which a microorganism can degrade or assimilate a pesticide. These factors include toxicity of the pesticide, pH, sunlight, temperature, and endogenous metabolism (Subashchandrabose et al. 2011, 2013). These factors can be easily controlled in a lab setting but need to be taken into consideration when designing in situ remediation.

6.4 Genetic Modifications in Bioremediation

A molecular understanding of the consortium metabolism can be used to genetically manipulate members of the microbial consortium to improve the remediation and augmentation. The enzyme family cytochrome P450 monooxygenase (CYPs) have been used for the degradation of polyaromatic hydrocarbons with wild-type enzymes from *Bacillus* and *Pseudomonas* species being modified to improve degradation activities (Gaur et al. 2018). The use of genetically altered bacteria and/or algae can provide fitness increases or differential stress responses through upregulating or modifying enzymatic activity, alleviating the rate-limiting steps in a metabolic pathway, or even enhancing energy production inside communities (Ortiz-Hernandez et al. 2013).

An algal strain of *Sphingobium japonicum* capable of HCH degradation was genetically engineered to display an organophosphate hydrolase enzyme from the bacterium *Pseudomonas syringae* (Cao et al. 2013). This study observed that the engineered organism could degrade parathion amounts of 100 and 10 mg/kg of lindane (γ-HCH) completely within 15 days (Cao et al. 2013). This example of genetic engineering completely circumvents microalgae-bacterial consortium use through combining one capable organism with the enzyme capabilities of bacteria. While these methods show immense potential for designing an organism to fit individual pollution sites, it is costly and time-consuming. Furthermore, the induced mutations may be energetically unfavorable to the organism and may lead to fitness decreases outside the laboratory (Gaur et al. 2018). Therefore, it is often easier and more cost-effective to use a symbiotic consortium of microalgae and bacteria.

6.5 Factors Affecting Pesticide Removal by Bacterial-Microalgal Consortium

There are many factors affecting the pesticide remediation of bacteria-microalgae consortium. The toxicity of the pesticide to members of the consortium, concentration, the site of contamination, temperature, pH, sunlight, and water availability all affect how a microbial consortium will accumulate or degrade a pesticide pollutant (Fang et al. 2010; Zhang et al. 2012). Furthermore, it is important to understand how the autochthonous organisms will interact with any inoculated organisms to avoid predation and competition (Cycoń et al. 2017).

Many pesticide pollutants may be toxic to the bacteria or algae species being used for remediation. Therefore, the concentration of the contaminant should be taken into consideration when designing the remediation. An example of this involves the uptake and transformation of the organophosphate nematicide fenamiphos into metabolites that were more toxic than the original compound (Cáceres et al. 2008a). While some species of *Chlorella* and *Anabaena* were able to oxidize fenamiphos, the oxidized by-product fenamiphos sulfoxide and bacterial metabolites fenamiphos phenol, fenamiphos sulfoxide phenol, and fenamiphos sulfone phenol were more toxic to *Pseudokirchneriella subcapitata* and *Chlorococcum* sp. (Cáceres et al. 2008a, b). Therefore, the microbial consortium should be designed to handle the uptake and transformation of all metabolic by-products of each species in the consortium. While these algae wouldn't be able to fully mineralize fenamiphos due to the toxicity of its transformation metabolite, pairing it with a bacterial species capable of degrading the partially oxidized phenols may prove effective, including strains from *Microbacterium*, *Sinorhizobium*, *Brevundimonas*, *Ralstonia*, and *Cupriavidus genera* (Cabrera et al. 2010). The wide variety of organisms capable of degradation have immense potential for combining bacteria and algae in bioremediation attempts.

The pH of soils has been shown to influence the degradation capabilities of some microbial consortiums. One study observed that the degradation of fenamiphos was improved as the pH of that soil increases, the more successful pHs being between 7.7 and 8.4 (Singh et al. 2003). It is hypothesized that more alkaline soils allow for higher expression of enzymes and higher total biomass (Singh et al. 2003). This may be exploited during ex situ bioremediation or bioaugmentation to enhance the efficacy of degradation, while changing the pH of a large contamination site may not be feasible. However, many strains of cyanobacteria and algae have been shown to produce extra polymeric substrates, sugars, proteins, and lipids under alkaline conditions which can enhance the growth and proliferation of heterotrophic bacteria (Subashchandrabose et al. 2011). This is just one example of the synergistic potentials of bacterial-algal consortiums. Another important factor affecting the bioremediation potentials is temperature.

Zhang et al. (2012) observed that γ-HCH (Lindane) was degraded faster at higher temperatures where 67.3% was degraded at 35 °C and 56.2% at 30 °C (Zhang et al. 2012). Fang et al. (2010) showed that DDT and associated metabolites were optimally degraded at 30 °C compared to 20 and 40 °C (Fang et al. 2010). This parameter would be nearly impossible to control in large in situ remediation attempts but should be considered when deciding on species consortiums and application types.

6.6 *Limitations of Bacterial-Microalgal Consortium*

One of the main limitations to the applied remediation consortium involves the specificity of strain to pollutant, as some strains of algae respond to different pesticides differently even resulting in toxicity (Subashchandrabose et al. 2011). This is not a one-size-fits-all solution. There is still a modicum of investigation needed to carry out effective remediation. Furthermore, some algal species are too sensitive to the toxicity of certain pesticides such as diazinon (Tien et al. 2011). This chapter was unable to find any species of microalgae capable of degrading or withstanding dieldrin or glyphosate.

The stability of a constructed consortium is only effective if there is division of labor and effective chemical communication between the species (Subash chandra bose et al. 2011). Inter-kingdom quorum sensing controls biofilm formation, co-metabolism, and stress responses but may not be compatible among some algae and bacteria. Furthermore, some nutrient requirements and physiochemical needs may not be compatible within every consortium. It is also hard to predict remediation outcomes based on laboratory and small pilot scale studies. The scalability of algal-bacteria consortiums may be a hindering aspect of its use and application.

In some instances, the metabolites produced during a biodegradation are more toxic than their precursors. For instance, one study observed that *Chlorella vulgaris* mediate degradation of diclofop-methyl (DM) to a less toxic metabolite diclofop (DC) (Cai et al. 2009). However, DC was then further degraded to 4-(2,4-dichlorophenoxy) phenol (DP) which was the most toxic metabolite of the three intermediates (Cai et al. 2009; Subashchandrabose et al. 2013). This was also discussed earlier in the case of fenamiphos by-products of bacterial metabolism being toxic to certain algal species. These are potential setbacks to applied remediation attempts due to the production of metabolites more harmful than the pesticide which was originally contaminated. The metabolites of all components of the microbial community should be taken into account during the engineering of a bacterial-algal consortium.

Not all algae-bacteria interactions are commensal or mutualistic, which is vital to engineering effective consortiums. These include quorum sensing inhibitors, algicidal metabolites produced by bacteria, and limiting nutrient competition (Amin et al. 2012). However, by identifying already present microbes in polluted areas, these types of interactions can be avoided during the remediation application.

7 Synergistic Potentials of Combined Remediation

Algae-bacteria relations have been studied heavily for decades primarily focusing on symbiosis of nutrient exchange, chelation, bacterial attachment, co-metabolism, and chemical communication (Rengifo-Gallego and Salamanca 2015; Subashchandrabose et al. 2011). Much of the symbiosis between microalgae and bacteria is based on nutrient exchange of vitamins, iron, and fixed nitrogen (Cooper and Smith 2015; Ramanan et al. 2016). In one example, some algae lacking a methionine synthase gene cannot produce vitamin B_{12} and require an exogenous source, which is produced by mutualistic bacteria species who in turn benefit from organic matter produced by the algae (Amin et al. 2012). Co-cultures of algae and bacteria have been shown to be more robust in the event of environmental flux and provide resistance to outside invasion or competition (Subashchandrabose et al. 2011). Furthermore, in the post omics age, mutualisms are being defined more closely and are providing insight into the application potentials of the mutualistic organisms (Cooper and Smith 2015). Biofilms conferring mutual advantage are described in Fig. 1. The relationship between archaea and microalgae is less well-known, but there is significant evidence for the interaction between the two kingdom based on chemical markers in marine sediment (Amin et al. 2012).

The relationship between the two kingdoms provides a unique opportunity to exploit the mutualism and synergy developed over millennia for the application of polymicrobial consortium for the bioremediation of pesticides. More comparative studies are needed to elucidate these relationships. Commonly, microbial-algal symbiosis occurs through the formation of biofilm. This association of consortiums through a biofilm matrix enhances the mutualisms previously discussed by bringing microbial populations closer together. This spatial organization makes for more efficient chemical communication, accumulation of limiting metals, and nutrient exchange. It has been shown that the heterotrophic bacteria have high 0_2 demand during degradation of complex organic molecules and therefore thrive in the upper layers of cyanobacteria and algal mats (Abed 2010). Furthermore, some cyanobac-

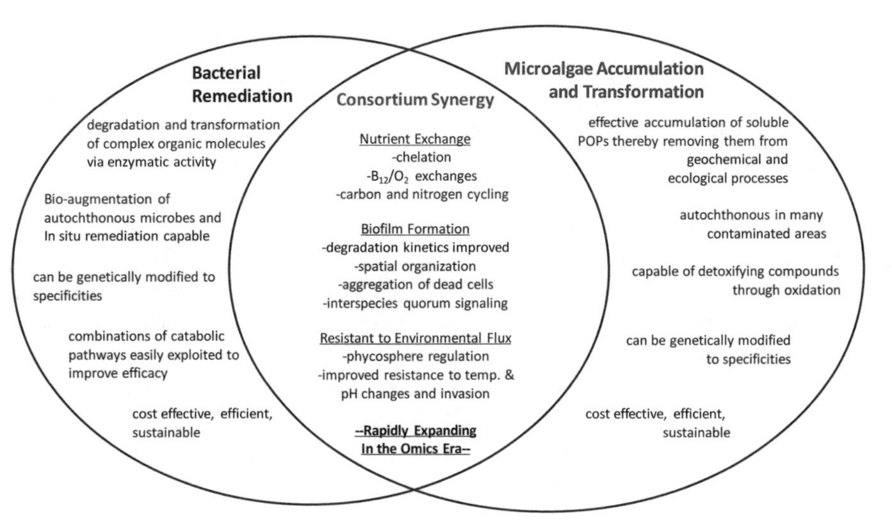

Fig. 1 Description of the benefits of biofilm formation between microalgae and bacteria. Various genera of bacteria and microalgae from selected studies included

teria even seem to regulate the eubacteria associated within their phycosphere by releasing certain carbon sources utilized by said eubacteria (Amin et al. 2012).

Biofilm formation may improve the remediation potentials of microalgal-bacterial consortiums. One study found that by immobilizing cells, they were able to increase the efficiency of the biodegradation of a polymicrobial consortium. This methodology immobilized cells on alginate beads, similar to the naturally occurring biofilms, which in turn improved the catalytic activity of the enzymes which increased the degradation efficiency (Yañez-Ocampo et al. 2009). This is further supported by an investigation into how biofilm formation improves the kinetics of degradation of diazinon, an organophosphate insecticide. It was shown that diazinon removal was 99.9% by algal-bacterial consortiums in biofilm compared to only 27% removal by the same species not in biofilm (Tien et al. 2011). This change in remediation was likely caused by the increased biomass associated with naturally occurring biofilms. This study also suggested that the remediation of diazinon was more efficient in springtime biofilms (99.9% removal) than wintertime biofilms (77% removal) due to higher levels of cellular absorption and adsorptions by extracellular polymeric substrate as well as increased levels of microbial degradation (Tien et al. 2011). Furthermore, cell aggregation and attachment will yield population increases in bacteria as algal cells die and begin to decompose. These dead and still attached algal cells provide alternative carbon sources thereby increasing bacterial populations, which has been shown to improve degradation potentials (Cycoń et al. 2017; Pino and Peñuela 2011).

The overall degradation of bacterial consortiums can be improved through the addition of an additional carbon source. For instance, one study using a fairly diverse consortium (Fig. 2) improved the degradation efficiency of methyl parathion by 28% and chlorpyrifos by 64% just by adding glucose to the medium (Pino and

Peñuela 2011). They reasoned that this additional carbon source greatly enhances the number of organisms in media thereby radically improving the overall degradation. One of the benefits reaped by bacteria through an algal mutualism is the addition of alternative carbon sources from the microalgae, thus further improving the remediation potential through increasing the number of bacteria.

The pH requirement of microbial consortiums can be matched and even manipulated to improve the degradation of pesticide pollutants. Fenamiphos degradation by bacteria was enhanced in alkaline soils, while *Scenedesmus*, *Chlamydomonas*, *Stichococcus*, *Chlorella*, *Nostoc*, and *Anabaena* species were all shown to accumulate and partially oxidize the same compound (Singh et al. 2003). Furthermore, bacterial degradation increased with the pH, while all the previous microalgae species are known to produce extracellular polymeric substrates, sugars, lipids, and vitamins that can be used as growth substrates by bacteria (Singh et al. 2003; Subashchandrabose et al. 2011). These interactions are described in Fig. 2.

8 Future Prospects

The future potentials of using a microalgae-bacterial consortium revolve around mitigating the limitations and can be improved upon in different ways. One such improvement includes the detection and selection of algal strains that are capable of withstanding larger environmental variation. In finding strains capable of growing at higher or lower pH, temperature, light availability, and pesticide levels, consortiums can be used in a wider range of applications. The toxicity of pesticides to

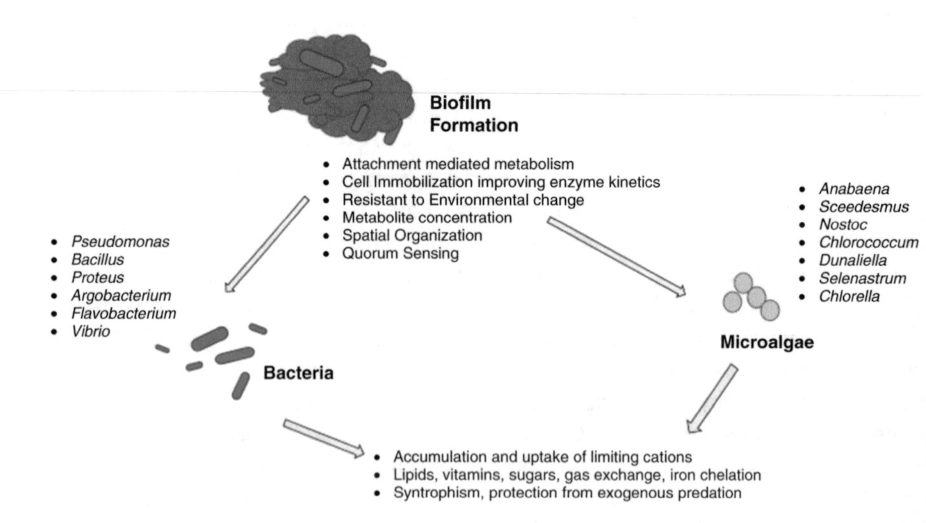

Fig. 2 Synergistic potentials of a bacterial-microalgal consortium to the application of bioremediation of pesticides

certain strains of algae remains a limitation in the application of microalgae-bacterial consortiums and may be improved by finding strains in extreme pollution settings.

Another future potential of engineering microalgae-bacterial consortium revolves around the long-term maintenance of homeostasis between species (Brenner et al. 2008). Engineering strains that can be selected for or against to control consortium ratios can be a potential fix. Finding ways to maintain the balance of organisms to optimize the remediation could greatly benefit from further study. Using alternative carbon sources and/or differential antibiotic regimes could be ways to control these populations.

Bacterial consortium engineering can exploit external chemical cues, like isopropyl β-D-1-thiogalactopyranoside (IPTG), to induce genetic circuits to promote commensalism and cooperation (Brenner et al. 2008). Inducible circuits should be further investigated to apply this concept to microalgae-bacterial populations. By elucidating the mechanisms that can confer interspecies communication and quorum sensing, these microalgae-bacterial consortiums can be tightly regulated with outside signals.

Auxotrophic mutants can be generated and used to make two species completely reliant upon one another. Research should be directed in ways to expand the ability to remediate pesticides that may be toxic to one or more members of the consortium and to control and augment the consortium ratios to improve the efficiency of remediation. Constructing consortiums that collapse when one or more members of the consortium expire can mitigate any ecological effects or imbalances associated with in situ remediation.

More research into the biochemical pathways involving the catabolism of consortiums would allow for more efficient remediation and novel applications. New "omics" tools and computational systems approaches can be employed for the development of consortium-based remediation.

Nanotechnologies including nano-adsorbents, nano-membrane-based filtration, and nanoparticle catalysts can be used in every stage of a remediation pipeline to improve a wide variety of processes (Gaur et al. 2018).

One study used a microalgae-bacterial consortium to anaerobically digest the microalgal biomass to produce methane for use as a biofuel. Using activated sludge from a wastewater treatment plant in Spain, they observed that low phosphorus levels of the incoming wastewater led to increases in lipids found in algal biomass (Hernández et al. 2013). More research can be done to improve the collection of biomasses to be used in fertilizers, pigments, animal feed, and nutrition supplements (Spolaore et al. 2006; Subashchandrabose et al. 2011). These future directions are described in Fig. 3 with various areas of research showing potential improvements in many places among the bioremediation pipeline. The general pipeline is described, and the areas of future improvement are demarcated with a yellow lightning strike and include nano-based technologies, consortium engineering, in situ discovery of novel strains, and improvements in the usage of incidental algal biomass in industrial applications.

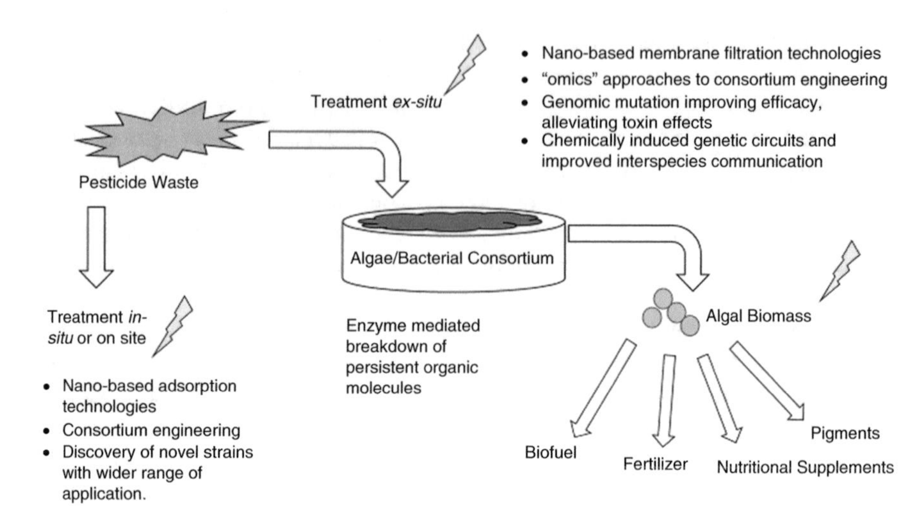

Fig. 3 Future potentials in the improvement of the bioremediation of pesticides using a microalgae-bacterial consortium

9 Conclusion

The use of pesticides has evolved and steadily increased since their first uses in the middle of last century. The production and application of pesticides is justified through overall increases in crop production and protection as well as through control of disease vector pests. The application of pesticides can be harmful to those involved and often leads to their deposit in aquatic and soil environments where many of them are resistant to degradation and accumulate over time. They may even enter the food web through primary producers becoming slowly magnified eventually becoming a risk to human health. The current levels of pesticides in the environment, water supply, and food stuffs have led to the need for developing efficient and cost-effective methods for their remediation and removal from these ecosystems. Bioremediation has been developed and used for these very reasons, and the bacterial-microalgae consortium applications were discussed in this chapter. By combining bacterial degradation with the bioaccumulation and degradation potentials of microalgae and cyanobacteria, the overall efficacy of bioremediation is improved. This is dependent upon endogenous characteristics of the consortium as well as the physiochemical aspects of the polluted site. While there are many limitations to the application of the bacterial-microalgal consortium, it remains wholly feasible and easily exploitable while ripe for further study and analysis. With over 200 million years of coevolution, these microbial consortiums can be used as an effective tool for the bioremediation and bioaugmentation of pesticide pollution (Table 1).

Table 1 Potential consortiums based on the degradation potentials of bacteria and algae by pesticide and type

Pesticide	Bacteria	References	Algae	References
HCH (Lindane)	*Pseudomonas*, *Burkholderia*, *Flavobacterium*, and *Vibrio*	Velázquez-Fernández et al. (2012).	*Anabaena azotica*	Zhang et al. (2012)
			Scenedesmus intermedius	González et al. (2012).
DDT	*Pseudomonas*, *Neisseria*, *Moraxella*, and *Acinetobacter*	Velázquez-Fernández et al. (2012)	*Chlorococcum* sp., *Anabaena* sp., *Nostoc* sp., *Aulosira fertilissima*	Subashchandrabose et al. (2013)
	Sphingobacterium	Fang et al. (2010)		
Endosulfan	*Pseudomonas spinosa*, *Pseudomonas aeruginosa*, and *Burkholderia*	Velázquez-Fernández et al. (2012)	*Chlorococcum* sp., *Scenedesmus* sp.	Sethunathan et al. (2004)
Dieldrin and endrin	*Pseudomonas*, *Bacillus*, *Trichoderma*, *Aerobacter*, *Mucor*, *Micrococcus*, and *Burkholderia*	Velázquez-Fernández et al. (2012)	*Dunaliella* sp. and *Agmenellum quadraplicatum*[a]	Matsumoto et al. (2009); Patil et al. (1972)
Atrazine	*Bacillus*, *Pseudomonas*, and *Burkholderia*	Dutta et al. (2016)	*Ankistrodesmus* sp. *Selenastrum* sp.	Geed et al. (2017)
Dimethomorph/ pyrimethanil	–	–	*Scenedesmus obliquus*, *Scenedesmus quadricauda*	Dosnon-Olette et al. (2010)
Glyphosate	*Pseudomonas* spp. strains GA07, GA09, and GC04	Zhao et al. (2015)	*Isochrysis galbana, Emiliania huxleyi, Skeletonema costatum, Phaeodactylum tricornutum*	Wang et al. (2016)
	Agrobacterium radiobacter, Burkholderia (Pseudomonas) pseudomallei, Sinorhizobium meliloti, Ochrobactrum anthropi	Hove-Jensen et al. (2014)	*Leptolyngbya boryana*, *Microcystis aeruginosa*, and *Nostoc punctiforme*	Hove-Jensen et al. (2014)
Methyl parathion	*Proteus vulgaris*, *Acinetobacter* sp., *Flavobacterium* sp., *Pseudomonas putida*, *Citrobacter freundii*, *Pseudomonas aeruginosa*, *Bacillus* sp., *Stenotrophomonas* sp., *Pseudomonas* sp., and *Proteus* sp.	Pino and Peñuela (2011)	*P. foveolarum*, *O. animalis*, *N. muscorum*, *N. linckia*, *S. bijugatus*, *C. vulgaris*	Megharaj et al. (1994)

continued

Table 1 continued

Pesticide	Bacteria	References	Algae	References
Fenamiphos	*Brevibacterium* sp.	Cáceres et al. (2008a)	*P. subcapitata*, *Chlorococcum* sp.	Cáceres et al. (2008b)
	Microbacterium, *Sinorhizobium*, *Brevundimonas*, *Ralstonia*, *Cupriavidus*	Cabrera et al. (2010)	*Scenedesmus* sp. *Chlamydomonas* sp. *Stichococcus* sp. *Chlorella* sp. *Nostoc* sp.	
Prometryne	*Ochrobactrum* sp., *Bacillus* sp. degradation was greatly enhanced by inoculation with nematodes	Zhou et al. (2012)	*Chlamydomonas reinhardtii*	Jin et al. (2012)
Fluroxypyr	Unknown consortium of soil bacteria	Tao and Yang (2011)	*Chlamydomonas reinhardtii*	Jin et al. (2012)

[a]*Dunaliella* sp. and *Agmenellum quadraplicatum* were shown to metabolize aldrin and dieldrin, but further study is suggested as the by-products; photodieldrin and ketoendrin are more toxic than the parent compounds

References

Abed RMM (2010) Interaction between cyanobacteria and aerobic heterotrophic bacteria in the degradation of hydrocarbons. Int Biodeter Biodegr 64(1):58–64. https://doi.org/10.1016/j.ibiod.2009.10.008

Aktar MW, Sengupta D, Chowdhury A (2009) Impact of pesticides use in agriculture: their benefits and hazards. Interdiscip Toxicol 2(1):1–12. https://doi.org/10.2478/v10102-009-0001-7

Alavanja MCR (2009) Pesticides use and exposure extensive worldwide. Rev Environ Health 24(4):303–309

Amin SA, Parker MS, Armbrust EV (2012) Interactions between diatoms and bacteria. Microbiol Mol Biol Rev 76(3):667–684. https://doi.org/10.1128/MMBR.00007-12

Atwood D, Paisley-Jones C (2017) Pesticides industry sale and usage: 2008–2012 market estimates. Retrieved from Online: https://www.epa.gov/sites/production/files/2017-01/documents/pesticides-industry-sales-usage-2016_0.pdf

Benbrook CM (2016) Trends in glyphosate herbicide use in the United States and globally. Environ Sci Eur 28(1):3. https://doi.org/10.1186/s12302-016-0070-0

Bonner MR, Alavanja MCR (2017) Pesticides, human health, and food security. Food Energy Secur 6(3):89–93. https://doi.org/10.1002/fes3.112

Brenner K, You L, Arnold FH (2008) Engineering microbial consortia: a new frontier in synthetic biology. Trends Biotechnol 26(9):483–489. https://doi.org/10.1016/j.tibtech.2008.05.004

Cabrera JA, Kurtz A, Sikora RA, Schouten A (2010) Isolation and characterization of fenamiphos degrading bacteria. Biodegradation 21(6):1017–1027. https://doi.org/10.1007/s10532-010-9362-z

Cáceres T, Megharaj M, Naidu R (2008a) Toxicity and transformation of fenamiphos and its metabolites by two micro algae Pseudokirchneriella subcapitata and Chlorococcum sp. Sci Total Environ 398(1):53–59. https://doi.org/10.1016/j.scitotenv.2008.03.022

Cáceres TP, Megharaj M, Naidu R (2008b) Biodegradation of the pesticide fenamiphos by ten different species of green algae and cyanobacteria. Curr Microbiol 57(6):643–646. https://doi.org/10.1007/s00284-008-9293-7

Cai X, Ye J, Sheng G, Liu W (2009) Time-dependent degradation and toxicity of diclofop-methyl in algal suspensions. Environ Sci Pollut Res 16(4):459–465. https://doi.org/10.1007/s11356-008-0077-1

Cao X, Yang C, Liu R, Li Q, Zhang W, Liu J et al (2013) Simultaneous degradation of organophosphate and organochlorine pesticides by Sphingobium japonicum UT26 with surface-displayed organophosphorus hydrolase. Biodegradation 24(2):295–303

Carvalho FP (2017) Pesticides, environment, and food safety. Food Energy Secur 6(2):48–60. https://doi.org/10.1002/fes3.108

Cimino AM, Boyles AL, Thayer KA, Perry MJ (2017) Effects of neonicotinoid pesticide exposure on human health: a systematic review. Environ Health Perspect 125(2):155–162. https://doi.org/10.1289/EHP515

Coats JR (1990) Mechanisms of toxic action and structure-activity relationships for organochlorine and synthetic pyrethroid insecticides. Environ Health Perspect 87:255–262

Cooper MB, Smith AG (2015) Exploring mutualistic interactions between microalgae and bacteria in the omics age. Curr Opin Plant Biol 26:147–153. https://doi.org/10.1016/j.pbi.2015.07.003

Cycoń M, Żmijowska A, Wójcik M, Piotrowska-Seget Z (2013) Biodegradation and bioremediation potential of diazinon-degrading Serratia marcescens to remove other organophosphorus pesticides from soils. J Environ Manag 117:7–16. https://doi.org/10.1016/j.jenvman.2012.12.031

Cycoń M, Mrozik A, Piotrowska-Seget Z (2017) Bioaugmentation as a strategy for the remediation of pesticide-polluted soil: a review. Chemosphere 172:52–71. https://doi.org/10.1016/j.chemosphere.2016.12.129

Damalas CA, Eleftherohorinos IG (2011) Pesticide exposure, safety issues, and risk assessment indicators. Int J Environ Res Public Health 8(5):1402–1419. https://doi.org/10.3390/ijerph8051402

Dosnon-Olette R, Trotel-Aziz P, Couderchet M, Eullaffroy P (2010) Fungicides and herbicide removal in Scenedesmus cell suspensions. Chemosphere 79(2):117–123. https://doi.org/10.1016/j.chemosphere.2010.02.005
Dutta A, Vasudevan V, Nain L, Singh N (2016) Characterization of bacterial diversity in an atrazine degrading enrichment culture and degradation of atrazine, cyanuric acid and biuret in industrial wastewater. J Environ Sci Health B 51(1):24–34. https://doi.org/10.1080/03601234.2015.1080487
Eddleston M (2016) Pesticides. Medicine 44(3):193–196. https://doi.org/10.1016/j.mpmed.2015.12.005
Fang H, Dong B, Yan H, Tang F, Yu Y (2010) Characterization of a bacterial strain capable of degrading DDT congeners and its use in bioremediation of contaminated soil. J Hazard Mater 184(1):281–289. https://doi.org/10.1016/j.jhazmat.2010.08.034
FAO (2018) Prevention and disposal of obsolete pesticides. Retrieved from http://www.fao.org/agriculture/crops/obsolete-pesticides/what-dealing/obs-pes/en/
Fukuto TR (1990) Mechanism of action of organophosphorus and carbamate insecticides. Environ Health Perspect 87:245–254
Gaur N, Narasimhulu K, PydiSetty Y (2018) Recent advances in the bio-remediation of persistent organic pollutants and its effect on environment. J Clean Prod 198:1602–1631. https://doi.org/10.1016/j.jclepro.2018.07.076
Geed SR, Shrirame BS, Singh RS, Rai BN (2017) Assessment of pesticides removal using two-stage Integrated Aerobic Treatment Plant (IATP) by Bacillus sp. isolated from agricultural field. Bioresour Technol 242:45–54. https://doi.org/10.1016/j.biortech.2017.03.080
González R, García-Balboa C, Rouco M, Lopez-Rodas V, Costas E (2012) Adaptation of microalgae to lindane: a new approach for bioremediation. Aquat Toxicol 109:25–32. https://doi.org/10.1016/j.aquatox.2011.11.015
Hernández D, Riaño B, Coca M, García-González MC (2013) Treatment of agro-industrial wastewater using microalgae–bacteria consortium combined with anaerobic digestion of the produced biomass. Bioresour Technol 135:598–603. https://doi.org/10.1016/j.biortech.2012.09.029
Hove-Jensen B, Zechel DL, Jochimsen B (2014) Utilization of glyphosate as phosphate source: biochemistry and genetics of bacterial carbon-phosphorus lyase. Microbiol Mol Biol Rev 78(1):176–197. https://doi.org/10.1128/MMBR.00040-13
Hu R, Huang X, Huang J, Li Y, Zhang C, Yin Y et al (2015) Long- and short-term health effects of pesticide exposure: a cohort study from China. PLoS One 10(6):e0128766–e0128766. https://doi.org/10.1371/journal.pone.0128766
Hvězdová M, Kosubová P, Košíková M, Scherr KE, Šimek Z, Brodský L et al (2018) Currently and recently used pesticides in Central European arable soils. Sci Total Environ 613-614:361–370. https://doi.org/10.1016/j.scitotenv.2017.09.049
Jayaraj R, Megha P, Sreedev P (2016) Organochlorine pesticides, their toxic effects on living organisms and their fate in the environment. Interdiscip Toxicol 9(3–4):90–100. https://doi.org/10.1515/intox-2016-0012
Jin ZP, Luo K, Zhang S, Zheng Q, Yang H (2012) Bioaccumulation and catabolism of prometryne in green algae. Chemosphere 87(3):278–284. https://doi.org/10.1016/j.chemosphere.2011.12.071
Kamel F, Hoppin JA (2004) Association of pesticide exposure with neurologic dysfunction and disease. Environ Health Perspect 112(9):950–958. https://doi.org/10.1289/ehp.7135
Katagi T (2010) Bioconcentration, bioaccumulation, and metabolism of pesticides in aquatic organisms. In: Whitacre DM (ed) Reviews of environmental contamination and toxicology. Springer New York, New York, pp 1–132
Kim K-H, Kabir E, Jahan SA (2017) Exposure to pesticides and the associated human health effects. Sci Total Environ 575:525–535. https://doi.org/10.1016/j.scitotenv.2016.09.009
Kniss AR (2017) Long-term trends in the intensity and relative toxicity of herbicide use. Nat Commun 8:14865. https://doi.org/10.1038/ncomms14865

Lawler SP (2017) Environmental safety review of methoprene and bacterially-derived pesticides commonly used for sustained mosquito control. Ecotoxicol Environ Saf 139:335–343. https://doi.org/10.1016/j.ecoenv.2016.12.038

Li Z, Jennings A (2017) Worldwide regulations of standard values of pesticides for human health risk control: a review. Int J Environ Res Public Health 14(7):826. https://doi.org/10.3390/ijerph14070826

Matsumoto E, Kawanaka Y, Yun S-J, Oyaizu H (2009) Bioremediation of the organochlorine pesticides, dieldrin and endrin, and their occurrence in the environment. Appl Microbiol Biotechnol 84(2):205–216. https://doi.org/10.1007/s00253-009-2094-5

Megharaj M, Madhavi DR, Sreenivasulu C, Umamaheswari A, Venkateswarlu K (1994) Biodegradation of methyl parathion by soil isolates of microalgae and cyanobacteria. Bull Environ Contam Toxicol 53(2):292–297. https://doi.org/10.1007/bf00192047

Mrozik A, Piotrowska-Seget Z (2010) Bioaugmentation as a strategy for cleaning up of soils contaminated with aromatic compounds. Microbiol Res 165(5):363–375. https://doi.org/10.1016/j.micres.2009.08.001

Muñoz R, Alvarez MT, Muñoz A, Terrazas E, Guieysse B, Mattiasson B (2006) Sequential removal of heavy metals ions and organic pollutants using an algal-bacterial consortium. Chemosphere 63(6):903–911. https://doi.org/10.1016/j.chemosphere.2005.09.062

Ortiz-Hernandez L, Enrique S-S, Dantan E, Castrejón-Godínez M (2013) Pesticide biodegradation: mechanisms, genetics and strategies to enhance the process. https://doi.org/10.5772/56098

Osteen CD, Fernandez-Cornejo J (2016) Herbicide use trends: a backgrounder. Choices 31(4):1–7

Patil K, Matsumura F, Boush G (1972) Metabolic transformation of DDT, dieldrin, aldrin, and endrin by marine micro-organisms. Environ Sci Technol 6:629–632

Pimentel D (1995) Amounts of pesticides reaching target pests: environmental impacts and ethics. J Agric Environ Ethics 8(1):17–29. https://doi.org/10.1007/bf02286399

Pino N, Peñuela G (2011) Simultaneous degradation of the pesticides methyl parathion and chlorpyrifos by an isolated bacterial consortium from a contaminated site. Int Biodeter Biodegr 65(6):827–831. https://doi.org/10.1016/j.ibiod.2011.06.001

Ramanan R, Kim B-H, Cho D-H, Oh H-M, Kim H-S (2016) Algae–bacteria interactions: evolution, ecology and emerging applications. Biotechnol Adv 34(1):14–29. https://doi.org/10.1016/j.biotechadv.2015.12.003

Rengifo-Gallego AL, Salamanca EJP (2015) Interaction algae–bacteria consortia: a new application of heavy metals bioremediation. In: Ansari AA, Gill SS, Gill R, Lanza GR, Newman L (eds) Phytoremediation: management of environmental contaminants, vol 2. Springer International Publishing, Cham, pp 63–73

Senthil Kumar P, Femina Carolin C, Varjani SJ (2018) Pesticides bioremediation. In: Varjani SJ, Agarwal AK, Gnansounou E, Gurunathan B (eds) Bioremediation: applications for environmental protection and management. Springer Singapore, Singapore, pp 197–222

Sethunathan N, Megharaj M, Chen ZL, Williams BD, Lewis G, Naidu R (2004) Algal degradation of a known endocrine disrupting insecticide, α-endosulfan, and its metabolite, endosulfan sulfate, in liquid medium and soil. J Agric Food Chem 52(10):3030–3035. https://doi.org/10.1021/jf035173x

Singh BK, Walker A, Morgan JAW, Wright DJ (2003) Role of soil pH in the development of enhanced biodegradation of fenamiphos. Appl Environ Microbiol 69(12):7035–7043. https://doi.org/10.1128/AEM.69.12.7035-7043.2003

Spolaore P, Joannis-Cassan C, Duran E, Isambert A (2006) Commercial applications of microalgae. J Biosci Bioeng 101(2):87–96. https://doi.org/10.1263/jbb.101.87

Subashchandrabose SR, Ramakrishnan B, Megharaj M, Venkateswarlu K, Naidu R (2011) Consortia of cyanobacteria/microalgae and bacteria: biotechnological potential. Biotechnol Adv 29(6):896–907. https://doi.org/10.1016/j.biotechadv.2011.07.009

Subashchandrabose SR, Ramakrishnan B, Megharaj M, Venkateswarlu K, Naidu R (2013) Mixotrophic cyanobacteria and microalgae as distinctive biological agents for organic pollutant degradation. Environ Int 51:59–72. https://doi.org/10.1016/j.envint.2012.10.007

Tang W, Wang D, Wang J, Wu Z, Li L, Huang M et al (2018) Pyrethroid pesticide residues in the global environment: an overview. Chemosphere 191:990–1007. https://doi.org/10.1016/j.chemosphere.2017.10.115

Tao L, Yang H (2011) Fluroxypyr biodegradation in soils by multiple factors. Environ Monit Assess 175(1–4):227–238. https://doi.org/10.1007/s10661-010-1508-2

Tien C, Chuang T, Chen CS (2011) The role of naturally occurring river biofilms on the degradation kinetics of diazinon. CLEAN–Soil, Air, Water 39(10):931–938

Uqab B, Mudasir S, Qayoom A, Nazir R (2016) Bioremediation: a management tool. J Bioremediat Biodegrad 7:331

Velázquez-Fernández JB, Martínez-Rizo AB, Ramírez-Sandoval M, Domínguez-Ojeda D (2012) Biodegradation and bioremediation of organic pesticides. In: Soundararajan RP (ed) Pesticides – recent trends in pesticide residue assay. InTech, Rijeka

Wang C, Lin X, Li L, Lin S (2016) Differential growth responses of marine phytoplankton to herbicide glyphosate. PLoS One 11(3):e0151633. https://doi.org/10.1371/journal.pone.0151633

Xia C, Ma X, Liu S, Fan P (2012) Studies on remediation of DDT-contaminated soil and dechlorination of DDT. Procedia Environ Sci 16:289–292. https://doi.org/10.1016/j.proenv.2012.10.040

Yadav IC, Devi NL, Syed JH, Cheng Z, Li J, Zhang G, Jones KC (2015) Current status of persistent organic pesticides residues in air, water, and soil, and their possible effect on neighboring countries: a comprehensive review of India. Sci Total Environ 511:123–137. https://doi.org/10.1016/j.scitotenv.2014.12.041

Yañez-Ocampo G, Sanchez-Salinas E, Jimenez-Tobon GA, Penninckx M, Ortiz-Hernández ML (2009) Removal of two organophosphate pesticides by a bacterial consortium immobilized in alginate or tezontle. J Hazard Mater 168(2–3):1554–1561. https://doi.org/10.1016/j.jhazmat.2009.03.047

Zhang H, Hu C, Jia X, Xu Y, Wu C, Chen L, Wang F (2012) Characteristics of γ-hexachlorocyclohexane biodegradation by a nitrogen-fixing cyanobacterium, Anabaena azotica. J Appl Phycol 24(2):221–225. https://doi.org/10.1007/s10811-011-9670-7

Zhao H, Zhu J, Liu S, Gao H, Zhou X, Tao K (2015) Bioremediation potential of glyphosate-degrading pseudomonas spp. strains isolated from contaminated soil. J Gen Appl Microbiol 61(5):165–170. https://doi.org/10.2323/jgam.61.165

Zhou J, Hu F, Jiao J, Liu M, Li H (2012) Effects of bacterial-feeding nematodes and prometryne-degrading bacteria on the dissipation of prometryne in contaminated soil. J Soils Sediments 12(4):576–585. https://doi.org/10.1007/s11368-012-0473-5

Potential of Blue-Green Algae in Wastewater Treatment

Pushan Bag, Preeti Ansolia, S. K. Mandotra, and Amit K. Bajhaiya

1 Introduction

Being the most primitive life form on earth, ranging from prokaryotes to eukaryotes, algae are a large group of photosynthetic multicellular/unicellular organisms characterized in two classes, macro- and microalgae. Large seaweeds such as green algae, red algae and brown kelp come under macroalgae, whereas microalgae are mainly freshwater-born single-cell organism, known to produce 70% of total atmospheric oxygen. Microalgae can be classified in several groups depending on cellular structure, life cycle and pigmentation. Blue-green algae (BGA) is the most important one among them due to its high growth rate and considered one of the most economic biomass-producing organisms. In general, the term blue green is used for the cyanobacteria (CB), which are group of gram-negative photosynthetic bacteria that have colonized earth over 3.5 billion years ago. Though they are bacteria, they have several features common with algae, and they can be naturally found in a wide variety of environments including river, ponds, lakes and streams. CB are considered as the predecessors of modern-day chloroplast. They are also known to possess great deal of morphological and metabolic diversity, which makes them

P. Bag
Department of Plant Physiology, Umeå Plant Science Centre, Umeå University, Umea, Sweden

P. Ansolia
Department of Pharmacy, Radharaman Institute of Technology & Science (RITS), Bhopal, India

S. K. Mandotra
Department of Botany, Punjab University, Chandigarh, Punjab, India

A. K. Bajhaiya (✉)
Department of Chemistry, Chemical Biological Centre (KBC), Umeå University, Umea, Sweden
e-mail: amit.bajhaiya@umu.se

S. K. Gupta, F. Bux (eds.), *Application of Microalgae in Wastewater Treatment*,
https://doi.org/10.1007/978-3-030-13913-1_17

extraordinary repertoire of different chemical compounds with application in food, feed, cosmetic, nutritional, pharmaceutical and even in biofuel industry. Although the BGA utilization is more than centuries old (*Nostoc* in Asia and *Spirulina* in Africa and Mexico), the purposeful cultivation of BGA has started only few decades ago. Apart from its increasing uses in agriculture, food and cosmetic industry, the emerging trend is to use BGA for wastewater treatment. In this chapter we will mainly focus on prospects of BGA in wastewater and its pros and cons.

2 Potential Applications of BGA

BGA has gained immense attentions due to its multivariant usage in biotechnology, specifically in agricultural biotechnology, natural products, cosmetics industry and production of numerous secondary metabolites including vitamins, enzymes and pharmaceuticals and most recently in wastewater treatment since the basic idea was given by Caldwell in the early 1940s.

2.1 Agricultural Biotechnology

Ability of BGA to photosynthesize and fix atmospheric N_2 gives inherent fertility to soil and explains how rice has be cultivated without any external supplies of N_2 even before invention of fertilizer. Field trails shows that N contribution by BGA is 20–30 kg/ha (Goyal 1993); thus farmers can get 20–30 kg/ha nitrogen without using any N supplements. Application of compost or dry algal mass in soil is more effective due to the availability of secondary N product in the field water. Effects have also been seen in other crop plants such as barley, oat, tomato, sugarcane, etc. BGA also helps in sustaining crop yield due to its multivalent capacity to produce vitamins, carbohydrates and growth hormones (Mishra et al. 1989; Kaushik 1998). Such N fixing cyanobacteria are listed in Table 1.

Apart from N fixation, BGA can be used as potential organisms to reclaim salinity-affected soils. *Anabaena torulosa* have been found to grow and enrich N status of saline coastal soils. Most of the sodium removed by BGA remains extracellularly trapped in their mucopolysaccharide sheaths (Apte and Thomas 1997); therefore, permanent salt removal from saline soils may not be possible, since Na^+ is released back into the soil subsequent to the death and decay of cyanobacteria. However, 25–38% of sodium can be removed by the removal of top soil.

Another major nutrient for plant growth is phosphate, which is very limiting in natural ecosystem, even if added externally; it is immediately converted in insoluble phosphate compounds which cannot be taken up by plants. BGA has the capability to solubilize phosphate compounds such as $(Ca)_3(PO_4)_2$ (tricalcium phosphate), $FePO_4$ (ferric orthophosphate), $AlPO_4$ (aluminium phosphate) and

Table 1 Nitrogen-fixing cyanobacterial species

Unicellular	Filamentous		References
	Heterocystous	Non-heterocystous	
Aphanothece, Chroococcidiopsis, Dermocarpa, Gloeocapsa, Myxosarcina, Pleurocapsa, Synechococcus, Xenococcus	*Anabaena, Anabaenopsis, Aulosira, Calothrix, Camptylonema, Chlorogloea, Chlorogloeopsis, Cylindrospermum, Fischerella, Gloeotrichia, Haplosiphon, Mastigocladus, Nodularia, Nostoc, Nostochopsis, Rivularia, Scytonema, Scytonematopsis, Stigonema, Tolypothrix, Westiella, Westiellopsis*	*Lyngbya, Microcoleus chthonoplastes, Myxosarcina, Oscillatoria, Plectonema boryanum, Pseudoanabaena, Schizothrix, Trichodesmium*	Vaishampayan et al. (2001), Pereira et al. (2009), Rana et al. (2012), Prasanna et al. (2013)

Mainly two types of cyanobacteria are mentioned here, i.e., unicellular and filamentous, which can be Heterocystous forming or non-Heterocystous forming

Table 2 Several agrochemical degrading cyanobacterial species which can be beneficial in wastewater treatment system to degrade mainly toxic pesticides used in cultivation

Cyanobacterial species	Pesticide degraded	Reference
Anabaena sp., *Microcystis novacekii, Nostoc linckia, N. muscorum, Oscillatoria animalis, Phormidium foveolarum.*	Methyl parathion	Fioravante et al. (2010)
Anabaena fertilissima, Nostoc muscorum	Monocrotophos, malathion, dichlorovos , phosphomidon	Subramanian et al. (1994)
Anabaena sp. *A. azotica, A. cylindrica, Cyanothece* sp., *Nodularia* sp., *Nostoc* sp., *Oscillatoria* sp., *Synechococcus* sp.	Lindane	El-Bestawy et al. (2007); Zhang et al. (2012)
Synechocystis sp. Strain PUPCCC 64	Anilofos	Singh et al. (2013)
Synechocystis sp. Strain PUPCCC 64	Chlorpyrifos	Singh et al. (2011)

Though specificity unknown but found to have significant effect under laboratory conditions

$Ca_5(PO_4)_3(OH)$ (hydroxylapatite) (Vaishampayan et al. 2001). Cyanobacteria such as *Anabaena doliolum*, *A. torulosa*, *Nostoc carneum* and *N. Piscinale* decompose and mineralize phosphate into soluble organic phosphates/orthophosphates and can also mobilize inorganic phosphates by means of extracellular phosphatases (Prasanna et al. 2013). Intensive dependence on agrochemicals has also brought significant pollution in soil ecosystems, and recent reports also suggest that cyanobacteria are capable to degrade agrochemicals to a certain extent (Subashchandrabose et al. 2013). Several agrochemicals degrading BGA have been listed in Table 2. Recent studies have shown that cyanobacteria are helpful in producing phytohormones, such as *Anabaena*, *Anabaenopsis*, *Calothrix*, *Chlorogloeopsis*, *Chroococcidiopsis*, *Cylindrospermum*, *Gloeothece*, *Nostoc*, *Oscillatoria*, *Plectonema*, *Phormidium* and *Synechocystis*, and help in production of auxins, whereas *Anabaena*, *Calothrix*, *Chlorogloeopsis*, *Chroococcidiopsis* and *Rhodospirillum* produce cytokines (Singh et al. 2016).

2.2 Food

BGA, specifically *Spirulina*, contains highest amount of proteins around 65% followed by soybean, dried milk (35%) and animal and fish flesh (15–25%). History of eating *Spirulina* by North African people and *Nostoc commune* by Chinese people goes back to 317–420 AD during the rule of Jin Dynasty. Apart from being rich in protein, *Spirulina* also contains significant amount of vitamins, lipids and other health-promoting substances, which make them commercially produced and sold in the names of Zyrulina, Recolina, etc. Dry powder containing capsules of *Aphanizomenon flos-aquae*, under the trade name of Klamath's Best® Blue Green Algae by Klamath Valley Botanicals LLC, USA, is famous in the USA, Germany, Canada, Korea, Japan and Austria due to having up to 20 antioxidants, 68 minerals and 70 trace elements, all amino acids, B vitamins and other important enzymes (Chakdar et al. 2012).

2.3 Natural Colours

Having the word 'green' in the names, BGA produces huge amount phycobilin and carotenoids apart from chlorophyll, which comprises up to 60% of total soluble proteins (Bogorad 1975). Phycocyanin (PC), phycoerythrin (PE) and long-chain terpenoids are among other phycobilins that have gained popularity as natural colourant for having nontoxic and environmental friendly effects. They are produced commercially from *Spirulina platensis* and *Anabaena*; several companies have incorporated them in their products containing natural colourant such as Dainippon Ink and Chemicals (Sakura, Japan) which developed a product called 'Lina blue' (PC extract from *S. platensis*), which is used in chewing gum, ice sherbets, popsicles, candies, soft drinks, dairy products and wasabi. High molar absorbance coefficients, high fluorescence quantum yield, large Stokes shift, high oligomer stability and high photostability properties make phycobiliproteins very powerful and highly sensitive fluorescent reagents. Purified native phycobiliproteins and their subunits fluoresce strongly; they have been widely used as external labels for cell sorting and analysis and a wide range of other fluorescence-based assays (Tooley et al. 2001).

2.4 Cosmetic Industry

BGA, namely, *Spirulina*, has gained major market uprising in cosmetic industry due to its natural colouring properties. Properties such as repairing signs of early skin aging, tightening effect, preventing stretch mark formation, improving moisturizing balance of skin, increasing skin's immunity naturally, lightning skin complexion and removing dead skin cells and photoprotective effect without having side effects have given *Spirulina* an edge over other synthetic products.

2.5 Bioactive Molecules and Antibiotics

The usage of BGA as medicine has long been established since 1500 BC, and compounds from *Anabaena*, *Nostoc* and *Oscillatoria* are known to produce an array of bioactive secondary metabolites, some of which are shown to have antibacterial and antifungal properties. A diterpenoid from *N. commune*, noscomin, showed antibacterial activity against *Bacillus cereus*, *Staphylococcus epidermidis* and *Escherichia coli* (Jaki et al. 1999). Natural products of *Nostoc* sp. are effective against *Cryptococcus* sp. as a causal agent of secondary fungal infections in patients with AIDS (Kuwaki et al. 2002). Anticancer properties have also been identified in *Scytonema* sp., *Phormidium tenue* and *Anabaena variables*. Cryptophycin-1, isolated from a *Nostoc* sp., has been found to have cytotoxic activity against nasopharyngeal carcinoma and human colorectal adenocarcinoma cell lines (Trimurtulu et al. 1994). Several modified bioactive compounds with reduced level of toxicity from *Spirulina* are also in second clinical trial phases as well (Tan 2010).

2.6 Biofuels

Cyanobacteria can be used for energy production, such as through production of hydrogen. Advantage of using natural energy produced by algae is its eco-friendly nature and almost no side effect or production of any pollutant (Dutta et al. 2005). Cyanobacteria mainly produces hydrogen as a secondary product of nitrogen fixation or by reversible activity of hydrogenase enzyme. More than 14 cyanobacterial genera including *Anabaena*, *Calothrix*, *Oscillatoria*, *Cyanothece*, *Nostoc*, *Synechococcus*, *Microcystis*, *Gloeobacter*, *Aphanocapsa*, *Chroococcidiopsis* and *Microcoleus* are known for their ability to produce hydrogen gas under various culture conditions. Several hydrogen producing BGA are mentioned in Table 3. Recently large-scale production of hydrogen in several bioreactors has been tried successfully and almost on its way to commercialisation (Dutta et al. 2005).

Table 3 Several hydrogen-producing blue green algae

Species of cyanobacteria	Growth conditions	Maximum hydrogen production	Reference
Anabaena sp. PCC 7120	Air, 20 μE m^{-2} s^{-1}	2.6 μmol mg^{-1} chl a h^{-1}	Masukawa et al. (2002)
Anabaena cylindrical lAMM-l	Air, 20 μE m^{-2} s^{-1}	2.1 μmol mg^{-1} chl a h^{-1}	Masukawa et al. (2002)
Anabaena variabilis AVMl3	Air and 1% CO_2, 100 μE m^{-2} s^{-1}	68 μmol mg^{-1} chl a h^{-1}	Happe et al. (2000)
Anabaena variabilis PK84	Air and 2 % CO_2, 113 μE m^{-2} s^{-1}	32.3 μmol mg^{-1} chl a h^{-1}	Tsygankov et al. (1999)
Anabaena variabilis ATCC 29413	73% Air, 25% N_2, 2 % CO_2, 90 μE m^{-2} s^{-1}	46.16 μmol mg^{-1} chl a h^{-1}	Sveshnikov et al. (1997)
Aphanocapsa montana	Air, photon fluence rate 290 μE m^{-2} s^{-1}	0.40 μmol mg^{-1} chl a h^{-1}	Howarth and Codd (1985)
Chroococcidiopsis thermalis	Ar and 1% CO_2	0.7 μmol mg^{-1} chl a h^{-1}	Serebryakova et al. (2000)
Gloeocapsa alpicola CALU 743	Sulfur free 4% CO_2; 25 μmol photons m^{-2} s^{-1}	0.58 μmol mg^{-1} protein	Antal and Lindblad (2005)
Gloeobacter PCC 7421	Air, photon fluence rate 20 μE m^{-2} s^{-1}	1.38 μmol mg^{-1} chl a h^{-1}	Moezelaar et al. (1996)
Microcystis PCC 7820	Air, photon fluence rate 20 μE m^{-2} s^{-1}	0.16 μmol mg^{-1} chl a h^{-1}	Moezelaar et al. (1996)
Nostoc commune lAMM-l 3	Air, 20 μE m^{-2} s^{-1}	0.25 μmol mg^{-1} chl a h^{-1}	Masukawa et al. (2002)
Synechococcus PCC 6803	Air, photon fluence rate 20 μE m^{-2} s^{-1}	0.26 μmol mg^{-1} chl a h^{-1}	Moezelaar et al. (1996)
Synechococcus PCC 6301	Air, photon fluence rate 20 μE m^{-2} s^{-1}	0.09 μmol mg^{-1} chl a h^{-1}	Howarth and Codd (1985)
Synechococcus PCC 6308	Air, photon fluence rate 20 μE m^{-2} s^{-1}	0.13 μmol mg^{-1} chl a h^{-1}	Howarth and Codd (1985)
Synechococcus PCC 6714	Air, photon fluence rate 20 μE m^{-2} s^{-1}	0.07 μmol mg^{-1} chl a h^{-1}	Howarth and Codd (1985)

Mainly tested in laboratory conditions and found to have significant hydrogen production rate. Mostly these species are from Anabaena and Synechococcus

2.7 *Wastewater Treatment*

The usage of BGA in wastewater treatment is the most recent activities of cyanobacteria, which are discussed below. Mostly BGA is used in combination with traditional wastewater treatment process.

Wastewater Composition and Related Hazards

Wastewater is a by-product of domestic, industrial, agricultural and commercial waste. By definition wastewater is 'used water from any combination of domestic, industrial, commercial or agricultural activities, surface runoff or stormwater and any sewer inflow or sewer infiltration' (Winfrey and Tilley 2016). Depending on the categories of sources, wastewater compositions are broadly classified into three categories which can come from all of the above-mentioned sources.

Chemical Compositions

Depending on sources, wastewater can contain a wide range of chemicals. Most harmful chemicals mainly come from industrial and commercial wastes which mainly contain heavy metals, including mercury, lead and chromium along with paints and other ammonium compounds from cosmetic industries. Some agricultural wastes such as urea, drugs, hormones, pesticides, fertilizers and primary and secondary nitrogenous and sulphur compounds are also there. On the other hand, faeces, hairs, food, vomit, paper fibres, plant material, humus, etc. come from domestic chemical wastes. Domestic wastewater is classified into two different classes such as grey water and black water. Grey water is all wastewater that is generated in household or office building sources without faecal contamination. Therefore, by definition, grey water does not include the discharge of toilets or highly faecally contaminated wastewater, which is designated sewage or black water and contains human waste.

Most of the times, chemical contaminates are from nitrogenous compounds, mainly nitrates. Main problems are related to conversion of nitrate to nitrite in the digestive system, which can cause severe problems due to its high absorption rate in the blood stream, where it binds to haemoglobin and forms methaemoglobin and eventually blocks the binding of oxygen creating an oxygen scarcity in blood.

Biological Compositions

Biological and chemical compositions of wastewater are very much well connected since most of the domestic wastewater contains biologically active organisms from human body or materials used in human households. There are mainly four types of major biological components of wastewater, which almost contains all possible disease-causing microorganisms, such as virus (hepatitis, rotavirus), bacteria (*Salmonella*, *Shigella*, *Campylobacter*, *Vibrio cholerae*) and protozoa (*Entamoeba*, *Giardia*). Apart from this, wastewater may also contain parasites such as Helminths (Ascaris).

There are several methods of wastewater treatment; most used and accepted on are the conventional processes by using disinfectant or by primary/secondary/tertiary treatment processes. But recent developments of industrial chemistry and ecological studies have also showed some emerging promises in natural ways of wastewater treatments.

Conventional Processes

Disinfecting Agents

The process of disinfection usually involves the injection of a solution of chlorine at the head end of a chlorine contact basin. The dosage of chlorine depends on the strength of the wastewater and other factors; however the dosages of 5–15 mg/l are commonly used. The ultraviolet (UV) and ozone irradiations can also be used for disinfecting wastewater; however these methods of disinfection are not in common use. The chlorine contact basins are mostly designed as rectangular channels, with obstructs to prevent short-circuiting and to give a contact time of about 30 minutes. However, in some specific conditions or to meet the advanced wastewater treatment requirements, a chlorine contact time can be increased to as long as 120 minutes so that it can meet the requirement for specific irrigation uses. In general, the bactericidal effects of chlorine and other commonly used disinfectants are dependent on the pH, organic content, contact time and effluent temperature (Singh 2017).

Preliminary Treatment

The idea of preliminary treatment is to remove large solid materials such as wood, pieces of glass, papers, plastic sand, etc. It helps to remove any floating or sedimented material and reduces the overall volume of the liquid. Pretreatment mainly includes a grit removal chamber where the flow of the liquid is controlled carefully to settle down the stones, sands and other solid materials from the liquid, but remaining suspended organic and inorganic material remains in the water. For this there are several screening processes such as coarse screening, fine screening, shredding, flow measuring, pumping and pre-aeration for further downstream process. Sometimes some disinfecting agents are also used to remove odour and to improve settling of grids (Fig. 1). Main goal of coarse screening is to remove materials that can damage the instrument; on the other hand, most of the fine screening is done to remove material that can block channels and other tubes in the machine; these also sometime couple with sedimentation process in the primary treatment.

Shredding is a culmination process where wastewater is prepared for sludge treatment. Culminators are loaded in a channel, and wastewater is passed through it where the blade of the culminator cuts down the rags until they can pass through the openings. Some advanced treatment plants also have specific shaped openings for more controlled shredding. After shredding is it send to a grit removal tank with a motive of removing inert material with a specific gravity of 2.65. Grit removal chambers are designed with specific size-dependent removal such as 0.011 inch with 65 meshes or sometimes 0.007 inches for activated sludge treatment. Most of the times, grit removal chamber is connected to an external sewer system and sanitary system to store excessive grit materials. There are several grit removing chambers such as (a) horizontal grit chamber, (b) detritus tanks, (c) aerated grit chambers, etc. After successful grit removal, raw sewage pumps take the liquid further to the primary treatment chamber and to sludge treatment chamber.

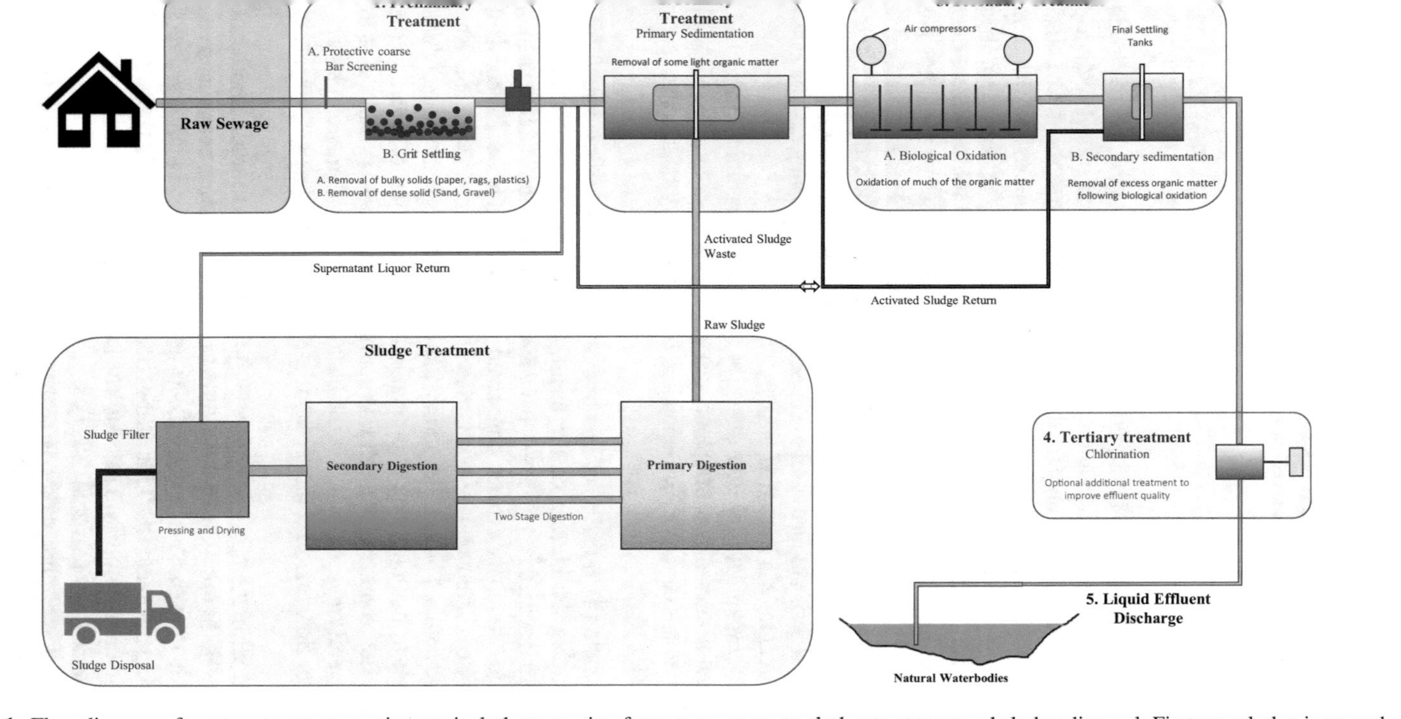

Fig. 1 Flow diagram of wastewater treatment in a typical plant, starting from raw sewage to sludge treatment and sludge disposal. First raw sludge is passed through preliminary treatment chamber where bulky and dense solids are removed, followed by treatment through primary treatment chamber to remove organic matter. Then, in the secondary treatment chamber, residual sludge is biologically oxidized and allowed for sedimentation. On the final step before releasing the liquid effluent to natural water bodies, it is treated with chlorine to improve the quality of the effluent. All residuals from primary and secondary sedimentation are released into sludge treatment chambers where it is passed through two digestion chambers and finally dried and disposed. Dense and bulky solids from preliminary treatment are directly moved for drying chamber in sludge treatment chambers, but sometimes depending on the type of sludge, it may be passed through sludge digestion chambers as well

Sometimes several additional preparation processes are added before sending it to primary treatment. This is mainly to improve wastewater treatability, providing grease separation, odour control and flocculation. Some pretreatment is also conjugated with primary treatment in case of domestic grey water treatments.

Primary Treatment

Primary treatment is mainly separating dissolved/colloidal organic and inorganic material by mainly filtration, sedimentation, phase separation or flotation. Previously primary treatment was considered only for domestic wastewater treatment. Colloidal suspensions of fine metals and organic materials are mainly removed by filtration through filters having pore size less than the particles. Particles of size more than colloids are mainly removed by gravity separation. Nonpolar organic substances are also separated by sedimentation. Containers like the API oil-water separator are specifically designed to separate nonpolar liquids (Weber 2004). Phase separation is mainly used to remove oils and grease by passing through a nonaqueous phase. Sometimes oils are saponified and then phase separated. Sometimes ion exchange and reverse phase osmosis are also used to separate nonpolar substances. The effluent from primary treatment is known as primary effluent.

Though primary treatment typically does not involve any chemical treatment, recently it has been observed that the plants use chemicals to coagulate colloidal materials (Grandclément et al. 2017). Depending on this primary treatment is classified into two types.

(a) Plain sedimentation is removal of heavy materials by gravitational field followed by clearing the bottom of the basin. Furthermore, several skimming devices are also installed here to remove the floatable substances such as scums, oil, grease, etc. which is further connected to sludge chamber. Successful removal from domestic water can comprise up to 40% of total BOD and 70% total suspended solids. The most important parameter for this is flow rate, which is very difficult to maintain due to contentious sedimentation of waste in the basin. Recent developments of high rate settlers provide better results due to addition of several trays and tubes in the basin for better settlement and maintaining a proper flow rate. But the problem of slime growth is never possible to remove fully.
(b) Sedimentation with chemical coagulant is introduced due to high amount of phosphorus waste in the industrial wastewater. These chemical coagulants are not at all used in domestic wastewater due to economic issues, but they are very effective to bring down the BOD for further treatments. Chemicals that are used singularly or in combination are salts of iron or aluminium, lime and synthetic organic polyelectrolytes (Yu et al. 2017).
(c) There are also some other methods such as extensive aeration or involving ponds or sometime with no primary treatment at all.

Secondary Treatment

Motive of secondary treatment is to remove soluble organic and inorganic substances mainly by chemical and biological-chemical oxidation; it may help in removing persistent organic and inorganic material mainly sulphur and phosphorus compounds. Sometimes chemical oxidation is carried out by adding ozone or chlorine to remove biological contaminants (virus and bacteria). Chemical oxidation is widely used for disinfection. Biological oxidation is mainly used for agricultural wastes and in sewage treatments mainly by using various microorganisms under controlled environments. Several aerobic microorganisms are used to breakdown organic materials and some inorganic compounds by means of anoxygenic photosynthesis. Most of the times, high rate biological oxidation is done in a very low volume under well-controlled environment which helps microorganisms to grow. In case of biological oxidation of organic material, it is necessary to remove microorganisms from wastewater by sedimentation to get secondary effluent. This sediment tank performs just like the primary treatment chamber. Followed by secondary sedimentation of microorganisms, this biologically degraded waste is known as biological sludge. The common high rate processes involve activated sludge treatment, biofilter or trickling filter, rotating biological contactors (RBC) or ditch filters. Mostly in case of municipal waste, activated sludge treatment is employed in combination with trickling filters to improve BOD.

Tertiary Treatment

Tertiary and/or advanced wastewater treatment is employed when specific wastewater constituents which cannot be removed by secondary treatment. For the tertiary treatment, individual treatment processes are necessary to remove phosphorus, nitrogen, additional heavy metals, suspended solids, dissolved solids and refractory organic waste. This advanced treatment is usually follows a high-rated secondary treatment and therefore sometimes called as tertiary treatment. However, the advanced treatment processes can be sometime combined with primary or secondary treatment (e.g. chemical addition to primary clarifiers or aeration basins to remove phosphorus) or used in place of secondary treatment (e.g. overland flow treatment of primary effluent). A flow diagram of stepwise treatment of wastewater is shown in Fig. 1.

Other Treatment Processes

Other treatment processes mainly include low-cost natural processes such as wastewater treatment ponds (which includes anaerobic ponds, facultative ponds, maturation ponds, etc.), overland treatment of wastewater, macrophyte treatment, nutrient film technique, etc.

3 Role of BGA in Wastewater Treatment

Recent developments in biotechnology and genetic engineering have opened a new way of treating wastewater with genetically modified microorganisms specifically blue-green algae. Role of BGA in human welfare is quite ancient as the earliest report available is almost 800 BC, but the use in wastewater treatment is very recent as the idea was proposed in 1945 (Caldwell 1946) and experimentally proved in 1957 (Oswald and Gotaas 1957).

The use of BGA in wastewater treatment has increased due to several reasons mentioned below.

1. Doesn't require nutrient-rich medium to grow; only enough amount of water is sufficient.
2. Since they are photosynthetic, so they can increase oxygen levels in water and also can utilize several organic and inorganic materials as a source of anoxygenic photosynthesis.
3. Cyanobacterial biomass is very easy to use in food and feed stock industry.
4. BGA do not produce any toxic substance rather can outperform the growth of other microorganisms

3.1 Water Quality Control

Water quality control can be monitored by monitoring cyanobacterial blooms due to eutrophication in water bodies. There are several strategies for monitoring cyanobacterial content in wastewater; among them the most used one is monitoring cyanobacterial pigment phycocyanin by spectroscopic methods from drones or spectroscope equipped air shuttles (Fig. 2) (Teta et al. 2017). BGA can also have several effects on human life due to its health related hazards coming from several sources mentioned in details below.

Drinking water: Due to the toxins and pigments produced by cyanobacteria, it can be harmful to be present in a certain amount in drinking water. Sometimes boiling water contaminated with high number of BGA may lead to production of more harmful chemicals and may lead to death.

Skin contact: Skin contact with BGA may lead to some irritation, rashes or maybe redness of eye or swelling of lips, etc., due to toxins present in cyanobacteria. Sometimes prolonged exposure to cyanobacteria may also lead to skin tumour formation.

Eating fishes infected with cyanobacteria: Eating fish or other marine seafoods infected with cyanobacteria may be harmful for our body due to toxins such as cylindrospermopsin; it has been identified in the Queensland freshwaters edible flesh of crayfish. Toxins such as 'Paralytic Shellfish Poisoning' (PSP) from the species of blue-green algae have highlighted concerns about possible neurotoxin bioaccumulation in edible mussels and other shellfish.

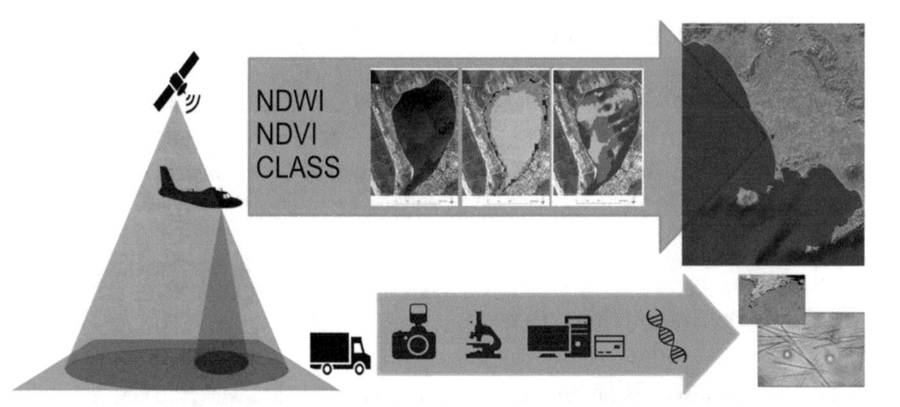

Fig. 2 Monitoring strategy based on a hierarchical approach, combining remote/proximal and in situ analytical/biotechnological data (https://doi.org/10.1088/1748-9326/aa5649)

But using proper detection and profiling techniques, it can be controlled and monitored for water quality, and this is actually very useful for fisheries and other agricultural fields.

3.2 Removal of Inorganic and Organic Toxins

Several recent studies have shown successful in removal of phosphorus and nitrogenous compounds from nutrient-rich wastewaters (Oswald et al. 1978; Chan et al. 1979). Biological processes such as suspended cultivation are very effective against nitrogenous compound removal; several species such as *Oscillatoria*, *Phormidium*, *Aphanocapsa* and *Westiellopsis* have been found to take up phosphorus and nitrogenous compounds very efficiently from the effluents which in turn reduce the pollution loads of environment (Vijayakumar 2012). There are several limitations of using suspending microalgae, such as perfect operating conditions that are hard to maintain, and it's difficult to maintain monospecificates. Secondly, there are not many effective processes available to separate microalgae from effluents before they can be discharged in the environment. Because of this sole reason, very limited number of stabilization ponds and high rate algal ponds are in use now.

Several immobilization processes such as entrapment of cyanobacteria in matrix (such as agarose, carrageenan, chitson, alginate and polyurethane foam) are used for microalgae immobilization. Process involving immobilized cells has been attempted for the treatment of effluents containing phenols, rubber press wastes, distillery waters, olive oil mill wastes, paper mill sludge, diary wastewaters and textile dye effluents. But these entrapment methods are found not to be very effective in terms of its activity to degrade discharges containing high amount of organic and inorganic compounds.

3.3 Maintaining Oxygen Levels in Water

Since BGA mainly uses oxygen to catalyse reactions to remove organic and inorganic wastes, so it is also a very good indicator of dissolved oxygen and hence helps in maintaining proper BOD and COD. Being photosynthetic organisms, they also tend to perform oxygenic photosynthesis and produce a lot dissolved oxygen.

3.4 Heavy Metal Treatments

In current situations of mining and mineral processing industries where excessive amount of metals and chemical are used for extra-metallurgical operations are raising concerns as it results in production and discharge of large amount of aqueous effluents with high metal contents which has a drastic effect on nearby water bodies (Vijayakumar 2012). The main concern of this era is to remove these toxic metallic compounds from effluents to an acceptable limit by using cost-effective and environment-friendly processes. Tiny cyanobacteria have really high metal absorption capacity with high doubling rate. These characters of cyanobacteria have encouraged their biomass usages in the detoxification of effluents. Moreover, as cyanobacteria are photosynthetic organisms, so it's more effective in removing heavy metals and detoxification of effluents. The interior pH of cyanobacterial cell is higher (approximately by two units) than the surrounding liquid, so it resists product transfer from the biofilm. Recent studies have revealed that cyanobacteria immobilized in matrix are more potent to remove heavy metals than free-living cyanobacteria. Some of the examples are:

1. An increased uptake of Cu and Fe by 45% and 23% seen in immobilized *Anabaena* compared to free counterpart
2. Another *Anabaena* species *A. Doliolum* showed 15–20% and 10–30 lower Cr and Ni removal by free-living cells when compared to that of immobilized cells.

Mechanism of metal removal is complex process which mainly occurs in two distinct phases: In phase I cations (positively charged ions) bind to negatively charged groups of the cell wall of cyanobacteria very rapidly which makes a negatively charged masking of the cell wall. This promotes the second phase, where the metal ions are taken up depending on the metabolic requirements and conditions of the cells (Pabbi, 2015). One intriguing fact was also proposed that this high metal uptaking property of immobilized cells over the free-living cells actually increases the photosynthetic energy productivity. This is maybe due to high amount cation pumping in the immobilized cells, which creates an H+ imbalance in the cell. This imbalance also increases H+ levels in chloroplast and mitochondria which is then used for ATP production while pumping out the protons to maintain the homeostasis. However, high metal uptaking property of the immobilized cells can also be due to increased permeability of cell wall (Khummongkol et al. 1982). It is also very much possible that immobilized cells have higher degree of successive collision with the metal ions than the free-living cells, which is also one of the major reasons of immobilization.

Mainly heavy metals like mercury, cadmium and lead pose the biggest hazard to human health, in addition to As, Be and Cr which are reported to be carcinogenic. These metals can cause serious damage to aquatic life due to accumulation through the trophic chain, production of toxic effects and teratogenic changes in plants, animals and human beings. This is also because of the remains of heavy metals in the sediments and release in freshwater or mixing of freshwater with heavy metal-contaminated wastewater (Wilde and Benemann 1993).

Traditional methods such as ion exchange, electrochemical treatment, precipitation, evaporation, reverse osmosis and sorption for heavy metal removal from waste streams are costly and not very much effective. Hence, biological approaches have emerged as an alternative remediation for heavy metal contamination. Since the last two decades, extensive study of microorganisms in bioremediation of heavy metals has shown a way to use BGA as the most effective remediating agents.

Most successful heavy metal remediation is dependent heavily on environmental conditions and mostly dependent on pH and temperature. Certain algae such as *Chlorella*, *Scenedesmus*, and *Hydrodictyon* can remove up to 90% of heavy metals from wastewater. Recent studies have found *Phormidium ambiguum* (*Cyanobacterium*), *Pseudochlorococcum typicum* and *Scenedesmus quadricauda var quadrispina* (Chlorophyta) to have high capacity of removing mercury and cadmium (Shanab et al. 2012). Another study has shown biosorption of different toxic heavy metals such as Pb, Cd, Co, Ni, Zn and Cu by exopolysaccharide (EPS) produced by *Paenibacillus jamilae* (Pérez et al. 2008). Another study showed rhizobia has significant roles of extracellular polysaccharides and biofilm formation (Nocelli et al. 2016). A study on *Spirulina platensis* showed significant uptake of chromium (Cr^{3+}) in free form rather than in an embayed form (Shashirekha et al. 2008).

These studies including other bioremediating properties give emerging promises for usage of BGA in traditional wastewater treatment plant to reduce cost-effectivity and better results, and large-scale experiments has also been started in the USA and Canada.

3.5 Coliform Removal

Ecotechnologies such as algal- and duckweed-based pond systems are becoming popular for wastewater treatments due to easy and cost-effective removal of pathogens in warm climatic conditions, though the mechanisms are not well understood. Several strategies based of basic physiological conditions are being used to remove coliforms by means of overproduction of BGA, such as increasing temperature (Brissaud et al. 2003; El-Shafai et al. 2007), nutrient deprivation (Van der Steen et al. 2000), sunlight, pH, dissolved oxygen (Davis-Colley et al. 2000) or algal toxins (Oudra et al. 2000).

Table 4 Comparison of roles of different blue green algae in duckweed and algal ponds in removing coliforms in different seasons during the year

Location	Season/temp. (°C)	DK	AL	CS	Reference
		Removal (log units)			
Accra, Ghanaa	Wet 24–29	3,8	4,8	4,3	Ansa (2013)
	Dry 30–33	3,5	4,6	4,3	
	Year-round	3,7	4,7	4,3	
Kumasi, Ghana	Year-round 24–27	4,0	5,0		Awuah et al. (2004)
West Bank, Palestine	Winter 7–13	1,0	3,1		Al-Sa'ed (2000)
	Summer 21–27	2,0	2,7		
Negev, Isreal	Winter 15–18		2,6	2,2	Van der Steen et al. (2000)
	Spring 18–27		2,7	2,3	
Belo Horizonte, Brazil	Yearround 20			6,7	Von Sperling and Mascarenhas (2005)

In nutrient-deprived conditions, algal growth not hampered very much due to their capability of photosynthesis, but on the other hand, coliforms lacking nutrient are easily outperformed. While increasing temperature, pH and less dissolved oxygen trigger the algal cultures to produce several super oxides and other free radicals which eventually kill possible coliforms. Again, selective growth of toxin producing algal stains also an effective idea of outperforming the growth of coliforms mainly in maturation ponds.

In both algal and duckweed ponds, faecal coliform levels have been found to be decreasing in different rates due to summer and winter conditions and availability of sunlight as well. Role of algae in duckweed and algal ponds in removing coliforms is represented in Table 4.

4 Concluding Remarks

Cyanobacteria being one of the primitive organisms are very simple for genetic engineering and have vast potential in environmental remediation. Several uses of BGA are mentioned here and furthermore are under research. Not only in remediation but also in energy production and usage in daily life make BGA more commercially valuable and frequently used microorganism due to its environment-friendly behaviour and ability to grow in almost any kind of conditions. Recently, it has been stated by NASA that the nutritional value of 1000 kg of fruits and vegetables equals to 1 kg of *Spirulina*. Even though it has all these advantages, cyanobacteria need to be explored more and more so that more fruitful results will come out.

Acknowledgement PB would like thank EU-funded Innovative Training Network (ITN) Solar Energy into Biomass (SE2B) Marie Skłodowska-Curie grant agreement no 675006. AKB would like to thank Sven and Lilly Lawski's Foundation, Sweden, for postdoctoral fellowship and Gunnar and Ruth Björkman's Foundation, Sweden, for research grant.

References

Al-Sa'ed R (2000) Wastewater management for small communities in Palestine

Ansa ED (2013) The removal of faecal coliforms in waste stabilization pond systems and eutrophic lakes. IHE Delft Institute for Water Education

Antal TK, Lindblad P (2005) Production of H2 by sulphur-deprived cells of the unicellular cyanobacteria *Gloeocapsa alpicola* and *Synechocystis* sp. PCC 6803 during dark incubation with methane or at various extracellular pH. J Ferment Bioeng 98:114–120

Apte SK, Thomas J (1997) Possible amelioration of coastal soil salinity using halotolerant nitrogen-fixing cyanobacteria. Plant Soil 189(2):205–211

Awuah E, Oppong-Peprah M, Lubberding HJ, Gijzen HJ (2004) Comparative performance studies of water lettuce, duckweed, and algal-based stabilization ponds using low-strength sewage. J Toxic Environ Health A 67(20–22):1727–1739

Bogorad L (1975) Phycobiliproteins and complementary chromatic adaptation. Annu Rev Plant Physiol 26(1):369–401

Brissaud F, Tournoud MG, Drakides C, Lazarova V (2003) Mixing and its impact on faecal coliform removal in a stabilisation pond. Water Sci Technol 48(2):75–80

Caldwell DH (1946) Sewage oxidation ponds: performance, operation and design. Sew Work J 18(3):433–458

Chakdar H, Jadhav SD, Dhar DW, Pabbi S (2012) Potential applications of blue green algae. J Sci Ind Res 71(1):13–20

Chan KY, Wong KH, Wong PK (1979) Nitrogen and phosphorus removal from sewage effluent with high salinity by Chlorella salina. Environ Pollut (1970) 18(2):139–146

Davis-Colley RJ, Donnison AM, Speed DJ (2000) Towards a mechanistic understanding of pond disinfection. Water Sci Technol 42(10–11):149–158

Dutta D, De D, Chaudhuri S, Bhattacharya SK (2005) Hydrogen production by cyanobacteria. Microb Cell Factories 4(1):36

El-Bestawy EA, Abd El-Salam AZ, Mansy HAR (2007) Potential use of environmental cyanobacterial species in bioremediation of lindanecontaminated effluents. Int Biodeterior Biodegrad 59:180–192. http://www.sciencedirect.com/science/article/pii/S0964830506002083

El-Shafai SA, El-Gohary FA, Nasr FA, Van Der Steen NP, Gijzen HJ (2007) Nutrient recovery from domestic wastewater using a UASB-duckweed ponds system. Bioresour Technol 98(4):798–807

Fioravante IA, Barbosa FAR, Augustic R, Magalhães SMS (2010) Removal of methyl parathion by cyanobacteria *Microcystis novacekii* under culture conditions. J Environ Monit 12:1302–1306. http://pubs.rsc.org/en/content/articlelanding/2010/em/b923288e#!divAbstract. https://doi.org/10.1039/b923288e

Goyal SK (1993) Algal biofertilizer for vital soil and free nitrogen. Proc Indian Natl Sci Acad B 59:295–295

Grandclément C, Seyssiecq I, Piram A, Wong-Wah-Chung P, Vanot G, Tiliacos N et al (2017) From the conventional biological wastewater treatment to hybrid processes, the evaluation of organic micropollutant removal: a review. Water Res 111:297–317

Happe T, Schütz K, Böhme H (2000 Mar 15) Transcriptional and mutational analysis of the uptake hydrogenase of the filamentous Cyanobacterium *Anabaena variabilis* ATCC 29413. J Bacteriol 182(6):1624–1631

Howarth DC, Codd GA (1985) The uptake and production of molecular hydrogen by unicellular cyanobacteria. J Gen Microbiol 131:1561–1569

Jaki B, Orjala J, Sticher O (1999) A novel extracellular diterpenoid with antibacterial activity from the cyanobacterium Nostoc commune. J Nat Prod 62(3):502–503

Kaushik BD (1998) Use of cyanobacterial biofertilizers in rice cultivation: a technology improvement. In: Subramanian G, Kaushik BD, Venkataraman GS (eds) Cyanobacterial biotechnology. Science Publishers, Enfield, pp 211–222

Khummongkol D, Canterford G, Fryer C (1982) Accumulation of heavy metals in unicellular algae. Biotechnol Bioeng 24:2643–2660

Kuwaki S, Ohhira I, Takahata M, Murata Y, Tada M (2002) Antifungal activity of the fermentation product of herbs by lactic acid bacteria against tinea. J Biosci Bioeng 94(5):401–405
Masukawa H, Mochimaru M, Sakurai H (2002) Disruption of the uptake hydrogenase gene, but not of the bidirectional hydrogenase gene, leads to enhanced photobiological hydrogen production by the nitrogenfixing cyanobacterium *Anabaena* sp. PCC 7120. Appl Microbiol Biotechnol 58:618–624
Moezelaar R, Bijvank SM, Stal LJ (1996) Fermentation and sulfur reduction in the mat-building cyanobacterium *Microcoleus chthonoplastes*. Appl Environ Microbiol 62:1752–1758
Mishra S, Parker JC, Singhal N (1989) Estimation of soil hydraulic properties and their uncertainty from particle size distribution data. J Hydrol 108:1–18
Nocelli N, Bogino PC, Banchio E, Giordano W (2016) Roles of extracellular polysaccharides and biofilm formation in heavy metal resistance of rhizobia. Materials 9(6):418
Oswald WJ, Gotaas HB (1957) Photosynthesis in sewage treatment. Trans Am Soc Civ Eng 122(1):73–105
Oswald WJ, Lee EW, Adan B, Yao KH (1978) New wastewater treatment method yields a harvest of saleable algae. WHO Chron 32(9):348–350
Oudra B, El Andaloussi M, Franca S, Barros P, Martins R, Oufdou K et al (2000) Harmful cyanobacterial toxic blooms in waste stabilisation ponds. Water Sci Technol 42(10–11):179–186
Pabbi S (2015) Blue green algae: a potential biofertilizer for rice. In: The Algae World. Springer, Dordrecht, pp 449–465
Pérez JAM, García-Ribera R, Quesada T, Aguilera M, Ramos-Cormenzana A, Monteoliva-Sánchez M (2008) Biosorption of heavy metals by the exopolysaccharide produced by Paenibacillus jamilae. World J Microbiol Biotechnol 24(11):2699
Pereira I, Ortega R, Barrientos L, Moya M, Reyes G, Kramm V (2009) Development of a biofertilizer based on filamentous nitrogen-fixing cyanobacteria for rice crops in Chile. J Appl Phycol 21:135–144. http://link.springer.com/article/10.1007/s10811-008-9342-4
Prasanna R, Chaudhary V, Gupta V, Babu S, Kumar A, Singh R et al (2013) Cyanobacteria mediated plant growth promotion and bioprotection against Fusarium wilt in tomato. Eur J Plant Pathol 136(2):337–353
Rana A, Joshi M, Prasanna R, Shivay YS, Nain L (2012) Biofortification of wheat through inoculation of plant growth promoting rhizobacteria and cyanobacteria. Eur J Soil Biol 50:118–126
Serebryakova LT, Sheremetieva ME, Lindblad P (2000) H2-uptake and evolution in the unicellular cyanobacterium *Chroococcidiopsis thermalis* CALU 758. Plant Physiol Biochem 38:525–530
Shanab S, Essa A, Shalaby E (2012) Bioremoval capacity of three heavy metals by some microalgae species (Egyptian isolates). Plant Signal Behav 7(3):392–399
Shashirekha V, Sridharan MR, Swamy M (2008) Biosorption of trivalent chromium by free and immobilized blue green algae: kinetics and equilibrium studies. J Environ Sci Health A 43(4):390–401
Singh H, Ahluwalia AS, Khattar JIS (2013) Induction of sporulation by selected carbon sources in *Anabaena naviculoides*, a diazotrophic strain capable of colonizing paddy field soil of Punjab (India). Phykos 43(2):18–25. http://phykosindia.com/current.html
Singh DP, Khattar JIS, Nadda J, Singh Y, Garg A, Kaur N, Gulati A (2011) Chlorpyrifos degradation by the cyanobacterium *Synechocystis* sp. strain PUPCCC 64. Environ Sci Pollut Res 18:1351–1359. http://link.springer.com/article/10.1007%2Fs11356–011-0472-x
Singh JS, Kumar A, Rai AN, Singh DP (2016) Cyanobacteria: a precious bio-resource in agriculture, ecosystem, and environmental sustainability. Front Microbiol 7:529
Singh RL (2017) Principles and applications of environmental biotechnology for a sustainable future. Springer
Subashchandrabose SR, Ramakrishnan B, Megharaj M, Venkateswarlu K, Naidu R (2013) Mixotrophic cyanobacteria and microalgae as distinctive biological agents for organic pollutant degradation. Environ Int 51:59–72
Subramanian G, Sekar S, Sampoornam S (1994) Biodegradation and utilization of organophosphorus pesticides by cyanobacteria. Int Biodeter Biodegr 33:129–143. http://www.sciencedirect.com/science/article/pii/0964830594900329

Sveshnikov DA, Sveshnikova NV, Rao KK, Hall DO (1997) Hydrogen metabolism of mutant forms of *Anabaena variabilis* in continuous cultures and under nutritional stress. FEMS Microbiol Lett 147:297–301

Tan LT (2010) Filamentous tropical marine cyanobacteria: a rich source of natural products for anticancer drug discovery. J Appl Phycol 22(5):659–676

Teta R, Romano V, Della Sala G, Picchio S, De Sterlich C, Mangoni A et al (2017) Cyanobacteria as indicators of water quality in Campania coasts, Italy: a monitoring strategy combining remote/proximal sensing and in situ data. Environ Res Lett 12(2):024001

Tooley AJ, Cai YA, Glazer AN (2001) Biosynthesis of a fluorescent cyanobacterial C-phycocyanin holo-α subunit in a heterologous host. Proc Natl Acad Sci 98(19):10560–10565

Trimurtulu G, Ohtani I, Patterson GM, Moore RE, Corbett TH, Valeriote FA, Demchik L (1994) Total structures of cryptophycins, potent antitumor depsipeptides from the blue-green alga Nostoc sp. strain GSV 224. J Am Chem Soc 116(11):4729–4737

Tsygankov AA, Borodin VB, Rao KK, Hall DO (1999) H2 photoproduction by batch culture of *Anabaena variabilis* ATCC 29413 and its mutant PK84 in a photobioreactor. Biotechnol Bioeng 64:709–715

Vaishampayan A, Sinha RP, Hader DP, Dey T, Gupta AK, Bhan U, Rao AL (2001) Cyanobacterial biofertilizers in rice agriculture. Bot Rev 67(4):453–516

Van der Steen P, Brenner A, Shabtai Y, Oron G (2000) The effect of environmental conditions on faecal coliform decay in post-treatment of UASB reactor effluent. Water Sci Technol 42(10–11):111–118

Van der Steen P et al (2000) Improved fecal coliform decay in integrated duckweed and algal ponds. Water science and technology 42.10–11:363–370

Vijayakumar S (2012) Potential applications of cyanobacteria in industrial effluents—a review. J Bioremed Biodeg 3:2

Von Sperling M, Mascarenhas LCAM (2005) Performance of very shallow ponds treating effluents from UASB reactors. Water Sci Technol 51(12):83–90

Weber WJ (2004) Optimal uses of advanced technologies for water and wastewater treatment in urban environments. Water Science and Technology: Water Supply 4.1:7–12

Wilde EW, Benemann JR (1993) Bioremoval of heavy metals by the use of microalgae. Biotechnol Adv 11(4):781–812

Winfrey BK, Tilley DR (2016) An emergy-based treatment sustainability index for evaluating waste treatment systems. Journal of cleaner production 112:4485–4496

Yu L, Han M, He F (2017) A review of treating oily wastewater. Arab J Chem 10:S1913–S1922

Zhang H, Hu C, Jia X, Xu Y, Wu C, Chen L, Wang F (2012) Characteristics of γ - hexachlorocyclohexane biodegradation by a nitrogen-fixing cyanobacterium, *Anabaena azotica*. J Appl Phycol 24:221–225. http://link.springer.com/article/10.1007%2Fs10811-011-9670-7

Photobioreactors for Wastewater Treatment

Vaishali Ashok, Sanjay Kumar Gupta, and Amritanshu Shriwastav

1 Introduction

Microalgae are commonly cultivated for wastewater treatment, range of food products, biofuels production, toxic wastes removal, biofertilizers, abatement of carbon dioxide, and other flue gases from atmosphere (Lopez et al. 2014). The technological understanding of many feasible species with their physical, chemical, and biological properties has already been understood, and many more are being researched every day. Major lag lies in the efficient design of PBR, which can provide efficient light utilization, uniform light distribution, uniform mixing, degasification, provision for aeration, temperature management due to diurnal and seasonal variation, cleaning, easy maintenance, and harvesting at lowest capital cost and energy input (Singh and Sharma 2012). There are myriad of PBR designs available for microalgal cultivation. Every reactor is unique in its own way with its own advantages and disadvantages (Singh and Sharma 2012). Despite such advantageous features and availability of PBR designs, utilization of these microorganisms in industrial scale is mainly constrained by the overall economics and complexity of operation.

V. Ashok
Civil Engineering Department, Indian Institute of Technology Kanpur, Kanpur, India

S. K. Gupta
Environmental Engineering, Department of Civil Engineering,
Indian Institute of Technology – Delhi, New Delhi, Delhi, India

A. Shriwastav (✉)
Centre for Environmental Science and Engineering, Indian Institute of Technology Bombay,
Mumbai, India
e-mail: amritan@iitb.ac.in

S. K. Gupta, F. Bux (eds.), *Application of Microalgae in Wastewater Treatment*,
https://doi.org/10.1007/978-3-030-13913-1_18

1.1 Historical Application of Microalgae for Wastewater Treatment

Earlier, oxidation, facultative, and high rate algal ponds treat a large amount of wastewater using natural illumination in open ponds (Palmer 1974). These open ponds were generally made up of concrete or earthen pits with plastic lining, with or without baffles (Singh and Sharma 2012). A schematic representation of high rate algal pond is shown in Fig. 1. These are open raceway ponds where algal consortia along with nutrients circulate around the raceway track. This provides necessary mixing as well as transportation of culture from input to output end (Arbib et al. 2013). In open configurations, there is lesser control over physical parameters of the culture, e.g., contamination, illumination losses, uneven illumination, evaporation losses, temperature variation, gaseous exchange, and species selection. In such systems, biomass growth rate has been reported to be lower as compared to closed systems (Chen et al. 2012). However, it has advantages of minimal maintenance, lower cost of construction, and lower running costs. Upgradability of open pond system has a major requirement of land. Therefore, such systems would be best suitable for wastewater treatment particularly in village areas with low land costs, least monitoring, low maintenance, and no to low power consumption.

1.2 Advantages of Photobioreactors over Natural Systems

- The natural open raceway ponds are shallow with depth in the order of 0.2–1 m (Park et al. 2011). This calls for highland footprint area. The PBR system, on the other hand, utilizes low space, extending vertically in three-dimensional space with higher depth.
- The natural system is dependent on sun and weather conditions for the availability of sunlight and temperature. This makes the treatment process slow and

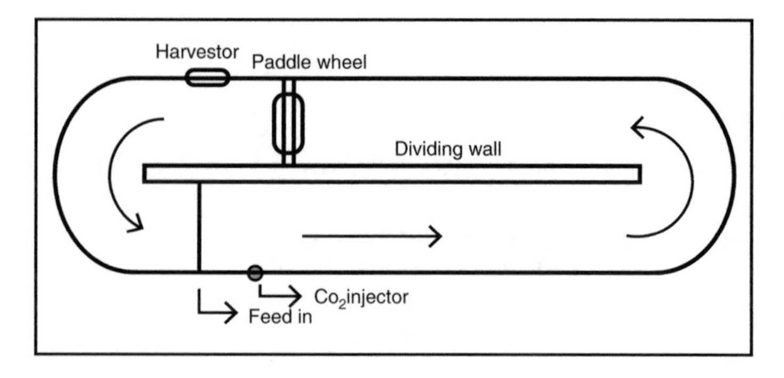

Fig. 1 Schematic diagram of open high rate algal pond with natural illumination. (Adapted and modified from Arbib et al. (2013))

highly variable due to climatic variations. The PBR system, on the other hand, with artificial illumination and temperature control conditions produces better and constant treatment in less time due to more control over parameters.
- The hydraulic retention times in natural systems are of the order of 4–10 days (Garcia et al. 2000), whereas in PBR system, it is as low as 1 or 2 days (Honda et al. 2012; Ashok et al. 2014).
- Open systems are susceptible to contamination, evaporation losses, and less control over species selection, whereas in PBR system more or full control over species, losses, and contamination is there. This makes the PBR system suitable to get desired results in less time.

2 Classifications of Photobioreactors

Broadly, PBRs have been classified as open and closed reactors. Other major subclassification types are based on their shape, mixing pattern, aeration provision, and feeding habits of species. Based on shape, PBRs can be categorized as tubular, rectangular or tank, columnar, cylindrical, annular, torus, flat panel, triangular, or other geometries. Based on their mixing patterns, they can be classified as stirred tank reactors, bubble column, rocking plate, and airlift. Chen et al. (2012) have classified the PBRs based on cultivation conditions as phototrophic, heterotrophic, mixotrophic, and photo-heterotrophic. Based on oxygen availability, PBRs have been classified as aerobic and anaerobic. Mallick (2002) has categorized bioreactors under five subheadings as fluidized bed bioreactors (FBR), packed bed bioreactors (PBR), parallel plate bioreactors (PPR), airlift bioreactors (ALR), and hollow fiber reactors (HFR). The following sections detail classification of PBR types based on shape and simultaneously will encompass other categorizations as well.

2.1 *Tank or Box Configurations*

A schematic diagram of closed stirred tank reactor is represented in Fig. 2 (Li et al. 2003; Gojkovic et al. 2013). Closed tank systems are generally made of glass, Plexiglas, or polyethylene. Mixing can be done either by stirring, mixing paddles, or air bubbling. A high biomass growth rate of 1.3 g/l had been reported in 15 l polycarbonate tank reactors using *Desmodesmus* sp. F2 using wastewater as feed (Huang et al. 2012). In both externally and internally illuminated tank reactors, light path length, intensity, and distribution need to be taken care of to avoid creation of dark zones (Kumar et al. 2013). A closed tank reactor however faces air sparging, degasification, mixing, and illumination costs. Uniform distribution of light becomes difficult for higher volumes, and numerous small units can pose difficulty in regular cleaning and maintenance of the reactor. Deposition of biomass and light distribution on edges of the tank can be avoided by opting for a circular design.

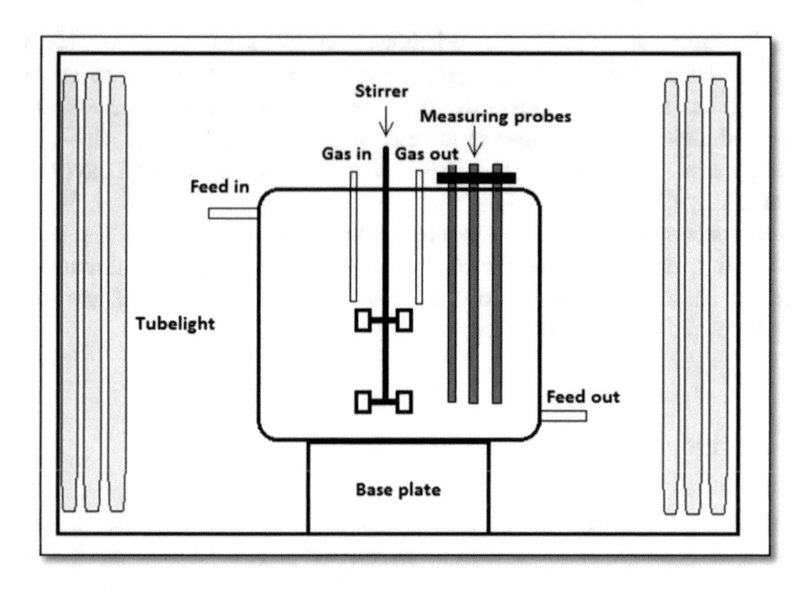

Fig. 2 Schematic diagram of stirred tank reactor with artificial illumination. (Adapted and modified from Gojkovic et al. (2013) and Li et al. (2003))

2.2 *Cylindrical or Columnar Configuration*

Algal systems are a boon to human development and the only kind of phototrophic organisms which doesn't require land to grow. This aspect of algal system can to be utilized by designing the treatment reactors along vertical axis. A general schematic of cylindrical airlift and cylindrical air bubbling reactors have been pictorially shown in Fig. 3. Several past researches on cylindrical reactor configurations with their salient features are listed in Table 1. Most of the studies conducted were in laboratory scale with very few on larger scales. Solovchenko et al. (2014) have tried to treat 50 l alcohol distillery wastewater and successfully removed more than 97% nitrates and 77% phosphates using *Chlorella sorokiniana*. Salas et al. (2013) using a 70 l cylindrical reactor have grown *Scenedesmus obliquus* species with a biomass productivity of 15.25 g/m^2/d. Optical fibers have also been experimented as an illumination source in a 15 l volume Pyrex cylindrical reactor (An and Kim 2000). Mixing is done by airlift, air bubbling, stirring, impeller, or paddle mixing.

Higher biomass growth rates have been reported in columnar reactors (Li et al. 2007; Yuan et al. 2011; Lopes and Franco 2013). Contamination problem can be avoided by covering the top portion of the reactor partially or fully. However, contamination in such reactors is higher than closed tubular reactors and lesser than open tank systems.

In such reactors, greater care needs to be taken for mixing provision by allowing central particles coming closer to the periphery to receive light. Partial aerobic and anaerobic portions may get created due to trapped gases, unavailability of proper lighting, and mixing conditions. The major advantage of such system is lesser land

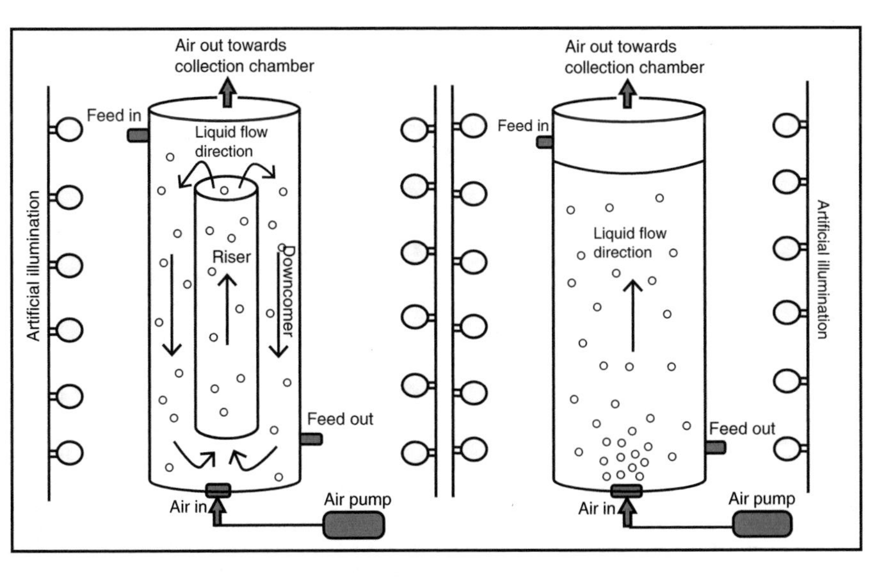

Fig. 3 Schematic diagrams for (**a**) cylindrical airlift reactor and (**b**) bubble column reactor

requirement due to its vertical extent and uniform distribution of light and mixing due to its circular shape. In addition, by virtue of its simple design, it allows ease in maintenance, cleaning, and upgradability.

2.3 *Tubular Configurations*

In tubular reactors, the culture flows through unidirectional tubular channels which allow high liquid flow rates with low shear. Polyethylene bags are commonly been used these days due to its low cost along with good light penetration efficiency (Trotta 1981; Cohen and Arad 1989). Such systems if developed on larger volumes can generate ample amounts of heat which needs to be abated by providing temperature control systems (Richmond 1987). Tubular reactors can be several meters long and can be arranged in different patterns like vertically coiled (Rorrer and Mullikin 1999; Travieso et al. 2001; Oncel and Kose 2014), horizontally coiled (Campo et al. 2001; Adessi et al. 2012), conical (Watanabe and Hall 1996; Morita et al. 2001), helical (Soletto et al. 2008), or 3D mesh layout (Giannelli and Torzillo 2012). A schematic diagram of vertically coiled, horizontally coiled, conical, and helical type tubular reactors is shown in Figs. 4, 5, 6, and 7, respectively. A list of past tubular PBR design and their salient features has been listed in Table 2.

Tubular reactors are generally a closed system arrangement, where maximum control over the physical and biological parameters is possible. Few researchers have reported a unique solution for temperature control by placing the tubular reactor inside a thermostatic water bath; 50 l Pyrex glass reactor (Adessi et al. 2012) and

Table 1 Comparison of past cylindrical or columnar PBR designs with their key features

Reactor configuration	Material	Species	Volume (liters)	Dimensions	Media	Mixing provision	Biomass productivity	Light intensity	Remarks	Reference
Cylindrical or columnar configurations										
Cylindrical	Pyrex	*C. thiosulfatophilum*	15	$D_i = 24$ cm $H = 37$ cm	Synthetic	Stirring	~750 mg/l HRT = 3 d	41–47 μE/m²/s	Biological desulfurization in optical fiber PBR Combined illumination of sunlight and metal halide lamps Physical scratching of optical fibers increased light availability five times	An and Kim (2000)
	Acrylic		17	$D = 15$ cm $T = 3$ mm		Airlift	7.68×10^6 cells/ml			

Cylindrical	Glass	*Chlorella vulgaris*	10	$H = 110$ cm $D_o = 11$ cm	Synthetic	Bubbling	5.8 × 10^7 cells/ml HRT = 1.75 d	24–236 μE/m²/s (12L/12D)	CO_2 fixation capacity enhanced from 80 to 260 mg/l/h with hollow fiber membrane module aeration Max. CO_2 fixation at 1% CO_2	Cheng et al. (2006)
Cylindrical	Borosilicate glass	*Protoceratium reticulatum* GG1AM	15	$D_i = 19.3$ cm	L1 media	Three-bladed marine propeller	25 × 10^3 cells/ml HRT = 8 d	242–766 μE/m²/s	PBR equipped with spin filters for cell retention Yessotoxin productivity 9.16 μg/l/d	Camacho et al. (2011)
Columnar	–	*Scenedesmus* sp., *N. salina*	18	$D_i = 7.2$ cm $H = 81$ cm $T = 0.5$ cm	–	Air and CO_2 bubbling	1.42 g/d	77–105 μE/m²/s	Internally illuminated PBR for algal cultivation Biomass productivity to energy input ratio of fresh water and marine algae 1.42 and 0.37 g/W/d, respectively	Pegallapati et al. (2012)

(continued)

Table 1 (continued)

Reactor configuration	Material	Species	Volume (liters)	Dimensions	Media	Mixing provision	Biomass productivity	Light intensity	Remarks	Reference
Cylindrical or columnar configurations										
Cylindrical	–	*Scenedesmus* sp.	20	D = 19 cm H = 1 m	–	Air and CO_2 bubbling	1.1 g/l HRT = 8 d	350–400 μmol/m²/s	Lutein extraction Increase in temperature increases impurity Ethanol as optimum cosolvent Lutein recovery yield of 76.7%	Yen et al. (2012)
Cylindrical	Opaque PVC	*Scenedesmus obliquus*	70	D_i = 0.3, 1.32 m H = 1.12 m	Modified BG11 media	Air and CO_2 bubbling	0.2 g/l HRT = 4.2–8.4 d	1000 μmol/m²/s (12L/12D)	Analyzing biomass cultivation and lipid production Dilution rates effects lipid productivity and composition	Salas et al. (2013)

Cylindrical	–	*Chlorella vulgaris*	25	–	Wright's cryptophytes (WC) media	Air bubble mixing	1 g/l	–	17 times higher biomass productivity than PBR 77% reduction in water footprint Salt accumulation increases conductivity Transparent exopolymeric particle accumulation due to recycling	Bilad et al. (2014), Discart et al. (2014)
Cylindrical	–	*Chlorella sorokiniana*	50	–	Alcohol distillery wastewater	Paddle mixing	2 g/l HRT = 4 d	180 μmol/m^2/s	Alcohol distillery wastewater remediation >97% nitrates, 35% sulfate and 77% phosphates removal	Solovchenko et al. (2014)

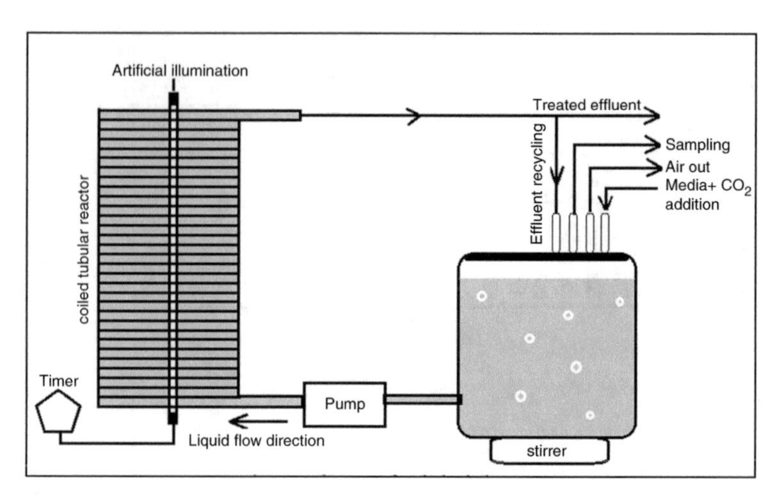

Fig. 4 Schematic diagram of vertical coiled tubular photobioreactor. (Modified from Rorrer and Mullikin (1999), Travieso et al. (2001) and Oncel and Kose (2014))

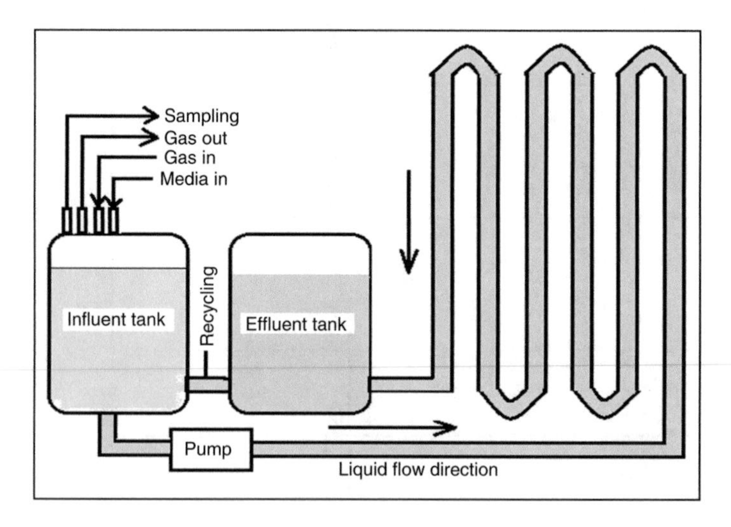

Fig. 5 Schematic diagram of horizontal tubular photobioreactor. (Adapted and modified from Campo et al. (2001) and Adessi et al. (2012))

110 l Plexiglas reactor (Giannelli and Torzillo 2012). Land requirement of such reactors is higher than columnar type but lower than tank, pond, or flat panel type reactors, if placed in stacks. Provision for atmospheric gaseous exchange needs to be employed by degasifiers and air suction pumps. Illumination for such systems can be provided by a combination of both artificial as well as natural sources (Briassoulis et al. 2010; Scoma et al. 2012). In conical helical reactor design, light radiations are distributed to a larger receiving area (Watanabe and Saiki 1997). Oncel and Kose (2014) reported 11% higher biomass productivity in tubular reactors when compared with panel reactors.

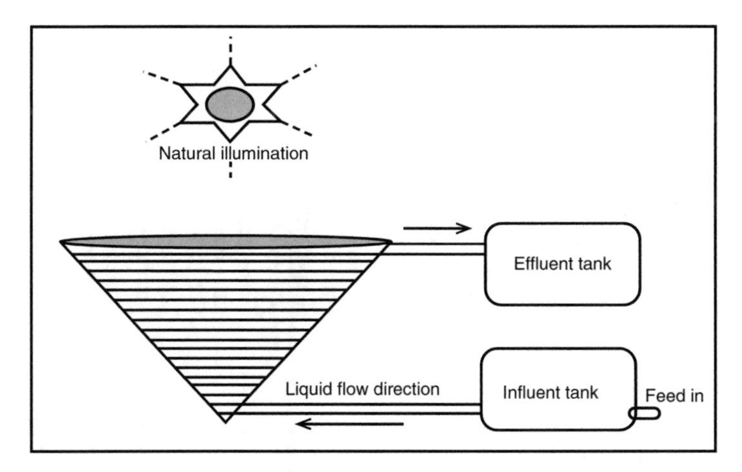

Fig. 6 Schematic diagram of conical tubular photobioreactor. (Adapted and modified from Morita et al. (2001))

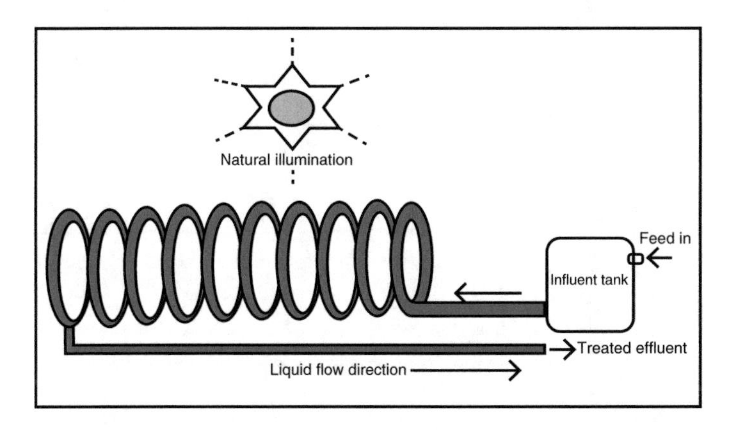

Fig. 7 Schematic diagram of helical shaped photobioreactor. (Adapted and modified from Soletto et al. (2008))

Tubular reactors are generally costlier to construct and difficult to clean. Also, they cannot be dismantled easily, thereby making it difficult for transportation and handling. Additional provisions for aeration, heat exchange, and degassing systems need to be provided for proper working. One of the major drawbacks with inclined tubular reactors is the loss of energy due to reflection (Lee 1986). Scaling up of vertical tubular reactors is not straightforward; however, scaling up of helical tube reactors might be easy by increasing the number of helical coils in the system (Borowitzka 1996). Although, pumping of water through the system can increase shear stress in the culture. Cost constraint in tubular reactors can be reduced by providing combination of artificial and natural light which can lower the running cost of the system. Above all, it has an unprecedented advantage of less contamination, high biomass productivity, high surface area to volume ratio, low to no evaporation losses, and high-value product returns.

Table 2 Comparison of past tubular PBR designs with their salient features

Reactor configuration	Material	Species	Volume (liters)	Dimensions	Media	Mixing provision	Biomass productivity	Light intensity	Remarks	Reference
Tubular configurations										
Tubular	Glass	*Rhodopseudomonas palustris* 42OL	53	$L = 2$ m $D_i = 4.85$ cm $N = 10$	Modified van Niel	Propeller shaft	1.45 g/l	900 W/m^2	Obtained biomass was rich in pigments and PHB Photosynthetic efficiency of 7.6% and 7.1% in winter and summer, respectively	Carlozzi and Sacchi (2001)
Tubular	PMMA	*Muriellopsis* sp.	55	$L = 90$ m $D_i = 2.4$ cm,	Synthetic	Airlift	40 g/m^2/d HRT = 2 d	1900 μmol/m^2/s	Outdoor reactor High biomass loss at night 20–25% 180 mg/m^2/d lutein production	Campo et al. (2001)
Tubular	–	*S. platensis* PCC 8005	77	–	Modified Zarrouk media	Airlift (sterile air)	–	50 W/m^2	Development of auto-regenerative life support system in space exploration 45–70% protein content	Morist et al. (2001)

Tubular conical	Transparent PVC	*Chlorella* sp.	~10	$D_i = 1.6$ cm $L = 49$ m $T = 2$ mm	–	CO_2 sparging	20 g/m^2/day	–	Analyzing biomass cultivation by heating and cooling effects due to temperature variation Proposed combination of strains for increasing productivity	Morita et al. (2001)
Tubular	Transparent PVC	*Spirulina platensis*	21	$D_i = 1.6$ cm $L = 0.9$ m	BG11 media	Airlift (CO_2)	5.82 g/l HRT = 5.3 d	156.6 µmol/m^2/s	Max. productivity of biomass was 0.4 g/l/d Biomass productivity in semicontinuous basis is higher than batch basis	Travieso et al. (2001)
Tubular	–	*Porphyridium cruentum*	100	$L = 1.5$ m $D_i = 4$ cm $N^* = 8$	Hemerick media	Air and CO_2 injection	3 g/l	206 µE/m^2/s	Short light-dark period is better mode of illumination than continuous lightening Specific growth rate at saturation 1.42/d	Feuga et al. (2003)

(continued)

Table 2 (continued)

Reactor configuration	Material	Species	Volume (liters)	Dimensions	Media	Mixing provision	Biomass productivity	Light intensity	Remarks	Reference
Tubular configurations										
Tubular	Glass	*Chlorella salina*	100	$D = 5.1$ cm $N^* = 16$	Seawater M5 media, Trebon media	Air and CO_2 airlift	10 g/l	60–380 μE/m²/s	100 l reactor requires 1 m² area Photoautotrophic conditions yielded 8 g/l and mixotrophic 10 g/l Productivity in 100 l higher than smaller scales	Walter et al. (2003)
Tubular	PVC & reinforced with steel	*Nannochloropsis* sp.	196	$D_i = 10$ cm $L = 25$ m $T = 5$ mm	Guillard's F2 media	Air and CO_2 bubbling	1.1–3.03 g/l/d	2364 μmol/m²/s Min. (18L/6D)	Natural plus artificial lighting Species prefer blue light over red Harvesting rates of 10% of the total volume Continuous cell monitoring sensor	Briassoulis et al. (2010)

Tubular	–	*Scenedesmus obliquus*	500	$D_i = 2.8$ cm	Synthetic	Water recycling	11.3 g/m²/d	250 μmol/m²/s	Plastic beads circulation to prevent biomass growth on surface; 2.18% photosynthetic efficiency Temperature control not necessary for long-term cultivation	Hulatt and Thomas (2011)
Tubular	Plexiglas	*Tetraselmis suecica*	30	$D = 5$ cm $L = 20$ m	Modified Walne media	CO_2 bubbling	2.04 g/l HRT = 14 d	282 μmol/m²/s	Analyzing volumetric productivity in outdoor day and light conditions High nightly biomass loss rate found due to prolonged respiration during dark	Michels et al. (2014)
Tubular triangular	Plexiglas	–	63	$D = 14$ cm $L = 2.34$ m	–	Airlift	–	–	Analyzing biomass growth by mixing and mass transfer variation	Pirouzi et al. (2014)

(continued)

2.4 Flat-Plate or Flat Panel Reactors

Such reactors promote better light penetration inside the culture and higher light utilization efficiency than any other reactor types. Flat panel design allows layer-by-layer arrangement of reactor and light source alternately (Fig. 8).

In case of artificial illumination, reactor is placed vertically in order to save on space. However, when illumination is a combination of both natural and artificial sources, the reactor plates are tilted toward the sun to receive maximum incident solar radiation; and with decrease in solar intensity, illumination is provided by artificial lightening with changing tilt angle. Hence for flat-plate reactor, moving base arrangement with adjustments in tilt angle should be provided to take advantage of both illuminations. A list of flat panel reactor designs with their salient features has been mentioned in Table 3.

Flat panel reactors can be expanded vertically as well as horizontally by increasing the number of panels (Shi et al. 2014). Land footprint of such reactor types is higher than tubular or columnar but lower than open tank systems. However, flat panel reactors are costlier to construct and operate. Being closed reactors, it needs to be provided with provisions of aeration, temperature control, pH control, and removal of accumulated gases. While scaling up the process, provision for temperature management needs to be implemented to counter high heat generation due to glass material itself. Also, regular cleaning of the surface can lead to the additional cost of chemical and water.

In spite of abovementioned limitations, flat panel reactors are most appropriate for higher light utilization efficiency (Borowitzka 1999). Being a closed type reactor, it has major advantages in higher biomass growth, lesser contamination, better

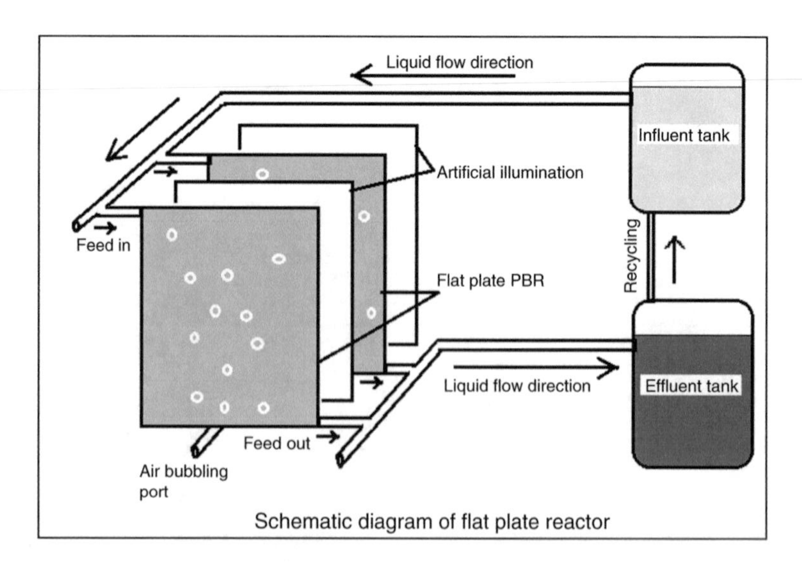

Fig. 8 Schematic diagram of flat-plate reactor

Table 3 Comparison of past flat panel PBR designs with their salient features

Reactor configuration	Material	Species	Volume (liters)	Dimensions	Media	Mixing provision	Biomass productivity	Light intensity	Remarks	Reference
Flat panel configurations										
Flat plate	Acrylic plastic	*Chlorella vulgaris* TISTR-8580, *Botryococcus braunii* NIES-2199, *Spirulina platensis*	9	$L = 10$ cm $W = 30$ cm $H = 50$ cm	Treated sewage	Stirrer, air sparging	0.9 g/l HRT = 1 d	50 W/m^2	Growth inhibitors like chlorellin at long SRTs Submerged membrane filtration Botryococcus became predominant species 91% total nitrogen removal achieved	Honda et al. (2012)
Flat-plate vertical	–	*Synechocystis* sp. PCC 6803	16	$L = 60$ cm $B = 60$ cm $T = 5$ cm	BG11P media	Air and CO_2 bubbling	3.8 g/l HRT = 28–32 d	209 W/m^2	Analyzing biomass cultivation by TDS variation Carbon utilization rate 32 mgC/l/d 97% N and 100% P removal efficiency	Kim et al. (2013)

(continued)

Table 3 (continued)

Reactor configuration	Material	Species	Volume (liters)	Dimensions	Media	Mixing provision	Biomass productivity	Light intensity	Remarks	Reference
Flat panel configurations										
Flat panel	Plexiglas	*Chlorella vulgaris*	1.5, 3	–	Modified DS media	Compressed air	~5 g/l HRT = 4.2 d	980 μE/m^2/s	Biomass productivity increases 1.7 and 2.5 times than randomly mixed bubble column and reduction in light path from 30 to 15 cm, respectively	Degen et al. (2001)
Flat panel	Acrylic	*Spirulina* sp.	50	L = 67 cm B = 15 cm H = 57 cm	Zarrouk media	Air bubbling	0.12 g/l/d	80 μmol/m^2/s	Max. mass transfer coefficient and gas holdup were 31.27/h and 0.065, respectively CO_2 utilization efficiency 30.57%	Velarde et al. (2010)

Flat plate	Polyethylene	–	250	$L = 2.5$ m $B = 0.07$ m $H = 1.5$ m	–	Gas sparging	–	–	53 W/m^3 power promote enough mass transfer rate to avoid DO accumulation 100 W/m^3 power can cause damage to cells, no damage in tubular PBR	Sierra et al. (2008)
Flat plate	Glass fiber	*Halochlorella rubescens CCAC* 0126	55	$L = 1$ m $B = 1$ m	Wastewater	–	6.3 g/m^2/d HRT = 8 d	220 μmol/m^2/s (14L/10D)	Twin-layer biofilm PBR 70–99% of N and P can be removed by immobilized biomass	Shi et al. (2014)

light penetration efficiency, greater control over parameters, and high-value end products. Flat panels are constructed specifically for enhancing light conversion efficiency due to high surface area to volume ratio. Such tank reactors can also be manufactured by industries as their trademark for derivation of microalgae-based products. However, such systems are less preferred for wastewater treatment, since there might be issues of contamination arising from the source and thereby the prime advantage of such systems will be suppressed.

2.5 *Other Design Types*

Some of the other design types which have been researched are torus (Degrenne et al. 2010; Ji et al. 2010), trapezoidal (Zhuang et al. 2014), dome (Sato et al. 2006), modular (Lucker et al. 2014), and bench (Ozkan et al. 2012) type. Torus type reactors have been researched mostly at the laboratory scale with a biomass growth up to 3.2 g/l (Ji et al. 2010) (Fig. 9).

Torus reactors being closed type can be used for the cultivation of pure cultures to have full control over various culture parameters. In the past, most of the studies were conducted for biohydrogen production (Degrenne et al. 2010; Ji et al. 2010). However, scaling up of such reactors can be cumbersome and expensive because of issues like limited illumination, nonuniform mixing, and regular maintenance. Other constraining issues for scaling up of a torus reactor are large space requirement and difficult to construct, transport, install, and maintain. However, such reactors have performed very well in laboratory scale and can be efficiently used in

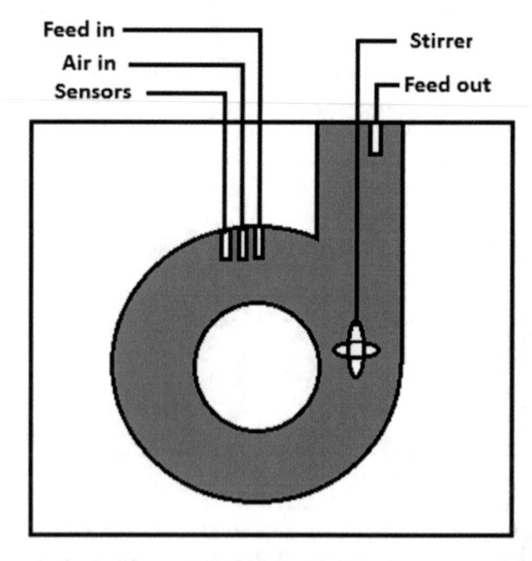

Fig. 9 Schematic diagram of torus reactor. (Adapted and modified from Ji et al. (2010) and Degrenne et al. (2010))

small scales. Designs like dome and trapezoidal type reactors can very well be used for the treatment of wastewater. However, such designs need to be experimented further before large-scale setup.

3 Major Parameters Affecting the Photobioreactor Operation

A generic list of factors emphasizing various parameters affecting PBR operation has been represented in Fig. 10. The process flow starts with the selection of the desired set of outputs. Based on the selected output, the specific reactor type and other related design parameters of the PBR (such as target population size, wastewater collection and distribution system, species, climatic conditions, harvesting, and economic considerations) should be determined. Population size will determine the capacity of PBR, while distance from the wastewater collection and distribution system will determine the transportation and other cost. Some of the design parameters like climate, species, and harvesting are interdependent, while parameters like population size, economic considerations, and distribution systems are dependent and will therefore jointly determine the efficient working of the reactor.

In order to fulfill all the necessary requirements for the optimal growth of biomass and desired products, an efficient PBR should take care of all the necessary requirements for the health of its dwellers. Carbon, nitrogen, and phosphorous are the most significant nutrients required for microalgal nutrition. Apart from these, other micronutrients required by them are S, Mg, Ca, Zn, Mn, Cu, and Mb (Song et al. 2012). Nutrients such as nitrates and phosphates remain untreated in the secondary wastewater treatment process and can be used as a feed for the algal-bacterial PBRs (Raouf et al. 2012). A very high concentration of nitrates and phosphates can lead to poor performance of the reactor, while lower nutrient concentrations are generally removed leading to higher reactor performance. Some microalgal species manage to survive in nutrient-limiting condition also by shifting from phototrophic growth pattern to that of heterotrophic or mixotrophic (Kumar et al. 2013). After secondary treatment, 90% of the carbon from wastewater gets removed (Arceivala and Asolekar 2007) and remaining can be utilized as carbon source by bacteria.

Nitrite assimilation inside the reactor is primarily governed by temperature, pH, light, and carbon availability, among others. The optimal temperature for growth varies from species to species. Light intensity, uniformity, and duration are other important parameters for photosynthetic microalgal growth.

Under light limitation condition, microalgal productivity depends directly upon the light conversion efficiency (Kumar et al. 2010). Most of the microalgae in natural or mixed culture favor pH values in the range of 5–12 (Dubinsky and Rotem 1974). Lower pH values can cause acidity; however, pH values greater than 9 can kill fecal coliform (Raouf et al. 2012). With higher pH values, dissolved oxygen

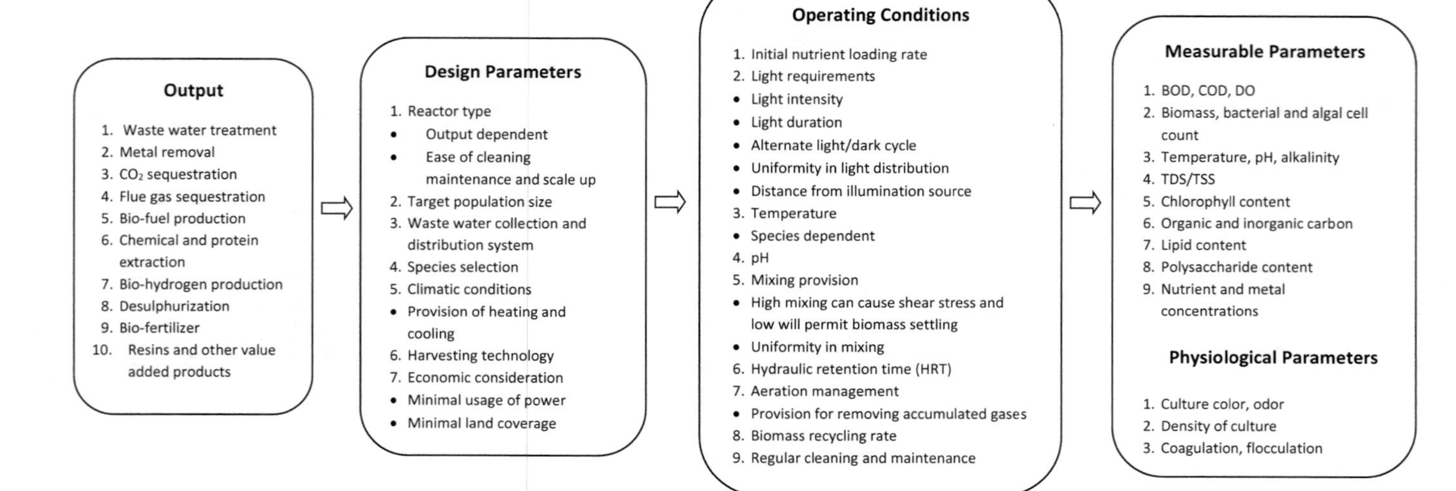

Fig. 10 Parameters affecting photobioreactor operation

concentration also rises which can lead to photooxidative damage to the cells. In order to avoid it, proper mixing and removal of accumulated gases from the reactor should be accompanied. Mixing should be done uniformly and optimally, so that shear stress to the algal cells could be minimized. For a suspended media reactor, mixing keeps the reactor medium in suspension, while in deep reactors, it also allows inner particles to move outward, thereby leading to uniform exposure to light source. Some of the common methods of mixing are shaking, bubbling or gas injection, pumping, and stirring. Aeration is a major aspect in closed reactors, especially in tubular reactors where uniform mixing can only be provided through aeration. CO_2 bubbling has shown to increase biomass productivity, thereby increased productivity of biofuels (Itoiz et al. 2012; Lopes and Franco 2013). Removal of accumulated gases such as oxygen can cause photooxidative stress to the algal cells and hence needs to be removed. It can be done by various means, like suction pump, circulation of gases, or an exhaust fan in open reactors. Other gas transfer equipments include mechanical systems such as blades, propellers, brushes and paddles, jet aerators, bubble diffusers, hollow fiber membrane unit, etc. (Kumar et al. 2013).

Optimum biomass recycling rates can increase biomass concentration inside the reactor, henceforth higher biomass productivities (Park and Craggs 2014). HRT of open pond system can be as high as 10–30 days (Arceivala and Asolekar 2007) which can be reduced eminently in an artificially illuminated PBR. Simpler the reactor design greater will be the ease of maintenance and regular cleaning of the reactor. Elected physical, chemical, biological, and physiological parameters should be monitored regularly to ensure the workability of the reactor. Physiological parameters of the culture help to have an intuitive guess over culture health in a PBR.

4 Recent Technical Advancements for Large-Scale Photobioreactor Application

The major constraint in going for industrial cultivation is the CAPEX and OPEX involved in PBR systems. In order to minimize the costs, algal cultivations are being carried on in polyethylene bags on tidal waves near seashore. This reduces the mixing and artificial illumination costs. These algal bags are ~4 m long and are seeded with wastewater and carbon dioxide (Omega 2009, NASA). This is an innovative method to grow algae, clean wastewater, capture carbon dioxide, and ultimately produce biofuel without competing with agriculture for water, fertilizer, or land. The algae use energy from the sun, carbon dioxide, and nutrients from the wastewater to produce biomass that can be converted into biofuels as well as other useful products such as fertilizer and animal food. The algae clean the wastewater by removing nutrients that otherwise would contribute to marine dead zone formation. Such system is employed in many places for commercial applications.

5 Conclusion

The current paper gives a brief overview of major past PBR designs, especially for wastewater treatment. The wastewater withholds a wide variety of beneficial end products which needs to be extracted. Large numbers of PBR designs have been experimented in the past, but unfortunately very few have been able to made up to the industrial scale. Each reactor has a different degree of performance which depends on the effective selection of parameters like mixing, light, climate, species, and reactor type. Major aspects which should be considered before construction of any reactor are required end products, ease of construction, simplicity, less power consumption, minimal footprint area, cost effective, and ease of cleaning and maintenance. Few key points that should be taken care of:

- Greater use of land for designing PBRs can create its shortage for other purposes like crop production. Therefore designing reactors vertically rather than horizontally creates smaller land footprint.
- Open system can affront high contamination, nonuniform illumination, and large land footprint constraints. Hence, it can be used for wastewater treatment in areas with low discharge and low land prices.
- Cylindrical or columnar reactors can very well be applied for wastewater treatment in high land price areas accompanying uniform distribution of light and mixing by virtue of its circular shape.
- Wastewater treatment using closed PBRs such as tubular and flat panels can be expensive, land, and labor intensive. Such systems should be preferred for contamination free biomass production for extracting high-value end products.

References

Adessi A, Torzillo G, Baccetti E, Philippis RD (2012) Sustained outdoor H_2 production with Rhodopseudomonas palustris cultures in a 50 L tubular photobioreactor. Int J Hydrogen Energy 37:8840–8849

An JY, Kim BW (2000) Biological desulfurization in an optical-fiber photobioreactor using an automatic sunlight collection system. J Biotechnol 80:35–44

Arbib Z, Ruiz J, Diaz PA, Perez CG, Barragan J, Perales JA (2013) Effect of pH control by means of flue gas addition on three different photo-bioreactors treating urban wastewater in long-term operation. Ecol Eng 57:226–235

Arceivala SJ, Asolekar SR (2007) Wastewater treatment for pollution control and reuse, 3rd edn. Tata McGraw-Hill Publishing Company Limited, New York

Ashok V, Shriwastav A, Bose P (2014) Nutrient removal using algal-bacterial mixed culture. Appl Biochem Biotechnol 174(8):2827–2838

Bilad MR, Discart V, Vandamme D, Foubert I, Muylaert K, Vankelecom IFJ (2014) Coupled cultivation and pre-harvesting of microalgae in a membrane photobioreactor (MPBR). Bioresour Technol 155:410–417

Borowitzka MA (1996) Closed algal photobioreactors: design considerations for large-scale systems. J Mar Biotechnol 4:185–191

Borowitzka MA (1999) Commercial production of microalgae: ponds, tanks, tubes and fermenters. J Biotechnol 70:313–321

Briassoulis A, Panagakis P, Chionidis M, Tzenos D, Laos A, Tsinos C, Berberidis K, Jacobsen A (2010) An experimental helical-tubular photobioreactor for continuous production of Nannochloropsis sp. Bioresour Technol 101:6768–6777

Camacho FG, Rodriguez JJG, Miron AS, Belarbi EH, Chisti Y, Grima EM (2011) Photobioreactor scale-up for a shear-sensitive dinoflagellate microalga. Process Biochem 46:936–944

Campo JAD, Rodriguez H, Moreno J, Vargas MA, Rivas J, Guerrero MG (2001) Lutein production by Muriellopsis sp. in an outdoor tubular photobioreactor. J Biotechnol 85:289–295

Carlozzi P, Sacchi A (2001) Biomass production and studies on Rhodopseudomonas palustris grown in an outdoor, temperature controlled, underwater tubular photobioreactor. J Biotechnol 88:239–249

Chen HW, Yang TS, Chen MJ, Chang YC, Lin CY, Wang EIC, Ho CL, Huang KM, Yu CC, Yang FL, Wu SH, Lu YC, Chao LKP (2012) Application of power plant flue gas in a photobioreactor to grow Spirulina algae, and a bioactivity analysis of the algal water-soluble polysaccharides. Bioresour Technol 120:256–263

Cheng L, Zhang L, Chen H, Gao C (2006) Carbon dioxide removal from air by microalgae cultured in a membrane-photobioreactor. Sep Purif Technol 50:324–329

Cohen E, Arad SM (1989) A closed system for outdoor cultivation of Porphyridium. Biomass 18:59–67

Degen J, Uebele A, Retze A, Staiger US, Trosch W (2001) A novel airlift photobioreactor with baffles for improved light utilization through the flashing light effect. J Biotechnol 92:89–94

Degrenne B, Pruvost J, Christophe G, Cornet JF, Cogne G, Legrand J (2010) Investigation of the combined effects of acetate and photobioreactor illuminated fraction in the induction of anoxia for hydrogen production by Chlamydomonas reinhardtii. Int J Hydrogen Energy 35:10741–10749

Discart V, Bilad MR, Marbelia L, Vankelecom IFJ (2014) Impact of changes in broth composition on Chlorella vulgaris cultivation in a membrane photobioreactor (MPBR) with permeate recycle. Bioresour Technol 152:321–328

Dubinsky Z, Rotem J (1974) Relations between algal populations and the pH of their media. Oecologia 16:53–60

Garcia J, Mujeriego R, Hernandez-Marine M (2000) High rate algal pond operating strategies for urban wastewater nitrogen removal. J Appl Phycol 12(3–5):331–339

Giannelli L, Torzillo G (2012) Hydrogen production with the microalga Chlamydomonas reinhardtii grown in a compact tubular photobioreactor immersed in a scattering light nanoparticle suspension. Int J Hydrogen Energy 37:16951–16961

Gojkovic Z, Nores IG, Jacinto VG, Barrera TG, Ariza JLG, Marova I, Lobato CV (2013) Continuous production of selenomethionine-enriched Chlorella sorokiniana biomass in a photobioreactor. Process Biochem 48:1235–1241

Honda R, Boonnorat J, Chiemchaisri C, Chiemchaisri W, Yamamoto K (2012) Carbon dioxide capture and nutrients removal utilizing treated sewage by concentrated microalgae cultivation in a membrane photobioreactor. Bioresour Technol 125:59–64

Huang CC, Hung JJ, Peng SH, Chen CNN (2012) Cultivation of a thermo-tolerant microalga in an outdoor photobioreactor: influences of CO_2 and nitrogen sources on the accelerated growth. Bioresour Technol 112:228–233

Hulatt CJ, Thomas DN (2011) Energy efficiency of an outdoor microalgal photobioreactor sited at mid-temperate latitude. Bioresour Technol 102:6687–6695

Itoiz ES, Grunewald CF, Gasol CM, Garces E, Alacid E, Rossi S, Rieradevall J (2012) Energy balance and environmental impact analysis of marine microalgal biomass production for biodiesel generation in a photobioreactor pilot plant. Biomass Bioenergy 39:324–335

Ji CF, Legrand J, Pruvost J, Chen ZA, Zhang W (2010) Chracterization of hydrogen production by Platymonas Subcordiformis in torus photobioreactor. Int J Hydrogen Energy 35:7200–7205

Kim HW, Vannela R, Rittmann BE (2013) Responses of Synechocystis sp. PCC 6803 to total dissolved solids in long-term continuous operation of a photobioreactor. Bioresour Technol 128:378–384

Kumar A, Ergas S, Yuan X, Sahu A, Zhang Q, Dewulf J, Malcata FX, Langenhove HV (2010) Enhanced CO_2 fixation and biofuel production via microalgae: recent developments and future directions. Trends Biotechnol 28:371–380

Kumar K, Sirasale A, Das D (2013) Use of image analysis tool for the development of light distribution pattern inside the photobioreactor for the algal cultivation. Bioresour Technol 143:88–95

Lee YK (1986) Enclosed bioreactors for the mass cultivation of photosynthetic microorganisms: the future trend. Trends Biotechnol 4:186–189

Li J, Xu NS, Su WW (2003) Online estimation of stirred-tank microalgal photobioreactor cultures based on dissolved oxygen measurement. Biochem Eng J 14:51–65

Li ZY, Guo SY, Li L, Cai MY (2007) Effects of electromagnetic field on the batch cultivation and nutritional composition of Spirulina platensis in an air-lift photobioreactor. Bioresour Technol 98:700–705

Lopes EJ, Franco TT (2013) From oil refinery to microalgal biorefinery. J CO2 Util 2:1–7

Lopez PP, Garcia SG, Jeffryes C, Agathos SN, McHugh E, Walsh D, Murray P, Moane S, Feijoo G, Moreira MT (2014) Life cycle assessment of the production of the red antioxidant carotenoid astaxanthin by microalgae: from lab to pilot scale. J Clean Prod 64:332–344

Lucker BF, Hall CC, Zegarac R, Kramer DM (2014) The environmental photobioreactor (ePBR): an algal culturing platform for simulating dynamic natural environments. Algal Res 6:242–249. https://doi.org/10.1016/j.algal.2013.12.007

Mallick N (2002) Biotechnological potential of immobilized algae for wastewater N, P and metal removal: a review. Biometals 15:377–390

Michels MHA, Slegers PM, Vermue MH, Wijffels RH (2014) Effect of biomass concentration on the productivity of Tetraselmis suecica in a pilot-scale tubular photobioreactor using natural sunlight. Algal Res 4:12–18

Morist A, Montesinos JL, Cusido JA, Godia F (2001) Recovery and treatment of Spirulina platensis cells cultured in a continuous photobioreactor to be used as food. Process Biochem 37:535–547

Morita M, Watanabe Y, Saiki H (2001) Instruction of microalgal biomass production for practically higher photosynthetic performance using a photobioreactor. Trans IChemE 79:176–183

Muller-Feuga A, Pruvost J, Le Guédes R, Le Déan L, Legentilhomme P, Legrand J (2003) Swirling flow implementation in a photobioreactor for batch and continuous cultures of Porphyridium cruentum. Biotechnology and bioengineering, 84(5), 544–551

OMEGA (2009–2012) https://www.nasa.gov/centers/ames/research/OMEGA/index.html

Oncel S, Kose A (2014) Comparison of tubular and panel type photobioreactors for bihydrogen production utilizing Chlamydomonas reinhardtii considering mixing time and light intensity. Bioresour Technol 151:265–270

Ozkan A, Kinney K, Katz L, Berberoglu H (2012) Reduction of water and energy requirement of algae cultivation using an algae biofilm photobioreactor. Bioresour Technol 114:542–548

Palmer CM (1974) Algae in American sewage stabilization's ponds. Rev Microbiol (S-Paulo) 5:75–80

Park JBK, Craggs RJ (2014) Effect of algal recycling rate on the performance of *Pediastrum boryanum* dominated wastewater treatment high rate algal pond. Water Sci Technol 70:1299–1306

Park JBK, Craggs RJ, Shilton AN (2011) Wastewater treatment high rate alagl ponds for biofuel production. Bioresour Technol 102(1):35–42

Pegallapati AK, Arudchelvam Y, Nirmalakhandan N (2012) Energy efficient photobioreactor configuration for algal biomass production. Bioresour Technol 126:266–273

Pirouzi A, Nosrati M, Shojaosadati SA, Shakhesi S (2014) Improvement of mixing time, mass transfer, and power consumption in an external loop airlift photobioreactor for microalgae cultures. Biochem Eng J 87:25–32

Raouf NA, Homaidan AAA, Ibraheem IBM (2012) Microalgae and wastewater treatment. Saudi J Biol Sci 19:257–275

Richmond A (1987) The challenge confronting industrial microagriculture: high photosynthetic efficiency in large-scale reactors. Hydrobiology 151/152:117–121

Rorrer GL, Mullikin RK (1999) Modeling and simulation of a tubular recycle photobioreactor for macroalgal cell suspension cultures. Chem Eng Sci 54:3153–3162

Salas LML, Castrillo M, Martinez D (2013) Effects of dilution rates and water reuse on biomass and lipid production of Scenedesmus obliquus in a two-stage novel photobioreactor. Bioresour Technol 143:344–352

Sato T, Usui S, Tsuchiya Y, Kondo Y (2006) Invention of outdoor close type photobioreactor for microalgae. Energy Convers Manag 47:791–799

Scoma A, Giannelli L, Faraloni C, Torzillo G (2012) Outdoor H_2 production in a 5-L tubular photobioreactor by means of a sulfur-deprived culture of the microalga Chlamydomonas reinhardtii. J Biotechnol 157:620–627

Shi J, Podola B, Melkonian M (2014) Application of a prototype-scale Twin-Layer photobioreactor for effective N and P removal from different process stages of municipal wastewater by immobilized microalgae. Bioresour Technol 154:260–266

Sierra E, Acien FG, Fernandez JM, Garcia JL, Gonzalez C, Molina E (2008) Characterization of a flat plate photobioreactor for the production of microalgae. Chem Eng J 138:136–147

Singh RN, Sharma S (2012) Development of suitable photobioreactor for algae production-a review. Renew Sust Energ Rev 16:2347–2353

Soletto D, Binaghi L, Ferrari I, Lodi A, Carvalho JCM, Zilli M, Converti A (2008) Effects of carbon dioxide feeding rate and light intensity on the fed-batch pulse-feeding cultivation of *Spirulina platensis* in helical photobioreactor. Biochem Eng J 39:369–375

Solovchenko A, Pogosyan S, Chivkunova O, Selyakh I, Semenova I, Voronova E, Scherbakov P, Konyukhov I, Chekanov K, Kirpichnikov M, Lobakova E (2014) Phycoremediation of alcohol distillery wastewater waitha novel Chlorella sorokiniana strain cultivated in a photobioreactor monitored on-line via chlorophyll fluorescence. Algal Res 6:234–241. https://doi.org/10.1016/j.algal.2014.01.002

Song L, Qin JG, Shengqi S, Xu J, Clarke S, Shan Y (2012) Micronutrient requirements for growth and hydrocarbon production in the oil producing green alga *Botryococcus braunii* (Chlorophyta). PLoS One 7(7):e41459. https://doi.org/10.1371/journal.pone.0041459

Travieso L, Hall DO, Rao KK, Benitez F, Sanchez E, Borja R (2001) A helical tubular photobioreactor producing Spirulina in a semicontinuous mode. Int Biodeter Biodegr 47:151–155

Trotta P (1981) A simple and inexpensive system for continuous monoxenic mass culture of marine microalgae. Aquaculture 22:383–387

Velarde RR, Urbina EC, Melchor DJH, Thalasso F, Villanueva ROC (2010) Hydrodynamic and mass transfer characterization of a flat-panel airlift photobioreactor with high light path. Chem Eng Process 49:97–103

Walter C, Steinau T, Gerbsch N, Buchholz R (2003) Monoseptic cultivation of phototrophic microorganisms—development and scale-up of a photobioreactor system with thermal sterilization. Biomolecular engineering, 20(4-6), 261–271

Watanabe Y, Hall DO (1996) Photosynthetic CO_2 conversion technologies using a photobioreactor incorporating microalgae-energy and material balances. Energy Convers Manag 37(6–8):1321–1326

Watanabe Y, Saiki H (1997) Development of a photobioreactor incorporating Chlorella sp. for removal of CO_2 in stack gas. Energy Convers Manag 38:S499–S503

Yen HW, Chiang WC, Sun CH (2012) Supercritical fluid extraction of lutein from Scenedesmus cultured in an autotrophical photobioreactor. J Taiwan Inst Chem Eng 43:53–57

Yuan X, Kumar A, Sahu AK, Ergas SJ (2011) Impact of ammonia concentration on Spirulina platensis growth in an airlift photobioreactor. Bioresour Technol 102:3234–3239

Zhuang LL, Hu HY, Wu YH, Wang T, Zhang TY (2014) A novel suspended-solid phase photobioreactor to improve biomass production and separation of microalgae. Bioresour Technol 153:399–402

Design Considerations of Algal Systems for Wastewater Treatment

Mahmoud Nasr

1 Introduction

Wastewater-grown algae are a promising approach for environmental remediation and sustainable production of animal feed and human nutritional requirements (Chen et al. 2011). Wastewater, as a culture medium, is used to provide the algal cells with essential nitrogen and phosphorus species (Craggs et al. 2014). Algal cultures can be acclimatized to receive various sources of wastewater such as raw and treated sewage from domestic areas and runoff from agricultural lands. Algal ponds can handle organic loading rates (OLRs) of 100–150 kg BOD/ha/d and assimilate nutrient species at rates of 24 kg N/ha/d and 3 kg P/ha/d (Fernandez et al. 2013). In addition, several algal strains have extensive capabilities to sorb toxic heavy metal ions such as Cu^{2+}, Ni^{2+}, Pb^{2+}, Cd^{2+}, Fe^{2+}, and Mn^{2+} from aqueous solutions (Mehta and Gaur 2005). In integrated algal-bacterial systems, algae release O_2 by the sequestration of CO_2 during the photosynthetic activity. In turn, bacteria utilize the produced O_2 gas for converting (oxidizing) the organic carbon into new cells and CO_2 gas. The obtained algal biomass can be further developed for the extraction of high-value substances such as polysaccharides (sugars) and triacylglycerides (fats) (Slade and Bauen 2013).

Algal systems are appropriately designed to attain high biomass productivity with low-cost and minimum energy inputs. Essential minerals/nutrients, radiative energy, and carbon source are required for an efficient cultivation process (Bhola et al. 2017). The photosynthetic production of 1 g of algal biomass consumes 1.8 g CO_2 and generates 1.3 g O_2 (Fernandez et al. 2013). An enriched gas mixture is supplied to the culture medium to provide the algal biomass with a CO_2 partial pressure of 0.2 kPa, i.e., equivalent to 0.076 mol/m^3 and 3.3 mg CO_2/L (Doucha et al. 2005). Mass transfer capacity is adopted by the supply of carbon dioxide along with the

M. Nasr (✉)
Sanitary Engineering Department, Faculty of Engineering, Alexandria University, Alexandria, Egypt

S. K. Gupta, F. Bux (eds.), *Application of Microalgae in Wastewater Treatment*, https://doi.org/10.1007/978-3-030-13913-1_19

removal of oxygen, in which an O_2 concentration of 7.2 mg/L at 20 °C can inhibit the photosynthetic activities of several algal strains (Fernandez et al. 2013).

Different designs and configurations of reactor systems have been used for the production of algae. Open pond reactors are the most commonly used systems for algal cultivation at a large-scale application due to their low costs of construction, operation, and maintenance (Gupta et al. 2015). This system is directly exposed to open air and uses a free source of energy from sunlight. However, it suffers from the possibility of contamination due to the limited control on environmental conditions (Markou and Nerantzis 2013). Other challenges of open ponds include high water evaporation, low quality and concentration of biomass production, and expensive downstream processes (Young et al. 2017). Open pond systems include waste stabilization ponds, circular/shallow ponds, and raceway ponds. The algal species of *Chlorella*, *Scenedesmus*, *Dunaliella*, and *Spirulina* are suitable for this mechanism.

Alternatively, closed photobioreactors have been developed to cope with the disadvantages of open pond systems. Photobioreactors do not allow for a direct exchange of gasses between the culture and atmosphere (Norsker et al. 2011). A typical photobioreactor is composed of four phases (Hincapie and Stuart 2015): (a) a solid phase containing algal biomass, (b) an aqueous phase for the growth medium, (c) a gaseous phase of CO_2 and O_2 gasses, and (d) a light-radiation field. Photobioreactors can be provided with a stirring unit to ensure mechanical agitation, heat and mass transfers, light dispersion, and homogeneous culture (Gupta et al. 2017). Photobioreactors safeguard suitable conditions, viz., nutrients and CO_2 supplies, optimal pH and temperature, adequate exposure to light, and sufficient mixing, for biomass growth. These reactors have been successfully used for the production of a wide variety of algal species such as *Chlorella*, *Dunaliella*, *Haematococcus*, *Phaeodactylum*, *Porphyridium*, *Spirulina*, and *Tetraselmis*. However, the high cost and power supply required to achieve the maximum biomass productivity are the main drawbacks of this system (Medipally et al. 2015). Based on the illuminated surface area, the common types of closed systems are tubular, flat plate, and column photobioreactors. Regarding the mode of culture flow, photobioreactors can be categorized as airlift reactor, bubble column, and stirred type.

The amount of solar energy received on the culture surface is influenced by the design and orientation of the cultivation system (Vejrazka et al. 2012). In addition, the reactor geometry affects the distribution of solar radiation on the surface, the irradiance propagation inside the culture, and the efficient utilization of light. Hence, this chapter presents the design considerations of various systems used for algal cultivation and wastewater treatment. The considered parameters included light dispersion, mixing, and temperature. The design, configurations, advantages, and limitations of several algal systems, i.e., open ponds and closed photobioreactors, were also discussed.

2 Designing Factors of Algal Culture Systems

2.1 Light Dispersion

Light intensity is an important factor that influences the cultivation of algal biomass. An adequate light operation should ensure high biomass productivity along with the reduction of both energy utilization and running costs (Singh and Singh 2015). However, a self-shading phenomenon may occur as the culture density/concentration increases over the threshold. Under this condition, the culture becomes subjected to an illuminated outer region and a relatively dark interior (Dalrymple et al. 2013). In dark zones, the algal cells are performing respiration rather than photosynthesis; hence, the algal activity decreases. The effect of self-shading can be minimized by using shallow or thin culture systems.

Photosynthesis rate is estimated as a function of the irradiance subjected to the culture surface (i.e., irradiance expresses the total amount of radiation that falls onto a unit area). The photosynthetically active radiation occurs at a spectral range of 400–700 nm, at which photosynthesis process occurs. Photosynthesis rate can be efficiently developed by flashing or intermittent light rather than by continuous illumination (Lee et al. 2015). In addition, an increase in the frequency of culture movement between illuminated and dark zones enhances the cell yield.

The light-inhibition model can be expressed by Eq. 1 (Nasr et al. 2017).

$$P_{O_2} = \frac{P_{O_2,\max} \cdot I}{I_k + I + \frac{I^2}{I_i}} \quad (1)$$

where P_{O_2} is the specific rate of oxygen production (mmol-O_2/m^3/s), $P_{O_2,\max}$ is the maximum photosynthesis rate (mmol-O_2/m^3/s), I is a given light intensity (μE/m^2/s), I_k is the light intensity half-saturation coefficient (μE/m^2/s), and I_i is irradiance at photoinhibition also known as the inhibition coefficient (μE/m^2/s).

The CO_2 consumption rate can be described by Eq. 2, assuming one-to-one molar ratio between CO_2 and O_2 (Fernandez et al. 2013).

$$P_{CO_2} = -P_{O_2} \quad (2)$$

The biomass production can also be estimated by Eq. 3.

$$P = \frac{P_{\max} \cdot I}{I_k + I} \quad (3)$$

where P is photosynthetic rate (mg C/m^3/h or mg C/mg chl-*a*/h) and $P_{\max}$ is photosynthetic rate (mg C/m^3/h or mg C/mg chl-*a*/h).

A static *P–I* model, which promotes both photoinhibition and photoadaptation, has been developed to mimic the realistic (dynamic) photosynthetic activity (Béchet et al. 2013). As shown in Fig. 1, a plot of photosynthesis rate against irradiance gives a hyperbolic curve that shows three irradiance points. These stages describe the photosynthesis rate as follows: (a) half-saturation constant (I_k) where the photosynthetic rate ensues at ½ P_{max}, (b) theoretical saturation irradiance (I_s) where the photosynthesis rate becomes saturated, and (c) inhibition irradiance (I_i) in which the photosynthesis rate initiates to decline. The irradiance values vary according to the culture conditions and enzymatic kinetic. For example, I_k, I_s, and I_i can be in ranges of <100, 100–500, and over 1000 μE/m²/s, respectively (Vejrazka et al. 2012).

The radiation at any position inside the reactor can be estimated by the Beer-Lambert law. An average irradiance (Eq. 4) is used to represent the amount of light delivered by algal cells that are randomly moving inside the medium (Grima et al. 1996).

$$I_{av} = \frac{I}{K_a \cdot p \cdot C}\left[1 - \exp\left(-K_a \cdot p \cdot C\right)\right] \tag{4}$$

where I_{av} is the average irradiance of the entire culture volume (kg/s³), K_a is the algal absorption coefficient (m²/g), p is the length of the light path (m), and C is the algae concentration (g/m³).

The exponential growth phase of algal cells is described by Eq. 5.

$$\mu = \frac{\mu_{max} \cdot I_{av}}{I_k + I_{av}} \tag{5}$$

where μ is the specific algae growth rate (1/h) and μ_{max} is the maximum specific growth rate (1/h).

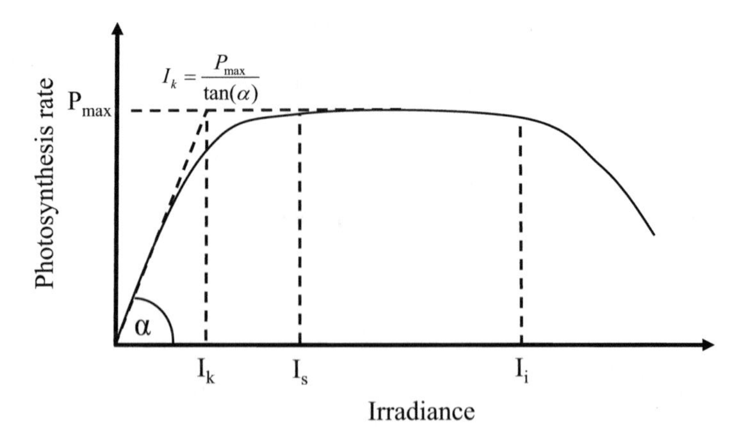

Fig. 1 Typical photosynthesis – irradiance (light intensity) curve

However, the estimation of average light intensity does not undertake the fluid dynamics or the light regime exposed to biomass cells. In the real application, bioreactors are affected by the culture concentration and mutual shading, and thus, each algal cell receives an irradiance value according to its position in the medium (Nasr et al. 2017). Light intensity declines exponentially from the irradiated surface to the photobioreactor center, where mutual shading of the algal cells may occur. In addition, algal biomass transfers according to the fluid dynamic in the reactor. Hence, the overall photosynthesis can be estimated by integrating local values of photosynthesis and growth rates over the total culture volume (Takache et al. 2012). Recently, computational fluid dynamic (CFD) simulation has been employed for an efficient reactor design by describing the potential light utilization of algae (Yang et al. 2016).

The decrease in light path length similar to tubular and flat panel photobioreactors is favorable for an efficient dispersion/utilization of light. Small tube diameter (<4 cm) enhances the biomass productivity by increasing the light utilization activity and ensuring adequate agitation. Norsker et al. (2011) reported photosynthetic efficiencies of 1.5% for open pond, 3% for tubular photobioreactor, and 5% for flat panel photobioreactor with algal productivities of 21, 41, and 64 ton dry weight per hectare. For outdoor applications, algae receive a high amount of sunlight throughout the day when the photobioreactors are well inclined with respect to the sun.

2.2 Agitation

The design of highly efficient photobioreactors should consider the mechanical part used for mixing and circulating the culture suspension. Mixing is beneficial for algal growth as it allows for a uniform nutrient distribution, avoids thermal stratification, enhances gas-liquid mass transfer, and maintains algal cells in suspension (Young et al. 2017). In addition, proper mixing ensures periodic dark-light cycles by the transfer of algal cells from the dark interior zone to the illuminated part near the reactor surface. However, high-speed mixing may destroy and damage the algal cells due to excessive shear forces, causing a reduction of biomass productivity (Chen et al. 2011). In addition, during aeration, the generated bubbles may break up or burst to cause harmful stress to cells. Apart from this, mixing rate has to be optimized to prevent excessive energy utilization and costs.

Mixing equipment differs according to the reactor type, being impeller (agitation) for stirred reactors, or sparger and baffles (aeration) for air-driven photobioreactors. Other mechanical devices can include pump, stirrer, or paddle wheel (Posten 2009). Circulation, which is achieved via the pressurized scheme, motor-driven pumping, and airlift, is a reliable method that can promote proper mixing (Gupta et al. 2015).

2.3 Temperature

The culture temperature is mainly influenced by natural convection via the surrounding environment (Singh and Singh 2015). In high-density culture, a large fraction of light is absorbed into algal cells as heat by radiation. For an optimal algal growth, the culture temperature should be maintained between 20 and 30 °C. The growth rate declines as temperature decreases below 20 °C, while at an increase in temperature over 30 °C, the algal biomass can be subjected to severe damage and probably death. It is recommended that marine algae are subjected to a temperature not exceeding 28 °C, whereas freshwater algae can accommodate a wide range of temperature between 25 and 35 °C (Fernandez et al. 2013).

Heat should be supplied or removed by proper amounts to avoid culture temperature variations and death of some cells. A heat balance model can be applied to estimate the power required for temperature control in closed and small systems (Huang et al. 2017). For outdoor culture systems, the heat mass balance should consider the amount of heat loss due to convection and evaporation. Moreover, some types of cooling system (water spray on the reactor surface or internal heat exchangers) can be used to control the temperature at hot-dry climatic zones.

3 Open Pond Culture Systems

3.1 Facultative Waste Stabilization Ponds

Facultative ponds are cost-effective waste stabilization ponds employed for the treatment of domestic wastewater due to their reliability and easy operation (Dalrymple et al. 2013). The ponds are composed of large and shallow basins (depth of 1.2–1.5 m) that use natural biological processes involving the activities of both growing algae and bacteria. In facultative ponds, several aerobic bacterial species including *Alcaligenes*, *Achromobacter*, *Flavobacterium*, and *Pseudomonas* oxidize organic matter and release CO_2 gas. In turn, algae utilize CO_2 during the photosynthetic activity to release O_2, which is consumed by bacteria during the oxidation of BOD (Nasr 2014). The amount of oxygen required for BOD removal depends on the algal photosynthesis performance. The algal concentration in facultative ponds is affected by sunlight, temperature, and nutrient loading and usually ranges between 500 and 2000 μg chlorophyll-*a* per liter (Mara 1987). Facultative ponds are designed based on a relatively low surface loading of 100–400 kg BOD/ha/d. Well-designed facultative ponds can achieve removal efficiencies of 70–90% of the influent BOD. The designed organic loading rate (OLR) of facultative ponds can be calculated from the empirical equations (Eqs. 6 and 7) (Mara and Pearson 1998).

$$\text{OLR} = 20T - 120 \tag{6}$$

$$\text{OLR} = 350\left[1.107 - 0.002T\right]^{(T-20)} \tag{7}$$

where OLR is organic loading rate (kg BOD/ha/d) and T is temperature (°C).

Based on OLR, the required surface area is estimated from Eq. 8 (Nasr 2014).

$$A = \frac{Q \times \text{BOD}}{\text{OLR}} \tag{8}$$

where Q is flow rate (m^3/d) and BOD is biological oxygen demand (g/m^3).

3.2 *Shallow Ponds*

The shallow pond system is used to naturally cultivate algae in lakes and reservoirs under high levels of solar radiation with low energy consumption (Posten and Walter 2013). Essential elements can be supplied to provide a nutrient-rich environment for the culture system. The shallow pond configuration is considered as a simple and cost-effective option for algal cultivation. It is easier to construct, implement, and operate than closed photobioreactors. It is suitable for remote, rural, and peri-urban areas where land availability is not a limiting factor (Craggs et al. 2014). However, the inadequate control of environmental conditions and the risk of contamination are the main drawbacks of this system. In addition, the low cell concentration in these ponds can negatively influence the microalgae harvesting process (Oswald and Golueke 1968). Moreover, insufficient contact time can occur between liquid and gas phases, causing a limited mass transfer of CO_2 inside the culture. Hence, some mechanical devices are required to ensure efficient mixing and enhance the biomass productivity.

The pond depth is determined based on the light penetration and culture volume that a unit can retain. The shallow ponds can be modified by flowing the culture through a titled surface (slope of 1–3%) using pumps. This configuration is known as thin layer reactor, and it has a maximum depth of 2 cm and a volume-to-surface area of 10–25 L/m^2 (Gupta et al. 2015). Pumps are used to lift the culture from a storage tank to the upper part (inlet) of the titled surface. The advantages of this system include low power consumption, increase in algal cells exposed to light, minimization of photoinhibition, and high biomass concentration (up to 30 g/L) (Doucha et al. 2005).

Raceway Ponds

The raceway pond has a rectangular cross section and horseshoe-shaped loops with channels, known as meanders (Fig. 2). The algal culture is continuously circulating in the whole system by means of mechanical devices such as paddle wheels (Posten

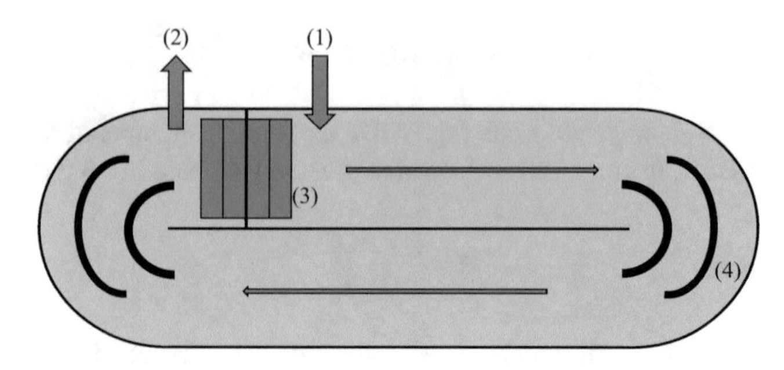

Fig. 2 Raceway (open culture system). In which, (1) culture feed, (2) culture harvest, (3) paddle wheel, and (4) deflector baffle

and Walter 2013). A turbulent flow regime (Reynolds number 8000) can be used to keep the algal cells in suspension, minimize thermal stratification, improve vertical mixing, and prevent O_2 accumulation (Chisti 2016). The pond can be implemented with a low depth between 10 and 15 cm, which allows for efficient light penetration and increased biomass productivity. The pond surface area is equivalent to 42–47% that of waste stabilization ponds. This flexible configuration provides a successful scale-up and achieves low energy requirements for mixing (Yang et al. 2016). The flow velocity in raceways varies between 0.2 and 0.3 m/s, which prevents solids precipitation or cell damage. Based on the climatic condition, the raceway system can receive a maximum OLR of 10–15 g BOD/m^2/d. Hydraulic retention time varies seasonally 7–9 d in winter and 3–4 d in summer (Craggs et al. 2014).

The channel bottom and walls are constructed from inexpensive materials, such as rubber sheet, concrete, or plastic. Young et al. (2017) reported that the cost of raceway pond (using paddle wheels for mixing) is 39.2–47.5% that of waste stabilization ponds. This system can be used for the growth of special microalgae species such as *Dunaliella salina*, *Chroococcus turgidus*, *Chlorella vulgaris*, *Scenedesmus*, *Haematococcus*, and *Nannochloropsis*. However, the raceway system suffers from the requirement of a large surface area, inadequate mass transfer of gas/liquid, poor control of environmental factors, the possibility of culture contamination, and low final algal productivity (Table 1).

The design of raceway algae reactors was previously reported in a study by Oswald and Golueke (1968). Their study assumed that the culture is flowing through a channel having a finite width and unspecified length. The length-to-width ratio is an important factor that should be considered during design (Young et al. 2017). The length-to-width ratio can be higher than or equal to 10/1, providing adequate mixing and mass transfer.

Head loss through the channel (Δd) provides the energy required to keep the flowing motion of algal biomass. It is influenced by bend loss due to the channel curvature (Eq. 9) and the friction of medium with bottom and side wall (Eq. 10).

Table 1 Advantages and disadvantages of algal cultivation systems

Cultivation system	Advantages	Drawbacks
Raceway open pond	High surface area-to-volume ratio Cost-effective technology Simple construction (mixing by paddle wheels) Flexible design	Lack of control to growth conditions Low gas transfer efficiency Possibility of contamination Low biomass productivity Large land area requirement Water loss due to evaporation
Tubular photobioreactor	High surface area-to-volume ratio Applicable for outdoor cultivation High biomass productivity Cost-effective Minimum effect of mutual shading	Accumulation of O_2 gas due to recirculation via pumps Possibility of photoinhibition Fouling due to biomass growth Large space requirement Difficult temperature control
Flat panel photobioreactor	High surface area-to-volume ratio Low space requirement Efficient photosynthetic activity Cost-effective technology Minimum O_2 buildup due to airlift agitation	Short light dispersion depth Not scalable Difficult temperature control Contains several components Periodic fouling and cleanup aspects
Airlift photobioreactor	High mass transfer via airlift Simple implementation (no internal moving parts) Sufficient agitation by bubble with minimum shear stress Low photosynthetic inhibition	Low surface area-to-volume ratio Costly construction material Limited scale-up application Self-shading effect may cause poor light distribution

$$\Delta d_{\mathrm{b}} = \frac{k \cdot U_{\mathrm{L}}^2}{2g} \tag{9}$$

where Δd_{b} is the head loss due to bends (m), k is kinetic loss factor (it varies between 10 and 40 according to the bend shape), U_{L} is average velocity of liquid culture (m/s), and g is the acceleration of gravity (assuming 9.8 m/s^2).

$$\Delta d_{\mathrm{c}} = U_{\mathrm{L}}^2 \cdot f_{\mathrm{M}}^2 \cdot \frac{L}{R^{4/3}} \tag{10}$$

where Δd_{c} is the head loss due to channel friction (m), f_{M} is manning channel roughness factor (s/m$^{1/3}$), R is the hydraulic radius of channel (m), and L is channel length (m).

The total head loss can be calculated by Eq. 11.

$$\Delta d = \Delta d_{\mathrm{b}} + \Delta d_{\mathrm{c}} \tag{11}$$

Further, the power required for paddle wheel can be estimated by Eq. 12 (Chisti 2016).

$$\text{Power} = \frac{Q \cdot \rho \cdot g \cdot \Delta d}{\eta} \quad (12)$$

where Power is the power consumption (kg·m^2/s^3), Δd is the change in depth also known as pump head (m), ρ is the culture density (kg/m^3), Q is pond flowrate (m^3/s), g is acceleration due to gravity (m/s^2), and η is efficiency of the paddle wheel (%).

4 Enclosed Culture Systems

4.1 Tubular Photobioreactors

Tubular photobioreactors (Fig. 3a) consist of transparent tubes (also known as solar collector tubes), which are arranged as horizontal and vertical rows, spirally wound around central support, or a helical structure (Markou and Nerantzis 2013). Horizontal tubular vessels having diameters between 1 and 6 cm and a total path length of several hundred meters are the most commonly used photobioreactors (Slade and Bauen 2013). The incident light is homogeneously distributed on the circumference of tubes, i.e., a phenomenon known as "lens" or "focusing effect." This configuration minimizes mutual shading by algae and improves the internal irradiance levels and radiation intensity (Dalrymple et al. 2013). Algal biomass grown in this system flows by means of mechanical pumps or aeration. Power consumption, in kg·m^2/s^3 per cubic meter of culture, varies from 100 for an airlift-driven reactor to 500 for a pump-driven fence system. The system is provided by a source of light, CO_2 supply to ensure carbonation, and a heat exchanger for temperature control.

The design of solar tubes should enhance the light profile inside medium and develop the transfer of culture between light and dark zones. A turbulent flow regime with Reynolds number over 3000 is recommended to avoid the stagnation of algae in the dark zone (interior) of the tube (Pagliolico et al. 2017). The volume-to-surface area ratio varies from 50 to 150 l/m^3, based on pipes diameters and arrangement. The design should also consider pH and temperature controls, carbon limitation, nonattachment of algae onto the inner surface of tubes, and proper circulation speed causing no damage to biomass cells. The culture pH and carbon applicability are simultaneously controlled by the proper injection of CO_2 gas, which propagates in the medium following the mass transfer capacity (Singh and Singh 2014). The CO_2 budget is regulated to neither exceed the inhibition level nor fall below the limited concentration.

The dimensions of fluid micro-eddies should be greater than those of algal cells to avoid cells stress and damage. The length of micro-eddies can be estimated by Eq. (13), following Kolmogorov's principle of locally isotropic turbulence (Chamecki and Dias 2004).

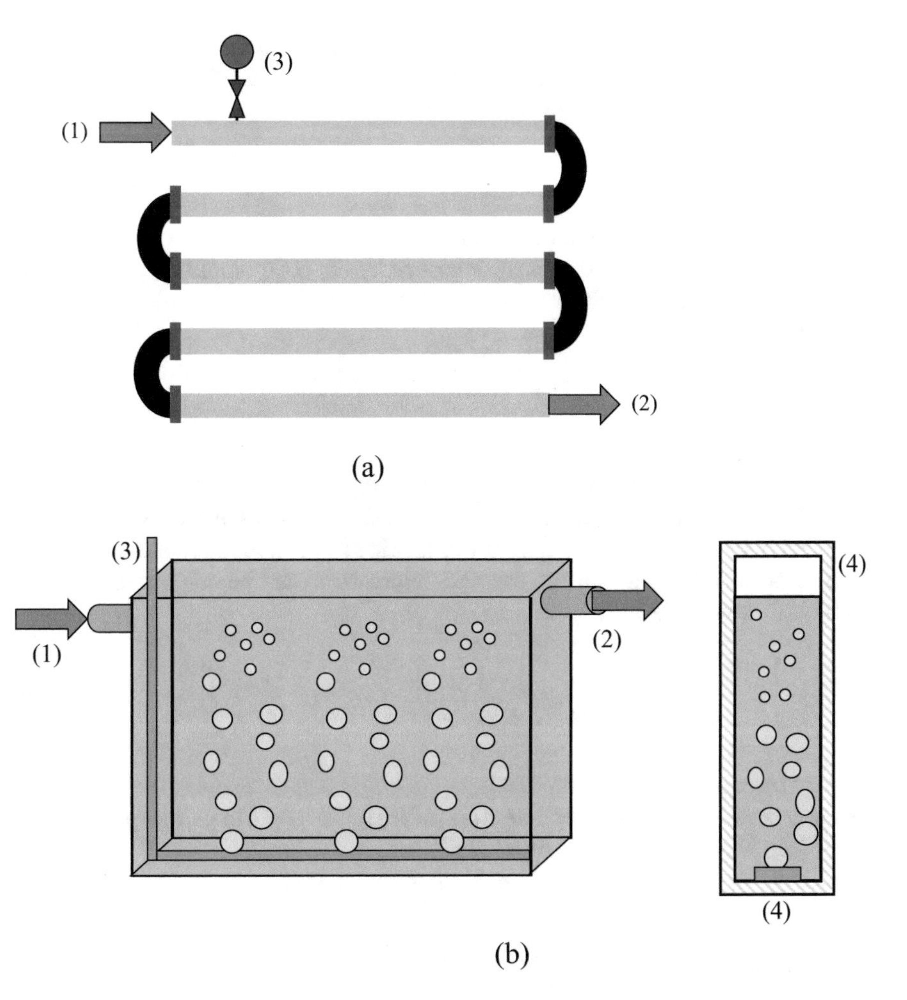

Fig. 3 Enclosed photobioreactors (**a**) tubular photobioreactor and (**b**) flat panel bioreactor. In which, (1) culture feed, (2) culture harvest, (3) flue gas supply, and (4) transparent glass

$$\lambda = \left(\frac{\mu_L}{\rho}\right)^{3/4} \cdot \xi^{-1/4} \tag{13}$$

where λ is the length of micro-eddy (m), μ_L is the fluid viscosity (kg/m/s), ρ is the fluid density (kg/m^3), and ξ is the specific energy dissipation rate (m^2/s^3) calculated from Eq. 14.

$$\xi = \frac{2C_f \cdot U_L^3}{d_t} \tag{14}$$

where U_L is the liquid velocity that is lower than the maximum velocity resulted from the micro-eddy length (m/s), d_t is tube diameter (m), and C_f is Fanning friction factor (dimensionless) estimated by Eq. 15.

$$C_f = 0.0791 R_e^{-1/4} \tag{15}$$

where R_e is Reynolds number (dimensionless).

The length of a solar collector, as described by Eq. (16), should be determined to avoid O_2 buildup in the culture (Posten and Walter 2013).

$$L = \frac{U_L\left([O_2]_{out} - [O_2]_{in}\right)}{R_{O_2}} \tag{16}$$

where L is the maximum length of solar tubes that prevents photosynthesis inhibition (m), $[O_2]_{in}$ and $[O_2]_{out}$ are the oxygen concentrations at the entrance and outlet of solar collectors (mg/L), and R_{O_2} is volumetric rate of oxygen generation (mg/l/s).

The power utilization in tubular photobioreactors can be calculated from the Bernoulli equation as follows (Eq. 17).

$$\text{Power} = \frac{\pi}{8V} d_t^2 U_L^3 + \frac{0.0791\pi}{8V} L \cdot d_t^{0.75} U_L^{2.75} \left(\frac{\mu_L}{\rho}\right)^{0.25} \tag{17}$$

where Power is power consumption (kg m^2/s^3), d_t is pipe diameter (m), U_L is culture velocity (m/s), V is volume per unit mass (m^3/kg), L is length of culture path or solar tubes (m), and ρ is culture density (kg/m^3).

4.2 Flat Panel Photobioreactors

Flat panel photobioreactors (Fig. 3b) consist of a rectangular frame covered by transparent/clear plates (Plexiglass alveolar plates with a thickness of 16 mm) on both sides (front/back). The plates are connected and joined to store the culture biomass, which transmits via aeration (Posten 2009). The distances between photobioreactors are identified to avoid the overlapping or shading among panels. Flue gas is supplied from the reactor bottom to ensure CO_2 transfer and O_2 release. No mechanical devices are situated in the culture medium, and the algal suspension is circulated by means of pumps (Pagliolico et al. 2017). The light penetration depth is minimized to provide a large illuminated surface area and an efficient photosynthetic activity. The design of this system considers the optimal position, angle of inclination, panels' orientation, and distance between panels. The dimensions are preferable as height <1.5 m and width <10 cm to prevent the use of expensive and mechanical resistant materials (Huang et al. 2017). Based on the culture depth and separation between panels, the surface area-to-volume ratio can vary between 6 and 20 m^2/m^3.

In flat panels, the power input is estimated as a function of aeration rate, as described by Eq. 18.

$$\text{Power} = \rho \cdot g \cdot U_{\text{G}} \tag{18}$$

where Power is power input per unit volume due to aeration in (kg m^2/s^3)/m^3, ρ is the density of culture (kg/m^3), g is gravitational acceleration (m/s^2), and U_{G} is superficial gas velocity in the aerated zone (m/s).

However, flat panel photobioreactors suffer from several issues including the requirement of support components, difficulty to control medium temperature, hydrodynamic stress associated with aeration, and algal attachment onto the wall of the system (Doucha et al. 2005). These drawbacks limit the development of flat panel photobioreactors for a commercial-scale application.

4.3 Airlift Photobioreactors

Airlift-type photobioreactors (Fig. 4) are composed of two interconnected sections (with no internal moving structure) that physically separate the upflow and downflow streams (Chen et al. 2011). The culture in the riser part moves upward by means of gas input that decreases the culture density. A portion of the released gas escapes from the culture phase to the top of the column, whereas the other part (composed of heavier bubble-free liquid) is allowed to circulate (Sadeghizadeh et al. 2017). The circulation pattern causes the algal cells to propagate within dark

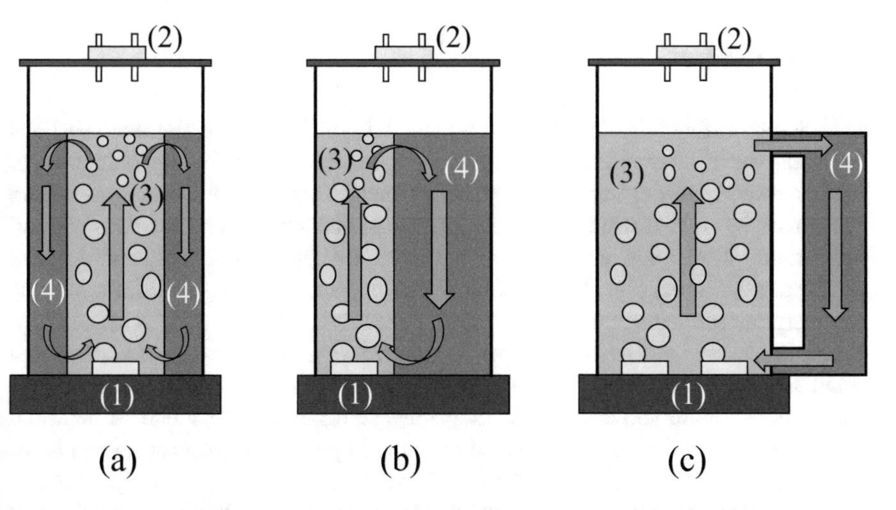

Fig. 4 Airlift photobioreactors (**a**) internal loop concentric, (**b**) internal loop, and (**c**) external loop. In which, (1) sparger for flue gas supply, (2) excess gas release, (3) culture upflow in illuminated zone, and (4) culture downflow in dark zone

and light phases continuously. The common airlift structures are (a) internal loop concentric (Fig. 4a), in which air is sparged by means of a concentric tube causing the algal culture to transfer from the riser (illuminated zone) to the downcomer (dark zone), (b) internal loop vessel (Fig. 4b) that employs an internal baffle to separate the riser (illuminated zone) from culture downflow (dark region), and (c) external loop vessel (Fig. 4c), in which an additional vertical section is externally connected to the main column by short horizontal tubes.

In airlift photobioreactors, the height is designed to be greater than twice the diameter. The sparging of compressed gas is used to achieve agitation and mixing as well as to remove O_2 gas from the culture by mass transfer (Bhola et al. 2017). The photobioreactor hydrodynamics, which affects the transfer of CO_2 from the gas phase to liquid phase, is controlled by the behavior of bubbles released from sparger. For example, when bubbles are uniformly distributed across the column cross section, the flow becomes homogeneous, and the back mixing of the gas phase can be neglected. However, an efficient heterogeneous flow occurs when both air bubbles and liquid tend to move upward from the center of the column, whereas an opposite direction (downward) of liquid ensues adjacent to the walls. This trend causes culture circulation from the central photic region to the external dark zone, causing additional back mixing of gas.

In a previous study, Hincapie and Stuart (2015) used an airlift reactor with a diameter of 20.3 cm (equivalent to 28 L volume), superficial gas velocity 1.80 cm/s, and gas flow 13.3 L/min for growing *Chlorella* sp. algae. Airlift photobioreactors have several advantages such as the absence of internal mechanical parts, continuous release of gas residues, efficient heat and mass transfers, high surface area-to-volume ratio, and low capital cost.

5 Conclusions

This chapter offered the design and fundamental principles of algal growth systems used for wastewater remediation. Algae can be developed for the reduction of organic matter, nutrient species, and heavy metals. Algal cultivation systems are influenced by light distribution, mixing, mass transfer rates, and medium pH and temperature. Raceways have been widely used as effective and inexpensive ponds for the production of algal biomass. However, this system may suffer from the surrounding environments, leading to relatively low biomass productivity. Alternatively, photobioreactors have been developed to achieve a high biomass growth along with small space requirements. Flat panel photobioreactors require lower energy inputs than tubular photobioreactors, but they may suffer from stress due to aeration. Future researches should be conducted to designing an efficient and scalable photobioreactor that can be operated with low energy. Moreover, essential parameters used for the improvement of algal performance at full scale should be identified in further studies.

References

Béchet Q, Shilton A, Guieysse B (2013) Modeling the effects of light and temperature on algae growth: state of the art and critical assessment for productivity prediction during outdoor cultivation. Biotechnol Adv 31:1648–1663

Bhola V, Swalaha F, Nasr M, Bux F (2017) Fuzzy intelligence for investigating the correlation between growth performance and metabolic yields of a Chlorella sp. exposed to various flue gas schemes. Bioresour Technol 243:1078–1086

Chamecki M, Dias N (2004) The local isotropy hypothesis and the turbulent kinetic energy dissipation rate in the atmospheric surface layer. Q J R Meteorol Soc 130:2733–2752

Chen C, Yeh K, Aisyah R, Lee D, Chang J (2011) Cultivation, photobioreactor design and harvesting of microalgae for biodiesel production: a critical review. Bioresour Technol 102:71–81

Chisti Y (2016) Large-scale production of algal biomass: raceway ponds. In: Algae biotechnology products and processes. Springer International Publishing, Cham

Craggs R, Park J, Heubeck S, Sutherland D (2014) High rate algal pond systems for low-energy wastewater treatment, nutrient recovery and energy production. NZ J Bot 52:60–73

Dalrymple O, Halfhide T, Udom I, Gilles B, Wolan J, Zhang Q, Ergas S (2013) Wastewater use in algae production for generation of renewable resources: a review and preliminary results. Aquat Biosyst 9:2

Doucha J, Straka F, Lívanský K (2005) Utilization of flue gas for cultivation of microalgae (*Chlorella* sp.) in an outdoor open thin-layer photobioreactor. J Appl Phycol 17:403–412

Fernandez F, Sevilla J, Grima E (2013) Photobioreactors for the production of microalgae. Rev Environ Sci Biotechnol 12:131–151

Grima E, Sevilla J, Pérez J, Camacho F (1996) A study on simultaneous photolimitation and photoinhibition in dense microalgal cultures taking into account incident and averaged irradiances. J Biotechnol 45:59–69

Gupta P, Lee S, Choi H (2015) A mini review: photobioreactors for large scale algal cultivation. World J Microbiol Biotechnol 31:1409–1417

Gupta S, Ansari F, Nasr M, Rawat I, Nayunigari M, Bux F (2017) Cultivation of Chlorella sorokiniana and Scenedesmus obliquus in wastewater: Fuzzy intelligence for evaluation of growth parameters and metabolites extraction. J Clean Prod 147:419–430

Hincapie E, Stuart B (2015) Design, construction, and validation of an internally lit air-lift photobioreactor for growing algae. Front Energy Res 2:1–7

Huang Q, Jiang F, Wang L, Yang C (2017) Design of photobioreactors for mass cultivation of photosynthetic organisms. Engineering 3:318–329

Lee E, Jalalizadeh M, Zhang Q (2015) Growth kinetic models for microalgae cultivation: a review. Algal Res 12:497–512

Mara D (1987) Waste stabilization ponds: problems and controversies. Water Qual Int 1:20–22

Mara D, Pearson H (1998) Design manual of waste stabilization ponds in mediterranean countries. Laggon Technology International, Leeds

Markou G, Nerantzis E (2013) Microalgae for high-value compounds and biofuels production: a review with focus on cultivation under stress conditions. Biotechnol Adv 31:1532–1542

Medipally S, Yusoff F, Banerjee S, Shariff M (2015) Microalgae as sustainable renewable energy feedstock for biofuel production. Biomed Res Int 2015:519513

Mehta S, Gaur J (2005) Use of algae for removing heavy metal ions from wastewater: progress and prospects. Crit Rev Biotechnol 25:113–152

Nasr M (2014) Application of stabilization ponds in the Nile Delta of Egypt. 2nd international conference on sustainable environment and agriculture. IPCBEE 76:1–5

Nasr M, Ateia M, Hassan K (2017) Modeling the effects of operational parameters on algae growth. In: Algal biofuels: recent advances and future prospects. Springer International Publishing, Cham, pp 127–139

Norsker N, Barbosa M, Vermuë M, Wijffels R (2011) Microalgal production--a close look at the economics. Biotechnol Adv 29:24–27

Oswald W, Golueke C (1968) Large-scale production of microalgae. In: Mateless RI, Tannenbaum SR (eds) Single cell protein. MIT Press, Cambridge, pp 271–305

Pagliolico S, LoVerso V, Bosco F, Mollea C, Forgia C (2017) A novel photo-bioreactor application for microalgae production as a shading system in buildings. Energy Procedia 111:151–160

Posten C (2009) Design principles of photo-bioreactors for cultivation of microalgae. Eng Life Sci 9:165–177

Posten C, Walter C (2013) Microalgal biotechnology: potential and production. Marine and freshwater botany. De Gruyter, Berlin/Boston

Sadeghizadeh A, Farhad Dad F, Moghaddasi L, Rahimi R (2017) CO_2 capture from air by Chlorella vulgaris microalgae in an airlift photobioreactor. Bioresour Technol 243:441–447

Singh S, Singh P (2014) Effect of CO_2 concentration on algal growth: a review. Renew Sust Energ Rev 38:172–179

Singh S, Singh P (2015) Effect of temperature and light on the growth of algae species: a review. Renew Sust Energ Rev 50:431–444

Slade R, Bauen A (2013) Micro-algae cultivation for biofuels: cost, energy balance, environmental impacts and future prospects. Biomass Bioenergy 53:29–38

Takache H, Pruvost J, Cornet J (2012) Kinetic modeling of the photosynthetic growth of Chlamydomonas reinhardtii in a photobioreactor. Biotechnol Prog 28:681–692

Vejrazka C, Janssen M, Streefland M, Wijffels R (2012) Photosynthetic efficiency of Chlamydomonas reinhardtii in attenuated, flashing light. Biotechnol Bioeng 109:2567–2574

Yang Z, Cheng J, Ye Q, Liu J, Zhou J, Cen K (2016) Decrease in light/dark cycle of microalgal cells with computational fluid dynamics simulation to improve microalgal growth in a raceway pond. Bioresour Technol 220:352–359

Young P, Taylor M, Fallowfield H (2017) Mini-review: high rate algal ponds, flexible systems for sustainable wastewater treatment. World J Microbiol Biotechnol 33:117

Index

S. K. Gupta, F. Bux (eds.), *Application of Microalgae in Wastewater Treatment*,
https://doi.org/10.1007/978-3-030-13913-1

E

Printed in the United States
By Bookmasters